D0909324

Undergraduate Texts in Mathematics

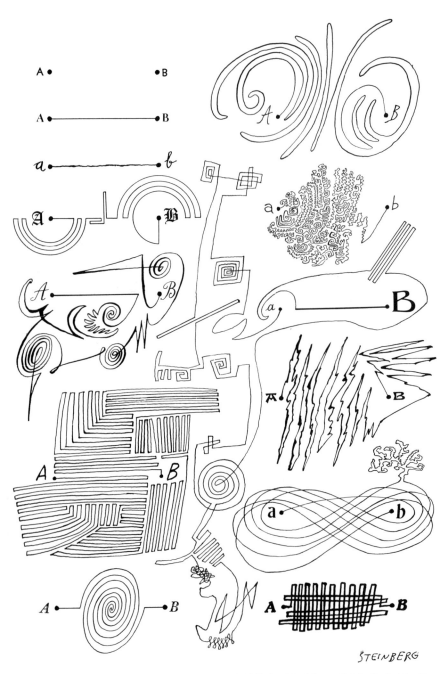

From THE LABYRINTH (Harper & Brothers). © 1960 Saul Steinberg. Originally in *The New Yorker*.

Walter Prenowitz
James Jantosciak

Join Geometries

A Theory of Convex Sets
and Linear Geometry

Springer-Verlag

New York Heidelberg Berlin

Walter Prenowitz (ret.)
Department of Mathematics
Brooklyn College
City University of New York
New York, NY 11210
USA

James Jantosciak
Department of Mathematics
Brooklyn College
City University of New York
New York, NY 11210
USA

AMS Subject Classifications: 50-01, 52-01

With 404 Figures

Library of Congress Cataloging in Publication Data

Prenowitz, Walter, 1906–
    Join geometries.

    (Undergraduate texts in mathematics)
    Includes index.
    1. Geometry.   I.   Jantosciak, James, joint
author.   II.   Title.
QA447.P73   516   79-1457

9 8 7 6 5 4 3 2 1

ISBN 0-387-90340-2 Springer-Verlag New York
ISBN 3-540-90340-2 Springer-Verlag Berlin Heidelberg

To Sophie

Contents

Introduction

The main object of this book is to reorient and revitalize classical geometry in a way that will bring it closer to the mainstream of contemporary mathematics. The postulational basis of the subject will be radically revised in order to construct a broad-scale and conceptually unified treatment.

The familiar figures of classical geometry—points, segments, lines, planes, triangles, circles, and so on—stem from problems in the physical world and seem to be conceptually unrelated. However, a natural setting for their study is provided by the concept of *convex set*, which is comparatively new in the history of geometrical ideas. The familiar figures can then appear as convex sets, boundaries of convex sets, or finite unions of convex sets. Moreover, two basic types of figure in linear geometry are special cases of convex set: *linear space* (point, line, and plane) and *halfspace* (ray, halfplane, and halfspace). Therefore we choose convex set to be the central type of figure in our treatment of geometry. How can the wealth of geometric knowledge be organized around this idea? By definition, a set is convex if it contains the segment joining each pair of its points; that is, if it is closed under the operation of joining two points to form a segment. But this is precisely the basic operation in Euclid. Our point of departure is to take the operation of joining two points to form a segment as fundamental, and to throw the burden of unifying the material on the consistent and relentless exploitation of this operation.

The postulates then will not involve complex ideas or complicated figures, but will state elementary properties of the join operation that can be grasped intuitively and verified concretely in planar diagrams.

The postulates are formulated as *universal* properties of points. Thus, there are no exceptional or degenerate cases to be excluded. This is in striking contrast with classical Euclidean postulates, such as: two *distinct* points determine a line; three *noncollinear* points determine a plane. As a result, proofs usually involve the application of the postulates as *general* principles and there is little or no need to consider the special or degenerate cases that arise so often in conventional treatments of Euclidean geometry.

A salient feature of the treatment is its freedom from the classical restriction to the study of geometries that are, at most, three-dimensional. Indeed, our postulates are dimension free—they involve no dimensionality assumption, explicit or implicit.

Consequently a major portion of the development is dimension free and is applicable to spaces of arbitrary dimension, finite or infinite. (This belies a widespread belief that the only effective way to study higher dimensional geometry is by the intervention of linear algebra.)

How does the theory compare with Euclidean geometry? The postulates are abstracted from Euclidean propositions, and the theory may be considered a generalization of Euclidean geometry. It is, however, much broader —the theory has been freed from constraints that arose naturally in the historical evolution of Euclidean geometry but now impede its development. Many familiar Euclidean propositions—in addition to the dimensional restriction mentioned above—are omitted from the postulate set. These include: (i) the Euclidean parallel postulate; (ii) the proposition that of three distinct collinear points, one is between the other two; and (iii) two distinct points determine a line. Moreover, the treatment is nonmetrical—no postulates for congruence have been assumed.

Can this brave new geometrical world be achieved merely by referring to a segment as the join of two points? Of course not. A reanalysis and reconstruction of classical geometry in terms of the join operation is required. First of all the join operation is not to be restricted artificially, it must apply equally well to all pairs of points, distinct *or* coincident. Even more important, the operation must be generalized to apply to all pairs of geometric figures.

In a Euclidean geometry, we define the *join* of points a, b, denoted by $a \cdot b$ or ab, to be the open segment with endpoints a and b if $a \neq b$ and define the join of a point a and itself to consist of a. The operation join is extended to apply to any two figures A and B in a natural way: The join $A \cdot B$ or AB, of A and B, is the union of all joins ab where point a ranges over A and point b over B.

There is, in Euclidean geometry, a second important operation—that of extending a segment indefinitely to form a ray. This can be treated as a sort of inverse operation to join and suggests the following

Definition. Let a and b be any points. Then a/b, the *extension* of a from b, is the set of points x which satisfy the condition that bx contain point a.

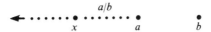

This operation is extended in the same way as join to define A/B, the *extension* of A *from* B, for any two figures A and B.

Chapter 1 provides an introduction to the abstract theory by studying the join and extension operations in Euclidean geometry in a concrete, intuitive, exploratory manner.

The formal development of the theory begins in Chapter 2, and is based on the idea of join operation. Let J be a set of elements (points) and $\cdot$ an operation which assigns to each ordered pair (a, b) of elements of J a uniquely determined subset of J, denoted by $a \cdot b$ or ab. Then the operation $\cdot$ is a *join operation* (in set J) and $a \cdot b$ is the *join* of a and b. We assume that the set J and the join operation $\cdot$ satisfy four postulates suggested by elementary properties of the Euclidean join operation. Convex sets are defined as closed under join, and their elementary properties deduced. The concepts of *geometrical* or *intrinsic* interior and closure of a convex set are defined. These ideas pervade the theory.

The convex hull of an arbitrary set is introduced in Chapter 3. Join theoretic formulas for convex hulls are derived. Polytopes are treated as convex hulls of finite sets.

In Chapter 4, the *extension* a/b of two elements a, b of J is defined, in the same way as Euclidean extension, in terms of join. Three new postulates involving extension are introduced to complete the basic postulate set for the theory. Speaking geometrically, the concept of *ray* (or *halfline*) is now available in the abstract theory. In formal terms, we have at our command an algebra, of strong deductive power, involving two "inverse" operations join $\cdot$ and extension$/$.

Chapter 5 introduces the idea of join geometry, our basic object of study. A *join geometry* is a model of the theory—it is a pair $(J, \cdot)$ composed of a set J and a join operation $\cdot$ in J, which satisfy the basic set of postulates. The notion of isomorphism of join geometries is studied. A collection of join geometries that are used as illustrative examples and counterexamples is presented. Real n-space $\mathbb{R}^n$ is converted into a join geometry by defining join in the natural manner. An infinite dimensional analogue of $\mathbb{R}^n$ is shown to contain a *pathological* convex set—a nonempty convex set whose interior is empty.

Chapter 6 studies *linear sets* (or *linear spaces*) defined as closed under join and extension. Among the topics considered are generation of linear sets, linear independence, and how *line* should be defined. It is interesting that linear sets of a join geometry bear analogies to subgroups of an abelian group.

Chapter 7 studies the idea of *extreme set of a convex set*. An extreme set of a convex set A is loosely a convex subset of A which is "peripheral" to A. The idea is suggested by, and is a generalization of, the classical notion of vertices, edges, and faces of a polyhedron. Two types of extreme sets, called components and faces, play important roles in the study of the structure of a convex set and are singled out for special study.

Chapter 8 deals with rays and halfspaces. A *ray* or *halfline* is defined as a set p/a, where p and a are points; p is its *endpoint*. Similarly, let L be a nonempty linear set, and a a point. Then L/a is a *halfspace* of L, or simply a *halfspace*; L is its *edge*. A study of rays is given, concentrating on rays with a common endpoint. This is generalized to an analogous treatment for halfspaces with a common edge. A halfspace of a linear set in a join geometry is analogous to a coset of a subgroup in an abelian group.

Chapter 9 presents a treatment of cones and hypercones based on the material of Chapter 8. A *cone* is the union of a family of rays that have a common endpoint. A *hypercone* is the union of a family of halfspaces that have a common edge.

In Chapter 10, the family of halfspaces of a linear space is converted into a geometrical system—called a *factor geometry*—by defining a join operation in it in a natural way. Factor geometries and join geometries share many common properties but differ markedly as algebraic systems, since a factor geometry has an identity element and its elements have inverses. The development has strong—though unforced—analogies with algebraic theories of congruence relations and factor or quotient systems.

Chapter 11 is devoted to the theory of *exchange geometries*, which are join geometries that satisfy a postulate equivalent to "two points determine a line." A theory of dimension is developed in an exchange geometry and the familiar incidence and intersection properties of lines and planes in Euclidean 3-space is generalized to finite-dimensional linear spaces.

Chapter 12 studies *ordered geometries*, which are join geometries that satisfy the Euclidean proposition: Of three distinct collinear points, one is between the other two. Among the results derived are basic geometric properties of polytopes; conditions for the separation of linear spaces by linear subspaces; the theorems of Radon, Helly, and Caratheodory on convex sets; and a striking formula for the linear space generated by a finite set of points.

In Chapter 13 various properties of polytopes in $\mathbb{R}^n$ are extended to ordered geometries. In particular, polytopes are related to intersections of halfspaces.

Since our approach is so different from the usual one, we felt compelled to develop the material slowly and deliberately with much concrete geometric motivation. This was done because of the unfamiliarity, not the inherent difficulty, of the treatment. The book assumes little formal knowledge of geometry and indeed little beyond high school algebra and some familiarity with intuitive set theory.

The book can be studied rather flexibly. Chapter 1 helps to provide a transition from intuitive informal geometry to an axiomatic formal development and can be read by able high school upperclassmen. The reader who has some degree of mathematical maturity need not begin with Chapter 1 but can use it as a source of supplementary material for the first few chapters. Chapters 2–6 form a basic course sequence in the abstract theory. (Some sections may be omitted in a first reading, for example: 2.20, 2.25, 2.26, 3.12–3.15, 4.20, 4.21, 4.23–4.26, and 6.20–6.24.) Except for the definition of join geometry, Chapter 5 can be skirted. But the reader is advised to make some contact with the models presented, since they shed so much light on the theory.

Here are some longer sequences with different emphases: Chapters 2–7; 2–6, 8; 2–6, 11; 2–6, 8, 9; 2–6, 8–10; 2–6, 8, 12; 2–8, 12, 13. A structure chart which indicates the interrelations of the chapters appears below. Footnotes to the titles of Chapters 7, 8, 10, and 12 provide more detail on the interrelations of the chapters.

The text is accompanied by a large and varied collection of exercises. They include simple exploratory exercises, verifying or testing a conjecture in a model, proofs that require only a few steps, difficult problems (the most difficult are indicated by an asterisk), and problems that involve extending the theory, labelled Projects.

Although the book was written as an undergraduate text, graduate students and mathematicians may find it of interest. There may be curiosity about a contemporary approach to the classical geometry which is our heritage from the Greeks. Those for whom geometry has little intuitive appeal may be attracted by the striking and unexpected analogies that appear between join geometries and algebraic structures, especially abelian groups. Specialists in the theory of convex sets may be interested both because of the broad vistas that seem to be opened up by the join theoretic axiomatization of the subject and the questions that arise on the extent to which the familiar theory of convexity in $\mathbb{R}^n$ can be extended to a join geometry, or to special types of join geometries.

Acknowledgments

The first acknowledgment is a personal one of Walter Prenowitz and concerns the origin of the book.

About ten years ago, Victor Klee suggested that I write a book on convex sets from the join theoretic point of view I had employed in an expository paper (Prenowitz [3]). I thought very well of the idea but felt the time was not ripe for so radical a departure in the treatment of geometry. The project was discussed when we met several times during the ensuing year and finally I changed my mind, feeling that Klee's initiative in proposing the project and persistence and courage in backing it were sufficient encouragement to take it on.

Prenowitz began by writing a preliminary version, which was published early in 1969 in a small soft-cover edition for classroom trial. Our thanks are extended to Jane W. DePaola, Branko Grünbaum, Walter Meyer, Paul T. Rygg, Seymour Schuster, and Ian Spatz for trying out this material. Also, we wish to thank William Barnier, George Booth, Melvin Hausner, Mark Hunacek, and Joseph Malkevitch for reading and criticizing portions of later versions of the manuscript.

Sophie Prenowitz was a mainstay during the whole period of evolution of the book from the preliminary version to the final manuscript. She typed, drew diagrams, proofread, and edited nontechnical portions of the manuscript. The cover drawing, which uses an oak leaf to illustrate the concept of a convex hull, was designed by her. She was a constant source of encouragment and support.

We are very grateful to the artist Saul Steinberg for permission to use his drawing as our frontispiece.

January 1979   Walter Prenowitz
                                                   James Jantosciak

xxi

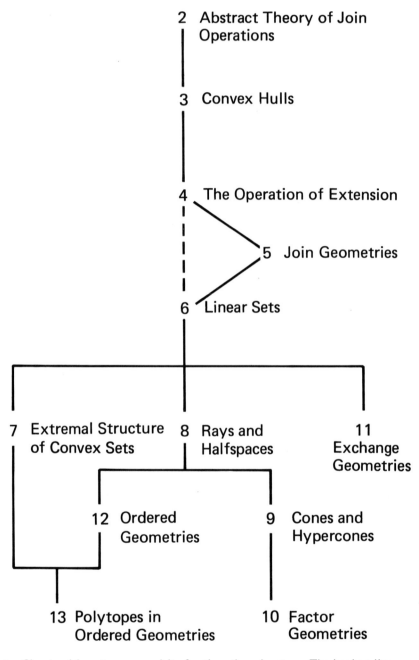

1   The Join and Extension Operations in
    Euclidean Geometry

2   Abstract Theory of Join
    Operations

3   Convex Hulls

4   The Operation of Extension

5   Join Geometries

6   Linear Sets

7   Extremal Structure   8   Rays and   11
    of Convex Sets   Halfspaces   Exchange
                                                    Geometries

12 Ordered   9   Cones and
    Geometries   Hypercones

13 Polytopes in   10 Factor
    Ordered Geometries   Geometries

Note. Chapter 1 is not a prerequisite for the other chapters. The broken line
indicates that it is possible to proceed directly from Chapter 4 to Chapter 6.

The Join and Extension Operations in Euclidean Geometry[1]

1

This chapter is intended to prepare the reader for the formal development of the theory of join operations which will begin in Chapter 2. It is devoted to the study, in concrete terms, of geometric operations which are suggested by two familiar ideas in Euclidean geometry:

(1) Joining two points to form a segment (Figure 1.1).

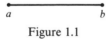

$$a \qquad\qquad b$$

Figure 1.1

(2) Extending or prolonging a segment endlessly beyond one of its endpoints (Figure 1.2).

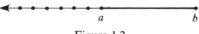

$$a \qquad\qquad b$$

Figure 1.2

The operations are made precise; they are generalized to be applicable to any two geometric figures; their elementary properties are studied. Diagrams and examples are used to discover and to verify basic properties of the operations. The treatment is concrete, intuitive, exploratory—it may be described, a bit loosely, as the intuitive geometry of the operations join and extension.

---

[1] Although Chapter 1 is not a prerequisite for the formal theory which begins in Chapter 2, some familiarity with it may help to prepare the reader for Chapter 2.

1.1 The Notion of Segment: Closed and Open

In order to define the join of two points it is necessary to clarify the notion
of segment. In high school geometry the term *segment* or *line segment*
stands for the portion of a (straight) line *L* which is bounded by two points

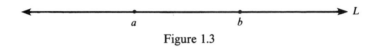

$$a \qquad\qquad b$$

Figure 1.3

of *L*, called the *endpoints* of the segment (Figure 1.3). What about the
endpoints: Are they to be included in the segment? Once this question is
posed it becomes evident that there are two notions of segment: *closed*
segment and *open* segment. A closed segment contains its endpoints—an
open segment excludes its endpoints. The two notions can be characterized
in terms of the idea of betweenness. Let *a* and *b* be distinct points. Then
the closed segment *ab* is the figure consisting of *a*, *b* and all points between
a and *b*. The open segment *ab* consists simply of all points between *a* and
b.

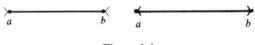

$$a \qquad\qquad b \qquad a \qquad\qquad b$$

Figure 1.4

You may have encountered a similar distinction on the number line in
the use of the terms closed interval and open interval. For example, the
closed interval [0, 1] consists of 0, 1 and all real numbers that are
numerically between 0 and 1, that is, all real numbers *x* which are greater
than 0 and less than 1. The open interval (0, 1) contains just the real
numbers that are numerically between 0 and 1—it has no least or greatest
number.

$$0 \qquad\qquad 1$$

Figure 1.5

In the study of joining operations we find it preferable to join points by
open rather than closed segments.[2] So for the sake of convenience we shall
use the term segment without qualification to refer to open segment.

Finally let us mention that a segment will be conceived as composed of
points, that is, a segment is a class or set of points. Geometric figures in
general will be conceived as sets of points.

---

[2] The basis for the preference will be indicated in the last section of the chapter.

1.2 The Join of Two Distinct Points

To begin we define the idea of the join of two distinct points.

Definition. Let *a* and *b* be distinct points. Then the *join of point a and point b* stands for the segment *ab* and is denoted operationally by *a·b* or simply *ab* (Figure 1.6).

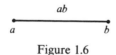

Figure 1.6

1.3 Two Basic Properties of the Join Operation

What can be said about this operation of joining two points? It has the following simple but very basic properties:

(1) The join *ab* of any two distinct points *a* and *b* is a definite nonempty set of points.
(2) Commutative law: *ab* = *ba* for any two distinct points *a* and *b*.

Can additional properties be found? Yes. Consider, for example, the principle that the segment joining two points of a given segment *S* is contained in *S* (Figure 1.7). This can be restated in join terminology as follows: If *p* and *q* are distinct points of the join *ab*, then their join *pq* is contained in *ab*.

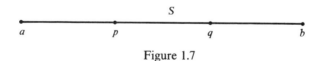

Figure 1.7

It is not hard to see that any geometric principle which is expressible in terms of the notion of segment or in terms of the notion of betweenness can be restated as a property of the join of pairs of points.

EXERCISES

1. Restate in join terminology: If *p* is between *a* and *b*, and *q* is between *a* and *p*, then *q* is between *a* and *b*.

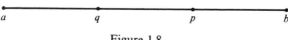

Figure 1.8

2. Restate in join terminology: If p is a point of segment S, then the segments that join p to the endpoints of S have no common point.

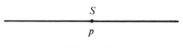

Figure 1.9

3. Restate in join terminology: If x is between a and a point p of segment bc, then x is between c and a point q of segment ab.

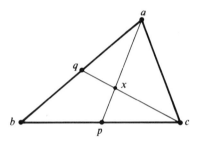

Figure 1.10

4. Find some other properties of segments and betweenness of points and state them in join terminology.

1.4 A Crucial Question

What has been achieved? A start has been made in the operation-centered treatment of elementary geometry: (1) the join operation has been defined for pairs of distinct points; and (2) two basic properties of the join operation have been formulated. You may feel that we have not achieved very much, that we are merely restating old ideas in a new vocabulary: Where Euclid says *segment* we say *join of two points*.

But our object is quite different. It is to change ideas rather than terminology. Our primary purpose is to develop a treatment of geometry in which the join operation plays a central role. It is hard to see how this can be done if we just stick to joins of two points. We should be able to construct more complicated figures than *segments*. To do this we should be able to join more complicated objects than *points*. Why shouldn't we be able to consider the join of a point and a segment or a point and a circle—or the join of any two figures for that matter?

Our approach to geometry cannot hope to be effective unless the scope of the join operation is significantly enlarged.

1.5 The Join of Two Geometric Figures

Now we show how the join operation can be extended to apply to two geometric figures, conceived as sets of points. Three examples are given before the formal definition is presented.

EXAMPLE I (The join of a point and a segment). Suppose points a, b and c are noncollinear (Figure 1.11). To get the join of point a and segment bc [denoted by $a(bc)$], join a to each point of bc and unite all the joins formed. The figure obtained is the interior of triangle abc. Thus the join of a and bc is the interior of triangle abc.

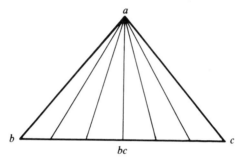

Figure 1.11 $a(bc)$ = Interior (triangle abc).

EXAMPLE II (The join of a point and a circle). Suppose point p is not in the plane of circle C (Figure 1.12). To get the join of point p and circle C (denoted by pC), join p to each point of C and unite all the joins formed. We obtain a circular cone (conical surface) with apex p and base C excluding p and all points of C.

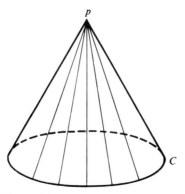

Figure 1.12 pC = circular cone.

EXAMPLE III (The join of two segments). Let ab and cd be two segments in a plane such that a, b, c and d are the vertices of a convex (nonreentrant) quadrilateral as indicated in Figure 1.13. The join of ab and cd, denoted by $(ab)(cd)$, is obtained by joining each point of ab to each point of cd and uniting all the joins formed. The join of ab and cd is the interior of quadrilateral $abcd$.

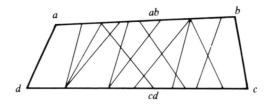

Figure 1.13 $(ab)(cd) =$ Interior (quadrilateral $abcd$).

Figure 1.14

These examples suggest the following definition (see Figure 1.14).

Definition. Let A and B be two geometric figures, which we conceive as sets of points. The *join of set A and set B* is the set of points obtained by joining each point a of A to each point b of B and uniting all joins ab formed in this way. The join of A and B is denoted by $A \cdot B$ or simply AB. If A consists of a single point a, we call the join of A and B simply the join of a and B and denote it by aB (Figure 1.15). Similarly, if B reduces to a single point b, we refer to the join of A and b and denote it by Ab.

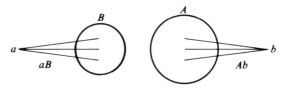

Figure 1.15

In simple terms the definition says: x is a point of the join AB of sets A and B if and only if there exists a point a of set A and a point b of set B such that x is a point of ab (Figure 1.16).

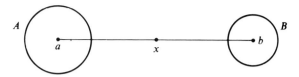

Figure 1.16

In this section and the next one the consideration of joins of two geometric figures is limited to the case where the figures do not intersect.

A Word on Notation. Systematically in our study of geometry, lower case letters $a, b, c, \ldots$ will be used to denote points and capitals $A, B, C, \ldots$ to denote geometric figures or sets of points.

This section began with the presentation of several examples of joins of geometric figures. The examples indicated that the join of two geometric figures could be a new and more complex figure. An important property of the join operation—which will be seen to pervade the theory—is that it affords a procedure for the construction of complex figures from simple ones. This property of join as a "figure generator" appears in the following set of exercises, some of which you may find surprising and stimulating—some, we hope, challenging.

EXERCISES (JOINS OF TWO FIGURES)

In these exercises only joins of nonintersecting figures are considered.

In Exercises 1–17, identify and describe the indicated joins. You may use diagrams or models. No proof is required.

1. The join of point p and I, the interior of circle C, if p is not in the plane of C. (Compare Example II in the text above.)

2. The join of point p and I, the interior of triangle abc, if p is not in the plane of triangle abc.

3. The join of point p and line L, if p is not on L.

4. The join of two parallel lines.

*5. The join of two skew lines, that is, two lines in space which do not lie in the same plane.

The Notion of Ray or Halfline. Suppose a and b are two distinct points. Then the *ray* or *halfline ab*, denoted by the symbol $\overrightarrow{ab}$, consists of all points of line ab that lie on the same side of point a as b (Figure 1.17).

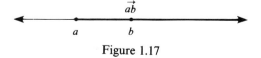

Figure 1.17

Point a is called the *endpoint* of $\overrightarrow{ab}$. Note that $\overrightarrow{ab}$ does not contain its endpoint a.

6. The join of point p and ray $\overrightarrow{ab}$ if p is not on line ab. (Compare Exercise 3 above.)

7. The join of ray $\overrightarrow{ab}$ and line L if a is on L and b is not on L. (Compare Exercise 3 above.)

8. The join of the distinct rays $\overrightarrow{ab}$ and $\overrightarrow{ac}$. Try not to miss any cases.

9. The join of segments ab and ac, if ab and ac do not intersect. Try not to miss any cases.

10. The join of segments ab and cd, if ab and cd are coplanar and do not intersect. Try not to miss any cases. (Compare Example III in the text above.)

*11. The join of two noncoplanar segments ab and cd.

12. The join of segment ab and circle C, if ab is perpendicular to the plane of C and a is the center of C.

13. The join of ray $\overrightarrow{ab}$ and circle C, if $\overrightarrow{ab}$ is perpendicular to the plane of C and a is the center of C.

14. The join of line L and circle C, if L is perpendicular to the plane of C at the center of C.

15. In Exercise 12, the join of ab and I, the interior of C.

16. In Exercise 13, the join of $\overrightarrow{ab}$ and I, the interior of C.

17. The join of line L and I, the interior of circle C, if L is perpendicular to the plane of C at a point of C.

18. Suppose a is not a point of segment bc. Will their join $a(bc)$ always be the interior of a triangle as in Example I in the text above? If not, what can it be? Try not to miss any cases.

19. Suppose point p is not a point of circle C. Will their join pC always be conical as in Example II in the text above? If not, what can it be? Have you considered all cases?

In Exercises 20 through 25 suppose that a Euclidean plane has been assigned a Cartesian coordinate system. Let p be the point $(0, 0)$ and C the graph of the given equation. Sketch the join pC and describe it as accurately as you can.

20. $y = x^2 + 1$.   21. $y = x^3 + 1$.   22. $y = 1/x$.

23. $y = e^x$.   24. $y = \ln x$.   25. $y = \cos x, \ 0 \leqslant x \leqslant 2\pi$.

26. Make up and solve some of your own problems on the join of two nonintersecting figures.

1.6 Joins of Several Points: Does the Associative Law Hold?

The join operation, which had been defined for pairs of points (Section 1.2), was extended in the last section to pairs of sets of points. This extension of the scope of the operation has two important consequences. First, as appeared in the last section (particularly in the exercises), it makes possible the construction of a wide variety of geometric figures starting with simple ones. Second, it enables us to define joins of *three* or more points and raises a host of questions about the figures formed by joining three or more points.

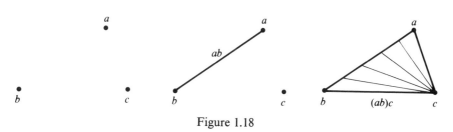

Figure 1.18

Our point of departure is quite simple. Suppose, for example, three noncollinear points a, b and c are given (Figure 1.18). Certainly ab and c are definite geometric figures. Then their join $(ab)c$ is a definite geometric figure (definition of join of two geometric figures, Section 1.5). Similarly $a(bc)$, $b(ac)$, $(ab)(ac)$, and so on, are definite geometric figures. It makes sense then to ask if these geometric figures are related. In particular, does the associative law

$$(ab)c = a(bc) \tag{1}$$

hold?

To answer the question recall that $a(bc)$ is the interior of triangle abc (Section 1.5, Example I). Consider $(ab)c$ (Figure 1.19). By definition $(ab)c$ is gotten by joining each point of ab to c and uniting all the joins formed. The result also is the interior of triangle abc, so that (1) holds.

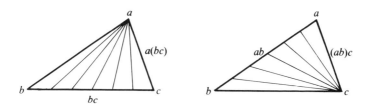

Figure 1.19

Summary. The associative law, $(ab)c = a(bc)$, is valid for any three noncollinear points a, b and c.

The result suggests the following definition.

Definition. If a, b and c are noncollinear points, each of the equal sets $(ab)c$, $a(bc)$ is denoted by abc, which is called the *join* of the points a, b and c.

A good sample of formal properties of join which supplement the commutative law (Section 1.3) and the associative law is presented in concrete form in the next exercise set. Some of the exercises involve triangles, and you may want to glance at the formal definition of the idea. You are not obliged to use it in solving a problem: you may find your intuitive notion of triangle sufficient. Here it is, expressed in join notation. (See Figure 1.20.)

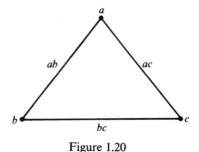

Figure 1.20

Definition. Let a, b and c be noncollinear points. Then the set consisting of a, b, c and all points of ab, ac and bc is called *triangle abc* (in symbols $\triangle abc$) or simply a *triangle*.

The definition of a tetrahedron (or triangular pyramid), which may be considered the analogue in 3-space of a triangle, appears below, following Exercise 1.

EXERCISES (PROPERTIES OF THE JOIN OPERATION)[3]

The following exercises involve the verification or the discovery of formal properties of join and are to be done with the aid of diagrams or models. Although proofs are not called for, try to clarify the intuitive geometric basis for your conclusion. In all cases only joins of nonintersecting figures are considered.

[3] The reader is advised to examine Exercises 1 and 2.

1. (a) If a, b and c are noncollinear points, verify that
$$abc = bac = \text{Interior}(\triangle abc).$$
(b) How many similar equalities can you find for abc?

Definition. Let a, b, c and d be noncoplanar points. Then the set consisting of a, b, c, d and all points of ab, ac, ad, bc, bd, cd, abc, abd, acd, bcd is called *tetrahedron abcd* or simply a *tetrahedron* (Figure 1.21).

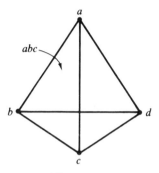

Figure 1.21

2. If a, b, c and d are noncoplanar points verify that
$$a(bcd) = (abc)d = \text{Interior}(\text{tetrahedron } abcd).$$

3. If a, b, c and d are noncoplanar points verify that
$$(ab)(cd) = \text{Interior}(\text{tetrahedron } abcd).$$
(Compare Exercise 11 at the end of Section 1.5.)

4. (a) If a, b, c and d are noncoplanar points, verify that
$$a(bcd) = c(dab) = \text{Interior}(\text{tetrahedron } abcd).$$
(b) How many similar equalities can you find for $a(bcd)$?

5. Let a and b be two distinct points.
   (a) Verify that $a(ab) = ab$.
   (b) Can you find a point x distinct from a which is not in ab and satisfies $x(ab) = ab$?
   (c) How many such points x are there?

6. Let a, b and c be noncollinear points.
   (a) Verify that $a(abc) = abc$.
   (b) Can you find a point x distinct from a which is not in abc and satisfies
   $$x(abc) = abc? \tag{1}$$
   (c) Can you find all points x which are not in abc and satisfy (1)? Can you describe the figure they form?

7. Try to make up and solve a problem similar to Exercises 5 and 6 for four noncoplanar points a, b, c, d.

8. (a) Identify the join $(\triangle abc)(abc)$.
   (b) Compare your result in (a) with that in Exercise 6(c).
   (c) Does the result in (a) extend to tetrahedrons?

9. Let I be the interior of a circle. Are there any points x that are not in I and satisfy $xI = I$?

10. (a) Let S be a closed segment. Are there any points x that are not in S and satisfy $xS = S$? Explain your answer.
    (b) The same as (a) if S is a closed circular region, that is, S is the set of all points on or interior to a circle.

11. (a) Identify pC if C is a circle and p is a point of its interior. The same if p is a point of the exterior of C. Compare your results.
    (b) Identify $p(\triangle abc)$ if p is a point of the interior of $\triangle abc$. The same if p is a point of the exterior of $\triangle abc$. Try not to miss any cases. Compare your results.

12. Suppose B is any figure (set of points) and point a is not in B.
    (a) Show that a cannot be a point of the join aB. Does this principle square with your answers to Exercise 11?
    (b) When will $a(aB) = aB$ hold?

13. If a, b and c are noncollinear points, verify that
    $$(ab)(ac) = (ba)(bc) = (ca)(cb).$$

14. Let a, b, c and d be the vertices of a convex quadrilateral, as indicated in Figure 1.22. Verify that
    $$(ab)(cd) = (ad)(bc) = \text{Interior(quadrilateral } abcd).$$
    (Compare Section 1.5, Example III.)

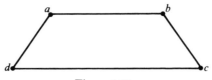

Figure 1.22

15. Does the conclusion in Exercise 14 hold if a, b, c and d are the vertices of a nonconvex (or reentrant) quadrilateral, as indicated in Figure 1.23? If not, can part of the conclusion be saved?

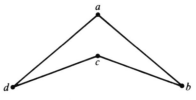

Figure 1.23

16. Let a, b, c and d be noncoplanar points. Verify that
    $$((ab)(ac))(ad) = (ab)((ac)(ad)).$$

1.7 The Join of Two Intersecting Geometric Figures

You may recall that a limitation was imposed in Section 1.5 (second paragraph following Definition) to consider only joins of nonintersecting figures, that is, figures that have no common point. Why was this done? To answer the question suppose two figures A and B do have a common point, say c (Figure 1.24). Then the definition of the join AB (Section 1.5) requires us to find the join cc. But no meaning has been assigned to the symbol cc: the join ab has been defined (Section 1.2) only if a and b are *distinct* points.

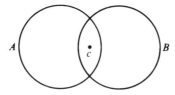

Figure 1.24

Thus consideration of the join of two intersecting figures was avoided specifically to postpone the problem of dealing with the special case of the join of a point and itself. We did this to focus attention on the *general* case of the join of two distinct points—to exhibit its essential role in constructing the join of two figures. (See Section 1.5, Examples I, II, III and Exercises.)

However, many nontrivial cases of the join AB arise where the figures A and B do intersect, for example (see Figure 1.25): the join of a segment and one of its points; the join of two intersecting segments, of two

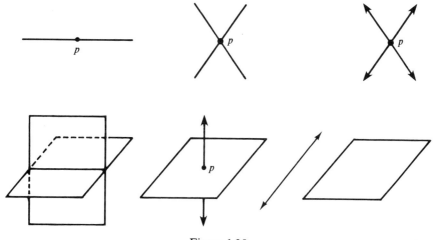

Figure 1.25

intersecting lines, or of two intersecting planes; the join of a line and a plane which intersect; and even the join of a line and itself or a plane and itself. Each of these calls for the join of a point and itself.

1.8 A Decision Must Be Made

We are confronted by a question which no longer can be put off: Should the join of a point and itself be assigned meaning? Should it be formally defined? If it is not to be defined, a question still remains: How shall the definition of the join AB of two figures A and B be interpreted if they intersect? The problem has not been manufactured artificially. It is not comparable, for example, to the question in arithmetic of defining the sum of an integer and itself. There, a uniform procedure exists for adding integers a and b independently of whether a and b are distinct or the same.

In an intuitive geometric sense the question before us is this: Does there exist a useful and reasonable special or degenerate form of the idea of segment?

1.9 The Join of a Point and Itself

We believe strongly that there does exist a suitable definition of the join of a point and itself, and present it with no more ado.

Definition. Let a be any point. We define the *join* of a and a, denoted by aa, to consist of point a, and write $aa = a$.

The definition of aa is simple—it is also, we feel, quite natural. In support of this consider the series of joins of points indicated in Figure 1.26. The first, $(abc)d$, is the interior of a tetrahedron (Exercise 2 at end of Section 1.6). Thus $(abc)d$ is a solid figure and is 3-dimensional or has *dimension number* 3. The second, abc, is a triangle interior and so is 2-dimensional or has dimension number 2. The third, ab, is a segment and

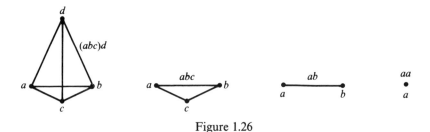

Figure 1.26

has dimension number 1. It seems reasonable to expect that aa, the final term in the series, should have dimension number 0—and it does, since aa is a, a point.

The definition of aa has a few simple but nontrivial consequences. First, the two basic properties of the join operation (Section 1.3) hold without restriction:

(1) The join ab of any two points a and b is a definite nonempty set of points.
(2) *Commutative law*: $ab = ba$ for any two points a and b.

Second, the definition of the join AB (Section 1.5) is now perfectly precise. It applies uniformly whether the figures A and B intersect or do not intersect. For the join ab of point a of A and point b of B is now precisely defined for the case $a = b$ as well as for the case $a \neq b$. In simple terms the definition of AB says: x is a point of the join AB of the figures A and B if and only if (1) x is between some point of A and some point of B, or (2) x is a point common to A and B (see Figure 1.27).

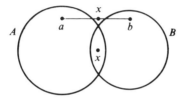

Figure 1.27

Summary Principle. The join AB of any two geometric figures A and B is a definite geometric figure.

EXERCISES (JOINS OF TWO INTERSECTING FIGURES)

In Exercises 1 through 12 identify and describe the indicated joins. You may use diagrams or models. No proof is required.

1. The join of point p and circle C if p is a point of C.

2. The join of point p and triangle T if p is a point of T and not a vertex; if p is a vertex.

3. The join of point p and line L if p lies on L.

4. The join of two distinct intersecting lines.

5. The join of two intersecting segments not necessarily distinct. Try not to miss any cases.

6. The join of a line with itself; a plane with itself.

7. The join of a circle with itself.

8. The join of circle C and line L if L lies in the plane of C and passes through its center.

9. The join of line L and I, the interior of circle C, if L is perpendicular to the plane of C at the center of C.

10. The join of two circles each of which passes through the center of the other.

11. The join of a line and a plane which intersect. Try not to miss any cases.

12. The join of two distinct intersecting planes.

In Exercises 13 through 17 suppose that a Euclidean plane has been assigned a Cartesian coordinate system. Let p be the point $(0, 0)$ and C the graph of the given equation. Sketch the join pC and describe it as accurately as you can.

13. $y = x^2$.

14. $y = x^3$.

15. $y = \sin x$, $0 \leqslant x \leqslant 2\pi$.

16. $y = \tan x$, $-\pi/2 < x < \pi/2$.

17. $y = xe^x$.

18. Make up and solve some of your own problems on the join of two intersecting figures.

1.10 The Unrestricted Applicability of the Join Operation

In ordinary numerical algebra exceptional cases sometimes arise in which an operation can't be performed. A familiar example is the application of the division operation to the numbers 1 and 0. Now that aa has been defined, this cannot happen for the join operation. There are no exceptional cases: the operation join is applicable without restriction to any finite number of points. As an illustrative example take the case of three points. Let a, b and c be any points, and consider $a(bc)$. The expression $a(bc)$ stands for the join of two geometric figures, namely, a and bc. By the Summary Principle (Section 1.9, last paragraph), $a(bc)$ is a definite geometric figure. Similar considerations apply to $(ab)c$, $(ab)(ac)$ and so on.

Suppose you want to see concretely what the figure $a(bc)$ is—or better, what it can be. Choose positions for the points a, b and c, find bc, and apply the definition of the join of two figures (Section 1.5) to a and bc. An example of this has already appeared: if a, b and c are noncollinear, we have seen (Section 1.5, Example I) that $a(bc)$ is a triangle interior (Figure 1.28).

What happens if a, b and c are collinear? First suppose they are distinct. If a is in bc (Figure 1.29), then $a(bc)$ involves the join aa and turns out to

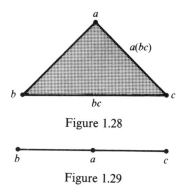

Figure 1.28

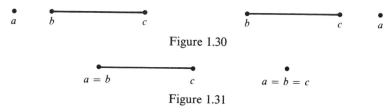

Figure 1.29

be the segment *bc*. If *a* is not in *bc* (Figure 1.30), the join *a(bc)* is still a segment, either *ac* or *ab*. If *a*, *b* and *c* are not distinct (Figure 1.31), you will find, by examining the cases, that *a(bc)* will always be a segment or a point.

Figure 1.30

Figure 1.31

Summary. The join *a(bc)* is a definite geometric figure for any three points *a*, *b* and *c*—regardless of their relative positions and regardless of whether or not they are distinct.

The same conclusion holds of course for *(ab)c*.

EXERCISES

1. Suppose the points *a*, *b* and *c* are collinear. List the possible relations of *a*, *b* and *c*, including coincidences, and identify *(ab)c* in each case.

2. Suppose the points *a*, *b*, *c* and *d* are distinct and coplanar, and no three of them are collinear. List the possible relations of *a*, *b*, *c* and *d*, and identify *(ab)(cd)* in each case.

3. The same as Exercise 2 for *((ab)c)d*.

1.11 The Unrestricted Validity of the Associative Law for Join

Recall (Section 1.6) that the associative law for join

$$(ab)c = a(bc) \tag{1}$$

holds with the restriction that *a*, *b* and *c* be noncollinear. In the last section we saw that the expressions $(ab)c$ and $a(bc)$ in (1) represent definite geometric figures for *all* choices of points a, b and c. We can hardly disregard the question: Can the restrictions be removed—does (1) hold for all choices of a, b and c?

Yes, it does. To justify the answer we argue by cases—of which there are many.

First suppose a, b and c are collinear and distinct. Then, by a basic principle of geometry, one of the three points must be between the other two. Consider first the case where b is between a and c (Figure 1.32). Then $(ab)c$ is simply the segment ac. Similarly $a(bc) = ac$, so that (1) holds. So far it has not been necessary to employ the definition of the join of a point and itself.

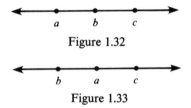

Figure 1.32

Figure 1.33

Next consider the case where a is between b and c. First note that $(ab)c = bc$. Then to get $a(bc)$ we must join a to each point of bc. Observe that bc falls into three parts: The points between b and a, the points between c and a, and a itself. So we can find $a(bc)$ by joining a to each of these parts. The results are, in order: the points between b and a, the points between c and a, and (by definition of aa) the point a itself. Uniting the three results, we get bc. Thus $a(bc) = bc$, so that $(ab)c = a(bc)$ and (1) holds in this case also.

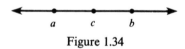

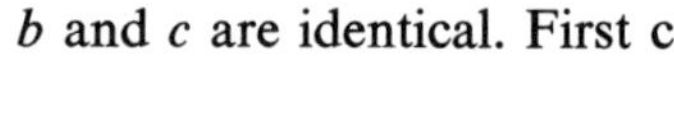

Figure 1.34

The case where c is between a and b (Figure 1.34) can be treated essentially in the same way as the preceding case.

Finally there remain the cases in which a, b and c are not distinct. Suppose just two of a, b and c are identical. First consider the case $a = b$ (Figure 1.35). Then

$$(ab)c = (aa)c = ac$$

and

$$a(bc) = a(ac) = ac,$$

so that (1) holds.

$$a = b \qquad\qquad\qquad\qquad c$$

Figure 1.35

Similar methods apply to the cases $b = c$ and $a = c$ (Figure 1.36).

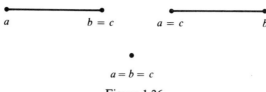

$$a = b = c$$

Figure 1.36

At long last we have the final case $a = b = c$. Now (1) reduces to $a = a$, and our proof is finished.

Summary. The associative law $(ab)c = a(bc)$ is valid for all points a, b and c.

As in Section 1.6, the associative law suggests a definition.

Definition. If a, b and c are any points, each of the equal sets $(ab)c$, $a(bc)$ is denoted by abc, which is called the *join* of the points a, b and c.

Remark 1. You recall the notion of an *identity* in classical algebra: an equation involving one or more variables which holds for all values of the variables. An example is the equation $(a + b)^2 = a^2 + 2ab + b^2$, which holds for all *numbers* a and b. The associative law $(ab)c = a(bc)$ is a *geometrical* identity which holds for all *points* a, b and c.

Remark 2. If you examine the proof of the associative law, you will find that it depends on the definition $aa = a$. Thus the associative law, which will play a key role in our theory, is a consequence of this definition.

EXERCISES

1. Verify by means of diagrams that $a(bcd) = (abc)d$, where a, b, c and d are any four distinct coplanar points, no three of which are collinear. Try not to miss any cases. (Compare Exercise 2 at the end of Section 1.6.)

2. Let a, b, c and d be the vertices of a convex quadrilateral as indicated in Figure 1.37. Verify that

$$(ac)(bd) = \text{Interior(quadrilateral } abcd).$$

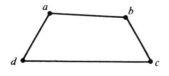

Figure 1.37

Did you apply the definition of join for a point and itself? Explain. (Compare Section 1.5, Example III.)

3. Let p be a point of segment ab. Verify that $p(ab) = ab$.

4. Let p be a point of triangle interior abc. Verify that $p(abc) = abc$.

1.12 The Universality of the Associative Law

The fact that the associative law holds universally for three points—not merely for three points in general position—is of great importance for the development of our theory. It indicates the possibility of constructing a theory of geometry in which the basic principles hold uniformly: they state universal properties of points that admit no exception. In algebra the desirability of this is taken for granted almost without discussion. The situation is quite different in geometry.

In the standard treatment of geometry the basic principles usually are not universal properties but hold for points in general position; for example: two distinct points determine a line, or three noncollinear points determine a plane. As a result the proofs of theorems sometimes require lengthy and involved discussion of special or degenerate cases.

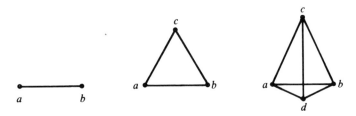

Figure 1.38

Moreover the standard treatment is figure-oriented in a rather rigid way. Consider the basic linear, planar and spatial figures: segment, triangle, tetrahedron (Figure 1.38). If you wish to refer to a segment, you must assign two distinct points; to a triangle, three noncollinear points; to a tetrahedron, four noncoplanar points. Why should we not be able to study a degenerate or squashed form of a triangle which would be determined by three collinear points, or a degenerate form of a tetrahedron determined by four coplanar points, as in Figure 1.39? We should be able to study four points assigned at random, without knowing whether they are in space, in a plane, in a line, or even coincident (Figure 1.40).

Our object in writing this book is to construct an operation-centered treatment of geometry of a *special* type: one in which the basic laws are *universal* properties of points. The associative law is our first nontrivial

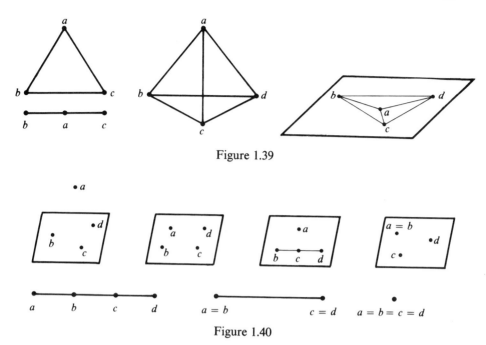

Figure 1.39

Figure 1.40

example of such a basic law. The validity of the associative law gives strong impetus to the search for other universal laws and foreshadows in a way the ultimate success of the project.

1.13 Alternatives to the Definition $aa = a$

Note to the Reader. This section may be omitted without loss of continuity. It is not a logical prerequisite to the remainder of the chapter but is designed to shed light on our choice of definition for aa. It indicates that the definition is the only one which is satisfactory for our purposes. If the definition did not seem quite right or you are curious about the issue, you might read the section.

The importance of the definition $aa = a$, and the ultimate reason for its adoption, lie in its utility—its effectiveness in organizing the geometric material we are studying. It has many nontrivial consequences, the most important of which is the unrestricted validity of the associative law for join.

Nevertheless you may feel that the definition is not quite right, and you may object: How do we know that an alternative definition won't work just as well?

Consider the following alternative.

Definition. For each point a, let $aa = \varnothing$, the empty set.

To define aa in this way may not seem unnatural. You may feel that the appropriate choice for a special or degenerate case of a *closed* segment should be a figure F consisting of a single point a. Then you may agree that the natural choice for a special or degenerate case of an *open* segment should be F with a deleted, that is, the empty set.

Whether the argument seems reasonable or not, the definition merits consideration as an alternative to $aa = a$.

Let us begin by asking a question: What effect does the definition have on the meaning of the join AB of two figures A and B? Suppose A and B do not intersect and so have no common point (Figure 1.41). Then the definition is not employed in forming the join AB and so has no effect on its meaning. Thus AB is obtained as usual, by uniting all the segments which join a point of A to a point of B.

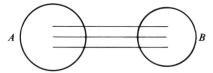

Figure 1.41

Now suppose A and B do have a common point, say c (Figure 1.42). Then $cc = \varnothing$, the empty set. Thus the join cc will contribute no point in forming the join AB. In other words, if c is common to A and B, the join cc can be disregarded in forming AB. Consequently the definition of AB is *now* equivalent to the following statement:

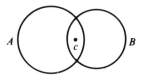

Figure 1.42

A point x is in AB if and only if x is in a segment which joins a point of A to a (distinct) point of B (Figure 1.43).

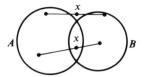

Figure 1.43

What consequences does the alternative definition have? In particular will the associative law

$$(ab)c = a(bc) \tag{1}$$

hold? Certainly (1) holds for the case where a, b and c are three noncollinear points—this was settled (Section 1.6) before the definition $aa = a$ was introduced (Section 1.9). But definitely (1) fails for the case where a is in bc (Figure 1.44).

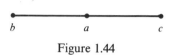

Figure 1.44

To justify this, suppose a is in bc. Then we show that a is in $(ab)c$ but a is not in $a(bc)$. Since c is not in ab, the join of ab to c does not involve the join of a point to itself. Hence $(ab)c = bc$. But a is in bc. Thus a is in $(ab)c$.

Is a a point of $a(bc)$? If this is so, then, by the principle above, a must be in a segment which joins a to a (distinct) point, say p, of bc (Figure 1.45). That is, a must be in the (open) segment ap. This is absurd. Consequently a is not in $a(bc)$, and the sets $(ab)c$ and $a(bc)$ cannot be the same. Thus (1) fails if a is in bc.

Figure 1.45

Summary. If we adopt the alternative definition $aa = \varnothing$, then the associative law fails—that is, it is not universally valid.

Will you conclude that it is illogical, or mathematically unsound, to adopt the alternative definition? It seems to us inconvenient rather than illogical. If the definition is adopted, a sound mathematical theory will result which is somewhat similar to our theory but has the disadvantage that the associative law will not hold for all triples of points.

EXERCISES (ALTERNATIVES TO THE DEFINITION $aa = a$)

In Exercises 1, 2, 3 assume the definition $aa = \varnothing$.

1. Let p be a point of segment ab. Identify $p(ab)$ and compare it with ab. Are they the same? Are they different? (Compare Exercise 3 at the end of Section 1.11.)

2. Let p be a point of triangle interior abc. Identify $p(abc)$ and compare it with abc. Are they the same? Are they different? (Compare Exercise 4 at the end of Section 1.11.)

3. Let *a*, *b*, *c* and *d* be the vertices of a convex quadrilateral as indicated in Figure 1.46. Identify $(ac)(bd)$ and compare it with the interior of quadrilateral *abcd*. (Compare Exercise 2 at the end Section 1.11.)

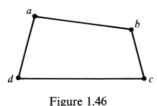

Figure 1.46

4. Let *a* be a particular point. Suppose that *aa* has been defined to be a definite set of points which does not contain *a*, for example, a circle of radius 1 centered at *a*. Choose *b* and *c* so that *a* is a point of segment *bc* (Figure 1.47). Show that

$$(ab)c \neq a(bc),$$

so that the associative law for join fails.

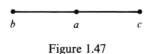

Figure 1.47

Remark 1. Suppose you want the associative law for join to hold. Exercise 4 shows that if you define *aa* for a point *a*, then *aa* must contain *a*.

5. Let *a* be a particular point. Suppose that *aa* has been defined to be a definite set of points, but that *aa* contains a point *p* which is distinct from *a*. Choose *b* and *c* so that *a* is a point of segment *bc* but *p* is not (Figure 1.48). Show that

$$(ab)c \neq a(bc),$$

so that the associative law for join fails.

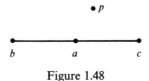

Figure 1.48

Remark 2. Suppose you want the associative law for join to hold. Exercise 5 shows that if you define *aa* for a point *a*, then *aa* cannot contain any point other than *a*.

Combining Remarks 1 and 2, we have: If the associative law for join is to hold and *aa* is to be defined for a given point *a*, then *aa* must inexorably be defined to consist of *a*. Moreover there is no loophole. You can't save the associative law by refusing to define *aa*. Suppose *aa* is not defined. Let *b* be a point distinct from *a* (Figure 1.49). Then the special case of the associative law

$$(aa)b = a(ab)$$

fails, since the left member is meaningless and the right member is *ab*.

a   b

Figure 1.49

1.14 Convex Sets

One of the most basic and important ideas in geometry is that of convex set. All of the familiar figures in high school geometry are either (1) convex sets:

or (2) figures formed by uniting convex sets:

or (3) boundaries of figures of types (1) and (2):

Intuitively, a convex set is a figure with no gaps

no holes

no indentations

The precise definition is even easier to state.

Definition. A set of points is *convex* if it contains the join of each pair (distinct or not) of its points.

Observe that a set consisting of a single point is convex.

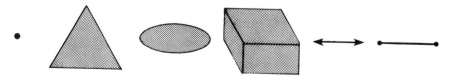

Figure 1.50

EXERCISES (EXPLORATORY)

1. If A and B are convex sets, must their union be convex? Can their union be convex? Justify your answer.

2. Let A and B be convex sets that have a common point. Must the intersection of A and B (that is, the set of all points which are common to A and B) be convex? Can their intersection be convex? Justify your answer.

3. (a) Is the join of two points convex?
   (b) The same for a point and a segment.
   (c) The same for two segments.
   (d) The same for two convex sets.

4. Suppose A is convex and p is a point not in A. Must the union of A and p be convex? Can it be convex? Explain.

5. Suppose A is convex and p is a point of A. Let set B be formed by deleting p from A. Must B be convex? Can B be convex? Explain.

6. (a) Suppose point p is not in convex set A but p is in qA for some point q. Can q be in pA?
   (b) Try to prove that your answer is correct.

7. (a) If p is a point of convex set A, must pA be contained in A? Must $pA = A$? Justify your answers.
   (b) If A is convex must AA be contained in A? Must $AA = A$? Justify your answers.

8. (a) Find a convex set A such that $pA = A$ for each point p in A.
   (b) Try to find several such sets A not all of the same type. Do you observe any common property?
   (c) Take your convex set A in (a) and try to find a point p not in A such that $pA = A$. Can you describe the figure formed by all such points?
   (d) The same as (c) for each of the sets in (b).

9. Can you find a convex set A such that $pA = A$ for each point p in A but $pA \neq A$ for each point p not in A? Try to find several. Can you find a common property?

Definition. Let A be a convex set and p a point. Suppose there is a join ab which contains p such that every point of pa is in A but no point of pb is in A (Figure 1.51). Then we call p a *boundary point* of convex set A. The set of all boundary points of A is called its *boundary*. Any point of A which is

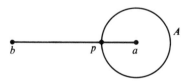

Figure 1.51

not a boundary point of A is called an *interior point* of convex set A. The set of all interior points of A is called its *interior*.

10. (a) If p is a boundary point of convex set A, must p be in A? Can p be in A? Explain.
   (b) If p is a point of convex set A, must p be a boundary point of A? Can p be a boundary point of A? Explain.

11. Suppose p is a boundary point of convex set A.
   (a) Can pA be contained in A? Must pA be contained in A? Explain.
   (b) Can $pA = A$? Must $pA = A$? Explain. [Compare Exercise 7(a).]

12. (a) Can you find a convex set whose boundary consists of a single point; two points; three points; four points?
   (b) Can you find a convex set whose boundary is a line; a plane; a segment?

13. (a) Try to find a convex set which contains none of its boundary points. Try to find several such convex sets of different types. Do you observe anything of interest?
   (b) The same as (a) for a convex set that contains all of its boundary points.

14. Try to find a convex set which has no boundary points. Try to find several of different types. Do you observe anything of interest?

15. (a) Suppose A is a convex set and p is a boundary point of A. If you adjoin p to A, can the resulting set be convex? Must the resulting set be convex? Explain.
   (b) If you adjoin to a convex set A *all* of its boundary points, can the resulting set be convex? Must it be convex? Explain.

16. (a) Suppose A is a convex set and p is a point of A which is one of its boundary points. If you delete p from A, can the resulting set be convex? Must the resulting set be convex? Explain.
   (b) If you delete from a convex set A *all* of its points which are boundary points, can the resulting set be convex? Must it be convex? Explain.

17. Can the boundary of a convex set be convex? Must it be convex? Explain.

18. Study 1-dimensional convex sets, that is, those that are not single points but are contained in a line. Try to classify them in terms of: (1) the nature of their boundary sets; (2) whether they extend endlessly or not; (3) whether they contain all, some, or none of their boundary points. Compare your results.

19. Mark two points a and b on a sheet of paper.
   (a) Sketch a convex set A that contains a and b.
   (b) Can you find a convex set containing a and b which is "greater than" A, that is, one which *contains* A?
   (c) Can you find a convex set containing a and b which is "less than" A, that is, one which *is contained in* A?
   (d) Can you find a convex set containing a and b which is neither "greater than" nor "less than" A?
   (e) Can you find a "greatest" convex set that contains a and b, that is, one which contains *every* convex set that contains a and b?

(f) Can you find a "least" convex set that contains a and b, that is, one which is contained in *every* convex set that contains a and b?

20. (a) Find a least convex set that contains (1) three noncollinear points a, b and c; (2) four noncoplanar points a, b, c and d; (3) three distinct collinear points a, b and c; (4) four coplanar points a, b, c and d, no three of which are collinear.

(b) Is there a greatest convex set in each case? Is the concept interesting?

1.15 A Geometric Proof in Join Terminology

Note to the Reader. The present section is optional and may be omitted without loss of continuity. It shows how the join operation can be used in proving geometric theorems and discusses the nature of proofs in geometry. Study it if you wish for your own satisfaction.

The join operation is used not merely in systematically constructing and describing figures, but also in constructing geometric proofs. The type of reasoning is not difficult or complex, but it has a quite different aspect from conventional geometrical reasoning—so we discuss an example in great detail.

Problem. Prove

$$(ab)(cd) = a(bcd), \tag{1}$$

using the associative law for join (Section 1.11) and the definitions of the join of two points and the join of two sets of points.

To prove (1) we must show that each point in the set $(ab)(cd)$ is in the set $a(bcd)$, and conversely that each point in $a(bcd)$ is in $(ab)(cd)$. So we suppose

$$x \text{ is in } (ab)(cd) \qquad \text{(Figure 1.52)}, \tag{2}$$

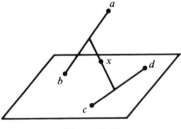

Figure 1.52

and we derive

$$x \text{ is in } a(bcd) \qquad \text{(Figure 1.53)}. \tag{3}$$

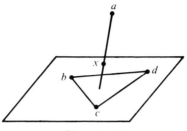

Figure 1.53

In order to get from (2) to (3) we must somehow break up the pair *ab* in (2) and associate *b* with *cd* as in (3). This "reassociation" will be justified by the associative law. To apply the associative law, we convert (2) to a form which asserts that *x* is in a join of three points. This takes two steps. Applying the definition of the join of the sets *ab* and *cd* to (2), we see that *x* must be in the join of a point in *ab* and a point in *cd*. Thus (2) implies

$$x \text{ is in } pq, \quad \text{where } p \text{ is in } ab \text{ and } q \text{ is in } cd \quad \text{(Figure 1.54)}. \quad (4)$$

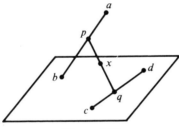

Figure 1.54

Comparing (4), which we have, and (3), which we want to get, we see that *x* must be related directly to *a*. Observe by (4) that *x* is in the join of a point *p* in *ab* and point *q*. Thus

$$x \text{ is in } (ab)q \quad \text{(Figure 1.55)} \quad (5)$$

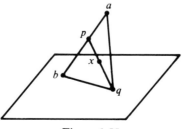

Figure 1.55

by definition of the join of set ab and point q. In effect we have eliminated p in the relations "x is in pq", "p is in ab", of (4) and have related x directly to ab. Thus (2) has been replaced by (5) which brings us closer to (3). Now the associative law is pleading to be applied in (5) and yields

$$x \text{ is in } a(bq) \qquad (\text{Figure 1.56}). \qquad (6)$$

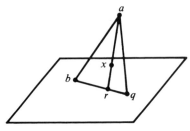

Figure 1.56

The relation (6) implies

$$x \text{ is in } ar, \quad \text{where } r \text{ is in } bq. \qquad (7)$$

Comparing (7), which we have, and (3), which we want, we conjecture r is in bcd. By (7) and (4), r is in bq, and q is in cd. These imply

$$r \text{ is in } b(cd) = bcd \qquad (\text{Figure 1.57}). \qquad (8)$$

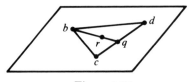

Figure 1.57

Finally by (7) and (8), x is in ar, and r is in bcd, which imply

$$x \text{ is in } a(bcd) \qquad (\text{Figure 1.58}).$$

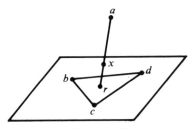

Figure 1.58

Thus we have shown that (2) implies (3).

To complete the proof, the converse, that (3) implies (2), must be shown. This can be proved essentially by reversing steps and is left as an exercise for the reader. (See Exercise 1 below.)

Query. Examine the proof. Can you verify each statement in the associated diagram? Are the diagrams helpful in constructing the proof? Are they necessary in testing the correctness of the proof? Is the formal proof valid if a, b, c and d are not in general position: for example, if they are coplanar and not the vertices of a tetrahedron as pictured?

Response. The proof as it stands is a formal logical argument which— though illustrated by diagrams—is independent of them. To verify this you can check the successive steps in the proof while keeping the diagrams covered. The fact that the proof is independent of the diagram is quite striking. It foreshadows a development of geometry based on formal properties of the join operation which, though possessed of a rich intuitive geometric content, will permit the construction of proofs that are verifiable without reference to that content. Often, of course, theorems will be *conjectured* and proofs *discovered* by employing diagrams. But the proofs themselves—to be considered valid—should not depend on diagrams. In high school geometry as usually developed, and in Euclid, steps sometimes appear in a proof which are justified only by appeal to a diagram.[4]

Observe, finally, that the relation $(ab)(cd) = a(bcd)$, though expressed in a cool algebraic formalism, carries several apparently unrelated items of concrete geometric information. This is illustrated in Figure 1.59.

EXERCISES (PROOFS IN JOIN TERMINOLOGY)

1. Complete the proof of equation (1) above by showing that (3) implies (2). Make diagrams to illustrate the steps.

In the following exercises prove the principle stated using the associative and commutative laws for join and the definitions of join of two points and join of two sets of points. In one or two exercises illustrate the steps in your proof by diagrams. Try to state the principles for the general or nondegenerate case, as properties of triangles or segments.

2. $(ab)(ac) = abc$.

3. $(ab)(ab) = ab$.

4. $a(abc) = abc$.

5. $(ab)(abc) = abc$.

6. $(abc)(abc) = abc$.

[4] See Prenowitz and Jordan [1], Chapter 1.

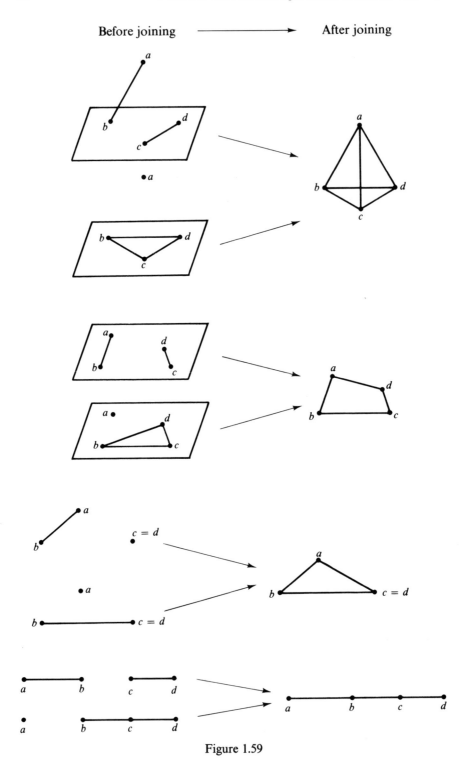

Figure 1.59

1.16 A New Operation: The Idea of Extension

Although the join operation is the most important operation in elementary geometry, it is not the only important one. For we often encounter the notion of extending or prolonging a segment, as indicated by the phrase: Extend segment *ab* beyond its endpoint *a* to a point *p* (Figure 1.60).

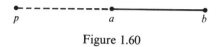

Figure 1.60

Suppose *ab* is extended endlessly beyond point *a*. Then the new points obtained, namely, the points that are situated beyond *a* from *b* (Figure 1.61), form a new type of linear figure—one which is not a segment and not a line. Observe that the figure is determined if we know the location of points *a* and *b*. Consequently there is no need to refer to segment *ab* in describing the figure—and it may be called simply the *extension* of *a* from *b*. This new idea is easily characterized in terms of the notion of betweenness.

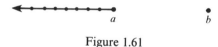

Figure 1.61

If *a* and *b* are two distinct points the extension of *a* from *b* consists of all points *x* such that *a* is between *x* and *b* (Figure 1.62).

Figure 1.62

1.17 The Notion of Halfline or Ray

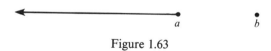

Figure 1.63

The extension of *a* from *b* is a portion of line *ab* which is bounded by point *a* and extends endlessly away from point *b*. Such figures occur in high school geometry and are called rays or halflines. A side of an angle (Figure 1.64) is a familiar example.

The notion of ray or halfline can be described informally as follows. Let *a* and *b* (Figure 1.65) be two distinct points. The *ray* or *halfline ab*, denoted by the symbol $\overrightarrow{ab}$, consists of all points of line *ab* that lie on the same side

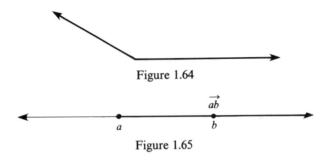

Figure 1.64

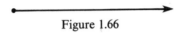

Figure 1.65

of point a as b. Point a is called the *endpoint* of $\overrightarrow{ab}$. Observe that a ray does not contain its endpoint and is, in a sense, an "open" figure like a segment. Sometimes it is necessary to study the figure consisting of a ray and its endpoint (Figure 1.66): this is called a *closed* ray. In contrast, a ray is sometimes referred to as an *open ray*.

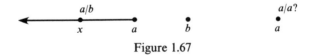

Figure 1.66

1.18 Formal Definition of the Extension Operation

In the last two sections we have been discussing somewhat informally the operation of extension and the related notion of ray. Now we proceed to give a *formal* definition of the operation of extension—and we frame it in terms of the join concept. The definition—in contrast with the informal treatment of the extension of a from b in Section 1.16—applies to any two points, distinct or coincident; it does not require the restriction $a \neq b$.

Definition. Let a and b be arbitrary points, not necessarily distinct. Then (Figure 1.67) the set of all points x such that bx contains a is called the *extension of a from b* or simply the *extension of a and b*, and is denoted by a/b (read "*a over b*" or "*a stroke b*").

Figure 1.67

1.19 Identification of Extension as a Geometrical Figure

First we show that the extension of two distinct points is a ray. Consider a/b, where $a \neq b$ (Figure 1.68). Choose a point c such that a is between b and c. Then a/b can be identified as the ray $\overrightarrow{ac}$, that is,

$$a/b = \overrightarrow{ac}.$$

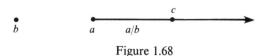

Figure 1.68

To justify this observe that each point of a/b is in $\overrightarrow{ac}$, and reversely, each point of $\overrightarrow{ac}$ is in a/b. Thus a/b is a ray.

Remark. In order to link the new concept of extension with a familiar notion in high school geometry, we have identified the extension a/b ($a \neq b$) with the idea of ray or halfline. However, a *prior* knowledge of the ray concept is not needed to grasp the idea of extension. Just the reverse! Once extension has been formally introduced (Section 1.18), it is perfectly feasible to define a ray as an extension a/b, where a and b are distinct points.[5]

Consider now the extension of a point and itself. The definition of a/b does not impose the condition $a \neq b$ and so assigns to a/a a definite meaning. a/a is simply the set of points x that satisfy the condition

$$ax \text{ contains } a \qquad \text{(Figure 1.69)}. \qquad (1)$$

$$\bullet \atop x \qquad\qquad a/a$$
$$\bullet \atop a$$

Figure 1.69

Certainly $x = a$ satisfies (1), since $aa = a$. Does any point other than a satisfy (1)? The answer is no—since if $x \neq a$, then ax is a segment and cannot contain the endpoint a. Thus a/a consists of point a, and we write $a/a = a$. We may think of a/a as a "degenerate" ray (or halfline) which consists of a single point and has no associated direction in space.

Summary. a/b is either a ray or a single point.

1.20 Properties of the Extension Operation

First we show, by reasoning directly from the definition of extension, that

(1) a/b does not contain a or b, provided $a \neq b$.

We are given $a \neq b$. Suppose a/b contains a. Then by definition of a/b (Section 1.18) the join ba must contain a. But this is impossible, since the segment ba can't contain one of its endpoints. Thus a/b can not contain a. Similarly, but more easily, we can show a/b does not contain b.

Note that extension, like join (Section 1.9) is an operation which associates with two points, in a given order, a definite nonempty set of

---

[5] See Prenowitz and Jordan [1], Chapter 10, particularly Sections 5, 8.

points. It differs from join, however, in not being commutative—indeed (Figure 1.70),

$$a/b \neq b/a \quad \text{if } a \neq b.$$

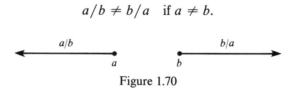

Figure 1.70

Next observe that the definition of a/b (Section 1.18) says simply

(2) x is in a/b if and only if bx contains a (see Figure 1.71).

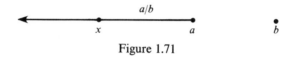

Figure 1.71

The condition (2) for x to be a point of a/b can be reformulated in the language of high school geometry:

(3) x is in a/b if and only if (i) a is between b and x (Figure 1.71), or (ii) $x = a = b$ (Figure 1.72).

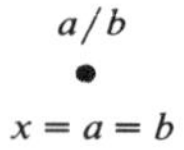

$$a/b$$
•
$$x = a = b$$

Figure 1.72

To justify (3) we show first that if x is in a/b then (i) or (ii) must hold. Suppose x is in a/b. Then by (2), a is a point of bx. First suppose $b \neq x$. Then bx is a segment and we have (i) a is between b and x. Next suppose $b = x$. We show $a = b$. Since $b = x$, the join bx equals the join bb, which consists of point b. But a is in bx. Hence $a = b$. This with $x = b$ justifies (ii). Thus in any case (i) or (ii) holds.

Conversely suppose (i) or (ii) holds. In either case a is in the join bx. Then by (2) above x is in a/b. Thus the justification of (3) is complete.

Finally, an application of the extension operation gives a neat and symmetrical formula for the line determined by two distinct points a and b:

(4) Line $ab =$ the union of ab, a/b, b/a, a and b.

Observe in Figure 1.73 that no two of the components of line ab in (4) have a common point.

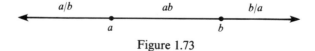

Figure 1.73

1.21 The Extension of Two Geometric Figures

To make further progress we generalize the operation extension from points to geometric figures, or sets of points, just as we did for join (Section 1.5).

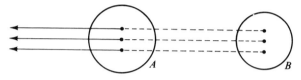

Figure 1.74

Definition. Let A and B be two sets of points (Figure 1.74). The *extension of set A from set B* is the set of points obtained by extending each point a of A from each point b of B and uniting all extensions a/b formed in this way. The extension of A from B is denoted by A/B (read "A over B" or "A stroke B"). If A consists of a single point a, we write A/B as a/B. Similarly, if B reduces to a point b, we write A/B as A/b.

There is no difficulty with the definition of A/B if the sets A and B intersect, since a/b has been defined for all a and b, distinct or identical.

In simple terms the definition says: x is a point of A/B if and only if there exists a point a of set A and a point b of set B such that x is in a/b. (See Figure 1.75.)

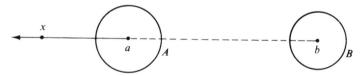

Figure 1.75

This can be reformulated, by applying the relation (3) of Section 1.20 above, in the following way:

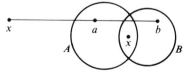

Figure 1.76

x is a point of A/B if and only if (i) some point of A is between some point of B and x or (ii) x is common to A and B.

Illustrations of A/B. Let A be a point a, and B a circle whose plane does not contain a (Figure 1.77). Then $A/B = a/B$ is a conical surface with apex a composed of all the rays a/b for all points b of B. The conical surface does not contain the apex a, since a/b does not contain point a when $a \neq b$ [Section 1.20, relation (1)].

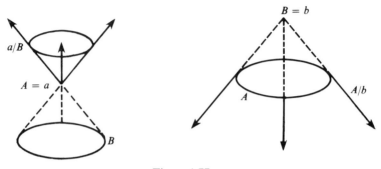

Figure 1.77

Suppose similarly B is a point b and A is a circle whose plane does not contain b. Then $A/B = A/b$ is the union of all the rays a/b for all points a of A. Thus A/B is a portion of the conical surface which has apex b and contains circle A: namely, the unbounded portion of the conical surface which is cut from it by the plane of A.

1.22 The Generation of Unbounded or Endless Figures

If the operation join is applied to two or three or more points $a_1, a_2, \ldots, a_n$ (Figure 1.78), the result is a "bounded" or "limited" figure F —one that does not extend endlessly. F will be contained, for example, in the interior of some circle or of some sphere. Moreover, if two figures A and B are bounded, their join AB will always be bounded (Figure 1.79).

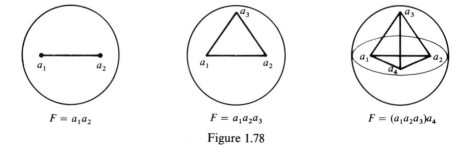

$F = a_1a_2$ $F = a_1a_2a_3$ $F = (a_1a_2a_3)a_4$

Figure 1.78

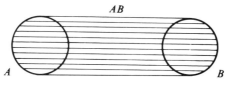

Figure 1.79

The extension operation behaves in a very different way. If A and B are (nonempty) figures (Figure 1.80), the extension A/B will be unbounded with a single exception: the case where A and B are the same point. The operation of extension then enables us to construct new types of geometric figures, not attainable by the application of join to points, segments or other familiar bounded figures. Indeed lines, planes and even 3-dimensional spaces are constructable by applying extension to very simple bounded figures. (See Exercise 14 below.)

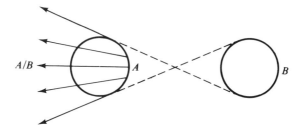

Figure 1.80

A final observation. You will find, we think, that the extension operation plays an important role in the theory. Not merely is extension an essential supplement to join in the construction of geometric figures, but the two operations are so strongly interwoven in the theory that the importance of each is enhanced by the other.

EXERCISES (EXTENSIONS OF TWO FIGURES)

In Exercises 1–14 sketch diagrams for, and describe, the figures defined by the given expressions. Unless the contrary is indicated, assume the points involved are in general position: that is, given a and b, assume them distinct; a, b and c, assume them noncollinear; a, b, c and d, assume them noncoplanar.

Notation. $\{a, b\}$ denotes the set whose points are a, b; $\{a, b, c\}$ the set whose points are a, b, c; and so on.

1. (a) $a/\{b, c\}$; (b) $\{b, c\}/a$; (c) $a/\{b, c, d\}$; (d) $\{b, c, d\}/a$; (e) $a/\{a, b\}$; (f) $a/\{a, b, c\}$.

2. $a/(bc)$. What happens if a, b, c are not in general position? Classify the cases and state the results.

3. $(ab)/c$.   4. $a/(bcd)$.   5. $(abc)/d$.

6. $a/(ab)$ and $(ab)/a$.   7. $a(a/b)$ and $b(a/b)$.

8. $a/(a/b)$ and $b/(a/b)$.

9. $a/(abc)$ and $(abc)/a$.

10. $a/(a(bcd))$ and $(a(bcd))/a$.

11. $a/(\triangle abc)$ and $(\triangle abc)/a$.

12. a/C and C/a, where a is a point of circle C.

13. The same as Exercise 12, where a is an interior point of C; an exterior point of C.

14. A/A where A is (a) $\{a, b\}$; (b) a circle; (c) a triangle; (d) a sphere; (e) a tetrahedron; (f) $\{a, b, c\}$; (g) a segment ab; (h) a circle interior; (i) a triangle interior abc; (j) the interior of a sphere; (k) a tetrahedron interior $a(bcd)$; (l) the union of two concentric circles; (m) the ring-shaped region between two concentric circles; (n) a line; (o) a plane; (p) a 3-space; (q) an open semicircle; (r) a closed semicircle.

In Exercises 15–20 suppose that a Euclidean plane has been assigned a Cartesian coordinate system. Let p be the point $(0, 0)$ and C the graph of the given equation. Sketch the extensions p/C and C/p, and describe them as accurately as you can.

15. $y = x^2$.   16. $y = x^3$.   17. $y = 1/x$.

18. $y = e^x$.   19. $y = \ln x$.

20. $y = \sin x$, $0 \leqslant x \leqslant 2\pi$.

21. Make up and solve some of your own problems on the extension of two figures.

EXERCISES (MISCELLANEOUS)

1. (a) Prove: AA contains A for any set A, that is, each point of A is a point of AA.
   (b) The same for A/A.

2. (a) Under what conditions does $(ab)/a$ contain point a? Can you give a formal proof to justify your answer?
   (b) The same for $(ab)/a$ and point b.

3. (a) The same as Exercise 2 for $(a/b)b$.
   (b) The same for $(a/b)a$.

4. (a) What is L/a if L is a line and a is a point not in L?
   (b) In (a) suppose a is in L.

(c) The same as (a) and (b) for a/L.

5. In Exercise 4 replace line L by plane P.

6. Let L and M be two lines. What is L/M if
   (a) L and M intersect in a single point;
   (b) L and M are parallel;
   *(c) L and M are skew, that is, do not lie in the same plane?

7. In Exercise 6 compare L/M with M/L.

8. Let P be a plane and L a line. What is P/L if
   (a) P and L intersect in a single point;
   (b) P and L are parallel?

9. In Exercise 8 compare P/L with L/P.

10. Can you find examples of sets A such that $A/A = A$? Can you find *all* such sets A?

11. Verify in a diagram, assuming that a, b and c are noncollinear:
    (a) $a(b/c)$ is contained in $(ab)/c$;
    (b) $a/(b/c)$ is contained in $(ac)/b$;
    (c) $(a/b)/c = a/(bc)$.

12. Obtain a formula for plane abc in terms of a, b, c and expressions formed by applying the operations of join and extension to a, b and c taken 2 or 3 at a time. Observe as indicated in Figure 1.81 that the three lines separate the plane into 7 regions. (Compare the formula for line ab in the relation (4) of Section 1.20.)

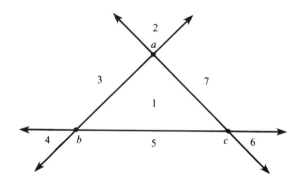

Figure 1.81

*13. Generalize Exercise 12 to four noncoplanar points a, b, c, d and obtain a formula for a 3-space. (There are 71 terms in the formula.)

14. Suppose no three of the points a, b, c and d are collinear. Suppose a/b and c/d intersect. What can you say about ad and bc? Can you state your result as a property of triangles?

1.23 Is There an Alternative to Join as Open Segment?

In this final section we consider an alternative definition of join.

Recall the definition of the join of two distinct points *a* and *b* as the (open) segment *ab* (Section 1.2; see Figure 1.82). You may have felt it would be more natural to define the join of *a* and *b* as the *closed* segment *ab*, consisting of *a*, *b* and all points between *a* and *b* (Section 1.1). In any case you may have wondered what *would* happen if the suggested definition were adopted. Would the theory be changed radically? Would the resulting treatment be preferable to the one presented?[6]

Figure 1.82

To begin, we formally define the new operation.

Definition. If *a* and *b* are distinct points, the *closed join* of *a* and *b* is the closed segment *ab*. The closed join of *a* and *a* consists of *a*.

To avoid confusion the original operation will be called *open* join.

The operation closed join is extended to two sets of points in the usual way (Section 1.5).

Definition. Let *A* and *B* be two sets of points (Figure 1.83). The *closed join* of *A* and *B* is the set of points obtained by forming the closed join of each point *a* of *A* and each point *b* of *B* and uniting all such closed joins.

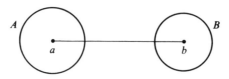

Figure 1.83

How do the two operations compare? Consider first the examples of the operation open join applied to: two distinct points, three noncollinear points, and four coplanar points that are the vertices of a convex quadrilateral, as shown in Figure 1.84.

---

[6] In high school algebra a similar question arises: Is there a satisfactory alternative to the definition "minus times minus is plus"?

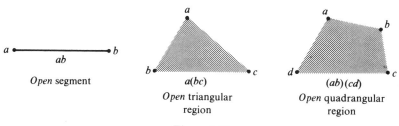

Figure 1.84

Observe that the result in each case is a convex set, but a convex set of special type: one which is *open* in the sense that it contains *none* of its boundary points. The examples illustrate a general principle: The open join of any finite set of points $a_1, \ldots, a_n$ is a convex set which is open in the indicated sense.

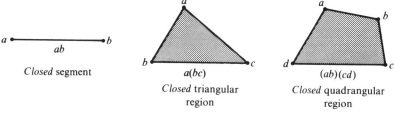

Figure 1.85

The closed join operation presents an interesting contrast. In Figure 1.85 are shown the analogues, for the operation closed join, of the examples above. The result in each case still is a convex set—but one which is *closed* in the sense that it contains *all* of its boundary points. Again a general principle holds: The closed join of any finite set of points $a_1, \ldots, a_n$ is a convex set which is closed in the indicated sense.

If a finite set of points is given, how are the open join and closed join related as geometrical figures? The two figures have the same boundary and the same interior. The closed join is the open join united with its boundary (Figure 1.86). Reversely, the open join is the interior of the closed join and is obtained from it by deleting or subtracting its boundary.

A logically sound theory, not unrelated to the one presented in this book, can be based on the closed join operation. Why then have we chosen open join as the basic operation? There are several reasons.

Figure 1.86

First, open segments, and open convex sets in general, form a simpler type of geometric figure than closed convex sets. An open convex set is homogeneous: Each of its points is *embedded within* the set, is *interior* to the set. A closed convex set, in general, contains points of two types: *interior* points, but also *boundary* points which are not embedded within the set but are *peripheral* to it (Figure 1.87).

Figure 1.87

Second, it is easier to obtain a closed convex set from an open one by "adding" boundary points than an open convex set from a closed one by "subtracting" boundary points. It is easier to study unions of sets rather than differences.

Our final reason involves the extension operation. In the closed join theory, the extension a/b will be defined naturally as the set of all points x such that the *closed* join bx contains a (Figure 1.88). The case a/a is crucial. By definition a/a will be the set of all points x such that the closed join ax contains a. But the closed join ax *always* contains a (Figure 1.89).

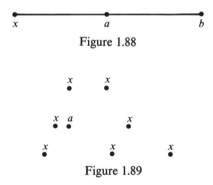

Figure 1.88

Figure 1.89

Thus a/a will contain *all* points—it will be the entire space that is studied. In a plane geometry a/a would be the base plane—in 3-dimensional geometry a/a would be 3-space. This does not seem attractive, to say the least, and clinches, in our view, the case for open join.

EXERCISES (CLOSED JOIN)

1. Solve some of the exercises at the end of Section 1.5 for the closed join operation. Compare the closed joins with the corresponding open joins.

2. The same as Exercise 1 for the exercises at the end of Section 1.9.

3. Does the commutative law hold for the closed join operation? Does the associative law?

4. In the closed join theory, verify that the extension a/b for $a \neq b$ is a closed ray (Section 1.17).

5. In the closed join theory, generalize the operation extension from points to sets in the usual manner (Section 1.21). Then solve some of the exercises on Extensions of Two Figures at the end of Section 1.22 in the closed join theory, and compare the results with the ones previously gotten.

2 The Abstract Theory of Join Operations

In this chapter we begin to develop an abstract theory of joining which is suggested by the operation of joining two points to form a segment in Euclidean geometry. The first section gives a brief treatment of the join operation in Euclidean geometry, defining it precisely and presenting its basic properties.[1] This "Euclidean join unit" motivates—becomes a model for—the abstract theory of join. The postulates for the operation join are given, and their elementary consequences deduced. The notion of convex set is then defined in terms of join and quickly takes on a central role in the theory. With each convex set A there are associated two important subsidiary sets which are themselves convex: the *interior* of A and the *closure* of A. These ideas receive a major share of our attention.

2.1 The Join Operation in Euclidean Geometry

One of our principal aims in writing this book is to try to bring classical geometry closer to the mainstream of contemporary mathematics. How is this aim to be achieved? May there not be something in the very nature of classical postulational geometry that precludes success in the venture?

---

[1] Chapter 1 contains a detailed treatment of the Euclidean join operation.

It has become evident in the development of mathematics that a branch of the subject can sometimes be restructured—by a new choice of basic ideas and assumptions—so as to seem almost a new subject.

How then do we intend to restructure geometry? Our point of departure is simply this: choose as a fundamental idea the operation of joining two points to form a segment, and base upon it a geometrical theory which centers on operations rather than figures.[2]

To begin, we indicate the concrete basis for the theory in Euclidean geometry. First the term segment must be clarified to dispel any ambiguity in the notion of joining.

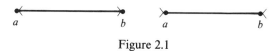

Figure 2.1

Definition. In a Euclidean geometry let a and b be distinct points (Figure 2.1). Then the set of all points between a and b is called the *open segment ab* or simply an *open segment*. The *closed segment ab* consists of a, b and all points between a and b. The points a and b are called *endpoints* of open segment ab and of closed segment ab. The term *segment* by itself will refer to open segment.

Then the join of two *distinct* points a and b is taken to be the segment ab (Figure 2.2). Now if the operation join is not to be artificially restricted, it is essential that it apply equally well to the points a and b regardless of whether $a \neq b$ or $a = b$. Thus we must clarify the idea of the join of a and a, this is taken to consist of point a.

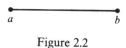

Figure 2.2

The discussion is restated and slightly sharpened in the definition which follows.

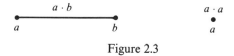

Figure 2.3

Definition. Let S be the set of points of a Euclidean geometry, planar or spatial. In S we define an operation $\cdot$ as follows (Figure 2.3). If a and b, $a \neq b$, are points of S, then $a \cdot b$ is the segment ab. For any point a of S,

---

[2] It is interesting that the first postulate in Euclid's *Elements* asserts in effect that there is a unique segment joining any two distinct points (see Euclid [1], p. 195, Postulate 1).

$a \cdot a$ consists of a.[3] The operation $\cdot$ is called the *Euclidean join operation* in S, or simply a *Euclidean join operation*; $a \cdot b$ will be called the (*Euclidean*) *join* of a and b for any points a, b of S. Usually $a \cdot b$ will be written simply as ab.

(The Euclidean join operation was introduced in Chapter 1, Sections 1.2, 1.9.)

The Euclidean Join of Two Geometric Figures

Finally the operation join must be extended to make it applicable, not just to pairs of points, but to any two geometric figures conceived as sets of points. As segments are generated from points, so complex figures will be constructed from simpler ones. Three illustrative examples of the idea are presented—then the definition will be formulated.

EXAMPLE I (The join of a point and a segment)[4]. Suppose points a, b and c are noncollinear (Figure 2.4). To get the join of point a and segment bc [denoted by $a(bc)$], join a to each point of bc and unite all the joins formed. The figure obtained is the interior of triangle abc. Thus the join of a and bc is the interior of triangle abc.

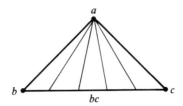

Figure 2.4 $a(bc) = $ Interior of triangle abc

EXAMPLE II (The join of a point and the interior of a triangle). Let p be a point which is not in the plane of triangle abc, and let I be the interior of triangle abc (Figure 2.5). To get the join of p and I (denoted by pI), join p to each point of I and unite the joins formed. The result is the interior of the pyramid whose vertices are p, a, b and c.

EXAMPLE III (The join of two intersecting segments). Let ab and cd be segments which intersect in a single point (Figure 2.6). Then the join of ab and cd, denoted by $(ab)(cd)$, is gotten by joining each point of ab to each point of cd and uniting the joins formed. The join of ab and cd is the interior of quadrilateral $acbd$.

---

[3] A discussion of alternatives to this definition appears in Section 1.13.

[4] This is the same as Example I, Section 1.5.

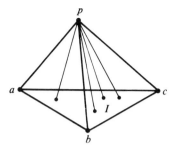

Figure 2.5   *pI* = Interior of a pyramid

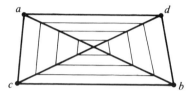

Figure 2.6   (*ab*)(*cd*) = Interior of quadrilateral *acbd*

Query. Was the definition of the join of a point and itself used in any of the Examples above?

Examples I, II and III suggest the following definition.

Definition. Let *A* and *B* be two geometric figures (Figure 2.7), which we conceive as sets of points. The (*Euclidean*) *join of set A and set B* is the set of points obtained by joining each point *a* of *A* to each point *b* of *B* and uniting all joins *ab* formed in this way. The join of *A* and *B* is denoted by *A* · *B* or simply *AB*. If *A* consists of a single point *a*, we call the join of *A* and *B* simply the join of *a* and *B* and denote it by *aB* (Figure 2.8).

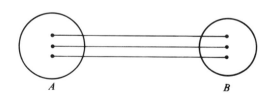

Figure 2.7

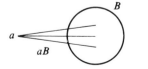
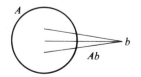

Figure 2.8

Similarly, if B reduces to a single point b, we refer to the join of A and b and denote it by Ab.

In simple terms the definition says: x is a point of the join AB of sets A and B if and only if there exist a point a of set A and a point b of set B such that x is a point of ab (Figure 2.9).

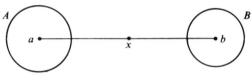

Figure 2.9

Basic Properties of the Euclidean Join Operation

The basic properties of the join operation in Euclidean geometry are now stated.

EJ1 (Existence Law). *If a and b are any points, ab is a nonempty set of points.*

EJ2 (Commutative Law). *If a and b are any points, ab = ba.*

EJ3 (Associative Law). *If a, b and c are any points, (ab)c = a(bc).*

EJ4 (Idempotent Law). *If a is any point, aa consists of a.*

It is not hard to see that EJ1, EJ2 and EJ4 are valid propositions of Euclidean geometry. In effect EJ1 says that any segment contains at least one point and EJ2 that segment ab is identical with segment ba. EJ4 is the content of a definition we have adopted. (EJ1 and EJ2 are covered by properties (1) and (2), Section 1.9.)

The verification of EJ3, the Associative Law, is more complicated for two reasons: It involves the operation of join of sets, and it requires the consideration of many cases. The "general" case, in which the points a, b and c are not collinear, is most important and very striking (Figure 2.10). Consider first $a(bc)$. By definition $a(bc)$ is gotten by joining a to each point of bc and uniting all the joins formed. The result is the interior of triangle abc. Consider now $(ab)c$. By definition this is gotten by joining each point of ab to c and uniting all the joins formed. The result is also the interior of triangle abc, and the associative law holds in the present case. (This was discussed in Section 1.6.)

The remaining cases of the associative law are indicated in Figure 2.11. Each of the seven cases must be considered to complete the verification of EJ3. (The question was treated in Section 1.11.)

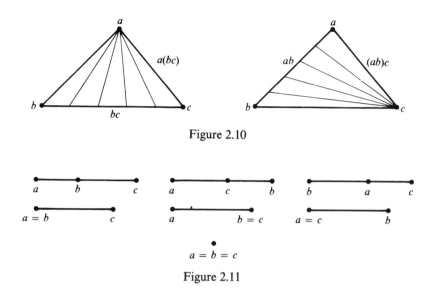

Figure 2.10

Figure 2.11

Remark. You may find it a burden to verify the seemingly trivial special cases of the associative law. But the verification yields an important benefit. The associative law is a nontrivial principle of Euclidean geometry which holds universally for the three points involved.

2.2 Join Operations in a Set—Join Systems

In this section we generalize the notion of Euclidean join operation (Section 2.1).

You have encountered in arithmetic and algebra operations, such as addition and multiplication, which assign to an ordered pair of numbers a uniquely determined *number*. We will be concerned with operations, such as the Euclidean join operation (Section 2.1), which assign to an ordered pair of elements a uniquely determined *set* of elements. Such an operation will be called a *join operation*. Here is the precise definition of the idea.

Definition. Let S be any set, not necessarily a set of points. Let $\cdot$ be an operation which assigns to each ordered pair of elements of S a uniquely determined subset of S. Then the operation $\cdot$ will be called a *join operation* in set S, or simply a *join operation*. S will be called the *domain* of the operation $\cdot$, and the pair $(S, \cdot)$ composed of the set and the operation will be called a *join system*. If (a, b) is an ordered pair of elements of S, the subset of S assigned to (a, b) by the operation $\cdot$ is called the *join* (or *product*) of a and b, and is denoted by $a \cdot b$ or ab. A specific operation $\cdot$ in a given context will usually be referred to as *join*.

Examples of Join Operations and Join Systems

EXAMPLE I. The Euclidean join operation, as defined in Section 2.1, is a *join operation* in the technical sense of the definition above. To tie it down to a specific domain, let S be a Euclidean plane or 3-space. Then the Euclidean join operation $\cdot$, applied to the points of S, is a join operation in S, and $(S, \cdot)$ is a join system, called not unnaturally a *Euclidean* join system.

EXAMPLE II. Let S be a Euclidean plane or 3-space. Define the operation $\cdot$ for points of S as follows: If $a \neq b$, let $a \cdot b$ be the Euclidean line ab; let $a \cdot a$ consist of a. Then $(S, \cdot)$ is a join system and may be called a *lineal* join system (Figure 2.12).

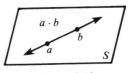

Figure 2.12

EXAMPLE III. Let S be the interior of a circle in a Euclidean plane P (Figure 2.13). Let $\cdot$ for points of S be the Euclidean join operation. Then $(S, \cdot)$ is a join system, which is, in a sense, a *subsystem* of the Euclidean join system $(P, \cdot)$.

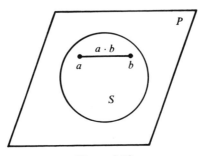

Figure 2.13

EXAMPLE IV. Let S be a Euclidean circle. Define $\cdot$ for points of S in this way: If a and b are distinct and not opposite, let $a \cdot b$ be the open minor arc of S with endpoints a and b; if a and b are opposite, $a \cdot b$ is the set consisting of a and b; $a \cdot a$ consists of a (Figure 2.14). Then $(S, \cdot)$ is a join system.

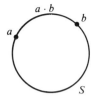

Figure 2.14

2.3 The Postulates for the Operation Join

In this section the formal basis for the theory—the postulates for the operation join—will be introduced and discussed.

The Genesis of the Theory

How did the theory arise? Some abstract mathematical theories arise by observation of common properties of several known (concrete) mathematical systems. In modern algebra the theory of fields (or generalized number systems) almost certainly arose from the observation that the rational, real, complex (and possibly other) number systems had many common properties. A suitable subset of the common properties was selected to form the basic assumptions or postulates of the theory of fields.

Sometimes an abstract theory has its genesis in one particular situation. The theory of join operations is a case at hand. Consider the four basic properties of the Euclidean join operation (Section 2.1).

EJ1. *If a and b are any points, ab is a nonempty set of points.*

EJ2. *If a and b are any points, ab = ba.*

EJ3. *If a, b and c are any points, (ab)c = a(bc).*

EJ4. *If a is any point, aa consists of a.*

EJ1–EJ4 are valid statements in Euclidean geometry that involve just two ideas: point and join of two points. Euclidean geometry is teeming with theorems involving a variety of ideas: lines, planes, circles, congruence, distance, similarity and so on. Suppose these four statements are singled out from Euclidean geometry. Can any significant deductions be made from them? The problem may be stated in this way: Suppose we try to build up geometry starting with the two ideas *point* and *join*—and suppose we assume EJ1–EJ4 as the basic properties of point and join. In effect, everything else in Euclidean geometry is to be washed out. How far can we hope to get?

To clarify the problem, observe that a new and autonomous theory is being conceived. *Point* now is not intended to refer to point in Euclidean geometry, nor *join* to the Euclidean join operation, defined in terms of Euclidean segments. Point and join are to be the basic terms in the theory —and they are to be characterized by four basic postulates corresponding to EJ1–EJ4.

Formulation of the Theory

The theory is suggested by, and has its origin in, Euclidean geometry, but it stands on its own: it is self-contained and could be studied by one who had never heard of Euclidean geometry.

Set Notation. To formulate the theory it is necessary to describe the set theoretic notation which will be employed. We will be concerned with the elements and subsets of a basic set J (the set of points). Elements of J will be denoted by $a, b, c, \ldots$, and subsets of J by $A, B, C, \ldots$. The empty or void set is denoted by $\varnothing$. The finite set whose elements are $a_1, \ldots, a_n$ is denoted by $\{a_1, \ldots, a_n\}$—the a's here are not necessarily distinct. (For example $\{a, b, a\}$ is significant and equals $\{a, b\}$.) As is customary, the symbols $\cup$, $\cap$ are used to denote union and intersection of two sets, and $\bigcup$, $\bigcap$ union and intersection of a family of sets. Set subtraction is indicated by a minus sign. Precisely stated: For any sets A and B, $A - B$ is the set of all elements that are in A but not in B. If sets A and B have a common element, we say A *meets* or *intersects* B. If A and B have no common element, they are *disjoint*.

Set containment is indicated by the familiar symbols $\subset$, $\supset$. Thus $A \subset B$ (or $B \supset A$) means that set A is contained in set B or A is a subset of B. If $A \subset B$ and $A \neq B$ we say A is a *proper* subset of B or is *properly* contained in B. It is convenient to represent element containment by using the same symbols. Suppose $A \subset B$ and $A = \{a\}$, the set whose only element is a. Then we have $\{a\} \subset B$, which is equivalent in conventional notation to $a \in B$, that is, a is an element of set B. We will suppress the braces in the statement $\{a\} \subset B$ and write it simply $a \subset B$. Similarly $B \supset \{a\}$ will be written $B \supset a$. These examples lead us to adopt the following convention: We shall when it is convenient write $\{a\}$ as a. In effect, we agree not to distinguish between the set $\{a\}$ and the element a. This permits us to dispense with the $\in$ notation, and affords the advantage of using the same symbols for element and set containment. There is an associated ambiguity, since element containment is not logically the same concept as set containment. But the ambiguity is resolved by our use of small letters for elements and capitals for sets.

Now consider an unspecified set J and an unspecified join operation $\cdot$ in J (Section 2.2). Let (a, b) be any ordered pair of elements of J. Then the subset of J assigned to (a, b) by the operation $\cdot$ is, as usual, called the *join* (or *product*) of a and b and is denoted by $a \cdot b$ or ab.

The operation · can be extended from pairs of elements of J to pairs of subsets of J, just as the Euclidean join operation was extended from pairs of points to pairs of geometric figures (Section 2.1).

Definition. Let A and B be any subsets of J. Then the *join* (or *product*) of *set A* and *set B*, denoted by $A \cdot B$ or AB, is defined by

$$A \cdot B = \bigcup_{a \subset A, \, b \subset B} (ab).$$

Thus the join of sets A and B is obtained by joining each element a of A to each element b of B and uniting all joins ab formed in this way. The definition says in simple terms:

$x \subset AB$ if and only if there exist elements a and b such that

$$x \subset ab, \qquad a \subset A \quad \text{and} \quad b \subset B.$$

The diagram in Figure 2.15 illustrates the idea of the join of two sets A and B when · is *interpreted* to be the Euclidean join operation.

If A consists of a single element a, we write AB for convenience simply as aB, instead of using the more exact notation $\{a\}B$. Similarly, if B consists of a single element b, we express AB as Ab.

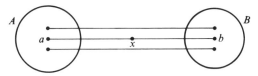

Figure 2.15

The postulates for the theory of join can now be stated. They are exact counterparts of the Euclidean join properties EJ1–EJ4 assumed for an abstract join system $(J, \cdot)$. The letters a, b and c denote any elements of J.

J1 (Existence Law). $ab \neq \varnothing$.

J2 (Commutative Law). $ab = ba$.

J3 (Associative Law). $(ab)c = a(bc)$.

J4 (Idempotent Law). $aa = a$.

Discussion of the Postulates

First consider the postulates individually. J1 asserts that the join of two (not necessarily distinct) elements of J is a nonempty subset of J, and so always contains an element of J. J2 and J3 are similar to the familiar commutative and associative laws in algebra—but observe that each asserts an equality of two *sets*, not of two elements. J3, however, is a more complex statement than its algebraic counterpart, since it involves the (defined) concept of the join of two sets. As in algebra, J3 enables us to omit parentheses and use the symbol abc to represent either of the equal sets $(ab)c$, $a(bc)$.

There is a difficulty in the statement of J4, since the left member is a set and the right member an element. A more precise but slightly awkward formulation of J4 would be $aa = \{a\}$. Since we have agreed not to distinguish between $\{a\}$ and a, it is permissible to employ the simpler form of J4.

J4 is analogous to a familiar property of union and intersection in set theory: $A \cup A = A$ and $A \cap A = A$ for any set A. It plays an important role in our "geometric" algebra and serves in a sense to distinguish it from the familiar algebras which had their origin in the study of numbers. J4 is called the *idempotent* law, since it implies that all powers of an element, under the given operation, are the same: $a^2 = a^3 = \cdots = a$.

The postulate set J1–J4 has a property which does not appear in the usual axiomatizations of geometry: Each postulate states a *universal* property of points; each brooks no exception; each covers all special or degenerate cases.

Finally note that J1–J4 state only the most elementary properties of the operation join. Meager as they may seem, they do provide enough deductive power to carry us through two chapters. In Chapter 4 additional postulates will be introduced to complete the basic set of assumptions for the theory.

A Possible Inconsistency

If our agreement to identify $\{x\}$ with x is applied to $\{a\}\{b\}$, it reduces to ab. But $\{a\}\{b\}$ has a meaning of its own determined by the definition above of the join of two sets. Thus it is possible that the agreement may lead to an inconsistency. This is easily settled. For by definition a point is in $\{a\}\{b\}$ if and only if it is in ab, so that $\{a\}\{b\}$ is the same set of points as ab. Thus $\{a\}\{b\} = ab$ indeed holds, and no difficulty arises.

2.4 Application of the Theory to Euclidean Geometry

Before starting to develop the abstract theory of the operation join, we show that it is applicable to Euclidean geometry.

What does it mean to make an application of an abstract mathematical theory such as the theory of join operations? First examine the postulates of the theory: J1–J4. They look like statements, they sound like statements, but they are not statements—not declarative sentences. For they involve two terms, "the set J" and "the join operation $\cdot$" to which no meaning has been assigned. J1–J4 become statements—true or false, valid or invalid—when suitable content is assigned to the terms J and join. To make an application of our theory we must assign to the basic terms J and join in the postulates J1–J4 specific meaning which *satisfies* the postulates —which converts them into true (or valid) statements.

There is no difficulty then in applying the theory to Euclidean geometry. Interpret J to be a Euclidean plane or 3-space (considered as a set of points) and interpret join to be the Euclidean join operation in that plane or 3-space (Section 2.1). Then J1–J4 are converted into EJ1–EJ4, which are valid propositions of Euclidean geometry (Section 2.1).

Summary Statement. We have constructed specific join systems $(J, \cdot)$ each consisting of a Euclidean plane or 3-space and an associated Euclidean join operation (Section 2.2). Each of these join systems *satisfies* the postulates J1–J4. In virtue of this they are called *models* (or *interpretations*) of the postulate set J1–J4 (or of the theory determined by the postulate set). Specifically, they will be called *Euclidean* models of the postulate set J1–J4. (There exist models different from the Euclidean ones. See Sections 2.21–2.24.)

EXERCISES

1. Let J be a Euclidean plane or 3-space. Define $\cdot$ for a, b in J as follows: If $a \neq b$, let $a \cdot b$ be line ab; $a \cdot a = a$. (Recall Example II, Section 2.2.)
   (a) Verify that $(J, \cdot)$ is a join system.
   (b) Does $(J, \cdot)$ satisfy the postulates J1–J4? Justify your answer.

2. The same as Exercise 1, given that J is the interior of a Euclidean circle, and $\cdot$ is the Euclidean join operation in J. (Recall Example III, Section 2.2.)

3. The same as Exercise 1 if J is a Euclidean circle and $\cdot$ is defined for a, b in J so: (i) if a and b are distinct and not opposite, $a \cdot b$ is the open minor arc of J that has endpoints a and b; (ii) if a and b are opposite points of J, $a \cdot b = \{a, b\}$; (iii) $a \cdot a = a$. (Recall Example IV, Section 2.2.)

4. The same as Exercise 1 if J is a Euclidean sphere and $\cdot$ is defined for a, b in J so: (i) if a and b are distinct and not opposite, $a \cdot b$ is the great circle of J that contains a and b; (ii) if a and b are opposite points of J, $a \cdot b = \{a, b\}$; (iii) $a \cdot a = a$.

5. The same as Exercise 1 given: J is the set of real numbers and $a \cdot b = (a + b)/2$ for any real numbers a and b.

6. In a Euclidean planar model find the join of the indicated figures:
   (a) A point and a line. (Exercises 3 at the end of Sections 1.5, 1.9.)[5]

---

[5] References in parentheses that follow an exercise refer to the appearance of the exercise or a related one in Chapter 1.

(b) Two parallel lines. (Exercise 4 at the end of Section 1.5.)
(c) Two distinct intersecting lines. (Exercise 4 at the end of Section 1.9.)
(d) A line and itself. (Exercise 6 at the end of Section 1.9.)

7. In a Euclidean 3-space model find the join of the indicated figures:
   (a) A point and a plane.
   *(b) Two skew (that is, noncoplanar) lines. (Exercise 5 at the end of Section 1.5.)
   (c) Two parallel planes.
   (d) Two distinct intersecting planes. (Exercise 12 at the end of Section 1.9.)
   (e) A plane and itself. (Exercise 6 at the end of Section 1.9.)
   (f) A line and a plane. (Exercise 11 at the end of Section 1.9.)

8. In a Euclidean 3-space model find the join of a line L and a circle C if L is perpendicular to the plane of C at its center. (Exercise 14 at the end of Section 1.5.)

In Exercises 9–14 suppose that a Euclidean plane has been assigned a Cartesian coordinate system. Let p be the point $(0, 0)$ and C the graph of the given equation. Sketch the join pC and describe it as accurately as you can.

9. $y = x^2$. (Exercise 13 at the end of Section 1.9.)

10. $y = x^3$. (Exercise 14 at the end of Section 1.9.)

11. $y = e^x$. (Exercise 23 at the end of Section 1.5.)

12. $y = \ln x$. (Exercise 24 at the end of Section 1.5.)

13. $y = x^2 + 1$. (Exercise 20 at the end of Section 1.5.)

14. $y = x^3 + 1$. (Exercise 21 at the end of Section 1.5.)

2.5 Elementary Properties of Join

The formal development of the theory begins with the deduction of several properties of the join operation from the postulates J1–J4.

It must be underscored that we are constructing an *abstract* theory. That the set J and the operation join in J1–J4 are literally undefined: No meaning is assigned to these terms in the *formal* development. They are not to refer to chalk dots on a blackboard—or to a process of drawing straight streaks to join pairs of such dots.

Therefore, for a proof to be valid, the justification of a step must be the application of a postulate, or of a definition adopted in the *formal* development. But this does not mean that you must become a logic-chopping machine—that when studying or trying to construct a proof you should not draw diagrams or make models, and should immediately try to suppress geometrical images or intuitive impressions that arise in your mind. Diagrams, models, mental images, intuitive impressions can be an important aid in grasping the point of a step, or suggesting what the next step should be, or in trying to scan a proof as a whole. They just carry no weight in the *logical* justification of an inference.

Recall as indicated above (Section 2.3) that elements of J will be denoted by $a, b, c, \ldots$ and subsets of J by $A, B, C, \ldots$. For the sake of convenience elements of J will usually be referred to as points.

The first theorem, which may seem long overdue, is a very useful analogue of the monotonic law for addition in algebra: $a < b$ implies $a + c < b + c$.

Theorem 2.1 (Monotonic Law for Join). *$A \subset B$ implies $AC \subset BC$ and $CA \subset CB$, for any set C.*

PROOF. The statement $AC \subset BC$ means that each point of AC is a point of BC. So to prove $AC \subset BC$ we suppose $x \subset AC$ and deduce $x \subset BC$. By definition of the join of two sets (Section 2.3), $x \subset AC$ implies the existence of a and c satisfying

$$x \subset ac, \qquad a \subset A, \qquad c \subset C.$$

The hypothesis, $A \subset B$, asserts that each point of A is a point of B. Thus $a \subset A$ implies $a \subset B$. Now we know $x \subset ac$, $a \subset B$, $c \subset C$; that is, point x is contained in the join of a point of B and a point of C. Hence $x \subset BC$ by the definition of the join of two sets. Thus we have justified $AC \subset BC$. Similarly we can show $CA \subset CB$. □

Remarks on the Theorem. The monotonic law for join does not seem comparable to ordinary algebraic principles since they are expressed in equations involving numbers, while it involves containment relations of sets. Such containment relations $A \subset B$ are very common in our work. The monotonic law says, in effect, that if a containment relation $A \subset B$ is given, a valid containment relation is obtained by "multiplying" both terms in $A \subset B$ by C. It is convenient to say that $AC \subset BC$ is obtained from $A \subset B$ by *multiplying $A \subset B$ by C.*

The monotonic law is often used in the case where A or C consists of a single point. If $A = \{a\}$, for example, the law may be stated: $a \subset B$ implies $aC \subset BC$ and $Ca \subset CB$. Similarly, if $C = \{c\}$ it takes the form: $A \subset B$ implies $Ac \subset Bc$ and $cA \subset cB$.

Theorem 2.1 easily yields a principle for multiplying (or joining) two containment relations term by term.

Corollary 2.1. *$A' \subset A$, $B' \subset B$ imply $A'B' \subset AB$.*

PROOF. Multiplying $A' \subset A$ by B' (on the right), we obtain

$$A'B' \subset AB'. \tag{1}$$

Similarly, multiplying $B' \subset B$ by A (on the left) yields

$$AB' \subset AB. \tag{2}$$

The relations (1) and (2) imply the conclusion $A'B' \subset AB$. □

The corollary holds of course if A' and B' are one-element sets, say $A' = \{a\}$, $B' = \{b\}$. Thus we may write

$$a \subset A, \quad b \subset B \quad \text{imply} \quad ab \subset AB.$$

This result is a direct consequence of the definition of AB. It is introduced here, as a special case of Corollary 2.1, merely for the sake of convenience.

A Few Simple Proofs

At this point we present a few illustrations of deductive proofs based on postulates J1–J4 and Theorem 2.1.

Suppose

$$p \subset ab. \tag{1}$$

Multiplying (1) by a, we have

$$ap \subset a(ab). \tag{2}$$

By J3 the associative law and J4 the idempotent law

$$a(ab) = (aa)b = ab. \tag{3}$$

Relations (2) and (3) imply

$$ap \subset ab. \tag{4}$$

Thus (1) implies (4) and we have proved:

(i) If $p \subset ab$, then $ap \subset ab$.

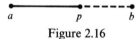

$$a \qquad p \qquad b$$

Figure 2.16

[Figures 2.16–2.19 each illustrates the corresponding result in a Euclidean model (Section 2.4).]

Similarly, using b as a multiplier in (1), we can obtain

(ii) If $p \subset ab$, then $pb \subset ab$.

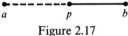

$$a \qquad p \qquad b$$

Figure 2.17

(i) and (ii) imply an interesting result:

$$\text{If} \quad p \subset ab \quad \text{then} \quad ap \cup p \cup pb \subset ab.$$

Next we prove

(iii) If $p \subset ab$ and $q \subset ab$, then $pq \subset ab$.

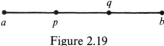
Figure 2.18

PROOF. By hypothesis

$$q \subset ab. \tag{5}$$

Multiplying (5) by p, and applying J3, we have

$$pq \subset p(ab) = (pa)b.$$

Thus

$$pq \subset (pa)b. \tag{6}$$

By J2 the commutative law, $pa = ap$. Then (6) yields

$$pq \subset (ap)b. \tag{7}$$

Now we let (7) stand and return to the other part of the hypothesis: $p \subset ab$. This implies, by (i) above,

$$ap \subset ab. \tag{8}$$

Multiplying (8) by b and applying J3 and J4 we get

$$(ap)b \subset (ab)b = a(bb) = ab.$$

Thus

$$(ap)b \subset ab. \tag{9}$$

The relations (7) and (9) imply $pq \subset ab$, and (iii) is justified. □

Finally we prove

(iv) If $p \subset ab$, then there is a point q such that

$$p \subset aq \quad \text{and} \quad q \subset ab. \tag{10}$$

Figure 2.19

Analysis. Observe that if a point q satisfies the relation (10), then

$$p \subset a(ab) \tag{11}$$

by the definition of the join of two sets (Section 2.3). Conversely, if (11) holds, then there is a point q that satisfies (10), again by the definition of the join of two sets. But (11) is easy to establish, since $p \subset ab$ is given, and as indicated above in (3)

$$a(ab) = ab. \tag{12}$$

Now it is not hard to write a proof of (iv).

Proof. By hypothesis, J4 and J3,

$$p \subset ab = (aa)b = a(ab).$$

Thus $p \subset a(ab)$. Then there is a point q such that

$$p \subset aq \quad \text{and} \quad q \subset ab$$

(by the definition of the join of two sets), and the proof is finished. ☐

Remark. The idempotent law has very little scope in itself, since it refers only to one point. However, in conjunction with other principles, particularly J3, it has strong deductive power. The relation (12) above, $a(ab) = ab$, is an illustration. Algebraically it may seem almost trivial—but it carries important geometric significance. In a Euclidean model it asserts that the join of an endpoint of a segment with the segment is precisely the segment. This can be used to characterize the endpoints of a segment in Euclidean geometry: The endpoints of a segment S are the only points x that are *not* in S and satisfy

$$xS = S.$$

(See Figure 2.20.)

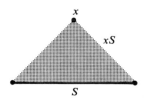

Figure 2.20

In the following exercises you may use J1–J4, Theorem 2.1 and Corollary 2.1 to prove the given statements.

EXERCISES

1. Prove: If $p \subset ab$, then there is a point q such that

$$p \subset aq \quad \text{and} \quad q \subset pb.$$

   [*Hint*. The conclusion is equivalent to $p \subset a(pb)$.]

2. Given $p \subset abc$. Prove:
   (a) There is a point d such that

   $$p \subset ad \quad \text{and} \quad d \subset bc.$$

   (b) There is a point e such that

   $$p \subset ae \quad \text{and} \quad e \subset abc.$$

3. (a) Prove: If $p \subset ab$ and $q \subset ac$, then $pq \subset abc$.
   (b) Interpret the result in a planar Euclidean model. Observe that it is verified.

4. (a) Prove: If $p \subset ab$ and $q \subset cd$, then $pq \subset (ab)(cd)$.
   (b) Interpret the result in a planar Euclidean model, if a, b, c and d are distinct and no three of them are collinear. Is the result verified?
   (c) Is the result verified if $c = a$; if $d = b$; if $c = a$ and $d = b$?

2.6 Generalizations of J1–J4 to Sets

An essential theme in our treatment is the study of general geometric figures, that is, arbitrary sets of points. This theme has appeared in the formulation of the concept of join of sets and the statement of the Monotonic Law. We develop it further by generalizing J1, J2 and J3 (the existence, commutative and associative laws) from points to sets of points. An extension of J4 is also obtained.

First the generalization of J1.

Theorem 2.2 (Existence Law). *Suppose $A \neq \varnothing$ and $B \neq \varnothing$. Then $AB \neq \varnothing$.*

PROOF. Let $a \subset A$ and $b \subset B$. Then $ab \subset AB$ by Corollary 2.1. But $ab \neq \varnothing$ by J1. Thus $AB \neq \varnothing$.   □

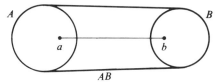

AB

Figure 2.21

NOTE. Figures accompanying *proofs* are Euclidean interpretations which may help you to grasp the proofs—they are not to be considered parts of the proofs. Similar considerations apply whenever figures appear in formal development.

Theorem 2.2 asserts that the join of two sets is nonempty, provided the two sets are not empty. One might wonder what the join would be if one of the sets *were* empty. The question is not of prime importance, but this possibility *is* permitted, since the definition of the join AB imposes no restriction on the sets A and B.

Consider then $A \cdot \varnothing$. In view of the definition of join of two sets, $A \cdot \varnothing$ is the set of all points x that have the following property:

(1) $x \subset ab$ for some point $a \subset A$ and some point $b \subset \varnothing$.

But $b \subset \varnothing$ is false for every choice of b. Hence no point x has property (1), that is, $A \cdot \varnothing = \varnothing$. Similarly $\varnothing \cdot A = \varnothing$.

Theorem 2.3 (Commutative Law). $AB = BA$.

PROOF. The set equality $AB = BA$ asserts by definition that $x \subset AB$ implies $x \subset BA$ and conversely. This is not hard to prove. Suppose $x \subset AB$. Then

$x \subset ab$, where $a \subset A$, $b \subset B$ by definition of join of sets (Section 2.3). By J2, $ab = ba$. Hence $x \subset ba$. Thus $x \subset BA$, again by definition of the join of sets. The converse follows by symmetry and the proof is complete. □

Theorem 2.4 (Associative Law). $(AB)C = A(BC)$.

PROOF. We are required to show that every element of each member of the equality is in the other member. Suppose $x \subset (AB)C$. By definition of the join of sets this implies

$$x \subset yc, \tag{1}$$

where $y \subset AB$, $c \subset C$. Similarly $y \subset AB$ implies

$$y \subset ab, \tag{2}$$

where $a \subset A$, $b \subset B$. By the Monotonic Law (Theorem 2.1) we may multiply both members in (2) by c, getting

$$yc \subset (ab)c. \tag{3}$$

Then (1), (3) and J3 imply

$$x \subset (ab)c = a(bc). \tag{4}$$

We have

$$a \subset A, \quad b \subset B, \quad c \subset C.$$

But $b \subset B$ and $c \subset C$ imply $bc \subset BC$ by Corollary 2.1. Similarly $a \subset A$ and $bc \subset BC$ imply

$$a(bc) \subset A(BC). \tag{5}$$

The relations (4), (5) imply $x \subset A(BC)$.

Thus we have shown: $x \subset (AB)C$ implies $x \subset A(BC)$. The converse, $x \subset A(BC)$ implies $x \subset (AB)C$, can be proved in a similar manner, and we conclude $(AB)C = A(BC)$. □

In view of the theorem we can employ the symbol ABC, as in high school algebra, to stand for either member of the equality $(AB)C = A(BC)$.

The idempotent law, $aa = a$, does not generalize to arbitrary sets (see Exercise 11 below), but we can obtain a weak generalization of it.

Theorem 2.5. $AA \supset A$.

PROOF. We must show that each point of A is a point of AA. Let $a \subset A$. Then using J4, we have

$$a = aa \subset AA$$

and the proof is complete. □

EXERCISES

In proofs use any combination of the postulates and preceding theorems you please.

1. (a) Interpret $(ab)(cd)$ in a Euclidean planar model, assuming a, b, c and d are distinct. Draw and briefly describe the figures you obtain.
   (b) The same as (a) for a Euclidean 3-space model, assuming a, b, c and d are not coplanar.

2. (a) Prove: $a(b(cd)) = (ab)(cd)$. (This was treated in Section 1.15. It can now be treated in a much more powerful fashion.)
   (b) Interpret the principle in Euclidean planar and spatial models. Is the result true or false?

3. Prove: $a(b(cd)) = ((ab)c)d$.

4. Prove: $a(b(cd)) = (a(bc))d$.

5. Prove: $(ab)(ab) = ab$.

6. Prove: $a(abc) = abc$.

7. Prove: $(ab)(abc) = abc$.

8. Interpret Exercises 6 and 7 in a Euclidean model, taking a, b and c to be noncollinear. Do your results suggest a geometrical principle?

9. Prove: $x \subset ABC$ if and only if there exist a, b, c such that $x \subset abc$ and $a \subset A$, $b \subset B$, $c \subset C$.

10. Prove: If $x \subset AB$ then $x \subset aB$ for some $a \subset A$, and $x \subset Ab$ for some $b \subset B$.

11. (a) Show that the analogue of J4 for sets, $AA = A$, is not true in a Euclidean model.
    (b) Could $AA = A$ be deduced from J1–J4? Explain.

12. (a) In a Euclidean plane choose a specific set of points A. (Represent A on a sheet of paper if you wish.) Interpret $\cdot$ to be the Euclidean join operation. Find $A^2 = A \cdot A$, $A^3 = (A \cdot A)A$, and so on. What do you observe? Make another choice or two for A and repeat the process.
    (b) The same as (a) for a Euclidean 3-space.
    (c) The same as (a) for a Euclidean line.
    (d) Can you find a general principle that covers your results in (a), (b) and (c)?

13. Consider the following proposition in Euclidean geometry, where a, b, c are vertices of a triangle: If x is between p and q, where p is between a and b and q is between a and c, then there is a point y such that x is between a and y and y is between b and c.
    (a) Convert the proposition into a statement about points and joins of points in Euclidean geometry.
    (b) Justify the proposition by proving that the statement in abstract form is a theorem in our theory.
    (c) Does the proposition hold if a, b, c are not the vertices of a triangle?
    (d) The same as (a), (b) and (c) for the converse of the proposition.

14. (a) Prove: $AB \supset A \cap B$.
    (b) Can $AB \subset A \cap B$? Must $AB \subset A \cap B$? Explain.
    (c) Can $AB = A \cap B$? Must $AB = A \cap B$? Explain.

15. (a) Can $AB \supset A \cup B$? Must $AB \supset A \cup B$? Explain.

(b) Can $AB \subset A \cup B$? Must $AB \subset A \cup B$? Explain.
(c) Can $AB = A \cup B$? Must $AB = A \cup B$? Explain.
(d) Can the containment relations in (a) and (b) be proper? Explain.

16. If $AA = BB$, must it follow $A = B$? Explain.

2.7 Extension of the Join Operation to n Terms

Since join has been defined for three terms and its basic associative and commutative properties are known, it would be natural now to move directly to the study of geometric figures. This will be done soon, but not quite yet. If we restrict our study of joins to three points, we can construct —at least in the Euclidean interpretation—figures which are at most 2-dimensional. The *intrinsic* study of classical geometry has too long been shackled by restrictions to the plane and 3-space which today are quite anachronistic. An important step in the removal of these restrictions, in our treatment, is the extension of the join operation to an arbitrary number of terms. This will facilitate the study of higher dimensional analogues of familiar lower dimensional figures.

We give an informal sketch of the development, since the formal treatment in algebra for an associative and commutative multiplication operation can be carried over without essential changes.[6] As in algebra, we can extend the join operation to an arbitrary finite sequence of elements inductively. Similarly the operation can be extended to an arbitrary finite sequence of sets. But it turns out to be easier to deal with sets rather than elements. For the fundamental associative and commutative laws hold for sets *and* for elements—but the join of two sets is a set, while the join of two elements is in general not an element. Thus the proofs employed in algebra can be applied without change to obtain associative and commutative laws for the join of n sets.

Consequently the following treatment is adopted. We associate with each finite sequence of sets $A_1, \ldots, A_n$ a *join* or *product* denoted by $A_1 \ldots A_n$. The join is characterized by the induction or recursion relations

$$A_1 \ldots A_i = A_1 \qquad\qquad \text{for } i = 1,$$
$$A_1 \ldots A_{i+1} = (A_1 \ldots A_i)A_{i+1} \quad \text{for } i \geq 1.$$

Then associative and commutative laws for n-term joins can be proved as in classical algebra. Here are statements of such laws.

The Basic Associative Law.
$$A_1 \ldots A_n = (A_1 \ldots A_i)(A_{i+1} \ldots A_n) \quad \text{for } 1 \leq i < n.$$

---

[6] See Chevalley [1], Jacobson [1].

The Generalized Commutative Law.

$$A_1 \ldots A_n = A_{i_1} \ldots A_{i_n},$$

where $i_1, \ldots, i_n$ is any permutation of the integers $1, \ldots, n$.

How are joins of elements taken care of? Quite easily. If the sets $A_1, \ldots, A_n$ consist of single elements $a_1, \ldots, a_n$, we write

$$A_1 \ldots A_n = a_1 \ldots a_n.$$

A bit more precisely, $a_1 \ldots a_n$ is defined to be $\{a_1\} \ldots \{a_n\}$. Then associative and commutative laws for the join of n elements are obtained as special cases of the associative and commutative laws for the join of n sets.

Joins of Points. Joins of finite sequences of points play a more important role here than products $a_1 \ldots a_n$ of numbers in algebra. For they denote sets of points or geometric figures and, as you will see, a rather important type of geometric figure. In general, the greater the number of points in the sequence the more complex the join will be.

Definition. The join $a_1 \ldots a_r$ of the finite sequence of points $a_1, \ldots, a_r$ is called a *join of r points*, an *r-point join*, a *finite join of points* or simply a *join of points*. Similarly the join $A_1 \ldots A_r$ of the finite sequence of sets $A_1, \ldots, A_r$ is called a *join of r sets*, a *finite join of sets* or a *join of sets*.

Remarks on the Definition. Note that the points $a_1, \ldots, a_r$ need not be distinct. Thus aaa is an example of a join of 3 points. A 1-point join is simply a point. A 2-point join is given a special name.

Definition. The join ab of a and b is called *segment ab*, or a *segment*, and is said to *join a and b*. The points a and b are called *endpoints* of ab. If $a \neq b$ the segment is said to be *proper* or *nondegenerate*. The segment $aa = a$ is said to be *degenerate* or *improper*.

This usage is often convenient for stating principles in English and is graphically suggestive. The degenerate case has been included to make "segment ab" correlative with "join ab".

We conclude the section with three theorems that extend previous results to n terms.

First the extension of Theorem 2.2.

Theorem 2.6. *Suppose $A_i \neq \varnothing$ for $1 \leqslant i \leqslant n$. Then*

$$A_1 \ldots A_n \neq \varnothing.$$

PROOF. We illustrate a proof, by successive cases, for $n = 4$.

By hypothesis, A_1, A_2, A_3, A_4 are not empty. Theorem 2.2 implies $A_1 A_2 \neq \varnothing$. Again by Theorem 2.2, $A_1 A_2 A_3 \neq \varnothing$. Similarly $A_1 A_2 A_3 A_4 \neq \varnothing$.   □

A formal proof can be given, of course, by mathematical induction.

Taking the special case where each A_i consists of a point, we have the following generalization of J1.

Corollary 2.6. *Any join of points $a_1 \ldots a_n$ is nonempty.*

Next, Corollary 2.1, the principle for multiplying containment relations, is generalized.

Theorem 2.7. *Suppose $A'_1 \subset A_1, \ldots, A'_n \subset A_n$. Then*

$$A'_1 \ldots A'_n \subset A_1 \ldots A_n.$$

PROOF. Use the method of Theorem 2.6 or mathematical induction. □

Consider the case where $A'_1, \ldots, A'_n$ reduce to single elements, say $a_1, \ldots, a_n$. Then the theorem asserts:

If $a_1 \subset A_1, \ldots, a_n \subset A_n$, then $a_1 \ldots a_n \subset A_1 \ldots A_n$.

Finally, the condition that a point x be in the join AB of two sets is extended to n sets.

Theorem 2.8.

$$x \subset A_1 \ldots A_n \tag{1}$$

if and only if there exist points $a_1, \ldots, a_n$ such that

$$x \subset a_1 \ldots a_n, \qquad a_1 \subset A_1, \ldots, a_n \subset A_n. \tag{2}$$

PROOF. That (1) implies (2) can be shown by the method of Theorem 2.6 or mathematical induction.

The converse, (2) implies (1), is an easy consequence of Theorem 2.7. □

2.8 Comparison with the Conventional View of Geometry

The extension of join and its basic properties to n terms may not seem world-shaking. But consider for a moment the usual treatment of geometry. In classical plane and solid geometry the basic figures such as segments, triangles, tetrahedrons and so on are studied essentially on a visually or pictorially motivated basis. When higher dimensional geometry is to be treated, it is hard to see how to generalize this treatment.

How, for example, can we hope to visualize or intuitively perceive points or figures in Euclidean 4-space? The conventional assumption is that it is not feasible to construct a theory of higher dimensional geometry as a *natural* outgrowth of two and three dimensional geometry. How then is higher dimensional geometry to be developed? Historically the key was found in two and three dimensional *coordinate* geometry. It was assumed that whatever Euclidean 4-space might be, its points could be represented by quadruples of real numbers (x_1, x_2, x_3, x_4). Then by analogy with Euclidean two and three dimensional coordinate geometry, algebraic relations were found to represent lines, segments, planes and other geometric figures.

This means that Euclidean four dimensional geometry is conceived as a different type of subject from Euclidean two and three dimensional geometry. It does not proceed from basic properties of points, lines and so on, stated as postulates. In effect it defines a point as an algebraic entity (x_1, x_2, x_3, x_4) and treats four dimensional geometry as a branch of algebra. What is the advantage of doing this? It is precisely that no unbridgeable chasm separates the algebra of 2 or 3 variables from the algebra of 4 or more variables. The algebraic methods employed in two or three dimensional coordinate geometry *can* be extended to four or to n-dimensional coordinate geometry.

In our treatment there *is* no chasm between lower dimensional and higher dimensional geometry. The trivial-seeming extendibility of the join operation to n terms takes care of this. It enables us to deal with the join of 5 or 6 or more points essentially as easily as 2 or 3 or 4 points. It fosters the direct study of geometric material of higher dimension, while helping to liberate us from excessive dependence on visual intuition. To make the point concrete, consider Figure 2.22, which indicates the typical (or nondegenerate) Euclidean interpretation of some joins listed. We have appended no diagram for the typical join of 5 elements. It is represented by the interior of the 4-dimensional analogue of a triangle or a tetrahedron, called a 4-dimensional simplex. The graphical representation in the second row breaks down at this stage—but the symbolic expression in the first row can be studied formally and abstractly, independent of the number of factors and so of dimension.[7]

Now you may ask: Are we asserting that a 4-dimensional space exists? In the theory we have introduced no existence assumptions at all. We are not asserting that a 3-space exists, or a plane or a line or even one lonely

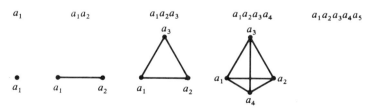

Figure 2.22

[7] This suggests a questionable piece of humor to make a serious point: 4-dimensional space is only 25% harder to study than 3-dimensional space.

point. But the fact that we can define the join $a_1a_2a_3a_4a_5$ of five points leaves open the possibility that the theory is applicable to 4-dimensional space: that it has a 4-dimensional model. In Chapter 5 it will be indicated that the theory has an n-dimensional model, of Euclidean type, for each integer $n \geqslant 1$.

EXERCISES

1. Without assuming the generalized associative law for join, prove it for four sets:
$$((AB)C)D = (AB)(CD) = (A(BC))D = A((BC)D) = A(B(CD)).$$
   Infer the law for four points. (Compare Exercises 1–4 at the end of Section 2.6.)

2. Prove: $(ab)(ac) = abc$. Interpret the result in a Euclidean model.

3. The same as Exercise 2 for each of these statements: $abcd = (ab)(cd)$; $abcd = (ab)(bcd)$; $abcd = ab(abcd)$; $abcd = (abc)(bcd)$; $abcd = (abc)(abcd)$.

4. Consider the following expressions which equal $abcde$: $(ab)(cde)$; $a(bcde)$; $(ab)(abcde)$; $(abc)(abcde)$. Interpret each in a Euclidean planar model and verify their equality. Try to do the same in a Euclidean 3-space model.

5. Without assuming the generalized commutative law for join, prove it for three sets:
$$ABC = ACB = BAC = BCA = CAB = CBA.$$
   Infer the law for three points.

6. Prove: $(ab)(cd) = (ac)(bd)$. Interpret the result in a Euclidean model.

2.9 Convex Sets

Finally we are prepared to study convex sets. It is hard to overrate the importance of the concept in classical geometry or contemporary linear geometry, and its importance is matched by its extreme simplicity. In Euclidean geometry a figure is convex if it contains the segment joining each pair of its points. The idea would demand attention in a join-centered theory of geometry even if it had not been known earlier. Some illustrations of convex and nonconvex sets are shown in Figure 2.23.

Most geometric figures encountered in mathematics and its applications are convex sets, boundaries of convex sets or unions of familiar convex sets.

Notation. A pair of containment statements of the form $A \supset B$, $A \supset C$ is abbreviated $A \supset B, C$. Similarly n statements
$$A \supset B_1, \ldots, A \supset B_n$$
are telescoped into
$$A \supset B_1, \ldots, B_n.$$

The same convention will be employed for similar sets of statements which involve the symbol $\subset$.

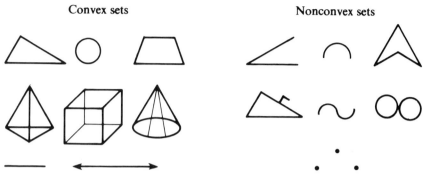

Figure 2.23

Definition. Let set A satisfy the condition

$$A \supset x,y \quad \text{implies} \quad A \supset xy.$$

Then we say A is *convex* or *closed under* the operation *join*.

Observe that J, the basic set, is convex; and that each point a forms a convex set, since by J4, $a = aa \supset aa$. Note that $\varnothing$ is convex, since the definition imposes no restriction on it.

In Euclidean geometry a so-called convex polygon (P in Figure 2.24) is not a convex set. It is however the boundary of a convex set, namely, its interior I.

Figure 2.24

A convex set A may be described in visual terms: All points of A are visible from any point of A.

Conditions that a set be convex can be formulated in terms of join of sets.

Theorem 2.9. *Each of the following conditions is equivalent to the convexity of A:* (a) $A \supset AA$; (b) $A = AA$.

PROOF. (a) Suppose A is convex. We show

$$A \supset AA. \tag{1}$$

Let $x \subset AA$. By definition of the join of sets,

$$x \subset pq, \tag{2}$$

where $p, q \subset A$. Thus since A is convex, $pq \subset A$. This with (2) yields $x \subset A$. Thus (1) is justified.

Conversely suppose (1). We show A is convex. Let $A \supset x, y$. Then

$$AA \supset xy \tag{3}$$

by Corollary 2.1. The relations (1) and (3) imply $A \supset xy$, and A is convex by definition.

(b) Observe that for any set A, $A \supset AA$ is equivalent to $A = AA$, since $AA \supset A$ by Theorem 2.5. $\square$

Remark on the Theorem. Part (b) calls to mind J4, the idempotent law. Let us say a set S is *idempotent* if $SS = S$. Then J4 asserts that each point is idempotent. Part (b) of the theorem says that a set is convex if and only if it is idempotent.

Using Theorem 2.9, we obtain easily the result that any join of points (Section 2.7) is convex.

Theorem 2.10. *Any join of points $a_1 \ldots a_n$ is convex.*

PROOF. By the generalized associative and commutative laws and J4,

$$(a_1 \ldots a_n)(a_1 \ldots a_n) = (a_1 a_1) \ldots (a_n a_n) = a_1 \ldots a_n.$$

Hence $a_1 \ldots a_n$ is convex by Theorem 2.9(b). $\square$

In view of the definition of segment (Section 2.7) the case $n = 2$ of the theorem may be stated: Any segment is convex.

The theorem is quite important. It provides a simple and basic procedure for constructing convex sets—we have, in effect, a machine for turning them out. The machine, it is true, does not produce all convex sets —only a certain type, namely, joins of points. But it will become apparent that joins of points play an important role in the construction of all convex sets. (See Theorem 3.7.)

Remarks on the Proof. The proof is brief, elegant, cogent. But it gives us no geometrical insight into what is happening. It seems difficult if not impossible to illustrate the steps in the proof graphically, since they involve algebraic manipulations of sets. It is possible to write a proof, based on the definition of convexity rather than Theorem 2.9, which can be pictured step by step for a given value of n. Here it is for the case $n = 4$ (see Figure 2.25).

Suppose

$$p, q \subset a_1 a_2 a_3 a_4. \tag{1}$$

Then $p \subset a_1(a_2 a_3 a_4)$, so that

$$p \subset a_1 p', \tag{2}$$

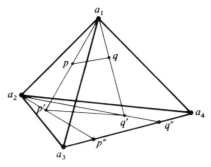

Figure 2.25

where
$$p' \subset a_2a_3a_4. \tag{3}$$
Similarly
$$q \subset a_1q', \tag{4}$$
where
$$q' \subset a_2a_3a_4. \tag{5}$$
By Corollary 2.1 we may multiply the relations (2) and (4) term by term, getting
$$pq \subset (a_1p')(a_1q') = a_1a_1p'q' = a_1p'q',$$
so that
$$pq \subset a_1p'q'. \tag{6}$$
Writing (3) and (5) together, we have
$$p', q' \subset a_2a_3a_4. \tag{7}$$
Observe that (7) has the same form as (1). So we can apply to (7) the argument which derived (2), (4) and (3), (5) from (1). We obtain
$$p' \subset a_2p'', \qquad q' \subset a_2q'', \tag{8}$$
where
$$p'', q'' \subset a_3a_4. \tag{9}$$
Then (8) implies
$$p'q' \subset (a_2p'')(a_2q'') = a_2p''q'',$$
so that
$$p'q' \subset a_2p''q''. \tag{10}$$
Similarly (9) implies
$$p''q'' \subset (a_3a_4)(a_3a_4) = a_3a_4. \tag{11}$$

Now (6), (10) and (11) yield

$$pq \subset a_1a_2p''q'' \subset a_1a_2a_3a_4.$$

Thus $a_1a_2a_3a_4$ is convex by definition.

A convex set, by definition, is closed under join of points. This implies it is closed under join of sets: that is, a convex set contains the join of any two of its subsets.

Theorem 2.11. *Let A be convex. Then $A \supset X, Y$ implies $A \supset XY$.*

PROOF. $A \supset X$, $A \supset Y$ can be combined to yield $AA \supset XY$ by Corollary 2.1. By Theorem 2.9(b), $AA = A$. Thus $A \supset XY$.   □

The result holds for n terms.

Corollary 2.11. *Let A be convex. Then $A \supset X_1, \ldots, X_n$ implies that $A \supset X_1 \ldots X_n$.*

PROOF. By successive cases (see Theorem 2.6) or by induction.   □

Note the special case: A convex set contains the join of any n of its points.

2.10 Joins, Intersections and Unions of Convex Sets

Is the join of two convex sets convex? The union? The intersection? We consider these questions now.

Theorem 2.12. *If two sets are convex, so is their join.*

PROOF. Let A and B be convex. Using Theorem 2.9(b) we have

$$(AB)(AB) = (AA)(BB) = AB.$$

Hence AB is convex by Theorem 2.9(b).   □

Alternative Proof of the Theorem. Here is a proof of Theorem 2.12 which is based on the definition of convexity and can be illustrated graphically. Suppose $p, q \subset AB$ (Figure 2.26). Then

$$p \subset a_1b_1, \qquad q \subset a_2b_2 \tag{1}$$

where

$$a_1, a_2 \subset A \quad \text{and} \quad b_1, b_2 \subset B.$$

Multiplying the relations in (1) term by term (Corollary 2.1) and applying the convexity of A and B, we obtain

$$pq \subset (a_1b_1)(a_2b_2) = (a_1a_2)(b_1b_2) \subset AB.$$

By definition AB is convex.   □

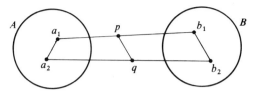

Figure 2.26

Observe that the germ of the proof is the relation

$$(a_1b_1)(a_2b_2) = (a_1a_2)(b_1b_2),$$

which in Euclidean geometry provides information (Figure 2.27) on the interior of a quadrilateral or a tetrahedron (triangular pyramid). (See Exercises 1 and 2 at the end of Section 2.6; Section 1.5, Example III; Exercise 3 at the end of Section 1.6.)

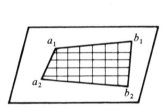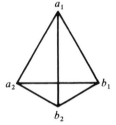

Figure 2.27

Corollary 2.12. *If n sets are convex, so is their join.*

PROOF. By induction, or by the method of Theorem 2.10. ☐

Theorem 2.13. *The intersection of two convex sets is also convex.*

PROOF. Let A and B be convex. We show $A \cap B$ is convex. Suppose $A \cap B \supset x, y$. Then $A \supset x, y$, so that $A \supset xy$. Similarly $B \supset xy$. Thus $A \cap B \supset xy$ and $A \cap B$ is convex by definition. ☐

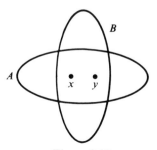

Figure 2.28

The theorem is easily extended to the case of n convex sets. The result holds, however, for any collection or family of convex sets, finite or infinite.

Theorem 2.14. *Let F be a family of convex sets. Then the intersection M of the sets of F is also convex.*

PROOF. Suppose $x, y \subset M$. We show $xy \subset M$. Let $p \subset xy$, and let K be any set of the family F. By definition M consists of all points which are in every set of the family F. Thus

$$x, y \subset K.$$

Since K is convex, $xy \subset K$. Thus $p \subset K$. Since K may be chosen arbitrarily as a set of F, p must be in every set of F, and so $p \subset M$. Consequently $xy \subset M$, and M is convex by definition.   □

Finally a few words on the union of convex sets. The union of two convex sets need not be convex. In Euclidean geometry, for example, choose two distinct points or two parallel lines. Several exercises in the following set pose problems on the convexity and nonconvexity of the union of convex sets.

EXERCISES (CONVEX SETS)

1. In Euclidean geometry let A be convex and b a point not in A.
   (a) Find examples in which $A \cup b$ is convex; is not convex.
   (b) Can you find a condition that will ensure $A \cup b$ is not convex? Is convex?
   (c) Can you find an A such that $A \cup b$ is convex (i) for no point b; (ii) for exactly one point b; (iii) for exactly two points b; (iv) for exactly n points b; (v) for infinitely many points b? What do you observe?

2. In Euclidean geometry let A be convex and b a point in A. Do the same as Exercise 1 for $A - b$ in place of $A \cup b$.

3. Represent on a sheet of paper the sets

$$a, \ b, \ c, \ ab, \ ac, \ bc, \ abc$$

   of a Euclidean plane, taking a, b, c to be noncollinear points.
   (a) Which of the unions of these sets, taken two at a time, are convex?
   (b) The same for unions of three of the sets.
   (c) The same for unions of four or more of the sets.

4. (a) Verify in a Euclidean model that if A contains exactly two points, $A^2 = AA$ is convex.
   (b) Similarly, if A contains exactly three points, $A^3 = A^2 \cdot A$ is convex.
   (c) Try to generalize the results of (a) and (b).

5. In Euclidean geometry suppose A and B are convex sets. Find examples in which $A \cup B$ is convex; is not convex. Can you conjecture a condition that will ensure the convexity of $A \cup B$?

6. Prove: $a \cup (ab)$ is convex. [*Hint*. Apply the definition of convexity: Suppose $p, q \subset a \cup (ab)$ and prove $pq \subset a \cup (ab)$ by considering cases.]

7. Prove: If A and B are convex, then $A \cup (AB)$ is convex.

8. Prove: If A and B are convex, then $A \cup (AB) \cup B$ is convex.

9. Prove: If A and B are convex, then $A \cup B$ is convex if and only if $A \cup B \supset AB$.

10. If $A \cup p$ is convex, can A be convex? Must A be convex? Explain.

11. If $A - p$ is convex, can A be convex? Must A be convex? Explain.

12. Can you prove Theorem 2.10 by using Corollary 2.12? Explain.

The Notion of Halfplane in Euclidean Geometry. Let L be a line of a Euclidean plane P (see Figure 2.29). Then the set H of points of P that are on a given side of L is called an *open halfplane* or simply a *halfplane*. The union $H \cup L$ is a *closed halfplane*. L is said to be the *edge* of H and also of $H \cup L$.

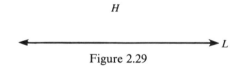

Figure 2.29

13. Let S be a square in a Euclidean plane. Consider the four closed halfplanes containing S whose edges each contain a side of S. Consider the corresponding four open halfplanes also.
    (a) What is the intersection of the four closed halfplanes? Is it convex?
    (b) The same as (a) for the four open halfplanes.

14. In a Euclidean plane let C be a circle with center p.
    (a) What is the intersection of all closed halfplanes containing p whose edges are tangent to C? Is it convex?
    (b) The same as (a) for open halfplanes.

EXERCISES (STAR-SHAPED SETS)

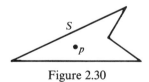

Figure 2.30

Definition. Let S be a set and p a point of S. Suppose $x \subset S$ implies $px \subset S$. Then we say S is *star-shaped relative to p* or *star-shaped from p*. The point p is called a *focal point* or *focus* of S. The set of all *focal points* of S is its *kernel*.

The idea is described cogently in visual terms: All points of S are visible from p.

1. Pick out focal points for each of the planar regions in Figure 2.31, and sketch or shade the kernel of the set.

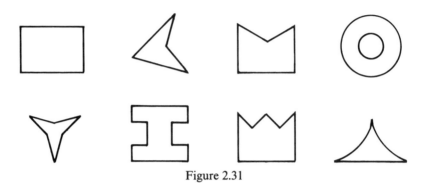

Figure 2.31

2. Prove: The kernel of any set S is convex.

3. Prove: If p is a focus of S and $p \subset T$, then p is a focus of ST.

4. Prove: If p is a focus of S and of T, then p is a focus of $S \cup T$ and of $S \cap T$.

5. Suppose $p \subset S$. Prove: S is star-shaped relative to p if and only if $S \supset pS$.

6. Let A and B be convex.
   (a) Prove: Each point of $A \cap B$ is a focus of $A \cup B$.
   (b) Can you prove that $A \cap B$ is the kernel of $A \cup B$? That $A \cap B$ is not the kernel of $A \cup B$? Explain.

2.11 Interiors and Frontiers of Convex Sets

The notion of the inside or interior of a physical object must have had its origins in prehistoric times. The idea of the interior of a geometrical figure such as a triangle or angle appears implicitly without formal definition in Euclid's *Elements* and also in treatments of classical geometry until comparatively recent times. Mathematicians, using metric or topological concepts, have constructed somewhat sophisticated definitions of the interior of an arbitrary set of points.

For our purposes it suffices to define the concept of interior solely for convex sets. In this case a simple definition can be framed which corresponds to familiar geometrical usage, as when we refer to an interior point of a triangular region or of a solid cube.

Motivation for the Definition. In Euclidean geometry consider the closed triangular region T with vertices a, b and c. It is evident from Figure 2.32 that T is a convex set. We are interested in the "fine structure" of this

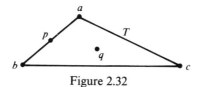

Figure 2.32

convex set. T contains points such as a, or a point p of ab, which are "peripheral" to the set—which lie, so to speak, on its "rim". These may be described as *frontier* points of T. Then T has points like q in abc which are embedded "within" the set and are naturally called *interior* points of T. The convex set T falls into two parts: its interior points fill out abc, its frontier points form $a \cup b \cup c \cup (ab) \cup (ac) \cup (bc)$.

Our immediate object is to replace the pictorial description of interior point by a simple mathematical criterion.

Consider the convex set A (Figure 2.33) in Euclidean geometry. Let a be an interior point of A in the intuitive-pictorial sense. Let x be any point of A distinct from a itself. Then there is a point y in A such that a is between x and y, or equivalently a is in the segment xy. Thus $a \subset xy$, the Euclidean join of x and y. Finally note that if $x = a$, we can choose $y = a$, and $a \subset xy$ in this case too.

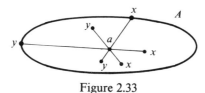

Figure 2.33

Summary. In Euclidean geometry let a be an "interior" point of a convex set A. Then a satisfies the following condition:

For each point $x \subset A$ there exists a point $y \subset A$ such that $a \subset xy$.

In the abstract join theory we adopt the above condition as the essential distinguishing characteristic of an interior point of a convex set.

Definition. Let A be a convex set. Then a point a is called a (*geometrical*) *interior point* of A if $a \subset A$ and satisfies the following condition:

(I) For each $x \subset A$ there exists $y \subset A$ such that $a \subset xy$.

The (*geometrical*) *interior* of A, denoted by $\mathcal{I}(A)$, is simply the set of all interior points of A.[8]

---

[8] Section 2.25 contains a discussion of the *metric* interior of a set in Euclidean geometry and its relation to the geometrical interior of a convex set.

Observe that $\mathcal{I}(A) \subset A$; also $\mathcal{I}(\emptyset) = \emptyset$ and $\mathcal{I}(p) = p$ for each point p.

Finally it can be noted that if the convex set $A \neq \emptyset$, the condition $a \subset A$ in the definition is dispensable and a is an interior point of A merely if condition (I) above holds. For then there exists $x \subset A$ and consequently $y \subset A$ such that $a \subset xy$, and the convexity of A implies $a \subset A$.

Remark on the Definition. Note that interior as defined above is quite different from "interior" as used in phrases such as "the interior of a triangle" or "the interior of a spherical surface." In the latter cases the interior is bounded by the given figure but is not a subset of it.

Once interior is defined, the notion of frontier falls right into place.

Definition. Let A be a convex set. Then a point of A which is not an interior point of A is called a *frontier point* of A. The *frontier* of A, denoted by $\mathcal{F}(A)$, is the set of its frontier points.   □

An equivalent definition:

$$\mathcal{F}(A) = A - \mathcal{I}(A) \quad \text{for any convex set } A.$$

Observe that $\mathcal{F}(A) \subset A$; also $\mathcal{F}(\emptyset) = \emptyset$ and $\mathcal{F}(p) = \emptyset$ for each point p.

2.12 Euclidean Interiors and Frontiers

The definitions of interior and frontier introduced in the last section are now applied to some simple Euclidean convex sets. The object is twofold: to show how the definitions are employed and to find out what the interiors and frontiers of a few specific convex sets actually are.

The results will not be justified by postulational reasoning as in the abstract theory, but by a mixture of geometric reasoning and intuitive observation of diagrams of the sort we have used in treating questions in Euclidean geometry.

EXAMPLE I (The closed segment ab). Let A be the closed segment ab (Section 2.1; see Figure 2.34). Then $A = a \cup b \cup (ab)$. The convexity of A can be verified by use of the diagram. We show informally that

$$\mathcal{I}(A) = ab. \tag{1}$$

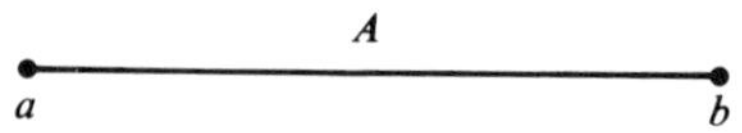

Figure 2.34

First we show that each point of ab is an interior point of A in accordance with the definition above (Section 2.11). Let $p \subset ab$. Suppose

Figure 2.35

$x \subset A$ (Figure 2.35). We must find a point $y \subset A$ such that

$$p \subset xy. \tag{2}$$

We argue by cases. If $x = a$, let $y = b$. Then (2) holds. Similarly if $x = b$, choose $y = a$. If $x = p$, take $y = p$. If $x \subset ap$ (Figure 2.36), let $y = b$.

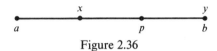

Figure 2.36

Similarly (and finally), if $x \subset bp$, let $y = a$. Then (2) is verified in each case, and we can assert that p is an interior point of A. Since p is any point of ab, we have shown

$$ab \subset \mathcal{I}(A). \tag{3}$$

To complete the proof we show that

$$\mathcal{I}(A) \subset ab, \tag{4}$$

that is, each interior point of A is in ab. So it remains to show only that a and b are not interior points of A. Suppose a is an interior point of A (Figure 2.37). By definition, since $b \subset A$, there exists $z \subset A$ such that

$$a \subset bz. \tag{5}$$

Certainly $z \neq b$: for $z = b$ implies, by (5), $a = b$. Then (5) implies that a is *between* b and z, which is impossible, since z is a point of A. Thus a is not an interior point of A. Similarly b is not an interior point of A. Thus (4) holds, and (3) and (4) imply (1).

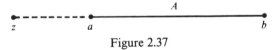

Figure 2.37

Note, in view of (1), that $\mathcal{I}(A)$ is convex (Theorem 2.10). Finally observe

$$\mathcal{F}(A) = \{a, b\}. \tag{6}$$

Summary. In Euclidean geometry the interior of the closed segment ab is the (open) segment ab, and its frontier consists of its endpoints.

EXAMPLE II (The segment ab). Note that ab (Figure 2.38) is convex by Theorem 2.10. We show as in Example I that

$$\mathcal{I}(ab) = ab. \tag{7}$$

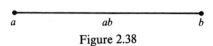

Figure 2.38

Let $p \subset ab$. Suppose $x \subset ab$. We show

$$p \subset xy \tag{8}$$

for some point $y \subset ab$. If $x = p$, let $y = p$. If $x \subset ap$ (Figure 2.39), choose $y \subset bp$. If $x \subset bp$, choose $y \subset ap$. In each case (8) is satisfied, and we conclude that p is an interior point of ab. Thus each point of ab is an interior point of ab, and (7) is justified.

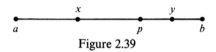

Figure 2.39

Finally we have

$$\mathscr{F}(ab) = \varnothing. \tag{9}$$

Summary. In Euclidean geometry the interior of a segment is the segment itself, and its frontier is the empty set.

Remark. Referring back to Example I above, let A be the closed segment ab. Then by (1),

$$\mathscr{G}(A) = ab. \tag{10}$$

Since $\mathscr{G}(A)$ is convex, we can form $\mathscr{G}(\mathscr{G}(A))$. Then using (10) and (7), we have

$$\mathscr{G}(\mathscr{G}(A)) = \mathscr{G}(ab) = ab = \mathscr{G}(A).$$

Thus

$$\mathscr{G}(\mathscr{G}(A)) = \mathscr{G}(A). \tag{11}$$

EXAMPLE III (A closed triangular region). Let B be the closed triangular region with vertices a, b, c (Figure 2.40). Then

$$B = a \cup b \cup c \cup (ab) \cup (ac) \cup (bc) \cup (abc).$$

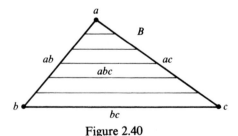

Figure 2.40

The convexity of B can be verified by using the diagram. We assert

$$\mathcal{I}(B) = abc. \tag{12}$$

No attempt will be made to give even an informal proof of (12). Rather, pictorial evidence will be presented to convince you of the correctness of the result.

First we indicate that each point of abc is an interior point of B in accordance with the definition. Let $p \subset abc$. Suppose $x \subset B$. Then we assert there is a point $y \subset B$ such that

$$p \subset xy. \tag{13}$$

Figure 2.41 indicates several choices for x and corresponding choices for y.

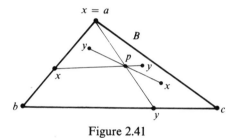

Figure 2.41

This "verifies" (13) and so yields

$$abc \subset \mathcal{I}(B). \tag{14}$$

Finally consider the reverse inclusion

$$\mathcal{I}(B) \subset abc. \tag{15}$$

This can be justified by showing that any point q of B which is not in abc cannot be an interior point of B. To verify the latter, consider first the case where $q \subset ab$ (Figure 2.42). Suppose q is an interior point of B. By definition, since $c \subset B$, there exists $z \subset B$ such that

$$q \subset cz. \tag{16}$$

If $c = z$, then (16) impliees $q = c$, which is impossible, since a, b and c are not collinear. Thus $c \neq z$, and (16) implies that q is between c and z. Then

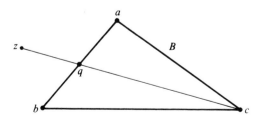

Figure 2.42

z is on the opposite side of line ab from c and cannot be in set B. Hence our supposition is false, and q can't be an interior point of B in the given case. Similar arguments can be employed for the other cases where q is in ac or bc or is a vertex of B. Thus (15) is verified and with (14) yields (12).

Finally we have

$$\mathcal{F}(B) = a \cup b \cup c \cup (ab) \cup (ac) \cup (bc). \tag{17}$$

Summary. In Euclidean geometry the frontier of the closed triangular region with vertices a, b and c is precisely triangle abc, and its interior is the interior of triangle abc.

EXAMPLE IV (The triangular region abc). Let a, b and c be three noncollinear points (Figure 2.43). Consider their join abc, which may be called a *triangular region* or an *open triangular region*.

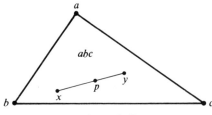

Figure 2.43

Note that abc is convex (Theorem 2.10). As in Example III, it can be seen that each point of abc is an interior point of abc. Thus

$$\mathcal{I}(abc) = abc. \tag{18}$$

It follows that

$$\mathcal{F}(abc) = \varnothing. \tag{19}$$

Summary. In Euclidean geometry the interior of a triangular region is precisely itself, and its frontier is the empty set.

Remark. In equation (12) above we showed $\mathcal{I}(B) = abc$, where B is the closed triangular region with vertices a, b and c. This relation together with (18) easily yields

$$\mathcal{I}(\mathcal{I}(B)) = \mathcal{I}(B). \tag{20}$$

(See the Remark at the end of Example II above.)

EXERCISES

1. Find the interior and the frontier in Euclidean geometry of the convex sets indicated in Figure 2.44.

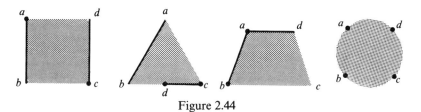

Figure 2.44

2. In Euclidean geometry suppose A is convex and $a, b \subset A$. Can $ab \subset \mathcal{I}(A)$? Must $ab \subset \mathcal{I}(A)$? Explain.

3. In Euclidean geometry can you find a convex set A such that $\mathcal{I}(A)$ is convex; is not convex? Explain.

4. The same as Exercise 3 with $\mathcal{I}$ replaced by $\mathcal{F}$.

5. In Euclidean geometry let A and B be convex sets.
   (a) Will $\mathcal{I}(AB) = \mathcal{I}(A) \cdot \mathcal{I}(B)$ hold ever; always?
   (b) The same for $\mathcal{I}(A \cap B) = \mathcal{I}(A) \cap \mathcal{I}(B)$.
   (c) The same for $\mathcal{I}(A \cup B) = \mathcal{I}(A) \cup \mathcal{I}(B)$, assuming $A \cup B$ is convex.

6. The same as Exercise 5 with $\mathcal{I}$ replaced by $\mathcal{F}$.

7. In Euclidean geometry can you find a convex set A such that $\mathcal{I}(A) = A$? Can you find several such sets? Can you describe or characterize in a simple way the family of all such sets?

8. The same as Exercise 7 for the following relations: (a) $\mathcal{I}(A) \neq A$; (b) $\mathcal{I}(A) = \varnothing$; (c) $\mathcal{F}(A) = A$; (d) $\mathcal{F}(A) \neq A$; (e) $\mathcal{F}(A) = \varnothing$.

9. In Euclidean geometry find a convex set which cannot be the interior of any convex set. Explain.

10. In Euclidean geometry find a set which cannot be the frontier of any convex set. Explain.

11. Let a, b and c be noncollinear points in Euclidean geometry. Find all points p such that $p \not\subset abc$ but $(abc) \cup p$ is convex. Find $\mathcal{I}((abc) \cup p)$, $\mathcal{F}((abc) \cup p)$. Can you express precisely the relation of p to the set abc?

12. In Euclidean geometry can you find a convex set A such that $\mathcal{F}(A)$ is a point; a line; a plane?

13. Consider in Euclidean geometry the equation
$$\mathcal{I}(X) = \mathcal{I}(A), \tag{1}$$
where A is a fixed convex set and X a variable one. Find A such that (1) has:
   (a) infinitely many solutions X;
   (b) more than one but only a finite number of solutions X;
   (c) exactly one solution, namely, $X = A$. Find several such A's. Can you describe or characterize all such A's?

14. (a) Find a convex set in Euclidean geometry whose frontier is a circle; the set of vertices of a rectangle; the set of vertices of a cube; the surface of a cube.

(b) In each case, can you find a second convex set whose frontier is the same figure?

Definition. Let A be a nonempty convex set in Euclidean geometry. Then A is *lineal* (or 1-*dimensional*) if it is contained in a line but is not a point; A is *planar* (or 2-*dimensional*) if it is contained in a plane but not in a line; A is *solid* (or 3-*dimensional*) if it is not contained in a plane; A is 0-*dimensional* if it is a point.

15. (a) Can you find in Euclidean geometry a lineal convex set A such that $\mathscr{F}(A)$ is $\varnothing$; a point; a point pair; a point triple? Explain.
    (b) Classify the lineal convex sets in Euclidean geometry.

16. (a) In Euclidean geometry if A is a lineal convex set, is $\mathscr{F}(A)$ convex? If so, what can you say about the dimensional class of $\mathscr{F}(A)$?
    (b) The same if A is planar; is solid.

2.13 Interiority Properties of Convex Sets

A series of properties of the interior of a convex set are now derived. All are easily verified in Euclidean geometry.

First we have: The join of an interior point of convex set K and a point of K is always contained in the interior of K.

Theorem 2.15. *Let K be convex. Then $a \subset \mathscr{F}(K)$ and $b \subset K$ imply that $ab \subset \mathscr{F}(K)$.*

PROOF. To prove the theorem we assume

$$p \subset ab \tag{1}$$

and derive

$$p \subset \mathscr{F}(K). \tag{2}$$

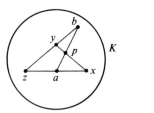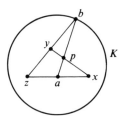

Figure 2.45

Note that $p \subset K$. To justify (2) we show that for each $x \subset K$ there exists y such that

$$p \subset xy \quad \text{and} \quad y \subset K. \tag{3}$$

Suppose $x \subset K$. By definition $a \subset \mathcal{I}(K)$ implies the existence of $z \subset K$ such that

$$a \subset xz. \tag{4}$$

The relations (1) and (4) imply (by the definition of the join of two sets, Section 2.3)

$$p \subset (xz)b = x(zb). \tag{5}$$

The relation (5) implies the existence of y such that

$$p \subset xy, \quad y \subset zb. \tag{6}$$

But $z, b \subset K$ and K is convex. Hence $zb \subset K$ and (6) implies

$$y \subset K. \tag{7}$$

The relations (6) and (7) yield (3). Then (2) holds (by definition) and the proof is complete. ☐

The theorem has two corollaries. The first generalizes the theorem to sets.

Corollary 2.15.1. *Let K be convex. Then $A \subset \mathcal{I}(K)$ and $B \subset K$ imply $AB \subset \mathcal{I}(K)$.*

PROOF. Let $p \subset AB$. Then $p \subset ab$, where $a \subset A$ and $b \subset B$. By hypothesis $a \subset \mathcal{I}(K)$ and $b \subset K$. By the theorem, $p \subset \mathcal{I}(K)$ and the corollary is established. ☐

The second corollary asserts the convexity of interiors.

Corollary 2.15.2. *For any convex set K, $\mathcal{I}(K)$ is convex.*

PROOF. Let $a, b \subset \mathcal{I}(K)$. By the theorem, $ab \subset \mathcal{I}(K)$. By definition $\mathcal{I}(K)$ is convex. ☐

The theorem can be strengthened.

Theorem 2.16. *Let K be a convex set. Then*

(i) $a \subset \mathcal{I}(K)$ *implies* $aK = \mathcal{I}(K)$;
(ii) $a \subset K$ *implies* $a \cdot \mathcal{I}(K) = \mathcal{I}(K)$.

Figure 2.46

Proof. (i) Suppose $a \subset \mathcal{I}(K)$. Corollary 2.15.1 implies

$$aK \subset \mathcal{I}(K). \tag{1}$$

To prove the converse inclusion to (1), let $x \subset \mathcal{I}(K)$. Certainly $a \subset K$. Thus by definition of interior point,

$$x \subset ay, \quad \text{where } y \subset K.$$

Hence $x \subset aK$, so that $\mathcal{I}(K) \subset aK$ and the result holds.

(ii) $a \cdot \mathcal{I}(K) \subset \mathcal{I}(K)$ can be shown as (1) was. For the converse let $x \subset \mathcal{I}(K)$. Then $a \subset K$ implies $x \subset ay$ where $y \subset K$. Hence using Theorem 2.15, the idempotent law and the monotonic law for join (Theorem 2.1), we have

$$x = xx \subset xay = a(xy) \subset a \cdot \mathcal{I}(K).$$

Thus $\mathcal{I}(K) \subset a \cdot \mathcal{I}(K)$ and the result holds. □

Corollary 2.16. $\mathcal{I}(K) \cdot K = \mathcal{I}(K)$ *for any convex set K.*

Proof. $x \subset \mathcal{I}(K) \cdot K$ if and only if $x \subset aK$ for some $a \subset \mathcal{I}(K)$. Apply (i) of the theorem. □

The Interior as a Function or Operation. It is sometimes convenient to think of the relation between a convex set K and its interior $\mathcal{I}(K)$ as a *function* or *operation*, naturally denoted by $\mathcal{I}$, which assigns to each convex set K, the set $\mathcal{I}(K)$. By Corollary 2.15.2 the "functional value" $\mathcal{I}(K)$ is always a convex set.

The function $\mathcal{I}$ satisfies a restricted monotonic law.

Theorem 2.17 (Monotonic Property of Interior). *Let A and B be convex sets. Then $A \subset B$ implies $\mathcal{I}(A) \subset \mathcal{I}(B)$, provided A meets $\mathcal{I}(B)$.*

Proof. By hypothesis there is a point o such that $o \subset A, \mathcal{I}(B)$. Suppose $p \subset \mathcal{I}(A)$. Since $o \subset A$, by definition of $\mathcal{I}(A)$ there is a point $q \subset A$ such that $p \subset oq$. Note that $q \subset B$. But $o \subset \mathcal{I}(B)$. Hence using Theorem 2.15,

$$p \subset oq \subset \mathcal{I}(B),$$

and we conclude $\mathcal{I}(A) \subset \mathcal{I}(B)$. □

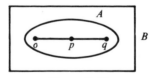

Figure 2.47

Corollary 2.17.1. *If A and B are convex, then $A \subset B$ implies $\mathcal{I}(A) \subset \mathcal{I}(B)$ or $A \subset \mathcal{I}(B)$.*

Corollary 2.17.2. *Let A and B be convex sets and $A \subset B$. Then*

$$\mathcal{I}(A) \subset \mathcal{I}(B) \tag{1}$$

if and only if (a) *A meets $\mathcal{I}(B)$ or* (b) *$\mathcal{I}(A) = \varnothing$.*

PROOF. If (a) or (b) holds, then (1) follows.

Conversely, suppose (1). If $\mathcal{I}(A) \neq \varnothing$, then $\mathcal{I}(A)$ meets $\mathcal{I}(B)$ and (a) follows. □

Finally we show that the composition of the interior function with itself yields itself.

Theorem 2.18. $\mathcal{I}(\mathcal{I}(K)) = \mathcal{I}(K)$, *for any convex set K.*

PROOF. $\mathcal{I}(K)$ is convex (Corollary 2.15.2) so that $\mathcal{I}(\mathcal{I}(K))$ is defined. Then $\mathcal{I}(\mathcal{I}(K)) \subset \mathcal{I}(K)$.

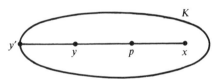

Figure 2.48

To prove the reverse inclusion

$$\mathcal{I}(K) \subset \mathcal{I}(\mathcal{I}(K)), \tag{1}$$

let $p \subset \mathcal{I}(K)$. To show $p \subset \mathcal{I}(\mathcal{I}(K))$ we let $x \subset \mathcal{I}(K)$ and show $p \subset xy$ for some $y \subset \mathcal{I}(K)$. Note $x \subset K$. Then certainly $p \subset xy'$ for some $y' \subset K$. Can we be sure $y' \subset \mathcal{I}(K)$? The diagram suggests a negative answer. However, we can be sure that

$$xy' \subset \mathcal{I}(K) \quad \text{(Theorem 2.15)} \tag{2}$$

and this is a clue to the proof. Using the idempotent law, we obtain

$$p \subset xy' = (xx)y' = x(xy').$$

Then $p \subset xy$ for some $y \subset xy'$. By (2) $y \subset \mathcal{I}(K)$, and $p \subset \mathcal{I}(\mathcal{I}(K))$ by definition. Hence (1) holds and the theorem is proved. □

Query on the Proof. The proof was suggested by the diagram. Are you sure the proof is logically valid and does not depend on the diagram?

In general a convex set is distinct from its interior, since it contains points which are not interior points. The theorem then places interiors of convex sets in a special category of convex sets—they are homogeneous, in a sense, since all of their points are interior points.

EXERCISES

1. Prove: If K is convex and $a \subset K$, then $\mathcal{I}(K) = \mathcal{I}(aK)$.

2. Suppose K is convex. If $\mathcal{I}(K) = \mathcal{I}(aK)$, does $a \subset K$? Explain.

3. (a) Prove: For a convex set K if $a \subset \mathcal{I}(K)$, then $a \subset xK$ for every $x \subset K$.
   (b) Is the converse valid? Justify your answer.

4. (a) Prove: If A and B are convex, then $\mathcal{I}(A \cap B) \subset \mathcal{I}(A) \cap \mathcal{I}(B)$, provided $\mathcal{I}(A)$ meets $\mathcal{I}(B)$.
   (b) Does the result hold if the proviso is dropped? Explain.

5. Prove: If A is convex and $a \subset \mathcal{I}(A)$, then for each $x \subset A$ there exists $y \subset \mathcal{I}(A)$ such that $a \subset xy$. (Compare the definition of interior of a convex set, Section 2.11.)

6. Prove:
   (a) If $a \cup (ab)$ is convex, then $\mathcal{I}(a \cup (ab)) = \mathcal{I}(ab)$.
   (b) If $a \cup b \cup (ab)$ is convex, then $\mathcal{I}(a \cup b \cup (ab)) = \mathcal{I}(ab)$.
   [*Note.* $a \cup (ab)$ and $a \cup b \cup (ab)$ are convex by Exercises 6 and 8 on Convex Sets at the end of Section 2.10.]

   In the following exercises define the interior of the set of points S exactly as we have defined it for a convex set in Section 2.11.

7. In Euclidean geometry let $S = \{a, b, p, q\}$, where p and q are distinct points of segment ab. Find $\mathcal{I}(S)$ and $\mathcal{I}(\mathcal{I}(S))$. What do you observe?

8. In Euclidean geometry let $S = (ab) \cup (cd)$ as indicated in Figure 2.49. Find $\mathcal{I}(S)$. Does the conclusion of Theorem 2.15 apply?

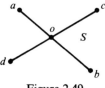

Figure 2.49

9. In Euclidean geometry let $S = (oac) \cup (obd) \cup o$ as indicated in Figure 2.50. Find $\mathcal{I}(S)$. Do you notice anything strange?

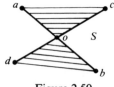

Figure 2.50

2.14 Absorption by Joining

The join of two sets of points usually has no simple relation to the individual sets. For example, AB may contain both A and B or neither of them. AB may be contained in A or B or neither. (See Exercises 14 and 15 at the end of Section 2.6.) In the last section we have encountered results such as

$$a \cdot \mathcal{G}(K) = \mathcal{G}(K) \quad \text{if } a \subset K \qquad \text{(Theorem 2.16)}$$

and

$$\mathcal{G}(K) \cdot K = \mathcal{G}(K) \qquad \text{(Corollary 2.16)}$$

in which the join of two terms is actually equal to one of the terms. These results may be described as "absorption principles", since the product of $\mathcal{G}(K)$ and a second factor lies within $\mathcal{G}(K)$. A related usage occurs in the theory of sets, where the principle "$A \supset B$ implies $A \cup B = A$" is called the *absorption law for union*, and its counterpart, "$A \subset B$ implies $A \cap B = A$", the *absorption law for intersection*.

Thus we introduce the following.

Definition. If the sets A and B satisfy $AB \subset A$, we say A *absorbs* B (*by joining*). Moreover, if $AB = A$, we say A absorbs B *properly* (by joining).

Examples of Absorption Properties. A convex set absorbs each of its subsets. The idempotent law may be stated: Each point absorbs itself properly. The interior of a convex set absorbs the set properly (Corollary 2.16). The null set absorbs every set, since $\varnothing A = \varnothing$ (see the second paragraph following Theorem 2.2).

EXERCISES (ABSORPTION OF SETS)

1. (a) Prove: If S absorbs a and b, then S absorbs $\{a, b\}$.
   (b) Prove: If S absorbs A and B, then S absorbs $A \cup B$.

2. Prove: The set of all points absorbed by set S is convex.

3. When can a set absorb itself? Explain.

4. Prove: If the convex set A absorbs p, then $\mathcal{G}(A)$ absorbs p.

5. Prove: If $p \subset S$ then S absorbs p if and only if S is star-shaped relative to p. (See Definition before the exercises on Star-Shaped Sets at the end of Section 2.10.)

6. Consider the set abc.
   (a) Find some points that are absorbed by abc.
   (b) Find some sets of points that are absorbed by abc.
   (c) What is the largest set of points you can find that is absorbed by abc?
   (d) Interpret your results in Euclidean geometry.

7. The same as Exercise 6 for the set $a_1 \ldots a_n$.

8. (a) Prove: If A absorbs B, then AC absorbs B for any set C.
   (b) Prove: If A absorbs B and A absorbs C, then A absorbs BC.
   (c) Prove: If A absorbs B and B absorbs C, then A absorbs BC.

9. In Euclidean geometry find examples of nonempty sets A, B and C such that A absorbs B and B absorbs C, but A does not absorb C.

2.15 Closures of Euclidean Convex Sets, Intuitively Treated

This section presents an introduction to a new concept: the closure of a convex set. The notion of closure is described intuitively and illustrated by applying it to the Euclidean convex sets studied in Examples I–IV of Section 2.12.

The closure of a convex set A, denoted by $\mathcal{C}(A)$, may be described intuitively as the set formed by adjoining to A its "peripheral" points: the points that lie on its "rim". The notion is complementary to that of interior, and you may find it interesting and helpful to compare the present results with the corresponding ones on interior in Section 2.12.

EXAMPLE I (The closed segment ab). Let A be the closed segment ab (Figure 2.51). Then

$$A = a \cup b \cup (ab).$$

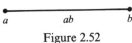

Figure 2.51

The peripheral points of A are a and b. Since a and b are already in A, the closure of A is A itself. In functional notation

$$\mathcal{C}(A) = A. \tag{1}$$

Conclusion. In Euclidean geometry the closure of a closed segment is the segment itself.

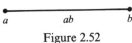

Figure 2.52

EXAMPLE II (The segment ab). The peripheral points of ab are a and b and do not belong to the set ab (Figure 2.52). Then

$$\mathcal{C}(ab) = (ab) \cup a \cup b = \text{the closed segment } ab. \qquad (2)$$

Conclusion. In Euclidean geometry the closure of the (open) segment ab is the closed segment ab.

Remark. Let A be the closed segment ab as in Example I. Then the conclusion can be written

$$\mathcal{C}(ab) = A. \qquad (3)$$

Then (3) and (1) of Example I above yield

$$\mathcal{C}(\mathcal{C}(ab)) = \mathcal{C}(A) = A = \mathcal{C}(ab),$$

so that

$$\mathcal{C}(\mathcal{C}(ab)) = \mathcal{C}(ab). \qquad (4)$$

Next we show

$$\mathcal{I}(\mathcal{C}(ab)) = ab. \qquad (5)$$

Recall [Section 2.12, Example I, equation (1)]

$$\mathcal{I}(A) = ab. \qquad (6)$$

Then (3) and (6) imply $\mathcal{I}(\mathcal{C}(ab)) = \mathcal{I}(A) = ab$, and (5) holds.

EXAMPLE III (A closed triangular region). Let B be the closed triangular region with vertices a, b, c (Figure 2.53), as in Section 2.12, Example III.

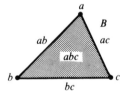

Figure 2.53

Then

$$B = a \cup b \cup c \cup (ab) \cup (ac) \cup (bc) \cup (abc).$$

The peripheral points of B form the set

$$B' = a \cup b \cup c \cup (ab) \cup (ac) \cup (bc).$$

Since $B' \subset B$, we have

$$\mathcal{C}(B) = B. \qquad (7)$$

Conclusion. In Euclidean geometry the closure of a closed triangular region is the region itself.

EXAMPLE IV (The triangular region abc). Consider abc the (open) triangular region (Figure 2.54), as in Section 2.12, Example IV. The peripheral points of abc form the set

$$a \cup b \cup c \cup (ab) \cup (ac) \cup (bc),$$

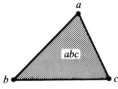

Figure 2.54

which is triangle abc. Thus

$$\mathcal{C}(abc) = (abc) \cup a \cup b \cup c \cup (ab) \cup (ac) \cup (bc).$$

Conclusion. In Euclidean geometry the closure of the triangular region abc is the closed triangular region with vertices a, b, c.

Remark. The following analogues of (4) and (5) in Example II above can be derived by the methods used there:

$$\mathcal{C}(\mathcal{C}(abc)) = \mathcal{C}(abc), \qquad \mathcal{I}(\mathcal{C}(abc)) = abc.$$

EXERCISES

1. Find the closure in Euclidean geometry of the convex sets that are indicated in Figure 2.55.

Figure 2.55

2. In Euclidean geometry suppose A is convex and $a, b \subset \mathcal{C}(A)$. Can $ab \subset \mathcal{I}(A)$? Can $ab \subset \mathcal{F}(A)$? Can ab be contained in neither? Explain.

3. In Euclidean geometry, can you find a convex set A such that $\mathcal{C}(A)$ is convex; is not convex? Explain.

4. In Euclidean geometry let A and B be convex sets.
   (a) Will $\mathcal{C}(AB) = \mathcal{C}(A) \cdot \mathcal{C}(B)$ hold ever; always?
   (b) The same for $\mathcal{C}(A \cap B) = \mathcal{C}(A) \cap \mathcal{C}(B)$.
   (c) The same for $\mathcal{C}(A \cup B) = \mathcal{C}(A) \cup \mathcal{C}(B)$, assuming $A \cup B$ is convex.

5. In Euclidean geometry can you find a convex set A such that $\mathcal{C}(A) = A$? Can

you find several such sets? Can you describe or characterize simply the family of all such sets? (Compare Exercise 7 at the end of Section 2.12.)

6. The same as Exercise 5 for the relation $\mathcal{C}(A) \neq A$.

7. The same as Exercise 5 for the relation $\mathcal{C}(A) = \mathcal{I}(A) = A$.

8. In Euclidean geometry find a convex set which cannot be the closure of any convex set. Explain. (Compare Exercise 9 at the end of Section 2.12.)

9. Consider in Euclidean geometry the equation

$$\mathcal{C}(X) = \mathcal{C}(A), \tag{1}$$

where A is a fixed convex set and X a variable one. Find A such that (1) has:
(a) infinitely many solutions X;
(b) more than one but only a finite number of solutions X;
(c) exactly one solution, namely, $X = A$. Find several such A's. Can you describe or characterize all such A's?
(Compare Exercise 13 at the end of Section 2.12.)

2.16 The Closure of a Convex Set

The idea of closure of a convex set, which is complementary to the notion of interior, is now formally defined. Closure, like its counterpart interior, can be defined for an arbitrary set of points in terms of metric or topological concepts. For convex sets it can be defined in elementary geometrical terms.

In the last section the closure of a convex set A was characterized intuitively as the set formed by adjoining to A all the points that are peripheral to it. This suggests that we define formally the idea of peripheral point of a convex set A and then define the closure of A to be the union of A and its set of peripheral points. Such a procedure is feasible—but it has the disadvantage that in dealing with closures, at least in proving basic theorems, we would have to consider two cases. So we prefer a different approach.

A point will be in the closure of A if and only if it is in A or is peripheral to A. Any point with this property will be called a *contact* point of A. It is not hard to define the notion of contact point by a *single* condition rather than the disjunction of two conditions. Then closure is definable in terms of contact point. So we adopt the following.

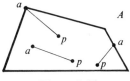

Figure 2.56

Definition. A point a is called a (*geometrical*) *contact point* of convex set A if $ap \subset A$ for some point p. The set of contact points of A is called the (*geometrical*) *closure* or *contact set* of A and is denoted by $\mathcal{C}(A)$.[9]

Observe that $\mathcal{C}(\varnothing) = \varnothing$.

By employing the notion of segment (see the last definition in Section 2.7) the definition of contact point can be restated in standard geometrical terms:

a is a contact point of convex set A if a is an endpoint of a segment contained in A.

In view of this the definition of closure can be restated so:

The closure of convex set A is composed of the endpoints of all subsegments of A.

Remark. The notion of closure does not have the high intuitive accessibility of interior—it seems a bit subtle, not forced on our attention by everyday experience. It will become evident in the development that the two ideas complement each other in a rather natural way. The basis for this complementarity is indicated in the following intuitive characterization of the ideas:

The interior (closure) of a convex set is formed by deleting (adjoining) the points peripheral to it (Figure 2.57).

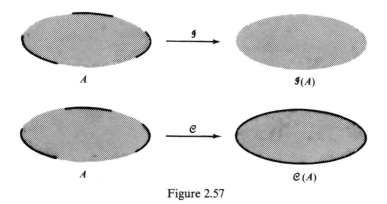

Figure 2.57

---

[9] Section 2.26 contains a discussion of the *metric* closure of a set in Euclidean geometry and its relation to the geometrical closure of a convex set.

2.17 Closure Properties of Convex Sets

The basic properties of the closure operation are now derived. Many of the results are analogous to properties of the interior operation that were studied in Section 2.13.

First a property with almost trivial proof.

Theorem 2.19. $A \subset \mathcal{C}(A)$ *for any convex set* A.

PROOF. If $a \subset A$, then $aa \subset A$ and $a \subset \mathcal{C}(A)$ by definition. ☐

Combining the theorem with $\mathcal{I}(A) \subset A$, the analogous property of the interior operation, we have

Corollary 2.19. $\mathcal{I}(A) \subset A \subset \mathcal{C}(A)$, *for any convex set* A.

By the theorem, $\mathcal{C}(J) = J$ and $\mathcal{C}(a) \supset a$. $\mathcal{C}(a) = a$ holds in Euclidean geometry but is not deducible from J1–J4. (See Exercise 9 below.) It will be proved in Chapter 4 after the postulate set has been strengthened (see Theorem 4.11).

The defining property of a contact point can be strengthened a bit.

Theorem 2.20. *Let* A *be a convex set. Then* a *is a contact point of* A *if and only if there exists a point* q *of* A *such that* $aq \subset A$.

PROOF. Suppose a is a contact point of A. By definition

$$ap \subset A \tag{1}$$

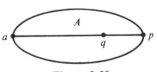

Figure 2.58

for some point p. Let $q \subset ap$. By the monotonic law for join (Theorem 2.1),

$$aq \subset a(ap) = ap. \tag{2}$$

The relations (1) and (2) yield $aq \subset A$, and clearly $q \subset A$. Thus the desired condition holds. The converse is trivial. ☐

In contrast with the interior operation, an unrestricted monotonic law holds for the closure operation.

Theorem 2.21 (Monotonic Property of Closure). *Let A and B be convex. Then $A \subset B$ implies $\mathcal{C}(A) \subset \mathcal{C}(B)$.*

(Compare Theorem 2.17.)

PROOF. Let $x \subset \mathcal{C}(A)$. Then $xp \subset A$ for some point p. By hypothesis $A \subset B$. Thus $xp \subset B$ and $x \subset \mathcal{C}(B)$ by definition.   □

Theorem 2.22. *For any convex set A, $\mathcal{C}(A)$ is convex.*

PROOF. Let $a, b \subset \mathcal{C}(A)$ and $x \subset ab$. Then for some p, q we have $ap \subset A$, $bq \subset A$. Thus

$$apbq \subset A \qquad \text{(Theorem 2.11)},$$

so that

$$(ab)(pq) \subset A. \tag{1}$$

Figure 2.59

Let r be a point of pq. Then $x \subset ab$ implies (Corollary 2.1)

$$xr \subset (ab)(pq). \tag{2}$$

The relations (2) and (1) yield $xr \subset A$, and $x \subset \mathcal{C}(A)$ by definition. Thus $ab \subset \mathcal{C}(A)$, and $\mathcal{C}(A)$ is convex by definition.   □

Now Theorems 2.15 and 2.16, the results of Section 2.13 involving $\mathcal{I}(K)$ and K, are extended by showing that they are valid for $\mathcal{I}(K)$ and $\mathcal{C}(K)$.

Theorem 2.23. *Let K be convex. Then $a \subset \mathcal{I}(K)$ and $b \subset \mathcal{C}(K)$ imply $ab \subset \mathcal{I}(K)$.*

(Compare Theorem 2.15.)

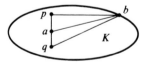

Figure 2.60

PROOF. First we show

$$ab \subset K. \tag{1}$$

Since $b \subset \mathcal{C}(K)$, we have

$$bp \subset K \tag{2}$$

for some point $p \subset K$ (Theorem 2.20). Since $a \subset \mathcal{I}(K)$ and $p \subset K$,

$$a \subset pq \tag{3}$$

for some point $q \subset K$. The relations (3), (2) and $q \subset K$ imply

$$ab \subset (pq)b = (bp)q \subset K.$$

Thus (1) is verified. In view of $a \subset \mathcal{I}(K)$ and (1), Corollary 2.15.1 implies

$$ab = a(ab) \subset \mathcal{I}(K),$$

and the theorem holds.   □

Corollary 2.23. *Let K be convex. Then $A \subset \mathcal{I}(K)$ and $B \subset \mathcal{C}(K)$ imply $AB \subset \mathcal{I}(K)$.*

(Compare Corollary 2.15.1.)

Theorem 2.24. *Let K be a convex set. Then*

(i) $a \subset \mathcal{I}(K)$ *implies* $a \cdot \mathcal{C}(K) = \mathcal{I}(K)$;
(ii) $a \subset \mathcal{C}(K)$ *implies* $a \cdot \mathcal{I}(K) = \mathcal{I}(K)$.

(Compare Theorem 2.16.)

PROOF. Adapt the method of proof of Theorem 2.16.   □

Corollary 2.24. $\mathcal{I}(K) \cdot \mathcal{C}(K) = \mathcal{I}(K)$ *for any convex set K.*

(Compare Corollary 2.16.)

Restatement. The interior of a convex set absorbs its closure properly.

Finally we prove an elementary property of the closure of a join.

Theorem 2.25. *Let A and B be convex and nonempty. Then $\mathcal{C}(AB) \supset A, B$.*

PROOF. Since AB is convex (Theorem 2.12), $\mathcal{C}(AB)$ is defined. We show $\mathcal{C}(AB) \supset A$. Let $a \subset A$. Since B is nonempty, there exists $b \subset B$. Then $ab \subset AB$. By definition $a \subset \mathcal{C}(AB)$. Thus $\mathcal{C}(AB) \supset A$. Similarly $\mathcal{C}(AB) \supset B$.   □

Corollary 2.25. *Let $A_1, \ldots, A_n$ be convex and nonempty. Then*

$$\mathcal{C}(A_1 \ldots A_n) \supset A_1, \ldots, A_n.$$

PROOF. By induction, making use of Theorem 2.6.   □

Remark. $\mathcal{C}(ab) \supset a, b$, that is, the closure of a segment contains its endpoints.

EXERCISES

1. Suppose A is convex and $\mathcal{I}(A) \neq \varnothing$. Prove: a is a contact point of A if and only if $ap \subset A$ for some $p \subset \mathcal{I}(A)$. (Compare Theorem 2.20.)

2. Let A and B be convex sets.
    (a) Prove: $\mathcal{C}(A \cap B) \subset \mathcal{C}(A) \cap \mathcal{C}(B)$.
    (b) Prove: $\mathcal{C}(A \cap B) \supset \mathcal{C}(A) \cap \mathcal{C}(B)$ provided $\mathcal{I}(A)$ meets $\mathcal{I}(B)$.
    (c) Show that the result in (b) fails if the proviso is dropped.

3. Prove: $\mathcal{C}(a_1 \ldots a_n)$ contains any join of points chosen from $\{a_1, \ldots, a_n\}$. Interpret the result in Euclidean geometry.

4. Prove: If A and B are convex and nonempty, then $\mathcal{C}(AB) \supset \mathcal{C}(A), \mathcal{C}(B)$.

5. (a) Prove: If A and B are convex, then $\mathcal{C}(AB) \supset \mathcal{C}(A)\mathcal{C}(B)$.
    (b) Find Euclidean convex sets A and B such that $\mathcal{C}(AB) \neq \mathcal{C}(A)\mathcal{C}(B)$.

6. Prove: If A is convex and $a \subset \mathcal{C}(A)$, then $\mathcal{C}(aA) = \mathcal{C}(A)$.

7. Prove: If A is convex and $\mathcal{C}(A)$ absorbs p, then $\mathcal{I}(A)$ absorbs p.

8. Let A be convex and $\mathcal{I}(A) \neq \varnothing$. Prove that if $\mathcal{I}(A)$ absorbs p, then $p \subset \mathcal{C}(A)$. Compare the result with Theorem 2.24(ii).

9. Let $J = \{0, 1\}$. Define $\cdot$ as follows: $0 \cdot 0 = 0 \cdot 1 = 1 \cdot 0 = 0$; $1 \cdot 1 = 1$. Verify that $(J, \cdot)$ is a join system (Section 2.2) which satisfies Postulates J1–J4 and so is a model of the postulate set J1–J4 (Section 2.4, Summary Statement). Show in this model that

$$\mathcal{C}(0) = \{0, 1\}.$$

Infer that the statement $\mathcal{C}(a) = a$ is not deducible from J1–J4.

2.18 Composition of the Interior and Closure Functions

The relation between a convex set and its closure may be considered as a function or operation $\mathcal{C}$—analogous to the function $\mathcal{I}$—which assigns to each convex set A the set $\mathcal{C}(A)$. We are interested in the functions obtained by combining or compounding the functions $\mathcal{C}$ and $\mathcal{I}$. $\mathcal{I}(A)$ and $\mathcal{C}(A)$ are convex for convex A, so that $\mathcal{I}(\mathcal{I}(A))$, $\mathcal{I}(\mathcal{C}(A))$, $\mathcal{C}(\mathcal{I}(A))$ and $\mathcal{C}(\mathcal{C}(A))$ are defined.

Theorem 2.18, $\mathcal{I}(\mathcal{I}(A)) = \mathcal{I}(A)$, asserts that $\mathcal{I}$ compounded with $\mathcal{I}$ yields $\mathcal{I}$. We begin with the analogous result for $\mathcal{C}$, which is restricted however by a proviso.

Theorem 2.26. $\mathcal{C}(\mathcal{C}(A)) = \mathcal{C}(A)$ *for any convex set A, provided $\mathcal{I}(A) \neq \varnothing$.*

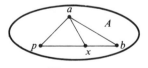

Figure 2.61

PROOF. $\mathcal{C}(A) \subset \mathcal{C}(\mathcal{C}(A))$, since any convex set is a subset of its closure (Theorem 2.19). It remains to prove the reverse inclusion

$$\mathcal{C}(\mathcal{C}(A)) \subset \mathcal{C}(A). \tag{1}$$

Let $a \subset \mathcal{C}(\mathcal{C}(A))$, that is, a is a contact point of $\mathcal{C}(A)$. By definition

$$ap \subset \mathcal{C}(A) \tag{2}$$

for some p. Since $\mathcal{I}(A) \neq \varnothing$, there is a point $b \subset \mathcal{I}(A)$. Using (2) and Corollary 2.24,

$$apb \subset \mathcal{C}(A) \cdot \mathcal{I}(A) = \mathcal{I}(A). \tag{3}$$

Let $x \subset pb$. Then using (3),

$$ax \subset apb \subset \mathcal{I}(A) \subset A.$$

Thus a is a contact point of A by definition, that is, $a \subset \mathcal{C}(A)$. This verifies (1) and the theorem follows. □

Discussion of the Theorem. The conclusion is not startling in view of our Euclidean experience. In forming $\mathcal{C}(A)$ we are enlarging A by adjoining those of its contact points that are not in A. Then we expect $\mathcal{C}(A)$ to be closed up—to contain all its contact points. This seems natural and reasonable. The sticking point is that we should have to assume $\mathcal{I}(A) \neq \varnothing$. For the result is trivial if $A = \varnothing$. And it seems impossible that there should exist a *nonempty* convex set A whose interior is empty. This negates all our experience with convex sets. There exist, nevertheless, examples of such convex sets A—and we can indeed find one which falsifies the conclusion of the theorem. The definition of the counterexample requires the construction of an infinite dimensional space and will be presented later. (See Section 5.13, Example II.)

Two theorems on the composition of the $\mathcal{C}$ and $\mathcal{I}$ functions are now proved.

Theorem 2.27. $\mathcal{C}(\mathcal{I}(A)) = \mathcal{C}(A)$ *for any convex set* A, *provided* $\mathcal{I}(A) \neq \varnothing$.

PROOF. $\mathcal{I}(A) \subset A$ implies $\mathcal{C}(\mathcal{I}(A)) \subset \mathcal{C}(A)$ by the monotonic property (Theorem 2.21).

To prove the reverse inclusion let $a \subset \mathcal{C}(A)$. Choose point p in $\mathcal{I}(A)$. By Theorem 2.23

$$ap \subset \mathcal{I}(A).$$

Hence $a \subset \mathcal{C}(\mathcal{I}(A))$ by definition. Thus $\mathcal{C}(A) \subset \mathcal{C}(\mathcal{I}(A))$ and the theorem holds. □

The condition $\mathcal{I}(A) \neq \varnothing$ is essential—for, as mentioned in the discussion of Theorem 2.26, there exist convex sets A that satisfy $A \neq \varnothing$ but $\mathcal{I}(A) = \varnothing$. (See Section 5.13.) Theorem 2.27 must fail for such sets A.

Theorem 2.28. $\mathcal{I}(\mathcal{C}(A)) = \mathcal{I}(A)$ *for any convex set* A, *provided* $\mathcal{I}(A) \neq \emptyset$.

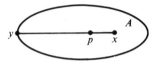

Figure 2.62

PROOF. Let $p \subset \mathcal{I}(\mathcal{C}(A))$. We show $p \subset \mathcal{I}(A)$. Fix x in $\mathcal{I}(A)$. Then $x \subset \mathcal{C}(A)$ by Corollary 2.19. Hence $p \subset \mathcal{I}(\mathcal{C}(A))$ implies $p \subset xy$ for some $y \subset \mathcal{C}(A)$. Since $x \subset \mathcal{I}(A)$ and $y \subset \mathcal{C}(A)$ we have $xy \subset \mathcal{I}(A)$ (Theorem 2.23). Hence $p \subset \mathcal{I}(A)$ and we have proved

$$\mathcal{I}(\mathcal{C}(A)) \subset \mathcal{I}(A).$$

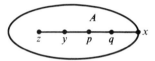

Figure 2.63

To prove the reverse inclusion, let $p \subset \mathcal{I}(A)$. Note that $p \subset \mathcal{C}(A)$. Let x be any point of $\mathcal{C}(A)$. We show $p \subset xy$ for some point $y \subset \mathcal{C}(A)$. By Theorem 2.23

$$px \subset \mathcal{I}(A) \subset A.$$

Choose q to satisfy

$$q \subset px. \tag{1}$$

Then $q \subset A$; and $p \subset \mathcal{I}(A)$ implies

$$p \subset qz \tag{2}$$

for some $z \subset A$. The relations (2) and (1) yield $p \subset (px)z = x(pz)$. This implies

$$p \subset xy, \quad \text{where} \quad y \subset pz. \tag{3}$$

Note that $pz \subset A \subset \mathcal{C}(A)$, so that $y \subset \mathcal{C}(A)$. Now (3) implies $p \subset \mathcal{I}(\mathcal{C}(A))$. Hence $\mathcal{I}(A) \subset \mathcal{I}(\mathcal{C}(A))$ and the theorem holds. ☐

The condition $\mathcal{I}(A) \neq \emptyset$ is essential here as in the last theorem (see Section 5.13, Example III and Exercise 3 at the end of Section 5.14).

Theorems 2.18, 2.28, 2.26 and 2.27 can be combined, with a slight loss of information, into the following

Summary. Any convex set A, its interior and its closure have the same interior and the same closure, provided $\mathcal{I}(A) \neq \emptyset$.

This summary principle provides a class of instances of convex sets with common interior and common closure. This class is significantly enlarged by the following theorem, the proof of which is left to the reader (Exercise 7 below).

Theorem 2.29. *Let A and B be convex with nonempty interiors. Then the following statements are equivalent:*

(a) $\mathcal{I}(A) = \mathcal{I}(B)$;
(b) $\mathcal{C}(A) = \mathcal{C}(B)$;
(c) $\mathcal{I}(A) \subset B \subset \mathcal{C}(A)$.

Finally a comparison table for the interior and closure operations:

$\mathcal{I}(A) \subset A$	$\mathcal{C}(A) \supset A$
	(Theorem 2.19)
$\mathcal{I}(A)$ is convex	$\mathcal{C}(A)$ is convex
(Corollary 2.15.2)	(Theorem 2.22)
$\mathcal{I}(\mathcal{I}(A)) = \mathcal{I}(A)$	$\mathcal{C}(\mathcal{C}(A)) = \mathcal{C}(A)$
(Theorem 2.18)	$[\mathcal{I}(A) \neq \varnothing]$
	(Theorem 2.26)
$\mathcal{I}(\mathcal{C}(A)) = \mathcal{I}(A)$	$\mathcal{C}(\mathcal{I}(A)) = \mathcal{C}(A)$
$[\mathcal{I}(A) \neq \varnothing]$	$[\mathcal{I}(A) \neq \varnothing]$
(Theorem 2.28)	(Theorem 2.27)
$A \subset B$ implies $\mathcal{I}(A) \subset \mathcal{I}(B)$	$A \subset B$ implies $\mathcal{C}(A) \subset \mathcal{C}(B)$
$[A \text{ meets } \mathcal{I}(B)]$	(Theorem 2.21)
(Theorem 2.17)	

EXERCISES

1. Prove: If A is convex, $p \subset \mathcal{I}(A)$ and $x \subset \mathcal{C}(A)$, then $p \subset xy$ for some $y \subset \mathcal{C}(A)$. Can y be chosen in A? in $\mathcal{I}(A)$?

*2. Prove: If A and B are convex $\mathcal{I}(A) \subset B \subset \mathcal{C}(A)$, and $\mathcal{I}(A) \neq \varnothing$, then $\mathcal{I}(B) \neq \varnothing$ and $\mathcal{I}(A) = \mathcal{I}(B)$, $\mathcal{C}(A) = \mathcal{C}(B)$.

3. (a) Prove: If K is convex, $a \subset \mathcal{C}(K)$ and $\mathcal{I}(K) \neq \varnothing$, then $\mathcal{I}(aK) = \mathcal{I}(K)$.
   (b) Interpret in Euclidean geometry.

4. Prove: If K is convex and $\mathcal{I}(aK) = \mathcal{I}(K) \neq \varnothing$, then $a \subset \mathcal{C}(K)$.

5. Let A be a fixed convex set such that $\mathcal{I}(A) \neq \varnothing$ and X a variable one, and suppose the equation $\mathcal{I}(X) = \mathcal{I}(A)$ has exactly one solution, namely, $X = A$. Prove: $\mathcal{C}(A) = \mathcal{I}(A) = A$. Interpret in Euclidean geometry. [Compare Exercises 13(c) and 7 at the end of Section 2.12, and Exercise 5 at the end of Section 2.15.]

6. The same as Exercise 5, replacing $\mathcal{I}(X) = \mathcal{I}(A)$ by $\mathcal{C}(X) = \mathcal{C}(A)$. [Compare Exercises 9(c) and 5 at the end of Section 2.15, and Exercise 7 at the end of Section 2.12.]

7. Prove Theorem 2.29.

2.19 The Boundary of a Convex Set

A boundary point of a convex set A is a point which "adheres" to A but is not in $\mathcal{I}(A)$. A formal definition is easily given.

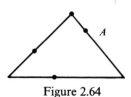

Figure 2.64

Definition. Let A be a convex set. Suppose $p \subset \mathcal{C}(A)$ but $p \not\subset \mathcal{I}(A)$. Then p is called a *boundary point* of A. The *boundary* of A, denoted by $\mathcal{B}(A)$, is the set of its boundary points.

The notion of frontier (Section 2.11) is closely related to that of boundary. Suppose $p \subset \mathcal{F}(A)$, A convex. By definition $p \subset A$, $p \not\subset \mathcal{I}(A)$. Certainly $p \subset \mathcal{C}(A)$. Then by definition $p \subset \mathcal{B}(A)$. Thus we have

$$\mathcal{F}(A) \subset \mathcal{B}(A).$$

The following theorem contains several simple set theoretic formulas involving $\mathcal{B}(A)$ and $\mathcal{F}(A)$.

Theorem 2.30. *For any convex set A,*

(a) $\mathcal{B}(A) = \mathcal{C}(A) - \mathcal{I}(A)$;

(b) $\mathcal{F}(A) = A - \mathcal{I}(A) = A \cap \mathcal{B}(A)$;

(c) $\mathcal{C}(A) = \mathcal{I}(A) \cup \mathcal{B}(A)$, $\mathcal{I}(A) \cap \mathcal{B}(A) = \varnothing$;

(d) $A = \mathcal{I}(A) \cup \mathcal{F}(A)$, $\mathcal{I}(A) \cap \mathcal{F}(A) = \varnothing$.

PROOF. (a) By definition $\mathcal{B}(A)$ consists of all points of $\mathcal{C}(A)$ that are not in $\mathcal{I}(A)$.

(b) Similarly the first equation is a restatement of the definition of $\mathcal{F}(A)$, Section 2.11.

We verify the second equation in the form

$$A \cap \mathcal{B}(A) = A - \mathcal{I}(A). \tag{1}$$

Using (a), we have

$$A \cap \mathcal{B}(A) = A \cap (\mathcal{C}(A) - \mathcal{I}(A)). \tag{2}$$

The right member of (2) calls for the set of points of A that are in $\mathcal{C}(A)$ and not in $\mathcal{I}(A)$. This is, since $A \subset \mathcal{C}(A)$, precisely the set of points of A that are not in $\mathcal{I}(A)$, that is, $A - \mathcal{I}(A)$. Thus (1) holds.

(c) The first equation follows from (a) and $\mathcal{I}(A) \subset \mathcal{C}(A)$. The second equation is an immediate consequence of (a).

(d) This follows from $\mathcal{F}(A) = A - \mathcal{I}(A)$ in (b), exactly as (c) from (a). $\square$

Remarks on the Theorem. (c) may be considered a transliteration of (a) from subtractive to additive form. In English it may be stated: $\mathcal{C}(A)$ is the union of the disjoint sets $\mathcal{I}(A)$ and $\mathcal{B}(A)$. Similar remarks hold for (d) and (b).

The theorem is easily verified in Euclidean geometry by using diagrams.

EXERCISES

1. In Euclidean geometry try to find examples of convex sets A such that
   (a) $\mathcal{B}(A) = \varnothing$;
   (b) $\mathcal{B}(A) = A$;
   (c) $\mathcal{B}(A)$ is not convex;
   (d) $\mathcal{B}(A)$ is nonempty and convex.

2. In Euclidean geometry try to find examples of two distinct convex sets which have
   (a) the same boundary;
   (b) the same frontier;
   (c) the same interior and the same boundary;
   (d) the same interior and the same frontier;
   (e) the same boundary and different interiors;
   (f) the same frontier and different interiors.

3. Prove: If $\mathcal{I}(X) = \mathcal{I}(Y)$ and $\mathcal{F}(X) = \mathcal{F}(Y)$ for convex sets X and Y, then $X = Y$.

4. Prove: Any convex set A, its interior and its closure have the same boundary, provided $\mathcal{I}(A) \neq \varnothing$.

5. Let X and Y be convex sets with nonempty interiors.
   (a) Prove: $\mathcal{I}(X) \subset Y \subset \mathcal{C}(X)$ implies $\mathcal{B}(X) = \mathcal{B}(Y)$; that is, if Y is "between" the interior and the closure of X, then X and Y have the same boundary.
   (b) Is the converse valid?
   (c) Does the conclusion hold for the frontier operation?

6. If A is convex, try to conjecture a formula for $\mathcal{F}(\mathcal{C}(A))$. Can you prove your formula without imposing restrictions on $\mathcal{I}(A)$?

7. (a) Does the monotonic law for the boundary operation [if X and Y are convex and $X \subset Y$, then $\mathcal{B}(X) \subset \mathcal{B}(Y)$] hold? Justify your answer.
   (b) Is a sort of reverse monotonic law [if X and Y are convex and $\mathcal{B}(X) \subset \mathcal{B}(Y)$, then $X \subset Y$] valid? Justify your answer.
   (c) The same as (a) and (b) for the frontier operation.

8. Suppose X and Y are convex sets.
   (a) Does $\mathcal{I}(X) \subset \mathcal{I}(Y)$ imply $X \subset Y$? Justify your answer.
   (b) Does $\mathcal{I}(X) \subset \mathcal{I}(Y)$ imply $\mathcal{C}(X) \subset \mathcal{C}(Y)$? Assume if needed $\mathcal{I}(X) \neq \varnothing$.

(c) The same as (b), replacing the closure operation by the boundary operation; by the frontier operation.

9. Suppose X and Y are convex sets.
   (a) Does $\mathcal{C}(X) \subset \mathcal{C}(Y)$ imply $X \subset Y$? Justify your answer.
   (b) Does $\mathcal{C}(X) \subset \mathcal{C}(Y)$ imply $\mathcal{I}(X) \subset \mathcal{I}(Y)$? Justify your answer. Assume if needed $\mathcal{I}(X) \neq \varnothing$.
   (c) The same as (b), replacing $\mathcal{I}(X) \subset \mathcal{I}(Y)$ by $\mathcal{B}(X) \subset \mathcal{B}(Y)$; by $\mathcal{F}(X) \subset \mathcal{F}(Y)$.

10. (a) In Euclidean geometry exhibit several convex sets A for which $\mathcal{B}(A) = \mathcal{F}(A)$; for which $\mathcal{B}(A) \neq \mathcal{F}(A)$.
    (b) Suppose A is convex. Is there any necessary relation between $\mathcal{F}(A)$ and $\mathcal{B}(A)$ other than $\mathcal{F}(A) \subset \mathcal{B}(A)$?
    (c) Does the following problem throw light on (b)? Find in Euclidean geometry examples of sets S such that if T is any preassigned subset of S there exists a convex set A such that $\mathcal{B}(A) = S$ and $\mathcal{F}(A) = T$.
    (d) Investigate in Euclidean geometry to what extent a convex set A can be reconstructed if $\mathcal{B}(A)$ is known: Is A uniquely determined? If not, what information can you obtain about A?

11. (a) Prove: If A is convex, $x \subset \mathcal{B}(A)$ and $y \subset \mathcal{I}(A)$, then $(xy) \cap \mathcal{B}(A) = \varnothing$.
    (b) Prove: If A is convex, $x, y \subset A$ and xy intersects $\mathcal{B}(A)$, then $x, y \subset \mathcal{F}(A)$.

12. Prove: For convex set A if $\mathcal{B}(A)$ is convex and $\mathcal{I}(A) \neq \varnothing$, then $\mathcal{C}(\mathcal{B}(A)) = \mathcal{B}(A)$. Interpret in Euclidean geometry.

2.20 Project: Another Formulation of the Theory of Join

Our formulation of the theory is based on two ideas, a set J and a join operation $\cdot$ in J, which satisfy postulates J1–J4. Most algebraic theories currently studied involve a basic set S and an operation $\circ$ which assigns to each ordered pair (a, b) of elements of S, a uniquely determined *element* of S, denoted by $a \circ b$. Such an operation is called a *binary operation* in or on S (or a *single-valued* binary operation in S). The question arises whether the theory of join can be given an *equivalent* formulation in terms of a binary operation in an appropriate set. We describe a way of doing this.

The new formulation is suggested by the fact that the operation join of sets of points *is* a binary operation in the family of all sets of points. So we retain the basic notion J (the set of points) and introduce as a second basic notion a binary operation $\circ$ in J', the family of subsets of J. Letters $a, b, c, \ldots$ will denote members of J; $A, B, C, \ldots$ members of J'. The postulates are suggested by familiar properties of join of sets in the theory.

Here they are:

J1′. $A \circ B = \emptyset$ *if and only if* $A = \emptyset$ *or* $B = \emptyset$.

J2′. $A \circ B = \displaystyle\bigcup_{\{a\} \subset A,\, \{b\} \subset B} (\{a\} \circ \{b\})$.

J3′. $A \circ B = B \circ A$.

J4′. $(A \circ B) \circ C = A \circ (B \circ C)$.

J5′. $\{a\} \circ \{a\} = \{a\}$.

How can the postulate sets J1–J4 and J1′–J5′ be shown to be equivalent? Since the basic terms in the two postulate sets are not the same, we cannot just say: Deduce the postulates in each set from those in the other set. To be specific, the basic term $\circ$ in J1′–J5′ does not occur in J1–J4, and it would be purposeless to try to derive, for example, J1′ from J1–J4. In order to go ahead, we shall have to *interpret* or *define* $\circ$ in terms of J and $\cdot$. The desired definition is at hand—it is suggested by the definition of the join of *sets* A, B in terms of the join of points (Section 2.3).

(I) *Definition.* If A and B are members of J',

$$A \circ B = \bigcup_{a \subset A,\, b \subset B} (a \cdot b).$$

In order to prove that J1–J4 and J1′–J5′ are equivalent we must show first:

(a) If $\cdot$ is a join operation in J, then $\circ$, as defined by (I), is a binary operation in J'.

(b) If $\circ$ is defined by (I), then J1–J4 imply J1′–J5′.

Reversely we define $\cdot$ in terms of J' and $\circ$:

(II) *Definition.* If a and b are members of J, then

$$a \cdot b = \{a\} \circ \{b\}.$$

Next we must show:

(c) If $\circ$ is a binary operation in J', then $\cdot$, as defined by (II), is a join operation in J.

(d) If $\cdot$ is defined by (II) then J1′–J5′ imply J1–J4.

Finally it must be shown that the definitions (I) and (II) satisfy the following conditions of *compatability*:

(e) Suppose the join operation $\cdot$ is given and $\circ$ is defined by (I). Then if we apply definition (II) to ∘ , the join operation obtained must be the same as · .

(f) Reversely, suppose the binary operation ∘ is given and · is defined by (II). Then if definition (I) is applied to · , the binary operation obtained must be the same as ∘ .

It is by now no secret that the project is to prove the postulate sets J1–J4 and J1′–J5′ equivalent.

The portion of the chapter devoted to the development of join theory based on postulates J1–J4 is now complete. The conclusion consists of two independent parts: the first a presentation of two new models or interpretations of the theory (Section 2.4), the second a discussion of the relation between our definitions of interior and closure and the conventional ones.

2.21 What Does the Theory Apply To?

It was shown in Section 2.4 that our theory is applicable to Euclidean geometry. Specifically, models of the theory (Euclidean models) were exhibited, each composed of a Euclidean plane or 3-space and its associated Euclidean join operation. It would be quite erroneous however to assume that these are the only models or interpretations of the theory. The postulates J1–J4 omit more of the basic principles of Euclidean geometry than they include: it is hard to believe that they provide an adequate foundation for the whole of Euclidean geometry. Actually they admit a wide variety of models different from the Euclidean ones—two of these will be studied below.

The study of such "non-Euclidean" models has definite advantages. Although Euclidean geometry is the basic model and is used as a touchstone for the theory, you may be misled if this is the only model you encounter. You may tend to lean too heavily on Euclidean intuition and develop a pro-Euclidean bias—a sort of implicit assumption that any familiar Euclidean principle must hold in the theory. Consequently, in solving problems or trying to grasp and assimilate the theory you may find yourself assuming and even trying to prove certain Euclidean propositions that cannot be deduced from J1–J4. Studying the non-Euclidean models or interpretations of the theory will help to foster an appreciation for the concrete significance of the theory and to develop a better balanced sense of what can and cannot be deduced from J1–J4.

2.22 The Triode Model

To form a model or interpretation of the theory we must present a pair $(J, \cdot)$, where J is a specific set and · a specific join operation in J which satisfy postulates J1–J4 (Section 2.4).

Our first non-Euclidean model involves the notion of closed ray in Euclidean geometry which is now characterized.

Euclidean Closed Ray. In Euclidean geometry, let L be a line and a a point of L (Figure 2.65). Let S be the set of points of L that lie on a given side of point a. Then $S \cup a$ is a *closed ray* (or *closed halfline*) and a is its *endpoint*.

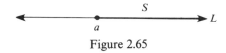

Figure 2.65

Let R_1, R_2 and R_3 be distinct closed rays with the same endpoint o in a Euclidean geometry (Figure 2.66). Let J, the basic set, be $R_1 \cup R_2 \cup R_3$. The join operation in J is defined piecemeal. If the points a and b of J are in the same ray R_1, R_2 or R_3, their join $a \cdot b$ is simply the Euclidean join of a and b (Section 2.1). If however a and b are not in the same ray, $a \cdot b$ is defined to be the union $(oa) \cup (ob) \cup o$, where oa and ob are the Euclidean joins of o to a and o to b respectively. It is interesting to note that $a \cdot b$ as defined is always the "natural path" that joins a and b in the geometric figure J.

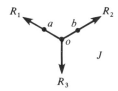

Figure 2.66

For convenience we omit the symbol $\cdot$ in the expression $a \cdot b$ for a join in J. Will this cause ambiguity? It should not. For if a and b are in the same ray R_1, R_2 or R_3, $a \cdot b$ denotes their Euclidean join—while if a and b are not in the same ray, we never refer to their Euclidean join and $a \cdot b$ denotes unambiguously $(oa) \cup (ob) \cup o$.

The operation $\cdot$ is a join operation in J (Section 2.2), since it assigns to each ordered pair (a, b) of elements of J a uniquely determined subset of J. Does $(J, \cdot)$ satisfy postulates J1–J4?

Clearly J1 is satisfied, since the join $a \cdot b$ always contains at least one point. J2 is easily seen to be satisfied, and J4 is immediate by the definition of join. The verification of J3 requires an examination of cases.

Verification of J3. (ab)c = a(bc). (Note that the diagrams below do not exhaust all cases but illustrate typical situations where the points a, b, c and o are distinct.)

Case I: a, b, c are in the same ray (Figure 2.67). Then J3 holds, since it holds in Euclidean geometry.

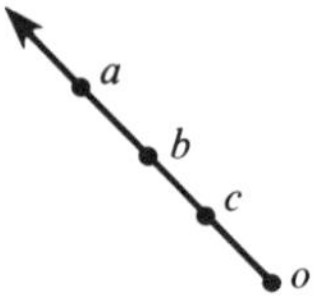

Figure 2.67

Case II: a, b, c are not in the same ray (Figure 2.68). It can be verified, for all positions of a, b and c, that

$$(ab)c = o \cup (oa) \cup (ob) \cup (oc) = a(bc).$$

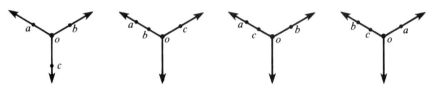

Figure 2.68

Thus $(J, \cdot)$ satisfies J1–J4 and is a model of the theory. We call it the *triode model*.

2.23 A Peculiarity of the Triode Model

We consider a property of the triode model which does not hold in Euclidean geometry and seems quite peculiar—indeed, almost bizarre.

Let a, b, c be points of the triode model that are distinct from o and in R_1, R_2, R_3 respectively (Figure 2.69). Then

$$a \not\subset bc, \qquad b \not\subset ac, \qquad c \not\subset ab, \tag{1}$$

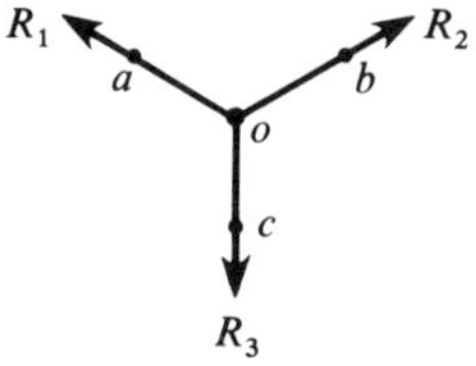

R_1 ⬉ ⬈ R_2
a b
o
c
R_3

Figure 2.69

that is, no one of the three points lies in the join of the other two. This is not remarkable. It occurs in Euclidean geometry all the time for three points that are noncollinear. Observe that

$$(ab) \cap (ac) = (ao) \cup o, \qquad (bc) \cap (ba) = (bo) \cup o,$$
$$(ca) \cap (cb) = (co) \cup o. \tag{2}$$

The relations in (2) say, in effect, that if we think of a, b and c as vertices of a triangle, then any two *sides* of the triangle have infinitely many common points.

This seems rather strange—at least in comparison with Euclidean geometry. Let us assume that the concept of line is significant in the triode model, without attempting now to clarify it. If we suppose a, b and c are contained in some line, then a, b and c are distinct collinear points that satisfy (1). This is impossible in Euclidean geometry—since if three distinct points are collinear, one of the three must be between the other two and so must be in their join.

If on the other hand we suppose there is no line containing a, b and c, we are confronted with a curious situation. For by (2) each two sides of triangle abc intersect in an infinite number of points.

EXERCISES

1. Consider the following proposition of Euclidean geometry: If the segments ab and ac have a common point, then $ab \subset ac$ or $ac \subset ab$. Does the proposition hold in the triode model?

2. The same as Exercise 1 for the proposition: If p is a point of segment ab, then $ab = (ap) \cup p \cup (pb)$.

3. The same as Exercise 1 for the proposition: If $p, q \subset ab$ and $p \neq q$, then $p \subset aq$ or $q \subset ap$.

4. (a) Does the following principle hold in a Euclidean model? If $p \subset ab$, then $ab = pab$.
   (b) The same for the triode model.

5. A set L is called a *linear set* if it satisfies the following conditions: (i) If $L \supset a, b$, then $L \supset ab$; (ii) If $L \supset p, q$ and $p \subset qx$, then $L \supset x$.
   (a) Find all types of linear set in a Euclidean 3-space model.
   (b) The same for the triode model.

6. (a) How many different types of convex set can you find in a Euclidean line? Briefly describe them.
   (b) The same for the triode model.

7. (a) Verify in the triode model, by choosing several examples, Corollary 2.15.2 that the interior of a convex set is convex.
   (b) The same for Theorem 2.22 that the closure of a convex set is convex.

8. In the triode model suppose $a \neq b$. Verify: $\mathcal{I}(ab) = ab$; $\mathcal{F}(ab) = \varnothing$; $\mathcal{C}(ab) = a \cup b \cup (ab)$. (Compare Section 2.12, Example II; Section 2.15, Example II.)

9. In the triode model suppose $a \neq b$. Verify:
   (a) $a \cup b \cup (ab)$ is convex.
   (b) $\mathcal{I}(a \cup b \cup (ab)) = ab$; $\mathcal{F}(a \cup b \cup (ab)) = \{a, b\}$; $\mathcal{C}(a \cup b \cup (ab)) = a \cup b \cup (ab)$.
   (Compare Section 2.12, Example I; Section 2.15, Example I.)

10. In the triode model let a, b, c be distinct from o and lie in R_1, R_2, R_3 respectively. Verify: $\mathcal{J}(abc) = abc$; $\mathcal{F}(abc) = \varnothing$; $\mathcal{C}(abc) = (abc) \cup a \cup b \cup c$. (Compare Section 2.12, Example IV; Section 2.15, Example IV.)

11. In the triode model let a, b, c be distinct from o and lie in R_1, R_2, R_3 respectively. Verify:
    (a) $(abc) \cup a \cup b \cup c$ is convex.
    (b) $\mathcal{J}((abc) \cup a \cup b \cup c) = abc$; $\mathcal{F}((abc) \cup a \cup b \cup c) = \{a, b, c\}$; $\mathcal{C}((abc) \cup a \cup b \cup c) = (abc) \cup a \cup b \cup c$.
    (Compare Section 2.12, Example III; Section 2.15, Example III.)

12. Let J be the union of four distinct closed rays with the same endpoint o. Define $a \cdot b$ for a, b in J as in the triode model. Will $(J, \cdot)$ satisfy J1–J4?

13. Try to find generalizations of the triode model. To what extent can you alter the model and retain the property that it satisfies J1–J4?

2.24 The Cartesian Join Model

We present a brief description of another non-Euclidean model of the theory.

Take J to be the set of all points of a Euclidean plane. To define join, choose a Cartesian coordinate system in J. Then if a is a point in J, a is represented by a uniquely determined ordered pair of real numbers (a_1, a_2). This relation will be indicated by writing $a = (a_1, a_2)$, and sometimes (a_1, a_2) will be used to denote the point a.

Define the operation $\cdot$ for $a = (a_1, a_2)$ and $b = (b_1, b_2)$ as follows (see Figure 2.70):

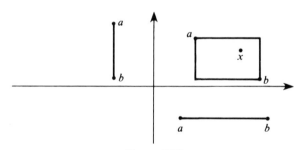

Figure 2.70

(1) If $a_1 = b_1$ or $a_2 = b_2$ (that is, if a and b are on the same vertical or horizontal line), then $a \cdot b$ is the familiar Euclidean join of points a and b.

(2) If $a_1 \neq b_1$ and $a_2 \neq b_2$ (that is, if a and b are not on the same vertical or horizontal line), then $a \cdot b$ is the set of all points $x = (x_1, x_2)$ such that

x_1 is (strictly) between a_1 and b_1 and x_2 is (strictly) between a_2 and b_2. In geometrical language $a \cdot b$ is the interior of the rectangle that has horizontal and vertical sides and segment ab as diagonal.

The operation $\cdot$ is a join operation in J (Section 2.2), and we call it the *Cartesian* join operation in J. The pair $(J, \cdot)$ satisfies J1–J4. Postulates J1, J2 and J4 are immediately verified; J3 requires a discussion of several cases. Thus $(J, \cdot)$ is a model of the theory—we call it the *Cartesian join* model or simply the *Cartesian* model.

It is worth noting that now we have two models with the same set J but radically different join operations.

EXERCISES

1. Verify J3 in the Cartesian model in several different cases by using diagrams.

2. Does the following proposition of Euclidean geometry hold in the Cartesian model? If the segments ab and ac have a common point, then $ab \subset ac$ or $ac \subset ab$. (Compare Exercise 1 at the end of Section 2.23.)

3. The same as Exercise 2 for the proposition: If p is a point of segment ab, then $ab = (ap) \cup p \cup (pb)$. (Compare Exercise 2 at the end of Section 2.23.)

4. The same as Exercise 2 for the proposition: If $p, q \subset ab$ and $p \neq q$, then $p \subset aq$ or $q \subset ap$. (Compare Exercise 3 at the end of Section 2.23.)

5. The same as Exercise 2 for the proposition: If $p \subset ab$, then $ab = pab$. (Compare Exercise 4 at the end of Section 2.23.)

6. In Euclidean geometry a segment has a unique pair of endpoints. This may be restated: If $ab = pq$, then $a = p$ and $b = q$ or $a = q$ and $b = p$. Does this principle hold in the Cartesian model?

7. Prove in the Cartesian model:

$$(a_1, a_2) \cdot (b_1, b_2) = (a_1, b_2) \cdot (b_1, a_2).$$

8. Find all types of linear set in the Cartesian model. For the definition of linear set see Exercise 5 at the end of Section 2.23. (Compare the result with that exercise.)

9. How many different types of convex set can you find in the Cartesian model? Briefly describe them or represent them by diagrams. (Compare Exercise 6 at the end of Section 2.23.)

10. (a) Verify in the Cartesian model, by choosing several examples, Corollary 2.15.2 that the interior of a convex set is convex.
    (b) The same for Theorem 2.22 that the closure of a convex set is convex.

11. In the Cartesian model suppose $a \neq b$. Verify: $\mathcal{I}(ab) = ab$; $\mathcal{F}(ab) = \varnothing$. (Compare Section 2.12, Example II; also Exercise 8 at the end of Section 2.23.)

12. (a) Find $\mathcal{C}(ab)$ for various choices of a, b ($a \neq b$) in the Cartesian model. Compare with the Euclidean case (Section 2.15, Example II).

(b) In the formal theory is it possible to prove: $\mathcal{C}(ab) = (ab) \cup a \cup b$?

13. In the Cartesian model suppose a, b and c are distinct, a and b are on a horizontal line, and b and c are on a vertical line. Find abc, $\mathcal{J}(abc)$, $\mathcal{C}(abc)$. How do your results for $\mathcal{J}(abc)$ and $\mathcal{C}(abc)$ compare with the Euclidean results in Section 2.12, Example IV, and Section 2.15, Example IV?

14. In the Cartesian model find the join $A \cdot A$, if A is (a) a line; (b) a circle; (c) the graph of $y = x^2$; (d) the graph of $xy = 1$, $x > 0$.

15. In the Cartesian model find a pair of sets whose Cartesian join and Euclidean join are identical. Avoid trivial cases such as two points on a horizontal line.

16. Try to find some generalizations of the Cartesian model.

2.25 The Metric Interior of a Set in Euclidean Geometry

This section and the next one—which are not required in the sequel—discuss the metric definitions of the interior and closure of an arbitrary set of points in a Euclidean geometry and their relation to our definitions of the interior and closure of a convex set (Sections 2.11, 2.16).

The metric concepts of interior and closure in Euclidean geometry are based on a rather simple idea. In a Euclidean geometry there is available a distance function $d(x, y)$ which gives the distance between the points x and y. The distance function is used to define *basic* regions (*circular* in a plane, *spherical* in a 3-space) which can be employed to characterize the interior points and the contact points of any given set.

Metric Interiors in a Euclidean Plane

We begin by introducing the idea of metric interior for a set of points in a Euclidean plane E_2. First circular regions are defined in E_2. Let a be a point of E_2 and ρ a positive real number (Figure 2.71). Then the set R of points x of E_2 for which $d(x, a) < \rho$ is called an *open circular region* of E_2 or simply a *circular region*; a is the *center* of R and ρ its *radius*. Now let S be any set of points of E_2, and p a point of S (Figure 2.72). Then p is a

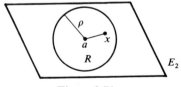

Figure 2.71

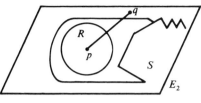

Figure 2.72

metric interior point of S (*relative* to E_2) if some circular region R of E_2 with center p is contained in S. The *metric interior* of S (*relative* to E_2) is the set of its metric interior points (relative to E_2) and is denoted by **m$\mathcal{I}$**($S : E_2$).

The point of the definition is that p is embedded within the region R, which is itself contained in S, so that p is well embedded within S. This can be viewed in another way. We ask: How far is p from a point *exterior* to S, that is, a point which is not in S but is in E_2? To make the issue specific, suppose the radius of R in the figure above is 2. Then the distance from p to any point q exterior to S is at least 2. Thus the exterior points of S cannot get close to p, and p cannot possibly be a "boundary" or "peripheral" point of S—indeed, p lies at least 2 units "deep" in S, and merits the title "metric interior point of S".

Our principal concern is this: If A is a convex set, how will the metric interior of A be related to $\mathcal{I}(A)$, the geometrical interior of A as defined in Section 2.11? To examine the question, we consider a few examples. First let A be a closed triangular region or a closed rectangular region in E_2 (Figure 2.73). Then p is a metric interior point of A (relative to E_2) if and only if p is an interior point of A in a familiar intuitive geometric sense. In other words, the concept of metric interior (relative to E_2) formalizes our intuitive geometric notion of interior for such *planar* convex sets A of E_2.

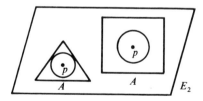

Figure 2.73

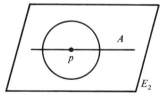

Figure 2.74

Now let A be a segment in E_2 (Figure 2.74). Clearly no circular region centered at a point p of A will be contained in A. Thus no point of A is a metric interior point of A (relative to E_2), and the metric interior of the segment A (relative to E_2) is $\varnothing$. In this case the metric definition of interior (relative to E_2) does not yield the familiar intuitive notion of interior of a segment.

The discussion shows that the metric interior of a convex set A relative to plane E_2 need not be the same as $\mathcal{J}(A)$. But it indicates that the two notions *will* be the same if A is a "planar" or "two dimensional" convex set.

Proposition I. *Suppose A is a planar convex set in a Euclidean plane E_2, that is, A is contained in E_2 but in no line of E_2. Then*

$$\mathbf{m}\mathcal{J}(A : E_2) = \mathcal{J}(A). \tag{1}$$

We indicate how to construct—but do not formalize completely[10]—a proof of Proposition I.

Suppose $p \subset \mathbf{m}\mathcal{J}(A : E_2)$ (Figure 2.75). By definition there exists a circular region R with center p that is contained in A. To show $p \subset \mathcal{J}(A)$, let $x \subset A$, $x \neq p$. Then the segment xp can be prolonged beyond p in A, since it can be prolonged beyond p in R. Thus there exists $y \subset A$ such that $p \subset xy$. By definition $p \subset \mathcal{J}(A)$.

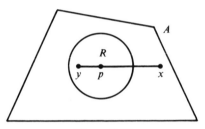

Figure 2.75

Conversely suppose $p \subset \mathcal{J}(A)$ (Figure 2.76). We show $p \subset \mathbf{m}\mathcal{J}(A : E_2)$. Thus we must show how to construct a circular region R centered at p and contained in A.

Choose $a \subset A$, $a \neq p$. Since A is planar, there exists a point b in A and not in line ap. Let $q \subset ab$. Since $p \subset \mathcal{J}(A)$, there exists $c \subset A$ such that $p \subset qc$. Then $p \subset abc$. Let R be a circular region of E_2 with center p and

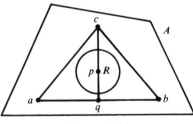

Figure 2.76

[10] A rigorous proof could hardly be given, since we have not introduced a set of postulates for Euclidean geometry.

radius less than the distances from p to the sides of triangle abc. Then

$$R \subset abc \subset A.$$

By definition p is a metric interior point of A relative to E_2. Thus $p \subset \mathbf{m}\mathcal{I}(A : E_2)$ and (1) holds.   □

A few words on the notion of metric interior in a Euclidean line before proceeding to 3-space.

Metric Interiors in a Euclidean Line

Let E_1 be a Euclidean line. First we define the analogue in E_1 of a circular region. Let a be a point and ρ a positive real number. Then the set R of points x of E_1 for which $d(x, a) < \rho$ is called an (*open*) *interval* of E_1 or simply an *interval*; a is the *center* of R, and ρ can be called its *radius*.

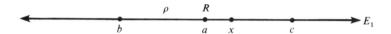

Let b and c be the points of E_1 whose distance to a is ρ. Then R is the (open) segment bc, and a is its midpoint.

Now let S be any set of points of E_1 and p a point of S (Figure 2.77). Then p is a *metric interior point* of S (*relative* to E_1) if some interval of E_1 with center p is contained in S. The *metric interior* of S (*relative* to E_1) is the set of its metric interior points (relative to E_1) and is denoted by $\mathbf{m}\mathcal{I}(S : E_1)$.

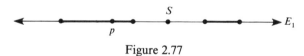

Figure 2.77

Let A be a convex subset of E_1. If A is not a point, it is not hard to see that

$$\mathbf{m}\mathcal{I}(A : E_1) = \mathcal{I}(A).$$

If A is a point, $\mathbf{m}\mathcal{I}(A : E_1) = \varnothing \neq \mathcal{I}(A)$.

Metric Interiors in a Euclidean 3-Space

Now we consider the concept of metric interior in a Euclidean 3-space. The formal definitions are exactly analogous to those for a plane or a line.

Let E_3 be a Euclidean 3-space, a a point of E_3 and ρ a positive real number. Then the set R of points x of E_3 for which $d(x, a) < \rho$ (Figure 2.78) is called an *open spherical region* of E_3 or simply a *spherical region*; a is the *center* of R and ρ its *radius*. Now let S be any subset of E_3, and p a point of S. Then p is a *metric interior point* of S (*relative* to E_3) if some spherical region of E_3 with center p is contained in S (Figure 2.79). The *metric interior* of S (*relative* to E_3) is the set of its metric interior points (relative to E_3) and is denoted by $\mathbf{m}\mathcal{I}(S : E_3)$.

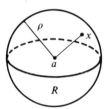

Figure 2.78

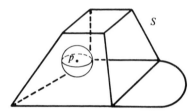

Figure 2.79

The definition ensures that p is well embedded within S as in the case of E_2 above.

Next we consider the relation between $\mathbf{m}\mathcal{I}(A : E_3)$ and $\mathcal{I}(A)$ if A is convex. First suppose A is a solid tetrahedron or a solid rectangular box in E_3 (Figure 2.80). Then p is a metric interior point of A (relative to E_3) if and only if p is "interior" to A in the familiar intuitive geometric sense. Thus we see intuitively that $\mathbf{m}\mathcal{I}(A : E_3) = \mathcal{I}(A)$ if A is one of these solid

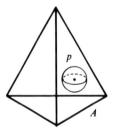
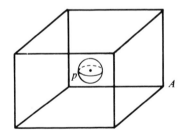

Figure 2.80

convex sets. Now let A be a nonsolid convex set in E_3, say a closed triangular or closed rectangular region or a segment (Figure 2.81). What is the metric interior of A now? In each case A contains no spherical region. So no point of A can be a metric interior point of A relative to E_3 and so $\mathbf{m}\mathcal{I}(A : E_3) = \varnothing$.

Now we state for E_3 an analogue of Proposition I on metric interiors of convex sets in E_2.

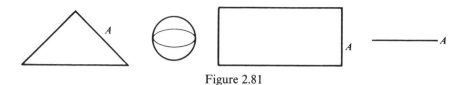

Figure 2.81

Proposition II. *Suppose A is a solid convex set in a Euclidean 3-space E_3, that is, A is contained in E_3 but in no plane of E_3. Then*

$$\mathbf{m}\mathcal{I}(A : E_3) = \mathcal{I}(A).$$

This can be justified by an argument analogous to that sketched above for Proposition I.

Dependence of Metric Interior on the Surrounding Space

In comparing metric interiors of figures relative to E_2 and E_3, we observe that the metric interior of a figure in E_2 may be different from its metric interior in E_3. This indicates a salient property of the metric interior concept, which we present here in a sharpened form:

Let A be a planar convex set, say a closed triangular region, in Euclidean space E_3. Let E_2 be the plane which contains A. Then the metric interior of A relative to E_2 is $\mathcal{I}(A)$, but its metric interior relative to E_3 is $\varnothing$.

Speaking informally, the metric interior of a geometric figure depends on the Euclidean space (line, plane, ...) in which it is considered to be immersed.

In view of this, the concept of metric interior is definitely deficient for general geometrical usage. The concept was not introduced for this purpose, but to study certain subtler properties of figures which do depend on the surrounding space. For example, the concept of metric interior in a 3-space gives important information about solid figures and is very useful in situations where such objects are of the essence. An example is the study of volume measure and integration in 3-space. Figures which have metric interior points will have positive volume; figures which have no metric interior points will have zero volume and will require little attention.

The notion of metric interior is unsuitable in elementary geometry because of the built-in rigidity involved in choosing a Euclidean space E and *then* assigning interiors to *all* geometrical figures in E relative to E. It is not hard to get around this. Consider, as an example, a rectangular region A. A is a *planar* convex set and is contained in a unique plane. Let E_2 be *that* plane. Then the metric interior of A relative to E_2 will be the familiar geometric interior, $\mathcal{I}(A)$.

This example indicates how to employ the metric interior concept to obtain the geometric interior of a convex set in a Euclidean space. The procedure is stated here for a Euclidean 3-space. Let A be a convex set in a Euclidean 3-space E. To avoid discussion of trivial cases, suppose A is not a point and not empty. Then A "determines" a Euclidean subspace E' (line, plane, 3-space) of E which can be characterized as the subspace of E of lowest dimension that contains A—or the least subspace of E that contains A. The metric interior of A *relative to* E' is the geometric interior of A. In symbols,

$$\mathbf{m}\mathcal{I}(A : E') = \mathcal{I}(A). \tag{1}$$

We conclude the section with a few remarks on the members of equation (1) and the names used in referring to them.

In the formal development of our theory $\mathbf{m}\mathcal{I}(A : E')$ does not appear—its role is played by $\mathcal{I}(A)$, which is defined (Section 2.11) without employing any notion of distance. $\mathcal{I}(A)$ may be called the geometrical or natural interior of the convex set A and is conceived as the intrinsic or absolute interior of A, since it depends on the internal structure of A and not on the surrounding space. In the literature of convex set theory (Rockafellar [1], p. 44), $\mathbf{m}\mathcal{I}(A : E')$ is called the *relative interior* of A. This sharp difference in terminology has a historical basis.

The notion of the metric interior of a set was developed before the more elementary idea of the geometric interior of a convex set. Thus for any convex set A in a Euclidean space E, $\mathbf{m}\mathcal{I}(A : E)$ was called *the* interior of A and conceived as its unique interior. When it became necessary to study the geometric interior of a convex set A, this was done by adjusting or "relativizing" the basic space from E to E'. Thus the *new* interior $\mathbf{m}\mathcal{I}(A : E')$ was called the *relative* interior of the convex set A in order to distinguish it from $\mathbf{m}\mathcal{I}(A : E)$. Actually $\mathbf{m}\mathcal{I}(A : E')$ can be defined independently of $\mathbf{m}\mathcal{I}(A : E)$, and it would seem desirable to assign to $\mathbf{m}\mathcal{I}(A : E')$ a name that indicates its intrinsic geometric quality.

2.26 The Metric Closure of a Set in Euclidean Geometry

Our attention now turns to the idea of metric closure of a set of points in a Euclidean geometry, which we take to be three dimensional.

Let E_3 be a Euclidean 3-space. Let S be a set of points of E_3 and p a point of S. Point p is a *metric contact point* of S if each spherical region of E_3 with center p contains a point of S. The *metric closure* of S is the set of its metric contact points and is denoted by $\mathbf{m}\mathcal{C}(S)$.

A Few Illustrations

In E_3 let S be a solid tetrahedron or a planar star-shaped region or an arc of a curve which is open in the sense of not containing its endpoints (Figure 2.82). For each choice of S we see that p_1 and p_2 are metric contact

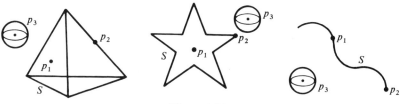

Figure 2.82

points of S, while p_3 is not. Thus a point is a metric contact point of S if and only if it is in "contact" with S in a natural intuitive–geometric sense. Observe that the idea of metric contact point is not biased in favor of solid figures as is the notion of metric interior point relative to E_3 (Section 2.25). Although phrased in terms of 3-dimensional regions, it yields, as indicated by the illustrations above, the same notion of contact for a planar region or a curve as for a solid tetrahedron.

Let us examine the notion of metric contact point a bit more closely. Let S be an arbitrary set of points of E_3. First suppose s is a point of S. Then each spherical region centered at s contains s. By the definition s is a metric contact point of S. Thus $S \subset \mathbf{m}\mathcal{C}(S)$. (Compare Theorem 2.19.)

What can we say about metric contact points of S in general? To answer this, suppose p is a metric contact point of S. Then each spherical region centered at p contains a point of S. Thus we can find points of S as close as we please to p. To be specific there is a point s_1 of S for which the distance $d(p, s_1) < 1$ (Figure 2.83). Similarly S contains a point s_2 for which $d(p, s_2) < \frac{1}{2}$ and, in general, a point s_n that satisfies $d(p, s_n) < 1/n$.

Summary Statement. Suppose $p \subset \mathbf{m}\mathcal{C}(S)$. Then there exists a sequence $s_1, \ldots, s_n, \ldots$ of points of S, which satisfies

$$d(p, s_1) < 1, \ldots, d(p, s_n) < 1/n, \ldots . \tag{1}$$

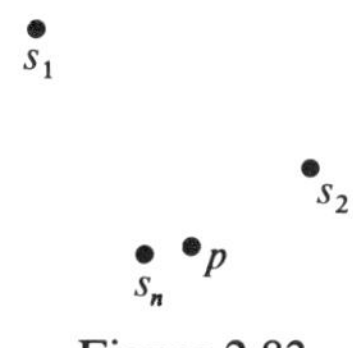

$$\bullet \atop s_1$$

$$\bullet \atop s_2$$

$$\bullet \atop s_n \quad \bullet \atop p$$

Figure 2.83

Note that if $p \subset S$ the result holds with $s_n = p$ for each n.
The result has a converse that is not hard to justify:

Let $p \subset E_3$. Suppose there exists a sequence $s_1, \ldots, s_n, \ldots$ of points of S which satisfies (1). Then $p \subset \mathbf{m}\mathcal{C}(S)$.

Speaking informally, p is a metric contact point of S if and only if p "adheres metrically" to S—or if you wish, the "distance" from p to S is zero.

Now we come to the question of the relation between the metrical closure of a convex set A and its geometrical closure as defined in Section 2.16. The answer is quite simple.

Proposition III. *Let A be any convex set in a Euclidean space E_3. Then*

$$\mathbf{m}\mathcal{C}(A) = \mathcal{C}(A). \tag{1}$$

(Compare Propositions I, II in Section 2.25 above.)

We sketch informally an argument to show this for the case where A is contained in a plane P. The case where A is contained in a line is not hard to dispose of, and we assume A is contained in no line.

Suppose $p \subset \mathbf{m}\mathcal{C}(A)$, that is, p is a metric contact point of A. First we show $p \subset P$. Suppose $p \not\subset P$ (Figure 2.84). Then there exists a spherical region centered at p which does not intersect P—and so not A. This contradicts the definition of metric contact point. Thus $p \subset P$ must hold.

Now let R be any spherical region with radius ρ and center p (Figure 2.85). Let $R' = R \cap P$. Then R' is the *circular* region of P with radius ρ and center p. But R intersects A. Hence R' must intersect A. It follows that every circular region of plane P with center p intersects A. From now on

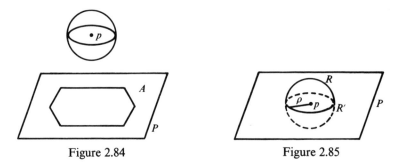

Figure 2.84   Figure 2.85

then we can dispense with spherical regions and operate with circular regions in P.

Our object is to show $p \subset \mathcal{C}(A)$. Suppose $p \not\subset \mathcal{C}(A)$ (Figure 2.86). Recall that A is not contained in any line. Hence A contains points a and b, $a \neq b$, that are not collinear with p. Choose point $c \subset ab$. Then $pc \not\subset A$. Hence there exists $q \subset pc$ such that $q \not\subset A$. Let a', b' satisfy $q \subset aa'$, bb'. Since $c \subset ab$, c is inside $\angle aqb$. Then $q \subset pc$ implies that p is inside $\angle a'qb'$, the angle vertical to $\angle aqb$. Let S be a circular region of P centered at p whose radius is less than the distances from p to the lines aa' and bb'. Then S also is inside $\angle a'qb'$. Since p is a metric contact point of A, the region S contains a point r of A. Certainly r is inside $\angle a'qb'$. Then the extension of segment rq beyond q will meet ab in a point s. Thus $q \subset rs \subset A$,

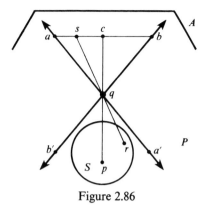

Figure 2.86

which is impossible. Hence our supposition is false and $p \subset \mathcal{C}(A)$. Thus $\mathbf{m}\mathcal{C}(A) \subset \mathcal{C}(A)$.

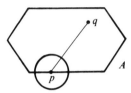

Figure 2.87

Conversely suppose $p \subset \mathcal{C}(A)$. By definition $pq \subset A$ for some point q (Figure 2.87). Then every spherical region centered at p contains a point in pq and so in A. By definition p is a metric contact point of A. Thus $p \subset \mathbf{m}\mathcal{C}(A)$ and (1) holds.

3 The Generation of Convex Sets—Convex Hulls

In this chapter a new operation—comparable in importance to the interior and closure operations—is introduced. It operates on each set S to produce a convex set—*the convex hull of S*—acting as a veritable machine for the generation of convex sets. In effect it expands an arbitrary set into a convex set in the simplest possible way.

The notion of polytope is introduced, a category of convex set which generalizes the closed convex regions of Euclidean geometry that are determined by certain polygons and polyhedrons. A start is made on the study of polytopes. It includes the representation of polytopes in terms of the join operation and preliminary results on their interiors and closures. The chapter concludes with the introduction of a generalized form of polytope and the derivation of some of its properties.

3.1 Introduction to Convexification: Two Euclidean Examples

Convex sets form a family of geometric figures of such great variety and complexity that it becomes essential to establish a systematic procedure for the determination or generation of convex sets.

Our approach is this. Given a set S, we try to *convexify* it—that is, to convert S into a convex set containing S in the simplest possible way. We introduce the process of convexification by considering examples in Euclidean geometry and begin with the simplest nontrivial one, the case of two distinct points.

EXAMPLE I. Let $S = \{a, b\}$, $a \neq b$ (Figure 3.1). If we adjoin to $\{a, b\}$ each point of ab, we get $a \cup b \cup (ab)$, the closed segment ab (Section 2.1), which is convex. Clearly this is the simplest way to enlarge $\{a, b\}$ to form a convex set—for any convex set that contains $\{a, b\}$ must contain the join ab. Thus in forming the set $S^* = a \cup b \cup (ab)$, we have found the simplest or "least" convex set that contains $\{a, b\}$, and so have convexified $\{a, b\}$. A way of describing the result is to say S^* is the convex set determined by a and b.

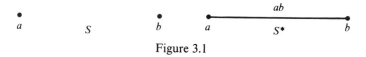

Figure 3.1

EXAMPLE II. Now let $S = \{a, b, c\}$, where a, b and c are noncollinear (Figure 3.2). Let K be any convex set that contains $\{a, b, c\}$. Then since K is closed under join, K contains ab, ac and bc. Similarly, since K contains a and bc, it must contain abc, since it is closed under join of sets (Theorem 2.11). Thus K contains

$$S^* = a \cup b \cup c \cup (ab) \cup (ac) \cup (bc) \cup (abc).$$

Figure 3.2

But S^* is the closed triangular region with vertices a, b, c (Section 2.12, Example III) and is convex. Thus S^* is the "least" convex set containing S, and we have convexified S in this case also. Here too S^* may be described as the convex set determined by a, b and c.

3.2 Convexification of an Arbitrary Set

In the last section two very simple Euclidean sets were convexified. Now, returning to the abstract theory, we prove that any set can be convexified. Stated formally this is an existence assertion:

For any set S there exists a "least" convex set containing S.

First, "least" needs a bit of attention.

Definition. Let F be a family or a collection of sets. If M is a set of F which is contained in every set of F, we say M is a *least* member of F. Similarly if N is a set of F which contains every set of F, N is a *greatest* member of F.

If F, a family of sets, has a least member M, it can have only one least member. For suppose M' is a least member of F. Then by definition $M' \subset M$ and $M \subset M'$. Thus $M' = M$. Similarly it can be shown that a family of sets can have only one greatest member.

Next, for convenience of expression, we introduce the following

Definition. If $A \supset S$ we call A an *extension* of S. Thus a convex set that contains S is described simply as a *convex extension* of S.

Now we can state the formal version of the convexification theorem.

Theorem 3.1 (Existence). *Let set S be given. Then the family F of convex extensions of S has one and only one least member.*

Figure 3.3

Analysis. Suppose the family F does have a least member L. Can we identify L? By definition L is contained in every member of F. Hence L is contained in the intersection of all the members of F. Call the intersection S^*. We have $L \subset S^*$. But L is a member of F, so that $S^* \subset L$. Hence $L = S^*$ and is identified.

PROOF. The family F is not empty, since it certainly has one member, namely, the basic set J. Let S^* be the intersection of all members of F. Then S^* contains S, since all members of F contain S. And S^* is convex, since it is the intersection of a family of convex sets (Theorem 2.14). Thus S^* is a convex extension of S—that is, S^* is a member of F. Since S^* is the intersection of all members of F, it must be contained in all members of F. Thus S^* is a least member of F, and F has a least member. But as we saw above (the paragraph following the first definition), F can have only one least member. Thus the theorem holds.   □

Restatement of Theorem 3.1. For each set there exists a unique least convex extension.

The following definition may now be stated.

Definition. Let S be any set. The least convex extension of S, denoted by $[S]$, is called the *convex hull* of S or the convex set *generated* (or *spanned* or *determined*) by S. If A is a given convex set and S a set for which $[S] = A$, we call S a *set of generators* of A (in the sense of *convexity*), and say S *generates* A (*convexly*) or A *is generated* (*convexly*) by S.

Note that $[A] = A$ if and only if A is convex; in particular $[J] = J$, $[\varnothing] = \varnothing$ and $[a] = a$ for each point a.

Theorem 3.1 asserts that any geometric figure can be convexified—any set of points has a convex hull. A better name for the idea might be convex cover or convex closure, but convex hull is widely used and does have the advantage of being brief. The term hull as used here comes from the German *Hülle*, which means cover or envelope and is related to the use of hull for husk or pod of a fruit, or body of a ship. This tends to be misleading, since it suggests that the convex hull bounds or envelopes the figure but doesn't contain it as a subset, which is contrary to the definition.

Remarks on the Notion of Convex Hull

The idea of convex hull may seem strange—unrelated to your previous knowledge of geometry. You may not have an intuitive feeling for the idea —as you did, for example, for the idea of the interior of a convex set in Chapter 2. This is not strange. The notion of convex hull has been assimilated into mathematics only relatively recently. Indeed, the concept of convex set itself is hardly more than a century old and has barely influenced the conventional treatment of geometry.

Actually the notion of convex hull has a simple physical interpretation (Figure 3.4). Let a geometric figure A be given in a Euclidean plane. Imagine that A is surrounded by a stretched rubber band—a mathematical

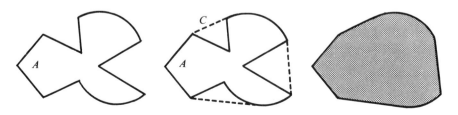

Figure 3.4

one, an elasticized simple closed curve. Suppose the band shrinks to form a closed curve C that just *envelopes* A. Then C united with its interior will form a convex set that just contains A—it will form the convex hull of A.

Similarly, imagine (Figure 3.5) a figure A in Euclidean 3-space enclosed in an inflated rubber balloon which shrinks to form a surface S which just envelopes A. Then the union of S and its interior will be the convex hull of A.

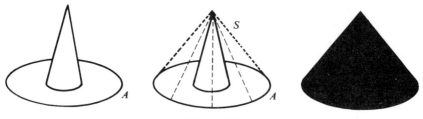

Figure 3.5

There is a graphical method for constructing the convex hull of a plane figure. Suppose a plane figure A is drawn on a sheet of paper and you want to construct its convex hull. By definition, the convex hull of A is convex and contains A—so it must contain the join of each pair of points of A. A natural approach then is to start with a "shading process": Draw segments joining pairs of points of A and try to determine the figure formed—or the part of the plane covered—by all such segments.

To illustrate the process, let A be an ellipse (Figure 3.6). Then the "shaded region" formed by the segments is I, the interior of A. Therefore the convex hull of A must contain $A \cup I$. But $A \cup I$ is convex. Hence $A \cup I$ is the convex hull of A.

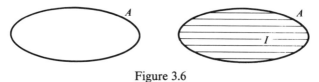

Figure 3.6

EXERCISES

1. Determine graphically the convex hulls of the figures given in Figure 3.7.

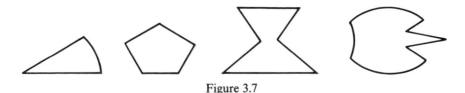

Figure 3.7

2. The same as Exercise 1 for the figures given in Figure 3.8, in which the indicated segments are closed.

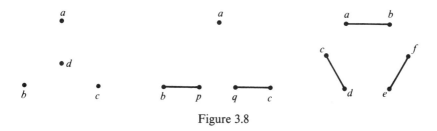

Figure 3.8

3. Find graphically the convex hull of each of these figures:

A, B, P, Q, S.

Now a few words on the importance and power of the convex hull concept. Euclidean 3-space is full of convex sets—linear ones, planar ones, solid convex sets—which in the conventional treatment of geometry are largely unrecognized. The notion of convex hull presents a systematic method for isolating and determining convex sets: for giving them mathematical existence. Choose any set S you please. The convex hull operation produces the simplest convex set containing S, tailored for the best fit.

As one of the simplest possible applications of the idea of convex hull, we indicate how it can be used to give a unified treatment of triangles and triangular regions. In a Euclidean geometry let a, b and c be noncollinear points (Figure 3.9). Let $T = [\{a, b, c\}]$. Then T is the closed triangular region with vertices a, b, c (Example II, Section 3.1 above). Since T is a convex set, it splits into its interior and its frontier. $\mathcal{I}(T)$ is the (open) triangular region abc, and $\mathcal{F}(T)$ is triangle abc. [See Section 2.12, Example III, relations (12) and (17).]

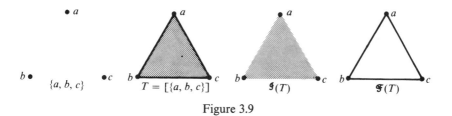

Figure 3.9

Euclidean Examples of Convex Hulls

Finally, in Figure 3.10 we give a list of convex hulls of some familiar figures.

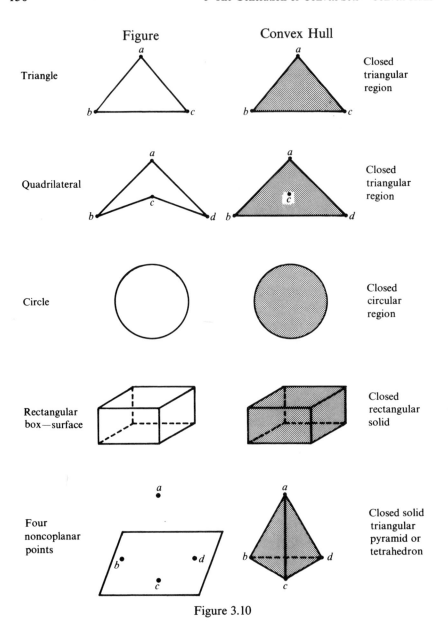

Figure 3.10

EXERCISES

1. (a) Does the family of all convex sets have a least member? a greatest?
   (b) Can you find a family of convex sets in a Euclidean geometry which has no least member? No greatest member?

2. (a) In Euclidean 3-space find the convex hull of $\{a, b, c\}$ assuming the points are distinct. List all types of figure that can arise.
   (b) The same for $\{a, b, c, d\}$.

3. (a) Interpret $[\{a, b\}]$ and $[\{a, b, c\}]$ in the triode model (Section 2.22) assuming in each case the points are distinct. Describe the figures you get.
   (b) The same for the Cartesian model (Section 2.24).

4. Prove:
   (a) If $A \supset B$, then $[A] \supset B$.
   (b) If $A \supset B$ and A is convex, then $A \supset [B]$.
   (c) If $A \supset B$, then $[A] \supset [B]$.

5. Prove: $[[S]] = [S]$.

6. Prove: $[A] \cup [B] \subset [A \cup B]$.

7. (a) Prove: $[A \cap B] \subset [A] \cap [B]$.
   (b) Show by example that equality need not hold in (a).

8. (a) If $[A] = [B]$, must $A = B$? Explain.
   (b) If $[A] \supset [B]$, must $A \supset B$? Explain.

9. Let a convex set A be given and suppose $[S] = A$. What can you say about the family F of sets S that satisfy this relation? Is F ever empty? Can you ever find a least member of F? Can you always find a least member of F? The same for a greatest member of F. Give examples in Euclidean geometry.

10. In a Euclidean plane, what sort of convex sets are generated by finite sets of points? The same for Euclidean 3-space.

11. Find convex sets in a Euclidean plane and 3-space which
    (a) cannot be generated by a finite set of points;
    (b) satisfy the condition in (a) and are generated by a denumerable infinite set, that is, one whose points can be represented by an infinite sequence $a_1, \ldots, a_n, \ldots$ where $a_i \neq a_j$ if $i \neq j$.

12. Find in Euclidean geometry a convex set A which is generated by $\mathscr{B}(A)$; by $\mathscr{F}(A)$ but not $\mathscr{B}(A)$; by $\mathscr{I}(A)$.

13. Find in Euclidean geometry a convex set that has a least set of generators which is finite; is infinite.

14. (a) Can you find a convex set A that contains a point p that belongs to no set of generators of A? Can you characterize such a point in simple geometrical terms?
    (b) The same if p belongs to some set of generators of A but not to all.
    (c) The same if p belongs to every set of generators of A.

15. (a) In the Cartesian model (Section 2.24) let $S = [\{a, b\}]$, where $a = (0, 0)$, $b = (1, 1)$. Find S, $\mathscr{I}(S)$, $\mathscr{F}(S)$, $\mathscr{B}(S)$, $\mathscr{C}(S)$.
    (b) The same in a Euclidean model, taking a and b to be distinct points. Compare your results with those in (a).

Exercises (Convex Hulls of Euclidean Sets)

1. In a Euclidean plane find the convex hull of the set described:
   (a) The union of point a and closed (open) segment bc, where a, b, c are noncollinear.

(b) The union of point a and line L, $a \not\subset L$. [Compare Exercise 6 (a) at the end of Section 2.4.]

(c) The union of two parallel lines. [Compare Exercise 6(b) at the end of Section 2.4.]

(d) The union of two distinct intersecting lines. [Compare Exercise 6(c) at the end of Section 2.4.]

(e) $(ab) \cup (cd)$, where $a \neq b$, $c \neq d$ and ab and cd have only one point in common.

(f) $(ab) \cup a \cup b \cup (cd) \cup c \cup d$, where $a \neq b$, $c \neq d$ and ab and cd have only one point in common.

2. In a Euclidean 3-space find the convex hull of the set described:

(a) The union of point a and plane P, $a \not\subset P$. [Compare Exercise 7(a) at the end of Section 2.4.]

(b) The union of two parallel planes. [Compare Exercise 7(c) at the end of Section 2.4.]

(c) The union of a plane and a line parallel to it. [Compare Exercise 7(f) at the end of Section 2.4.]

(d) The union of a plane P and an intersecting line L, $L \not\subset P$. [Compare Exercise 7(f) at the end of Section 2.4.]

(e) The union of two skew (that is, noncoplanar) lines. [Compare Exercise 7(b) at the end of Section 2.4.]

(f) The union of two noncoplanar segments.

(g) $\{a, b, c, d, e\}$, where a, b are distinct, c, d, e are noncollinear and ab and cde have a single point in common.

(h) The union of a point and a circle that are noncoplanar.

(i) The union of two congruent circles that lie in parallel planes.

(j) The union of two congruent noncoplanar triangles whose pairs of corresponding sides are parallel.

(k) The union of line L and circle C if L is perpendicular to the plane of C at its center. (Compare Exercise 8 at the end of Section 2.4.)

3. Suppose that a Euclidean plane has been assigned a Cartesian coordinate system. Find the convex hull of the set of points (x, y) which satisfy the given conditions:

(a) $y = x^2$.

(b) $y = x^2$, $0 \leqslant x$.

(c) $y = x^3$.

(d) $y^2 = x^3$.

(e) $y = 1/x$, $1 \leqslant x$.

(f) $x^2 - y^2 = 1$, $1 \leqslant x$ and $0 \leqslant y$.

(g) $x^2 + y^2 = 1$, where x and y are rational numbers.

3.3 Finitely Generated Convex Sets—the Concept of Polytope

Many of the familiar convex sets of Euclidean geometry are convex hulls of finite sets of points, for example, closed segments, closed triangular or rectangular regions, closed tetrahedral or cubical regions in 3-space. These seem to represent an especially simple type of convex set and suggest the following definitions.

Definition. A convex set is said to be *finitely generated* (*convexly*) if it is the convex hull of some finite set.

Definition. A convex set is called a *convex polytope*, or simply a *polytope*, if it is finitely generated and nonempty. Equivalently, a polytope is a convex set which is generated by a finite nonempty set.

The standard form for representing a polytope is $[\{a_1, \ldots, a_n\}]$, which usually will be written $[a_1, \ldots, a_n]$.

Note that each point p is a polytope, since $[p] = p$.

The class of polytopes may seem a small fragment to split off from the family of all convex sets, but it is not an unimportant one. Polytopes will be a recurrent theme in our work, both for their intrinsic importance as geometric objects and for their involvement in the study of more complex types of convex set.[1]

3.4 A Formula for a Polytope

Theorem 3.1, the existence theorem, assures us that if a set S is given, its convex hull $[S]$ exists. But it provides no simple procedure for constructing or representing $[S]$ in terms of the points of S. However, in Euclidean geometry (Example II, Section 3.1), $[S]$ for $S = \{a, b, c\}$ was obtained as the union of the joins of a, b, c taken one or more at a time. This type of construction can be used in the formal theory and extended to any nonempty set S. The case where S is finite is considered in the next theorem. The theorem involves the notion of a join of points, and we have to devote a few paragraphs to it.

The term *join of points* was introduced to refer to a generalized join $a_1 \ldots a_r$ (Section 2.7, the definition next to the last).

First an extension of the terminology.

Definition. Suppose $a_1, \ldots, a_r \subset S$. Then $a_1 \ldots a_r$ is called a *join of points* of S. For example, a_1, $a_2 a_4$, $a_4 a_1 a_2 a_3$ and $a_4 a_2 a_4 a_1 a_2$ are joins of points of the set $\{a_1, a_2, a_3, a_4\}$.

Next it is necessary to represent joins of points of $\{a_1, \ldots, a_n\}$ in simplified or standard forms. Consider, for example, $a_4 a_2 a_4 a_1$. In most situations we would express it in the form $a_1 a_2 a_4$. But if the context called for 5-point joins, we might write it as $a_1 a_2 a_4 a_4 a_4$. To take care of this the following lemma is introduced.

---

[1] Grünbaum [1] is a remarkably comprehensive work on the theory of polytopes.

Lemma 3.2 (Representation Principle for Joins of Points). *If $a_{i_1} \ldots a_{i_r}$ is a join of points of $\{a_1, \ldots, a_n\}$, it can be expressed in the following forms:*

(i) $a_{j_1} \ldots a_{j_s}$, *where* $1 \leqslant j_1 < \cdots < j_s \leqslant n$;

(ii) $b_1 \ldots b_n$, *where the b's are in* $\{a_{i_1}, \ldots, a_{i_r}\}$.

PROOF. (i) Repetitions of terms in $a_{i_1} \ldots a_{i_r}$ can be eliminated by using the idempotent law J4 and generalized associative and commutative laws. Then the terms can be arranged according to increasing subscripts by applying the generalized commutative law.

(ii) Observe that $s \leqslant n$ in (i). Let $b_1 = a_{j_1}, \ldots, b_s = a_{j_s}$. Then by (i)

$$a_{i_1} \ldots a_{i_r} = b_1 \ldots b_s. \tag{1}$$

If $s = n$, the result holds by (1). If $s < n$, let $b_{s+1} = \cdots = b_n = b_s$. Then (1) implies

$$a_{i_1} \ldots a_{i_r} = b_1 \ldots b_n,$$

and the lemma is proved.   □

Thus any join of points of $\{a_1, \ldots, a_n\}$ can be represented in two forms: (i) with increasing subscripts and (ii) as an n-term join.

Notation. We adopt the convention that in expressions involving the operations of join and union, portions separated by $\cup$ signs are to be considered enclosed in parentheses; for example, $AB \cup CDE \cup AF$ stands for $(AB) \cup (CDE) \cup (AF)$.

Now we are prepared to represent a polytope as a union of joins of points.

Theorem 3.2 (The Polytope Join Formula).

$$[a_1, \ldots, a_n]$$

$$= (a_1 \cup \cdots \cup a_n) \cup (a_1 a_2 \cup \cdots \cup a_{n-1} a_n) \cup \cdots \cup (a_1 \ldots a_n).$$

PROOF. By definition $[a_1, \ldots, a_n]$ is convex and

$$[a_1, \ldots, a_n] \supset a_1, \ldots, a_n. \tag{1}$$

Hence (1) implies

$$[a_1, \ldots, a_n] \supset a_1 a_2, a_1 a_3, \ldots, a_{n-1} a_n.$$

Similarly (1) implies

$$[a_1, \ldots, a_n] \supset a_1 a_2 a_3, \ldots, a_{n-2} a_{n-1} a_n,$$

since a convex set is closed under the extended join operation (Corollary 2.11). Continuing in this way, we get finally

$$[a_1, \ldots, a_n] \supset a_1 \ldots a_n.$$

Combining these results, we have

$$[a_1, \ldots, a_n]$$

$$\supset (a_1 \cup \cdots \cup a_n) \cup (a_1 a_2 \cup \cdots \cup a_{n-1} a_n) \cup \cdots \cup (a_1 \ldots a_n). \tag{2}$$

Let S denote the right member of (2),

$$S = (a_1 \cup \cdots \cup a_n) \cup (a_1 a_2 \cup \cdots \cup a_{n-1} a_n) \cup \cdots \cup (a_1 \ldots a_n).$$

We assert S is convex. To justify this let $x, y \subset S$. Then x and y belong to joins of the a's. Say

$$x \subset a_{i_1} \ldots a_{i_r}, \quad \text{where } 1 \leqslant i_1 < \cdots < i_r \leqslant n \tag{3}$$

and

$$y \subset a_{j_1} \ldots a_{j_s}, \quad \text{where } 1 \leqslant j_1 < \cdots < j_s \leqslant n. \tag{4}$$

The relations (3) and (4) imply

$$xy \subset (a_{i_1} \ldots a_{i_r})(a_{j_1} \ldots a_{j_s}) = a_{i_1} \ldots a_{i_r} a_{j_1} \ldots a_{j_s}. \tag{5}$$

By Lemma 3.2,

$$a_{i_1} \ldots a_{i_r} a_{j_1} \ldots a_{j_s} = a_{k_1} \ldots a_{k_t}, \quad \text{where } 1 \leqslant k_1 < \cdots < k_t \leqslant n. \tag{6}$$

Then (5) and (6) imply $xy \subset S$, and S is convex by definition. Certainly $S \supset \{a_1, \ldots, a_n\}$, so that S is a convex extension of $\{a_1, \ldots, a_n\}$. Thus S contains the least convex extension of $\{a_1, \ldots, a_n\}$, that is,

$$S \supset [a_1, \ldots, a_n]. \tag{7}$$

The theorem follows by (2) and (7).   □

The result is quite important. It gives body to Theorem 3.1 (the existence theorem for convex hulls) and initiates our study of polytopes by providing a join theoretic formula for a polytope in terms of a set of generators. In view of Lemma 3.2(i), the theorem can be restated in the following form: The convex hull of a finite nonempty set S is the union of all joins of points of S.

A Few Examples of the Polytope Join Formula

EXAMPLE I $(n = 2)$.

$$[a, b] = a \cup b \cup ab. \tag{1}$$

In (1) suppose $a \neq b$. Then (1) asserts in Euclidean geometry that $[a, b]$ is the closed segment ab (Section 3.1, Example I; see Figure 3.11). Thus (1) suggests the following definition in the abstract theory.

Figure 3.11

Figure 3.12

Definition. Let a and b be any points. Then $[a, b]$ is called a *closed segment*; a and b are called *endpoints* of $[a, b]$. If $a \neq b$ the closed segment is said to be *proper* or *nondegenerate*. The closed segment $[a, a] = a$ is said to be *improper* or *degenerate*.

The notion of closed segment supplements that of segment (Section 2.7, the last definition).

EXAMPLE II ($n = 3$).

$$[a, b, c] = a \cup b \cup c \cup ab \cup ac \cup bc \cup abc. \tag{2}$$

This calls to mind a closed triangular region in Euclidean geometry (Figure 3.12). Indeed, if a, b and c are noncollinear points in a Euclidean geometry, (2) asserts that $[a, b, c]$ is the closed triangular region with vertices a, b, c (Section 3.1, Example II). You may wonder whether we can define an analogue of a Euclidean closed triangular region in the abstract theory. To do this we would have to characterize noncollinearity of three points or an equivalent idea that three points are *independent* or lie in *general position*. (In Chapter 6, Section 6.14, a notion of linearly independent points is introduced and used to define the concept of *simplex*, which is a generalization of Euclidean closed segment and closed triangular region.)

EXAMPLE III ($n = 4$).

$$[a, b, c, d] = (a \cup b \cup c \cup d)$$

$$\cup (ab \cup ac \cup ad \cup bc \cup bd \cup cd)$$

$$\cup (abc \cup abd \cup acd \cup bcd) \cup (abcd). \tag{3}$$

In Euclidean geometry if a, b, c, d are noncoplanar, $[a, b, c, d]$ is a closed solid triangular pyramid or tetrahedron (Figure 3.13), and (3) gives the structure of this figure in terms of its vertices, edges, facial regions and interior. It should not be missed that (3) [as well as (2) and (1) above] holds for all special or degenerate cases. Thus in Euclidean geometry (3) holds if a, b, c, d are distinct and ab and cd have a single point in common. In this case $[a, b, c, d]$ is the closed quadrangular region with vertices a, b, c, d (Figure 3.14).

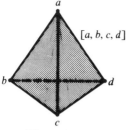

Figure 3.13

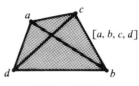

Figure 3.14

The Polytope Join Formula is a key result. But there still are many—apparently simple—questions to be answered. Can we find a formula for the interior, closure, frontier or boundary of a polytope? Is it identical with its closure; is its frontier identical with its boundary, as in the familiar Euclidean polytopes? Does it have vertices, edges and faces, like the polyhedra of Euclidean 3-space? Is there a simple way to represent a polytope as a finite intersection of convex sets? These and other questions on polytopes will involve our concern in the sequel.

EXERCISES (POLYTOPES)

1. (a) Interpret proper closed segment in the triode model (Section 2.22). What types of figure do you obtain?
   (b) The same for the Cartesian model (Section 2.24).
   (Compare Exercise 3 in the first set at the end of Section 3.2.)

2. (a) Suppose in the triode model a, b, c are distinct and no one of the points lies in the join of the other two. Verify the polytope join formula for $[a, b, c]$.
   (b) The same for the Cartesian model.
   (Compare Exercise 3 in the first set at the end of Section 3.2.)

3. Suppose in a Euclidean model a, b, c, d are distinct. Verify the polytope join formula for $[a, b, c, d]$. [Compare Exercise 2(b) in the first set at the end of Section 3.2.]

4. (a) In a Euclidean 3-dimensional model find polytopes whose join formula has the property that no two of its terms meet.
   (b) The same for the property that some two of the terms meet.
   (c) What do you conjecture about the family of polytopes in (a); in (b)?

5. (a) Suppose $A = [a_1, a_2, a_3]$, $B = [b_1, b_2]$, and $A \cup B$ is convex. Must $A \cup B$ be a polytope? Explain.
   (b) The same for arbitrary polytopes A and B.

6. If A and B are convex and $A \cup B$ is a polytope, must A and B be polytopes? Must one of A, B be a polytope? Explain.

7. (a) In a Euclidean model can you find a polytope that contains all its contact points? One that does not?

(b) The same for the triode model.

(c) The same for the Cartesian model.

8. Verify: In a Euclidean model a, b and c are collinear if and only if in the join formula for $[a, b, c]$ there are two terms which meet.

9. Generalize Exercise 8 to four points a, b, c, d.

10. (a) Prove: $[a_1, a_2, a_3] = [a_1, a_2] \cup [a_1, a_3] \cup [a_2, a_3] \cup a_1 a_2 a_3$.

(b) Try to generalize the result in (a) to $[a_1, \ldots, a_n]$.

11. Prove: $[a_1, \ldots, a_n] = [a_2, \ldots, a_n]$ if and only if $a_1 \subset [a_2, \ldots, a_n]$.

12. Verify, using diagrams, Radon's Theorem for a Euclidean plane: If a, b, c and d are four distinct points in a Euclidean plane, the four points can be split into two disjoint sets whose convex hulls have a common point.

13. Assuming Radon's Theorem (Exercise 12), prove: If a, b, c and d are in a Euclidean plane, then $[a, b, c]$, $[a, b, d]$, $[a, c, d]$ and $[b, c, d]$ have a common point.

14. (a) Assuming the result in Exercise 13, prove Helly's Theorem for a Euclidean plane: If A, B, C and D are convex sets in a Euclidean plane each three of which have a common point, then all four have a common point.

(b) Show that the result in Exercise 13 is deducible from Helly's Theorem.

15. (a) Let a, b, c and p be points in a Euclidean plane such that a, b and c are not collinear and

$$[p, a, b] \cap [p, b, c] \cap [p, a, c] = p.$$

What do you conjecture about the position of p relative to a, b, and c?

(b) Assuming Radon's Theorem or Helly's Theorem, try to prove your conjecture.

16. Assuming Radon's Theorem (Exercise 12), verify, using diagrams as needed: If p is in the Euclidean plane abc, then p lies on a line aq, where $q \subset [b, c]$, or a line br, where $r \subset [a, c]$, or a line cs, where $s \subset [a, b]$.

17. (a) Try to generalize Exercise 12 to Euclidean 3-space.

(b) The same for Exercises 13–16.

3.5 A Distributive Law

The following distributive law for join with respect to union of sets is quite useful.

Theorem 3.3 (Distributive Law for Join). $A(B \cup C) = AB \cup AC$.

PROOF. Let $x \subset A(B \cup C)$. Then $x \subset ay$, where

$$a \subset A, \qquad y \subset B \cup C.$$

If $y \subset B$, then $x \subset AB$; similarly, if $y \subset C$, then $x \subset AC$. In any case $x \subset AB \cup AC$.

Conversely let $x \subset AB \cup AC$. Then $x \subset AB$ or $x \subset AC$. In either case the monotonic law for join (Theorem 2.1) implies $x \subset A(B \cup C)$, and the theorem holds. $\qquad \qquad \square$

Generalized distributive laws follow from the theorem.

Corollary 3.3.1. $A(B_1 \cup \cdots \cup B_n) = AB_1 \cup \cdots \cup AB_n$.

Corollary 3.3.2. $(A_1 \cup \cdots \cup A_m)(B_1 \cup \cdots \cup B_n) = A_1 B_1 \cup \cdots \cup A_m B_1 \cup \cdots \cup A_1 B_n \cup \cdots \cup A_m B_n$.

EXERCISES

1. Prove: If A and B are convex, then $A \cup AB$ and $A \cup B \cup AB$ are convex.

2. (a) Given that A and B are convex, find and justify a formula for $[A \cup B]$.
   (b) Similarly for three convex sets A, B, C.

3. Let S^2 denote $S \cdot S$.
   (a) Prove: $([a, b] \cup [b, c])^2 = [a, b, c]$.
   (b) Verify (a) in Euclidean geometry.
   (c) Does the following analogue of (a) hold?

   $$([a, b] \cup [b, c] \cup [c, d])^2 = [a, b, c, d].$$

4. Prove: $A(B \cap C) \subset (AB) \cap (AC)$. Show that equality need not hold.

5. (a) Prove: $[a_1, a_2, a_3] = a_1 \cup [a_2, a_3] \cup a_1[a_2, a_3]$.
   (b) Try to generalize the result in (a) to $[a_1, \ldots, a_n]$.

3.6 An Absorption Property of Polytopes

Theorem 3.4 (Absorption). $a_1 \ldots a_n[a_1, \ldots, a_n] = a_1 \ldots a_n$.

Restatement. If polytope P is generated by $\{a_1, \ldots, a_n\}$, then the join $a_1 \ldots a_n$ absorbs P properly (Section 2.14).

PROOF. First observe that $a_1 \ldots a_n$ absorbs properly any one of the a's:

$$(a_1 \ldots a_n)a_i = a_1 \ldots a_n, \qquad 1 \leqslant i \leqslant n.$$

This relation implies that $a_1 \ldots a_n$ absorbs properly any *join* of the a's:

$$(a_1 \ldots a_n)a_{i_1} \ldots a_{i_r} = a_1 \ldots a_n, \qquad 1 \leqslant i_1 < \cdots < i_r \leqslant n. \quad (1)$$

Then the Polytope Join Formula (Theorem 3.2), the generalized distributive law (Corollary 3.3.1) and (1) imply

$$a_1 \ldots a_n[a_1, \ldots, a_n] = a_1 \ldots a_n(a_1 \cup \cdots \cup a_{i_1} \ldots a_{i_r} \cup \cdots$$

$$\cup a_1 \ldots a_n)$$

$$= a_1 \ldots a_n \cup \cdots \cup a_1 \ldots a_n$$

$$= a_1 \ldots a_n. \qquad \qquad \square$$

Corollary 3.4. $a_1 \ldots a_n$ *absorbs properly every join of points of* $\{a_1, \ldots, a_n\}$.

3.7 Interiors and Closures of Polytopes

In this section the interior and closure operations are applied to polytopes. Let P be a polytope. Can we find $\mathcal{I}(P)$ and $\mathcal{C}(P)$ if a finite set of generators of P is given? Consider $\mathcal{I}(P)$ first.

Notation. $\mathcal{I}([a_1, \ldots, a_n])$ is written as $\mathcal{I}[a_1, \ldots, a_n]$; parentheses are dispensed with in similar expressions involving the closure, frontier and boundary operators $\mathcal{C}$, $\mathcal{F}$ and $\mathcal{B}$.

In a Euclidean geometry there is a definitive result:

$$\mathcal{I}[a_1, \ldots, a_n] = a_1 \ldots a_n. \qquad \qquad \text{(A)}$$

To illustrate (A) suppose in a Euclidean geometry a, b and c are noncollinear (Figure 3.15). We show

$$\mathcal{I}[a, b, c] = abc. \qquad \qquad \text{(B)}$$

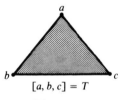

$$[a, b, c] = T$$

Figure 3.15

$[a, b, c]$ is the closed triangular region T with vertices a, b, c (Section 3.4, Example II). But $\mathcal{I}(T) = abc$ (Section 2.12, Example III). Thus (B) is verified.

The relation (A) cannot be deduced from the postulates J1–J4 (see Exercise 1 below). In the next chapter the postulate set J1–J4 will be enlarged and (A) will be proved (see Theorem 4.30). Now we are able to establish a weaker result—but one which carries a significant piece of information and will be used in the proof of (A).

Theorem 3.5. $\mathcal{I}[a_1, \ldots, a_n] = \mathcal{I}(a_1 \ldots a_n)$.

PROOF. First we show

$$\mathcal{I}[a_1, \ldots, a_n] \subset \mathcal{I}(a_1 \ldots a_n). \tag{1}$$

We suppose

$$p \subset \mathcal{I}[a_1, \ldots, a_n] \tag{2}$$

and show

$$p \subset \mathcal{I}(a_1 \ldots a_n). \tag{3}$$

Note $a_1 \ldots a_n \neq \varnothing$ by Corollary 2.6. Let $x \subset a_1 \ldots a_n$ (Figure 3.16). Then $x \subset [a_1, \ldots, a_n]$ by Theorem 3.2, the Polytope Join Formula. Hence (2) implies

$$p \subset xy, \tag{4}$$

where $y \subset [a_1, \ldots, a_n]$. In order to establish (3) we must find a point y' such that

$$p \subset xy' \quad \text{and} \quad y' \subset a_1 \ldots a_n.$$

It is not too hard to find such a y'. Applying the absorption property of a polytope (Theorem 3.4), we have

$$xy \subset a_1 \ldots a_n[a_1, \ldots, a_n] = a_1 \ldots a_n.$$

Then use the following device. Rewrite (4) in the form

$$p \subset xxy = x(xy).$$

Hence

$$p \subset xy', \quad \text{where} \quad y' \subset xy \subset a_1 \ldots a_n. \tag{5}$$

Note that $p \subset a_1 \ldots a_n$. Then by definition, (5) implies (3), and (1) is verified.

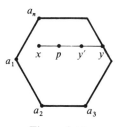

Figure 3.16

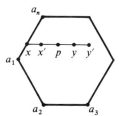

Figure 3.17

Conversely we show

$$\mathcal{I}(a_1 \ldots a_n) \subset \mathcal{I}[a_1, \ldots, a_n]. \tag{6}$$

Suppose

$$p \subset \mathcal{I}(a_1 \ldots a_n). \tag{7}$$

Let x satisfy (Figure 3.17)

$$x \subset [a_1, \ldots, a_n].$$

Then we shall show

$$p \subset xy, \quad \text{where} \quad y \subset [a_1, \ldots, a_n]. \tag{8}$$

The relation (7) implies

$$p \subset x'y', \quad \text{where} \quad y' \subset a_1 \ldots a_n \tag{9}$$

provided $x' \subset a_1 \ldots a_n$. How shall x' be chosen? Observe that

$$xp \subset [a_1, \ldots, a_n] a_1 \ldots a_n. \tag{10}$$

The relation (10) implies, by the absorption property of a polytope,

$$xp \subset a_1 \ldots a_n.$$

Let $x' \subset xp$. Then $x' \subset a_1 \ldots a_n$ and (9) holds. Now (9) and $x' \subset xp$ imply $p \subset xpy'$. Hence $p \subset xy$ where $y \subset py'$. But

$$p, y' \subset a_1 \ldots a_n \subset [a_1, \ldots, a_n].$$

Hence

$$y \subset [a_1, \ldots, a_n]$$

and (8) holds. By definition $p \subset \mathcal{I}[a_1, \ldots, a_n]$. Thus (6) is justified and the theorem follows. □

Remark. The theorem reduces the problem of finding $\mathcal{I}[a_1, \ldots, a_n]$ to the simpler one of finding $\mathcal{I}(a_1 \ldots a_n)$. This is solved in Theorem 4.28, which asserts in effect that $\mathcal{I}(a_1 \ldots a_n) = a_1 \ldots a_n$.

The theorem has two simple corollaries.

Corollary 3.5.1. $\mathcal{I}[a_1, \ldots, a_n] \subset a_1 \ldots a_n.$

Corollary 3.5.2. $\mathcal{F}[a_1, \ldots, a_n] \supset [a_1, \ldots, a_n] - a_1 \ldots a_n.$

PROOF. By Theorem 2.30(b) and the preceding corollary,

$$\mathcal{I}[a_1, \ldots, a_n] = [a_1, \ldots, a_n] - \mathcal{I}[a_1, \ldots, a_n]$$

$$\supset [a_1, \ldots, a_n] - a_1 \ldots a_n. \quad □$$

Now we briefly consider closures of polytopes, following more or less the treatment for interiors.

In a Euclidean geometry we have

$$\mathcal{C}[a_1, \ldots, a_n] = [a_1, \ldots, a_n]. \tag{C}$$

This is illustrated by the example used above: $[a, b, c]$, the closed triangular region T with vertices a, b, c. We have (Section 2.15, Example III)

$$\mathcal{C}[a, b, c] = \mathcal{C}(T) = T = [a, b, c].$$

The relation (C) is not deducible from J1–J4 (see Exercise 2 below). It does hold under a strengthened set of postulates (see Theorem 12.15). It is interesting that the analogue for closure of Theorem 3.5 is valid (see Exercise 3 below).

EXERCISES

1. Suppose $J = \{a, b, c\}$, and $\cdot$ is given by Figure 3.18.

$\cdot$	a	b	c
a	a	a	a
b	a	b	J
c	a	J	c

Figure 3.18

(a) Verify that $\cdot$ is a join operation in J, and postulates J1–J4 hold for $(J, \cdot)$.
(b) Show $\mathcal{J}[b, c] = a$ but $bc \neq a$.
(c) Infer that the relation (A) above (third paragraph of the section) is not deducible from J1–J4.

2. In the Cartesian model (Section 2.24), show that points a and b can be chosen so that $\mathcal{C}[a, b] \neq [a, b]$. Infer that the relation (C) above is not deducible from J1–J4. (Compare your solution of Exercise 7(c) at the end of Section 3.4.)

3. Prove: $\mathcal{C}[a_1, \ldots, a_n] = \mathcal{C}(a_1 \ldots a_n)$.

3.8 Powers of a Set and Polytopes

The convex hull of a set S is related to repeated joins, or "powers" of S. For by definition $[S] \supset S$ and is convex. Thus $[S] \supset SS$. Similarly $[S] \supset (SS)S$. Thus

$$[S] \supset S \cup SS \cup (SS)S.$$

This process can be continued without difficulty.

In this section we define powers of a set and prove that a polytope is expressible as a power of any finite set that generates it.

Definition. Let a set S be given. Let $S_1 = \cdots = S_n = S$. Then the n-term join $S_1 \ldots S_n$ is called the nth *power* of S and is denoted by S^n. In practice the n-term join S^n will usually be written as $S \ldots S$ (n terms).

As in ordinary algebra, we can prove

$$S^m \cdot S^n = S^{m+n} \quad \text{for any positive integers } m, n.$$

Theorem 3.6. $\left[\{a_1, \ldots, a_n\} \right] = \{a_1, \ldots, a_n\}^n$.

PROOF. Let $S = \{a_1, \ldots, a_n\}$. First we show

$$[S] \supset S^n. \tag{1}$$

By definition

$$[S] \supset S. \tag{2}$$

Since $[S]$ is convex, it is closed under n-term join (Corollary 2.11). Thus (2) implies

$$[S] \supset S \ldots S \, (n \text{ terms}) = S^n,$$

and (1) holds.

Conversely, we show

$$S^n \supset [S]. \tag{3}$$

Let $x \subset [S] = [a_1, \ldots, a_n]$. By the polytope join formula (Theorem 3.2),

$$x \subset a_{i_1} \ldots a_{i_r} \tag{4}$$

where the a_i's are in S. By Lemma 3.2(ii),

$$a_{i_1} \ldots a_{i_r} = b_1 \ldots b_n \tag{5}$$

where the b's are in $\{a_{i_1}, \ldots, a_{i_r}\} \subset S$. Then by (4) and (5),

$$x \subset b_1 \ldots b_n$$

where $b_1 \subset S, \ldots, b_n \subset S$. By Theorem 2.8,

$$x \subset S \ldots S \, (n \text{ terms}) = S^n.$$

Thus (3) holds and the proof is complete. $\qquad\square$

Corollary 3.6.1. *Let set S be finite, nonempty and have cardinal number n. Then $[S] = S^n$.*

Corollary 3.6.2. $\{a_1, \ldots, a_n\}^n = (a_1 \cup \cdots \cup a_n) \cup (a_1 a_2 \cup \cdots \cup a_{n-1} a_n) \cup \cdots \cup (a_1 \ldots a_n).$

PROOF. Compare the theorem with Theorem 3.2 $\qquad\square$

EXERCISES

1. (a) Find a formula for $\{a_1, a_2, a_3\}^2$ in terms of joins of points of $\{a_1, a_2, a_3\}$.
   (b) Conjecture a formula for $\{a_1, \ldots, a_n\}^2$ in terms of joins of points of $\{a_1, \ldots, a_n\}$. Try to prove your conjecture.
   (c) Conjecture a formula for $\{a_1, \ldots, a_n\}^m$ in terms of joins of points of $\{a_1, \ldots, a_n\}$ when $m < n$. Try to prove your conjecture.

2. Prove: $S^n \subset S^{n+1}$, $n \geqslant 1$, for any set S, not necessarily finite.

3. Can $[S] = S^m$ for $m < n$, where $S = \{a_1, \ldots, a_n\}$? Explain.

4. Prove: If S is nonempty and has cardinal number n, then

$$S^n = S^{n+1} = S^{n+2} = \cdots .$$

5. Prove: If S is any set and $S^{n+1} = S^n$ for some n, then S^n is convex and $[S] = S^n$.

6. If m and n are positive integers, does $(S^m)^n = S^{mn}$?

7. If n is a positive integer, does $(AB)^n = A^n B^n$?

3.9 The Representation of Convex Hulls

Theorem 3.2, the Polytope Join Formula, may be stated in this form: The convex hull of a finite nonempty set S is the union of all joins of points of S. The result holds for any set S.

Theorem 3.7 (Convex Hull Representation Theorem). *$[S]$ is the union of all joins of points of S.*

PROOF. Let S^* be the union of all joins of points of S. We show $[S] = S^*$. Let $a_1, \ldots, a_n$ be any points of S. Then

$$[S] \supset S \supset a_1, \ldots, a_n.$$

Since $[S]$ is convex, $[S] \supset a_1 \ldots a_n$, that is, $[S]$ contains *any* join of points of S. Thus $[S]$ contains the union of all joins of points of S, that is,

$$[S] \supset S^*. \tag{1}$$

Conversely we prove

$$S^* \supset [S]. \tag{2}$$

First we show

$$S^* \supset S. \tag{3}$$

Let $a \subset S$. Then a is a join of points of S (Section 2.7, the next to last definition). Thus $a \subset S^*$ and (3) holds.

Next we show S^* convex. Let $x, y \subset S^*$. By definition of S^*

$$x \subset a_1 \ldots a_m, \qquad y \subset b_1 \ldots b_n,$$

where the a's and the b's are in S. Then

$$xy \subset a_1 \ldots a_m \cdot b_1 \ldots b_n \subset S^*.$$

Thus S^* is convex. By (3), S^* is a convex extension of S, and (2) follows. This with (1) completes the proof. $\square$

Corollary 3.7.1. *$x \subset [S]$ if and only if x is in a join of points of S.*

Corollary 3.7.2. *$[S]$ is the union U of all polytopes generated by finite subsets of S.*

PROOF. Let $x \subset [S]$. By the preceding corollary, $x \subset a_1 \ldots a_n$, where $a_1, \ldots, a_n \subset S$. Thus $x \subset [a_1, \ldots, a_n] \subset U$ and $[S] \subset U$. To show $U \subset [S]$, let $x \subset U$. Then $x \subset [a_1, \ldots, a_n]$, where $a_1, \ldots, a_n \subset S$. Thus $[S] \supset \{a_1, \ldots, a_n\}$, so that $[S] \supset [a_1, \ldots, a_n]$. Then $x \subset [S]$, and we have justified $U \subset [S]$ and so the corollary. $\qquad\square$

Corollary 3.7.3 (Finiteness of Generation). $x \subset [S]$ *if and only if there exist finitely many points* $a_1, \ldots, a_n$ *of S such that* $x \subset [a_1, \ldots, a_n]$.

This corollary asserts that a point x is in the convex hull of set S if and only if x is in the convex hull of a finite subset of S, and so is described as a property of *finiteness of generation*. This is not to say that $[S]$ is finitely generated but each point in it is finitely generated, so to speak.

Theorem 3.7 indicates the importance of finite joins of points as building blocks of a sort for the construction of a convex set from a set of generators.

3.10 Convex Hulls and Powers of a Set

In this section we prove that the convex hull of any set S is representable as the union of all powers of S. The result is a sort of generalization of Theorem 3.6.

Theorem 3.8. $[S] = S^1 \cup S^2 \cup \cdots \cup S^n \cup \cdots$.

PROOF. $[S] \supset S$. This implies, since $[S]$ is convex, that $[S] \supset S^i$, $i \geqslant 1$. Hence

$$[S] \supset S^1 \cup S^2 \cup \cdots \cup S^n \cup \cdots.$$

To prove the reverse inclusion let $x \subset [S]$. By Corollary 3.7.1

$$x \subset a_1 \ldots a_j, \qquad (1)$$

where

$$a_1 \subset S, \ldots, a_j \subset S.$$

This implies, by Theorem 2.7,

$$a_1 \ldots a_j \subset S \ldots S \ (j \text{ terms}) = S^j. \qquad (2)$$

By (1) and (2)

$$x \subset S^j \subset S^1 \cup S^2 \cup \cdots \cup S^n \cup \cdots,$$

and the proof is complete. $\qquad\square$

Remark on the Theorem. The infinite series of powers of S,

$$S^1 \cup S^2 \cup \cdots \cup S^n \cup \cdots$$

may seem a bit strange to you. For you may have observed that each term is contained in its successor:

$$S^1 \subset S^2 \subset \cdots \subset S^n \subset S^{n+1} \subset \cdots . \qquad (1)$$

This follows readily from $S^1 \subset S^2$ (Theorem 2.5). In view of (1) we can write the theorem in the form

$$[S] = S^{1630} \cup S^{1631} \cup \cdots ,$$

or

$$[S] = S^2 \cup S^4 \cup \cdots \cup S^{2n} \cup \cdots .$$

This may seem strange, but it doesn't invalidate the theorem. If you wish, the theorem can be put in the form

$$[S] = S^1 \cup (S^2 - S^1) \cup \cdots \cup (S^{n+1} - S^n) \cup \cdots ,$$

in which each term is disjoint to its successor.

EXERCISES

1. Prove that if S is finite, nonempty and has cardinal number c then

$$S^c = S^1 \cup \cdots \cup S^n \cup \cdots .$$

2. Prove Theorem 3.8 without using any of the results of Section 3.9.

3. Deduce Theorem 3.7 from Theorem 3.8. (See Exercise 2.)

4. Suppose a family F of sets has the property that if A and B are members of F, then AB is contained in some member of F.
   (a) Prove that the union of the sets of F is convex.
   (b) Find some Euclidean examples of such a family F and verify the result in (a).
   (c) Use the result in (a) to prove that $S^1 \cup \cdots \cup S^n \cup \cdots$ is convex for any set S.

3.11 Bounded Sets

The notion of a bounded set has its origin in the familiar observation that some geometric figures are finite or limited and others are endless or unlimited. In a Euclidean plane a figure is said to be *bounded* if it is contained in the interior of a circle (Figure 3.19). Similarly, in a Euclidean 3-space a figure is *bounded* if it is contained in the interior of a sphere. Circles and spheres, in these definitions, can be replaced by other figures, for example, triangles and tetrahedrons.

Boundedness is not an unimportant idea, but we do not have occasion to make much use of it. The following definition is adequate for our purposes.

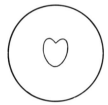 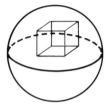

Figure 3.19

Definition. Let S be any set. If there is a polytope which contains S, we say S is *finitely bounded* or simply *bounded*.

Remarks on the Definition. In Euclidean geometry the definition is equivalent to the conventional one stated above. It has the advantage of being nonmetrical—it does not require the introduction of metrical ideas like distance (or congruence) which are used to define circles and spheres. The definition has the disadvantage of being too restrictive: it implies, under suitable assumptions, that a bounded set must be contained in a finite dimensional space.[2] Thus it rules out the possibility of a bounded set being intrinsically of infinite dimension. The definition is, however, suitable in Euclidean geometry and in general in finite dimensional geometries.

EXERCISES

1. Prove: If S and T are bounded, then $S \cap T$, $S \cup T$ and ST are bounded.

2. Prove: If S is bounded, then $[S]$ is bounded.

3. In Euclidean geometry, if a convex set S is bounded, must $\mathcal{C}(S)$ be bounded? Give some evidence to justify your answer.

4. (a) Find the bounded convex subsets of a Euclidean line.
   (b) Show that a Euclidean line is not bounded.

3.12 Project—the Closed Join Operation

Postulates J1–J4 were suggested by properties EJ1–EJ4 of the Euclidean join operation (Section 2.3), which was defined in terms of the notion of open segment (Section 2.1). However, you may be curious about how a "closed join" operation in Euclidean geometry would behave and what an associated abstract theory would be like. (The question for Euclidean geometry was discussed informally in Section 1.23.) Our theory can shed

---
[2] See Section 6.9, Remark.

some light on the issue, since the notion of *closed* segment has been formally defined (Section 3.4). We can easily define a corresponding *closed* join operation and proceed to study its properties.

Definition. Let a and b be points (elements of J). Then the *closed join* of a and b, denoted by $a * b$, is $ab \cup a \cup b = [a, b]$.

Observe that $*$ is a join operation in J (Section 2.2). One naturally asks: how does the closed join operation $*$ compare with $\cdot$ the (open) join operation? Some answers are indicated below.

EXERCISES

1. Prove that $(J, *)$ satisfies Postulates J1–J4.

2. Interpret the associative law for $*$ in a Euclidean geometry. Do the same for $\cdot$. What do you observe?

3. Prove: $A * B = A \cup B \cup AB$, provided $A, B \neq \emptyset$.

4. Prove $[a_1, \ldots, a_n] = a_1 * \cdots * a_n$.

3.13 Convex Hulls of Finite Families of Sets —Generalized Polytopes

The remainder of the chapter is devoted to a generalization of the idea of polytope and its properties as developed in Sections 3.4, 3.6 and 3.7.

There are, in Euclidean geometry, figures which are not polytopes but are polytope-like: They are generated, not by a finite collection of points, but by a finite collection of convex sets. To construct an example, consider the solid triangular prism $[a, b, c, a', b', c']$ in Figure 3.20, where the

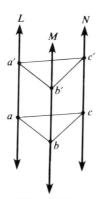

Figure 3.20

triangles *abc* and *a′b′c′* are congruent and the lines *L, M, N* which
join corresponding vertices of the triangles are parallel. Imagine the
prism stretched to form an endless solid triangular tube whose edges
are *L, M* and *N*. This endless prismatic solid is "generated by" the lines
L, M and *N*.

The first problem, then—and it is not very difficult—is to characterize
formally this new way of generating convex sets. Since it is no more
difficult to treat, we take up the case of a finite family $\{S_1, \ldots, S_n\}$ of sets
that are not necessarily convex. The treatment is an analogue of that for
the convex hull of a single set (Section 3.2). First a definition.

Definition. Let a family of sets $\{S_1, \ldots, S_n\}$ be given. Suppose M is
convex and $M \supset S_1, \ldots, S_n$. Then M is called a *convex extension* of
$\{S_1, \ldots, S_n\}$ or simply a *convex extension* of $S_1, \ldots, S_n$.

Thus M is a convex extension of $S_1, \ldots, S_n$ if and only if M is a convex
extension of each of these sets.

Now we can state the analogue of Theorem 3.1.

Theorem 3.9. *Let sets $S_1, \ldots, S_n$ be given. Then the family F of convex
extensions of $S_1, \ldots, S_n$ has one and only one least member*

(Compare Theorem 3.1.)

PROOF. Follow the proof of Theorem 3.1, merely taking into account that
F is the family of convex extensions of the sets $S_1, \ldots, S_n$ rather than that
of the set S.   □

Restatement of Theorem 3.9. For each finite family of sets $\{S_1, \ldots, S_n\}$
there exists a unique least convex extension.

Now we may state the

Definition. Let $\{S_1, \ldots, S_n\}$ be a finite family of sets. The least convex
extension of $\{S_1, \ldots, S_n\}$ is called the *convex hull* of $\{S_1, \ldots, S_n\}$ or of
$S_1, \ldots, S_n$ and is denoted by $[\{S_1, \ldots, S_n\}]$ or $[S_1, \ldots, S_n]$. The convex
hull of $S_1, \ldots, S_n$ is also called the convex set *generated* (or *spanned* or
determined) by $S_1, \ldots, S_n$. If A is a given convex set and sets $S_1, \ldots, S_n$
satisfy

$$[S_1, \ldots, S_n] = A,$$

we say $S_1, \ldots, S_n$ generate A (*convexly*), or A is *generated* (*convexly*) by
$S_1, \ldots, S_n$.

Definition. A convex set is called a *generalized polytope* if it is generated
(convexly) by a finite nonempty family of convex sets.

Thus A is a generalized polytope if and only if A is representable in the form $[B_1, \ldots, B_n]$ where the B's are convex.

Observe that

$$[S_1, \ldots, S_n] = [S_1 \cup \cdots \cup S_n],$$

since a set is a convex extension of $S_1, \ldots, S_n$ if and only if it is a convex extension of $S_1 \cup \cdots \cup S_n$.

Finally we assert that

$$[S_1, \ldots, S_m] \supset T_1, \ldots, T_n \tag{1}$$

implies

$$[S_1, \ldots, S_m] \supset [T_1, \ldots, T_n].$$

To justify this, observe that (1) implies $[S_1, \ldots, S_m]$ is a convex extension of $T_1, \ldots, T_n$ and so must contain the *least* convex extension $[T_1, \ldots, T_n]$.

3.14 Properties of Generalized Polytopes

The development now parallels closely that for polytopes in Sections 3.4, 3.6 and 3.7.

In order to derive an analogue of the Polytope Join Formula (Theorem 3.2) we need to extend the terminology of joins of sets (Section 2.7, the definition next to the last) as we did for joins of points (Section 3.4).

Definition. Suppose $S_1, \ldots, S_r$ are members of F, a family of sets. Then $S_1 \ldots S_r$ is called a *join of sets of F*.

Now we can state an analogue of Lemma 3.2.

Lemma 3.10 (Representation Principle for Joins of Sets). *Let $A_1, \ldots, A_n$ be convex and $A_{i_1} \ldots A_{i_r}$ a join of sets of $\{A_1, \ldots, A_n\}$. Then $A_{i_1} \ldots A_{i_r}$ can be expressed in the following forms*:

(i) $A_{j_1} \ldots A_{j_s}$, *where $1 \leqslant j_1 < \cdots < j_s \leqslant n$;*
(ii) $B_1 \ldots B_n$, *where the B's are in $\{A_{i_1}, \ldots, A_{i_r}\}$.*

(Compare Lemma 3.2.)

PROOF. Apply the method of proof of Lemma 3.2. Use Theorem 2.9(b) in place of J4. □

Now a generalization of the Polytope Join Formula (Theorem 3.2) can be proved.

Theorem 3.10.

$$[S_1, \ldots, S_n] = ([S_1] \cup \cdots \cup [S_n]) \cup ([S_1][S_2] \cup \cdots$$

$$\cup [S_{n-1}][S_n]) \cup \cdots \cup ([S_1] \ldots [S_n]).$$

(Compare Theorem 3.2.)

PROOF. Observe that

$$[S_1, \ldots, S_n] = [[S_1], \ldots, [S_n]], \tag{1}$$

for $[S_1, \ldots, S_n] \supset [S_i]$ and $[[S_1], \ldots, [S_n]] \supset S_i$, where $1 \leqslant i \leqslant n$. Now apply the argument of Theorem 3.2 to the right member of (1), replacing $a_1, \ldots, a_n$ by $[S_1], \ldots, [S_n]$. $\square$

Next the *analogue* of the Polytope Join Formula for a generalized polytope:

Corollary 3.10 (Generalized Polytope Join Formula). *Let* $A_1, \ldots, A_n$ *be convex. Then*

$$[A_1, \ldots, A_n]$$

$$= (A_1 \cup \cdots \cup A_n) \cup (A_1 A_2 \cup \cdots \cup A_{n-1} A_n) \cup \cdots \cup (A_1 \ldots A_n).$$

(Compare Theorem 3.2.)

The analogue, for a generalized polytope, of the absorption property of a polytope (Theorem 3.4) now follows.

Theorem 3.11 (Absorption). *Let* $A_1, \ldots, A_n$ *be convex. Then*

$$A_1 \ldots A_n[A_1, \ldots, A_n] = A_1 \ldots A_n.$$

(Compare Theorem 3.4.)

PROOF. Apply the argument of Theorem 3.4. $\square$

Corollary 3.11. *Let* $A_1, \ldots, A_n$ *be convex. Then* $A_1 \ldots A_n$ *absorbs properly every join of sets of* $\{A_1, \ldots, A_n\}$.

(Compare Corollary 3.4.)

Finally we have an analogue of Theorem 3.5 on the interior of a polytope.

Theorem 3.12. *Let* $A_1, \ldots, A_n$ *be nonempty convex sets. Then*

$$\mathcal{I}[A_1, \ldots, A_n] = \mathcal{I}(A_1 \ldots A_n).$$

(Compare Theorem 3.5.)

PROOF. Apply the method of theorem 3.5. $\square$

Corollary 3.12.1. *Let $A_1, \ldots, A_n$ be nonempty convex sets. Then*

$$\mathcal{G}[A_1, \ldots, A_n] \subset A_1 \ldots A_n.$$

(Compare Corollary 3.5.1.)

Corollary 3.12.2. *Let $A_1, \ldots, A_n$ be nonempty convex sets. Then*

$$\mathcal{GF}[A_1, \ldots, A_n] \supset [A_1, \ldots, A_n] - A_1 \ldots A_n.$$

(Compare Corollary 3.5.2.)

Remarks on the Notion of Generalized Polytope. First we point out that the convex hull $[S_1, \ldots, S_n]$, where the S's are arbitrary sets, is a generalized polytope. For by (1) in the proof of Theorem 3.10 above,

$$[S_1, \ldots, S_n] = [[S_1], \ldots, [S_n]],$$

which by definition is a generalized polytope. Thus the convexity condition in the definition of generalized polytope is redundant. The condition was imposed to underscore the analogy between polytopes as convex hulls of finitely many points $a_1, \ldots, a_n$ and generalized polytopes as convex hulls of finitely many convex sets $A_1, \ldots, A_n$.

Next observe that the family of generalized polytopes is identical to the family of convex sets. For any convex set A can be written $[A]$. The concept will be useful and produce convex sets with polytope-like structure when the set of generators is suitably chosen—in some sense they should be in general position. This is indicated in the following examples.

3.15 Examples of Generalized Polytopes

A few examples of generalized polytopes in Euclidean 3-space are given for their own sake and for the purpose of illustrating results proved in the last section.

EXAMPLE I. Let $L_1, \ldots, L_n$ ($n \geqslant 3$) be parallel lines no three of which are coplanar (Figure 3.21). Then $[L_1, \ldots, L_n]$ is an endless convex prismatic solid.

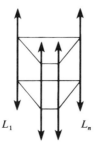

Figure 3.21

The remaining examples involve the notion of ray in Euclidean geometry.

Euclidean Ray. Let L be a line in a Euclidean geometry and a a point of L. Then the set S of points of L that lie on a given side of point a is an *open ray* or simply a *ray*, and a is its endpoint.

EXAMPLE II. Let point o and plane P be given, $o \not\subset P$ (Figure 3.22). Let $R_1, \ldots, R_n$ $(n \geqslant 3)$ be rays with endpoint o, which intersect P, no three of which are coplanar. Then $[R_1, \ldots, R_n]$ is an (endless) convex pyramidal solid.

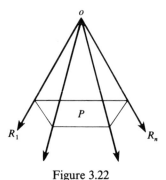

Figure 3.22

EXAMPLE III. Let $R_1, \ldots, R_n$ $(n \geqslant 3)$ be rays which satisfy these conditions: (1) the lines they determine are parallel; (2) their endpoints lie on a plane P; (3) they lie on a given side of P; (4) no three of them are coplanar (Figure 3.23). Then $[R_1, \ldots, R_n]$ may be called a semi-endless convex prismatic solid.

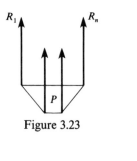

Figure 3.23

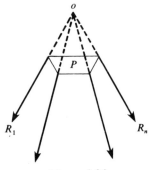

Figure 3.24

EXAMPLE IV. Let $R_1, \ldots, R_n$ ($n \geqslant 3$) be rays which satisfy these conditions: (1) their endpoints lie on a plane P; (2) they can be extended to meet in a point o; (3) no three of them are coplanar (Figure 3.24). Then $[R_1, \ldots, R_n]$ may be called a truncated convex pyramidal solid.

EXERCISES

1. Prove: $A \subset [B]$ implies $[A, B] = [B]$. This is a sort of absorption principle for the convex hull operation.

2. Prove: $[[A, B], C] = [A, [B, C]] = [A, B, C]$.

3. In Euclidean 3-space, identify each of the indicated generalized polytopes. Find its interior, closure, frontier, boundary. Verify the generalized polytope join formula. Use diagrams at will.
   (a) $[a, L]$, where L is a line and $a \not\subset L$. [Compare Exercise 1(b) on Convex Hulls of Euclidean Sets at the end of Section 3.2.]
   (b) $[a, R]$, where R is an open ray and a is not on the line that contains R.
   (c) $[L, M]$, where L and M are distinct lines. Consider all cases. [Compare Exercises 1(c), (d) and 2(e) on Convex Hulls of Euclidean Sets at the end of Section 3.2.]
   (d) $[R, S]$, where R and S are rays that do not meet. Consider all cases.
   (e) $[L, M, N]$, where the lines L, M, N are noncoplanar and parallel in pairs. (See Example I above.)
   (f) $[R, S, T]$, where the rays R, S, T are noncoplanar and have the same endpoint. (See Example II above.)
   (g) $[ab, ac, bc]$, where a, b, c are noncollinear. Is $[ab, ac, bc]$ a polytope?
   (h) The same as (g) for $[[a, b], [a, c], [b, c]]$.
   (i) $[ab, cd]$, where a, b, c, d are noncoplanar. [Compare Exercise 2(f) on Convex Hulls of Euclidean Sets at the end of Section 3.2.]

4. (a) Prove: $[A][B] = [AB]$.
   (b) Derive a new join formula from Theorem 3.10.

5. Write out proofs of (a) Theorem 3.10, (b) Theorem 3.11.

6. Write out a proof of Theorem 3.12. Where in the proof is the assumption that $A_1, \ldots, A_n$ are nonempty needed?

7. Prove: If $A_1, \ldots, A_n$ are nonempty convex sets then

$$\mathcal{C}[A_1, \ldots, A_n] = \mathcal{C}(A_1 \ldots A_n).$$

(Compare Exercise 3 at the end of Section 3.7.)

4 The Operation of Extension

In this chapter the theory of join operations is expanded by defining an operation of extension, which is a sort of inverse to join. Three postulates involving extension are introduced to complete the basic postulate set. These are employed to derive principles and formulas involving extension and join which supplement and enrich the formal theory of join in Chapters 2 and 3. The theory is applied to new ideas: extreme points of convex sets, linear order of points and two categories of convex set referred to as *open* and *closed*. New results are obtained on familiar ideas: Theorem 4.28—any join of points is an open convex set; Theorem 4.30— the interior of a polytope P is the join of the points of any finite set of generators of P; and Theorem 4.31—$\mathcal{I}(AB) = \mathcal{I}(A)\mathcal{I}(B)$, provided $\mathcal{I}(A), \mathcal{I}(B) \neq \varnothing$.

4.1 Definition of the Extension Operation

An operation of extension is defined now in terms of the concept of join.

Definition. Let a and b be any points (Figure 4.1). Then the *extension* of a *from* b (or the *quotient* of a and b), denoted by a/b, is the set of all points x which satisfy $bx \supset a$. (The expression a/b may be read "a stroke b" or "a over b".)

Remarks on the Definition. First note that a and b are arbitrary elements of the basic set J (the set of points) and are not necessarily distinct.

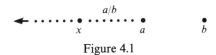

Figure 4.1

Next observe that the definition is equivalent to the statement

$$x \subset a/b \quad \text{if and only if} \quad bx \supset a. \tag{1}$$

This calls to mind the definition of the quotient a/b of two numbers a and b in algebra, which is equivalent to

$$x = a/b \quad \text{if and only if} \quad bx = a. \tag{2}$$

Statement (2) indicates a reciprocal relationship between the operations of multiplication and division which is described by saying that multiplication and division are *inverse* operations. Statement (1) asserts similarly that join and extension are inverse operations in a sense appropriate to our theory.

   This seems to be a good place to make a point about the abstract nature of definition. Practically all the preceding definitions in the theory were suggested by some familiar Euclidean notion which could be pictured easily. This may have fostered the impression that definitions in our theory have *intrinsic* meaning.

   The definition does not say specifically what the extension of a from b is. Rather it expresses the notion of extension in terms of the basic notions of point (element of J) and join. But point and join themselves are abstract —they have no *intrinsic* meaning.

   To put it differently: Do not read into the definition more than it asserts. It says in effect no more and no less than statement (1) says. Is extension then a "meaningless" notion? If so, how can we reason about it? The answer is not hard to give. To reason in mathematics we must know the *properties* of our terms—not their content or meaning. Any reference to extension is by its definition a reference to point and join. And we know scores of properties of point and join: for they satisfy J1–J4 and so all theorems of Chapters 2 and 3 that have been deduced from J1–J4.

   Can the term extension be assigned specific meaning? Of course it can. We have assigned meaning to the basic terms "point" and "join" in several different ways in setting up the Euclidean, triode and Cartesian join models (Sections 2.4, 2.22, 2.24). To obtain specific meaning for extension, just assign specific meaning to the terms "point" and "join" in the definition of extension.

Euclidean Interpretation of Extension

We proceed to interpret extension in a Euclidean model $(J, \cdot)$. To be specific, let J be a Euclidean 3-space. Let a and b be any points of J. The extension a/b will be the set of all points x of J such that the Euclidean join $bx \supset a$.

First suppose $a \neq b$. Then $x \neq b$, and "$bx \supset a$" is equivalent to "a is between x and b". Thus a/b will consist of all points x of J such that a is between x and b (Figure 4.2). Equivalently a/b is the set of all points of line ab that lie on the side of a *opposite* to b. This is an example of a figure called a ray.

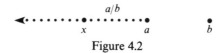

Figure 4.2

Euclidean Ray. In Euclidean geometry let L be a line and a a point of L (Figure 4.3). Let R be the set of all points of L that lie on a given side of point a. Then R is an *open ray* (*open halfline*) or simply a *ray*, and a is its *endpoint*. Observe that a is not a point of R. (The description of a closed ray in Euclidean geometry appears in Section 2.22.)

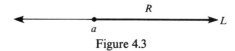

Figure 4.3

Thus a/b is a ray with endpoint a. It may be described as the ray with endpoint a that is opposite to point b (Figure 4.4).

Figure 4.4

Now suppose $a = b$, which is permitted by the definition of extension. By definition a/a is the set of all points x that satisfy

$$ax \supset a. \tag{1}$$

By J4, $x = a$ satisfies (1). Suppose $x \neq a$. Then the join ax is a Euclidean (open) segment and (1) is false. Thus the only solution of (1) is $x = a$, and we have $a/a = a$.

Summary. In a Euclidean model, if $a \neq b$, then a/b is the ray with endpoint a that is opposite to b; $a/a = a$.

Rays—in contrast with segments—form a class of unbounded linear figures. They play an essential role in Euclidean geometry in the study of angles and directions. Rays also are involved in the structure of lines and planes.

Here are two applications of the extension operation to yield formulas for lines (Figures 4.5, 4.6):

$$\text{Line } ab = (ab) \cup (a/b) \cup (b/a) \cup a \cup b \quad (a \neq b).$$

$$\text{Line } ab = (o/a) \cup (o/b) \cup o \quad (o \text{ is between } a \text{ and } b).$$

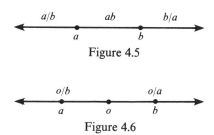

Figure 4.5

Figure 4.6

These formulas indicate the importance of the ray concept in elementary Euclidean geometry.

A few words are in order on the choice of the term *extension* in referring to a/b. As we saw above, in Euclidean geometry if $a \neq b$, then a/b is the ray with endpoint a that is opposite to point b or is directed away from b (Figure 4.7). If x is a point of a/b, segment ab can be extended beyond its endpoint a to x. Conversely, every such "extension point" of ab must lie on a/b. Hence a/b can be conceived as arising by endlessly extending ab beyond a. Thus a/b might be called the extension of ab beyond a. But there is no need to refer to segment ab—for a/b is determined by the points a and b. Thus we were led to call a/b the extension of a from b.

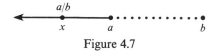

Figure 4.7

In conclusion we introduce a graphically suggestive name for the extension a/b in the abstract theory.

Definition. The extension a/b is called a *ray* or *halfline*; a is its *endpoint*. If $a \neq b$, the ray is said to be *proper* or *nondegenerate*. The ray a/a is said to be *improper* or *degenerate*.

Thus to say R is a ray with endpoint a means that R can be expressed in the form a/b for some point b.

Compare the definition with that of segment in Section 2.7.

4.2 The Extension Operation for Sets

The extension operation for points induces a corresponding extension
operation for sets just as join for points induced join for sets (Section 2.3).

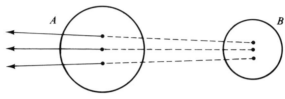

Figure 4.8

Definition. Let A and B be any sets of points (Figure 4.8). Then A/B, the
extension of *set A from set B* or the *quotient* of *sets A and B*, is defined by

$$A/B = \bigcup_{a \subset A,\, b \subset B} (a/b).$$

This means, in simple terms: $x \subset A/B$ if and only if there exist a, b
such that

$$x \subset a/b, \qquad a \subset A \quad \text{and} \quad b \subset B.$$

Euclidean Examples of Extension of Sets

EXAMPLE I (The extension of a point from a segment). Let a, b and c be
noncollinear points in a Euclidean model (Figure 4.9). Then $a/(bc)$ is the
portion of plane abc which is covered by the rays that emanate from a and
are directed away from the points of bc. Thus $a/(bc)$ is the interior of the
angle whose sides are the rays a/b and a/c, that is, the angle vertical to
$\angle bac$.

EXAMPLE II (The extension of a segment from a point). Choose a, b and c
as in Figure 4.10. Then $(ab)/c$ is the union of all rays x/c for all points x
of ab. Each of these rays is contained in the interior of $\angle acb$. Thus $(ab)/c$
is a portion of the interior of $\angle acb$: namely, the endless portion of it
which is cut off from it by the segment ab.

In Euclidean geometry the operation of extension of sets is very useful
in the representation of various unbounded figures such as cones, angle

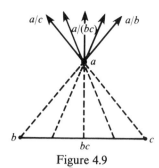

Figure 4.9

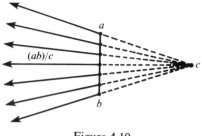

Figure 4.10

interiors, halfplanes, and so on, and in the framing of formulas for planes and 3-spaces (see both sets of exercises below). It will be found no less useful in the abstract theory.

EXERCISES (FIRST SET)[1]

In Exercises 1 through 17 sketch diagrams for and describe the figures defined by the given expressions in Euclidean geometry. Assume the points involved are in general position: that is, given a and b, assume them distinct; a, b and c noncollinear; a, b, c and d noncoplanar.

1. (a) $a/\{b, c\}$; (b) $\{b, c\}/a$; (c) $a/\{b, c, d\}$; (d) $\{b, c, d\}/a$; (e) $a/\{a, b\}$; (f) $a/\{a, b, c\}$.

2. $a/[b, c]$. Compare Exercise 1(a), Example I above.

3. $[b, c]/a$. Compare Exercise 1(b), Example II above.

4. $a/(bcd)$ and $a/[b, c, d]$. Compare Exercises 1(c), 2.

5. $(bcd)/a$ and $[b, c, d]/a$. Compare Exercises 1(d), 3.

6. $a/(ab)$ and $(ab)/a$.

7. $a/(a/b)$ and $(a/b)/a$. Compare Exercise 6.

8. $a/(abc)$ and $(abc)/a$.

9. a/C and C/a if a is a point of circle C.

10. The same as Exercise 9 if a is interior to C; exterior to C.

11. The same as Exercise 9 if a is not in the plane of C.

12. (a) L/a and a/L if L is a line and a is not on L.
    (b) In (a) suppose a is on L.
    (c) Compare L/a and a/L with La in (a) and in (b).

---

[1] Many of these exercises are the same as or similar to items appearing in the two exercise sets at the end of Section 1.22.

13. In Exercise 12 replace line L by plane P.

14. L/L and P/P if L is a line and P a plane.

15. L/M if L and M are lines that
(a) intersect in a single point;
(b) are parallel;
*(c) are skew, that is, are not coplanar.

16. P/L if P is a plane and L a line such that
(a) $P \supset L$;
(b) P and L intersect in a single point;
(c) P and L are parallel.

17. The same as Exercise 16 for L/P.

18. Assign a Cartesian coordinate system to a Euclidean plane. Let p be the point $(0, 0)$ and C the graph of the given equation. Sketch p/C and C/p and describe them as accurately as you can.
(a) $y = x^2$;
(b) $y = 1/x$;
(c) $y = e^x$.

EXERCISES (SECOND SET)

1. Suppose a, b and c are noncollinear points in a Euclidean plane.
(a) Sketch and describe the sets: $a/(bc)$, $(ab)/c$, $a(b/c)$, $(a/b)/c$, $a/(b/c)$, $(a/b)c$.
(b) Are any two of the sets equal? Does any one contain another?

2. Suppose R and S are rays in Euclidean geometry with the same endpoint o. Sketch and describe (a) oR; (b) o/R; (c) R/o; (d) RS; (e) R/S;
(f) aR, a/R, R/a if a is not the endpoint of R. Try not to miss any cases.

3. In Euclidean geometry identify: $(ab)/(ab)$; $(abc)/(abc)$; $(abcd)/(abcd)$. Assume in the respective problems $a \neq b$; a, b and c noncollinear; a, b, c and d noncoplanar.

4. (a) In the triode model (Section 2.22) find a/b, $a \neq b$. How many different types of figure do you get?
(b) The same for $(ab)/(ab)$, $a \neq b$.

5. The same as Exercise 4 in the Cartesian model (Section 2.24).

6. In Euclidean geometry identify
(a) $p(p/a)$, $a \neq p$;
(b) $L(L/a)$ where L is a line and $a \not\subset L$;
(c) $P(P/a)$ where P is a plane and $a \not\subset P$.

7. Obtain a formula for the Euclidean plane abc in terms of a, b, c and expressions formed by applying the operations of join and extension to a, b and c taken 2 or 3 at a time. Observe as indicated in Figure 4.11 that the three lines separate the plane into 7 regions. [Compare the first formula for line ab in Section 4.1, Euclidean Interpretation of Extension.] This is the same as Exercise 12 in the Miscellaneous exercises at the end of Section 1.22.

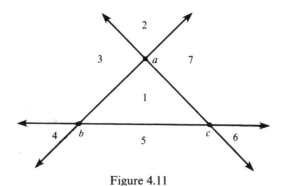

Figure 4.11

*8. Generalize Exercise 7 to four noncoplanar points a, b, c, d and obtain a formula for a Euclidean 3-space. (There are 71 terms in the formula.) This is the same as Exercise 13 in the Miscellaneous exercises at the end of Section 1.22.

4.3 The Monotonic Law for Extension

There is a monotonic law for extension analogous to the monotonic law for join (Theorem 2.1).

Theorem 4.1 (Monotonic Law for Extension). $A \subset B$ *implies* $A/C \subset B/C$ *and* $C/A \subset C/B$, *for any set C.*

(Compare Theorem 2.1.)

PROOF. The method is familiar. Assume $x \subset A/C$ and prove $x \subset B/C$. $x \subset A/C$ implies by definition $x \subset a/c$, where $a \subset A$, $c \subset C$. Then $A \subset B$ implies $a \subset B$. Thus $x \subset a/c$ implies $x \subset B/C$, and we conclude $A/C \subset B/C$. The second part of the conclusion is proved similarly. □

Corollary 4.1.1. $A \subset B$, $C \subset D$ *imply* $A/C \subset B/D$.

(Compare Corollary 2.1.)

PROOF. The method used in the proof of Corollary 2.1 applies. □

Corollary 4.1.2. $A/A \supset A$.

(Compare Theorem 2.5.)

PROOF. Let $a \subset A$. By J4, $a \subset aa$. By definition of extension and the preceding corollary,

$$a \subset a/a \subset A/A$$

and the result holds. □

4.4 Distributive Laws for Extension

The distributive law for join (with respect to union), Theorem 3.3, has two
easily proved analogues for extension. To state them more simply, a
notational convention is adopted.

Notation. In expressions involving the operations of join, extension
and union, portions separated by $\cup$ signs are to be considered enclosed in parentheses. For example, $(A/B \cup C)D \cup EF$ stands for
$(((A/B) \cup C)D) \cup (EF)$. This extends the corresponding convention for
the operations of join and union (Section 3.4, Notation).

Theorem 4.2 (Distributive Laws for Extension).

(a) $A/(B \cup C) = A/B \cup A/C$.

(b) $(B \cup C)/A = B/A \cup C/A$.

(Compare Theorem 3.3.)

PROOF. The method of proof of Theorem 3.3 applies.   □

Corollary 4.2.
$$(A_1 \cup \cdots \cup A_m)/(B_1 \cup \cdots \cup B_n)$$
$$= A_1/B_1 \cup \cdots \cup A_m/B_1 \cup \cdots \cup A_1/B_n \cup \cdots \cup A_m/B_n.$$

(Compare Corollaries 3.3.1, 3.3.2.)

4.5 The Relation "Intersects" or "Meets"

One of the commonest and simplest relations in geometry (indeed generally in mathematics) occurs if one set *intersects* or *meets* another (Section
2.3, Set Notation). The relation occurs so frequently in the sequel that a
compact and convenient notation for it should be introduced. The stan-
dard mathematical expression for "A intersects B" is $A \cap B \neq \varnothing$: their set
intersection is not empty. This is quite awkward for our purposes, since it
puts emphasis on the noun *intersection*, rather than the verb *intersects*.
Thus we introduce the following

Notation. $A \approx B$ means A intersects or meets B, that is, A and B have a
common element.

The notation is particularly useful because intersection relations imply
other intersection relations, somewhat as one equation in algebra implies
another.

Theorem 4.3.

(a) $A \approx A$ *if and only if* $A \neq \emptyset$.
(b) $A \approx B$ *implies* $B \approx A$.
(c) $A \approx B$ *and* $B \subset C$ *imply* $A \approx C$.
(d) $A \approx B$ *and* $C \neq \emptyset$ *imply* $AC \approx BC$.
(e) $A \approx B$ *and* $C \approx D$ *imply* $AC \approx BD$.

PROOF. (a), (b) and (c) are immediate consequences of the definition of the relation $\approx$.

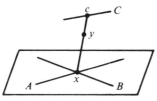

Figure 4.12

Consider (d). Assume $A \approx B$ and $C \neq \emptyset$ (Figure 4.12). Then there exist x and c such that $x \subset A$, B and $c \subset C$. Thus $xc \subset AC$, BC. By J1, $xc \neq \emptyset$. Let $y \subset xc$. Then $y \subset AC$, BC. By definition $AC \approx BC$.
Conclusion (e) is proved similarly. □

Remark on the Theorem. Conclusion (d) reminds one of the familiar algebraic principle for multiplying an equality by a term, while (e) is suggestive of the rule: "If equals are multiplied by equals, the results are equal".

Remark on the Intersection Relation. The relation $\approx$ may be considered a weak generalization of equality, since if A and B are one-element sets and $A \approx B$, then $A = B$. However, the relation does not have all the properties of equality. It does have the reflexive property that $A \approx A$ (*provided* $A \neq \emptyset$) and the symmetric property that $A \approx B$ implies $B \approx A$. But the transitive property, $A \approx B$ and $B \approx C$ imply $A \approx C$, does not hold. Can you find an example to show this?

The intersection notation has an additional and unforeseen advantage: It covers the idea of member of a set. For $a \subset A$ is equivalent to $\{a\} \approx A$, which we write simply $a \approx A$. This expression and its reverse $A \approx a$ are often more convenient than $a \subset A$ or $A \supset a$.

4.6 The Three Term Transposition Law

The relation between join and its inverse, extension, can be expressed, using the intersection relation, as a simple and useful transposition law.

Theorem 4.4 (The Three Term Transposition Law). $A \approx BC$ *if and only if* $A/B \approx C$.

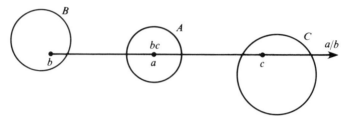

Figure 4.13

PROOF. Suppose $A \approx BC$. This means there exists a such that $a \subset A$, $a \subset BC$ (Figure 4.13). Observe that the latter implies $a \subset bc$ where $b \subset B$, $c \subset C$. Thus $bc \supset a$ and by definition of a/b we have $c \subset a/b$. This with $a \subset A$, $b \subset B$ implies, by definition of A/B, that $c \subset A/B$. Since we already know $c \subset C$, we conclude $A/B \approx C$.

For the converse we reason similarly. Suppose $A/B \approx C$. Then $c \subset A/B$, $c \subset C$ for some c. Thus $c \subset a/b$, where $a \subset A$, $b \subset B$. By definition of a/b we have $bc \supset a$. Thus $BC \supset a$, and since $A \supset a$, we conclude $A \approx BC$.   □

Permitting A, B, C to reduce to single elements a, b, c, we have

Corollary 4.4. $a \approx bc$ if and only if $a/b \approx c$.

Remark. The definition of extension implies: $a \subset bc$ if and only if $a/b \supset c$. Comparing this with the corollary, we observe the algorithmic advantage of using the intersection relation $\approx$ to express containment of elements.

EXERCISES

1. (a) Prove: If $p/a \approx bc$, then $p/b \approx ac$.
   (b) Interpret the result in Euclidean geometry.

2. Can you prove $ab \approx ab$; $a/b \approx a/b$?

3. Prove: If $c \subset a/b$, then ac, $ab \subset bc$.

4. Prove: (a) $(ab)/a \supset b$; (b) $(ab)/a \supset ab$; (c) $(ab)/a \supset b/a$.
   Interpret the results in Euclidean geometry. Do you have any conjectures about $(ab)/a$? Test them in the Cartesian model (Section 2.24).

5. Prove: If $c \subset a/b$, then $c/b \subset a/b$.

6. Prove:
   (a) $a(b/a) \supset b$, assuming $b/a \neq \varnothing$.
   (b) $a(b/a) \supset ab$, assuming $b/a \neq \varnothing$.
   (c) $a(b/a) \supset b/a$, assuming that the extension of any two points is not empty.
   Interpret the results in Euclidean geometry. Do you have any conjectures about $a(b/a)$? Test them in the Cartesian model (Section 2.24).

7. Find a model that satisfies J1–J4 in which $a/b = \varnothing$ for some pair of points a, b. (*Hint.* The segment ab would not be extendible beyond point a.) Compare your answer to Exercise 2. Do the results of Exercise 4 hold in the model? Do the containment relations of Exercise 6 hold in the model?

8. Prove:
   (a) $A/(B \cap C) \subset (A/B) \cap (A/C)$.
   (b) $(B \cap C)/A \subset (B/A) \cap (C/A)$.
   (c) Show that equality need not hold in (a) and (b).

9. (a) Prove in the formal theory that $a/a \supset a$.
   (b) Show that $a/a = a$ does not hold in the formal theory, in the following way: Let $J = \{0, 1\}$. Define $\cdot$ so: $0 \cdot 0 = 0 \cdot 1 = 1 \cdot 0 = 0$; $1 \cdot 1 = 1$. Show that $(J, \cdot)$ is a model of J1–J4. Show that $a/a = a$ does not hold in this model. (Compare Exercise 9 at the end of Section 2.17.)

4.7 The Mixed Associative Law

One of the most important properties of extension is an associative law for the operations of join and extension combined, which—rather remarkably—requires no new postulates and is deducible from J1–J4. First a word on notation.

Notation. In expressions involving $\cdot$ and $/$, we eliminate excess parentheses by the agreement that portions separated by $/$ signs are to be considered enclosed in parentheses. For example, a/bc stands for $a/(bc)$ and $(A/B)C/DE$ for $((A/B)C)/(DE)$.

Theorem 4.5 (The Mixed Associative Law). $(a/b)/c = a/bc$.

PROOF. We show every point of each member is in the other. Suppose $x \subset (a/b)/c$, that is, $x \approx (a/b)/c$. Then

$$xc \approx a/b \qquad \text{(Theorem 4.4),}$$

$$(xc)b \approx a \qquad \text{(Theorem 4.4),}$$

$$x(bc) \approx a \qquad \text{(J3, J2),}$$

$$x \approx a/bc \qquad \text{(Theorem 4.4).}$$

Conversely we suppose $x \approx a/bc$ and retrace the steps to complete the proof. ☐

Corollary 4.5 (The Mixed Associative Law for Sets). $(A/B)/C = A/BC$.

PROOF. You can apply the method of the theorem. Or else you can reduce the result to the theorem by proving that $x \subset (A/B)/C$ if and only if there exist $a \subset A$, $b \subset B$, $c \subset C$ such that $x \subset (a/b)/c$, and a similar condition for $x \subset A/BC$. [Compare the proof of $(AB)C = A(BC)$, Theorem 2.4.] ☐

EXERCISES

1. Verify Theorem 4.5 in a Euclidean model if a, b and c are noncollinear. Verify it in one or two other cases.

2. (a) Prove: $(a/b)/c = (a/c)/b$.
   (b) Verify in a Euclidean model the equation in (a).

3. (a) Prove: $(a/b)/b = a/b$.
   (b) Verify the equation in (a) in Euclidean and Cartesian models.

4. (a) Generalize Theorem 4.5 to obtain a formula for $((a/b)/c)/d$ and justify your result.
   (b) Interpret the result in a Euclidean model, taking a, b, c, d to be the vertices of a tetrahedron.

4.8 Three New Postulates

Postulates J1–J4 have yielded a substantial body of theorems—much more than one might have expected at first glance. Still, there are many principles of Euclidean geometry which we feel should be valid in the theory but cannot be deduced from J1–J4. Here are a few of them.

(1) Any segment is extendible beyond each of its endpoints.
(2) Any segment contains infinitely many points.

As a matter of fact, we cannot deduce from J1–J4 that a segment contains two distinct points.

(3) For any segment ab, $\mathcal{G}(ab) = ab$. (Section 2.12, Example II.)
(4) Any ray is convex.

To get these results—and a host of others—three new postulates are introduced.

The first postulate is suggested by (1) above.

J5 (Existence Law). $a/b \neq \varnothing$.

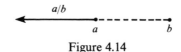

Figure 4.14

The second postulate is suggested by a principle in Euclidean geometry which is easily verified.

(A) Suppose p, q and r are noncollinear, $s \subset pq$, $t \subset pr$. Then $sr \approx tq$. (See Figure 4.15.)

Principle (A) is related to certain basic, but intuitively familiar, properties of opposite sides of a line. Here is a simple illustration.

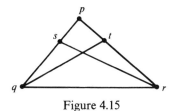

Figure 4.15

(B) Suppose a, b and c are noncollinear, a and b are on opposite sides of line L, $c \subset L$ and $d \subset ac$. Then b and d are on opposite sides of L. (See Figure 4.16.)

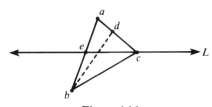

Figure 4.16

We indicate informally that (B) follows from (A).

Since a and b are on opposite sides of L, ab must meet L, say in e. Then (A) implies $bd \approx ce$. Thus $bd \approx L$, so that b and d are on opposite sides of L.

Principle (A) does not seem prepossessing as a candidate for postulate in our theory. As stated, it is not a universal property of points. It is hard to remember and employ as a formal principle. Rather it seems to be grasped and used through visual intuition as a pictorial property of a triangle. There is, however, a gem beneath its rough exterior.

Discarding the noncollinearity condition in (A), we get

(C) If $s \subset pq$, $t \subset pr$, then $sr \approx tq$.

Note that p does not appear in the conclusion of (C). Can we eliminate p from the hypothesis? The hypothesis can be rewritten

$$s/q \supset p \quad \text{and} \quad t/r \supset p \quad \text{for some point } p.$$

This is equivalent to

$$s/q \approx t/r.$$

Thus (C) is equivalent to:

(D) If $s/q \approx t/r$, then $sr \approx qt$.

We adopt (D) as a postulate.

J6 (The Four Term Transposition Law). If $a/b \approx c/d$, then $ad \approx bc$.

Our third postulate is the idempotent law for extension which as we have seen holds in Euclidean geometry (Section 4.1).

J7 (Idempotent Law). $a/a = a$.

Note as a consequence: An improper ray (Section 4.1, last definition) is a point.

The following exercises indicate how to show that principles (1)–(4) above are not deducible from J1–J4. The corresponding problem for Postulates J5, J6, J7 appears in Exercise 1 at the end of Section 5.3. See also Exercises 7 and 9 at the end of Section 4.6.

EXERCISES

1. (a) Let J be a closed circular region in a Euclidean plane and $\cdot$ be the Euclidean join operation applied to J. Show $(J, \cdot)$ is a model of J1–J4.
   (b) Show $(J, \cdot)$ does not satisfy principle (1) above, and infer that (1) is not deducible from J1–J4.

2. The same as Exercise 1 for principle (2), defining J to be $\{0, 1\}$ and $\cdot$ by $0 \cdot 0 = 0 \cdot 1 = 1 \cdot 0 = 0$, $1 \cdot 1 = 1$. (Compare Exercise 9 at the end of Section 2.17.)

3. The same as Exercise 1 for principle (3). Use the model in Exercise 1 at the end of Section 3.7, and show $\mathfrak{g}(bc) \neq bc$.

4. The same as Exercise 1 for principle (4), using the triode model (Section 2.22).

4.9 Discussion of the Postulates

Postulates J5, J6 and J7 greatly amplify the deductive power of the theory and deserve a section to themselves.

J5 says in effect that the extension of a point from a point is not existentially trivial: it always contains at least one point. In terms of the operation join, J5 asserts that for any points a and b there exists a point x such that $bx \supset a$. In Euclidean geometry J5 is equivalent to the principle that a segment can be extended beyond each of its endpoints.[2] The main importance of J5 lies in its consequence (Theorem 4.6 below) that $A/B \neq \varnothing$ if $A, B \neq \varnothing$. To illustrate this consider in Euclidean geometry L/p, where L is a line and $p \not\subset L$ (Figure 4.17). Then L/p, the "side of L" opposite to p, is not empty. In effect J5 enables us to construct a broad class of unbounded figures.

J6 is one of the most interesting and powerful of the postulates J1–J7. Although it stems from a subtle and strongly pictorial geometrical principle, it is easy to remember and use in formal argument and abstract

---

[2] Compare Euclid [1], p. 154, Postulate 2.

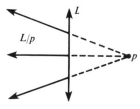

Figure 4.17

analysis, since it has the symmetric structure of the familiar transposition principle in elementary algebra: If $a/b = c/d$, then $ad = bc$. The importance of J6, in the formal development, lies in the fact that it provides a relation between join and extension which is not covered by the Three Term Transposition Law.

J7, in formal terms, is an attractive analogue of J4. But it has a deep substantive significance in the treatment. For J7 implies (Theorem 4.9 below)

$$ab \not\supseteq a, b \quad \text{provided } a \neq b, \tag{1}$$

that a proper segment does not contain its endpoints. In view of this, a join operation that satisfies (1) may be called an *open* join operation. The adoption of J7 then marks a decision to base the theory on the concept of an open join operation. The advantages, we feel, will be borne out amply in the subsequent development.

Still, you may be curious about an alternative theory that would be suggested by taking the join of a and b ($a \neq b$) in Euclidean geometry as the *closed* segment ab rather than the (open) segment ab. Would this make much difference in the abstract theory? We consider this question in concrete terms in a Euclidean geometry.[3]

In a Euclidean space J let $\cdot$ denote the Euclidean (open) join operation. In J we define an operation $*$ by

$$a * b = a \cdot b \cup a \cup b, \tag{1}$$

and call $a * b$ the *closed* join of a and b.

The relation (1) implies $a * b \supset a$ for every b. Let $/$ be the extension operation associated with $*$. Then $a/a \supset b$ for every b, and $a/a = J$. This is quite unattractive; the opulence of the result is embarrassing.

Another difficulty arises in treating $*$ because sometimes open joins will be needed—for example, $a \cdot b$ or $a \cdot b \cdot c$ in order to get the interior of $[a, b]$ or $[a, b, c]$. Although it is easy to form closed joins from open joins [as (1) attests], to do the reverse is not so easy. How can we invert (1) and express $a \cdot b$ in terms of $a * b$? It is natural to try

$$a \cdot b = a * b - \{a, b\}. \tag{2}$$

But (2) fails, since if $a = b$, it asserts $a \cdot a = a - a = \varnothing$.

---

[3] Sections 1.23 and 3.12 are concerned with closed join operations.

Finally consider the relation of the theory to Euclidean geometry. We know that the postulates J1–J4 are satisfied by the Euclidean join systems $(J, \cdot)$ in which J is a Euclidean plane or 3-space and $\cdot$ the Euclidean join operation in J (Section 2.4). It is important to observe that these systems also satisfy J5, J6 and J7—they are models of the postulate set J1–J7. The verification for J5 and J7 is almost trivial. To verify J6 is tedious but nonetheless important. Although J6 stems from an intuitively obvious triangle property, it is not at all obvious—but is indeed the case—that J6 holds for all possible positions of the points a, b, c, d: they may be collinear, two of them may coincide, we may even have $a = b = c = d$.

Although postulates J1–J7 are applicable to Euclidean geometry, they are much too weak to *characterize* Euclidean geometry. Many Euclidean properties have been omitted. We have not included a parallel postulate, postulates for congruence, or postulates for incidence, such as, two distinct points determine a line.[4] Moreover, J1–J7 do not imply that the points of a line are ordered (Section 4.24 below) i.e., that one of three distinct points of a line is in the join of the other two. (See Exercise 1 at the end of Section 6.10.) Finally note that we have not included any dimensionality restriction: J1–J7 are valid in Euclidean geometries of arbitrary dimension. (See Sections 5.6–5.9.)

EXERCISES

1. Verify J6 in a Euclidean geometry, making sure that you cover all special or degenerate cases.

2. (a) Prove: If $a/b \approx ab$, then $a = b$.
   (b) Prove: If $a \approx bc$, $b \approx ca$, then $a = b = c$.
   (c) Prove: If $a \approx bc$, $b \approx cd$, $c \approx da$, then $a = b = c = d$.

3. Prove: $a(b/a) \subset ab/a$.

*4. Prove: If $b \subset ac$, then $a/b = a/c$. Verify in a Euclidean model.

5. Prove: If $b \subset ac$, then $a/b \subset b/c$. Verify in a Euclidean model.

6. (a) Prove: If $a/b \approx b/a$, then $a = b$.
   (b) The same, but not using J6.

7. Prove: $a/(a/b) \subset ab/a$.

8. Prove: If A is convex and $p, q \not\subset A$, then $p \not\subset qA$ or $q \not\subset pA$.

4.10 Formal Consequences of the Postulate Set J1–J7

Now we proceed to deduce consequences of the full set of postulates J1–J7, beginning with formal properties of join and extension. Some of the theorems are simple and supplement previous results; some are powerful

---

[4] Line is defined in Section 6.10.

formal principles which will be used to extend the theory of convex sets. Postulate J5 can be generalized to sets exactly as J1 was in Theorem 2.2.

Theorem 4.6 (Existence Law). *Suppose $A \neq \varnothing$ and $B \neq \varnothing$. Then $A/B \neq \varnothing$.*

(Compare Theorem 2.2.)

Moreover, the argument (which follows Theorem 2.2) that $A \cdot \varnothing = \varnothing \cdot A = \varnothing$ can be applied to justify

$$A/\varnothing = \varnothing/A = \varnothing.$$

It is not surprising that J5 yields new intersection relations.

Theorem 4.7.

(a) $A \approx B$ and $C \neq \varnothing$ imply $A/C \approx B/C$ and $C/A \approx C/B$;
(b) $A \approx B$ and $C \approx D$ imply $A/C \approx B/D$.

[Compare Theorem 4.3(d), (e).]

PROOF. Adopt the method of Theorem 4.3(d), (e), using J5 instead of J1. □

Corollary 4.7. *Suppose $A \neq \varnothing$. Then (a) $A(B/A) \supset B$; (b) $AB/A \supset B$; (c) $A/(A/B) \supset B$.*

PROOF. Suppose $b \subset B$, that is, $b \approx B$. Then

$$b/A \approx B/A \qquad [\text{Theorem 4.7(a)}]; \tag{1}$$

$$Ab \approx AB \qquad [\text{Theorem 4.3(d)}]; \tag{2}$$

$$A/b \approx A/B \qquad [\text{Theorem 4.7(a)}]. \tag{3}$$

From (1), Theorem 4.4 gives $b \approx A(B/A)$, so that (a) holds. Similarly (b) follows from (2) and (c) from (3). □

Remark. The corollary has an analogue in school algebra. Consider the simple rules $a(b/a) = b$, $ab/a = b$ and $a/(a/b) = b$. This suggests as a memory aid (or at least a check) for the corollary, that the left member of each relation should reduce to the right member when simplified as in elementary algebra.

Now J6 is extended to sets.

Theorem 4.8 (The Four Term Transposition Law). *If $A/B \approx C/D$ then $AD \approx BC$.*

PROOF. By hypothesis there exists x such that

$$x \subset A/B, \qquad x \subset C/D.$$

Hence by definition

$$x \subset a/b, \qquad x \subset c/d,$$

where $a, b, c, d \subset A, B, C, D$ respectively. Thus $a/b \approx c/d$, and J6 yields $ad \approx bc$. Thus there exists y such that

$$y \subset ad, \qquad y \subset bc.$$

This implies $y \subset AD, y \subset BC$, so that $AD \approx BC$ and the theorem holds.   □

Next a few consequences of J7.

Theorem 4.9. *If $a \neq b$, then $ab \not\supset a, b$.*

PROOF. Suppose $ab \supset a$. Then $b \subset a/a$ by definition of extension. But $a/a = a$ by J7. Thus $b \subset a$, that is, $b = a$. This is contrary to hypothesis, and we conclude $ab \not\supset a$. Similarly $ab \not\supset b$.   □

Remark. The theorem says that a proper segment excludes its endpoints. In this regard ab $(a \neq b)$ is analogous to an (open) segment in Euclidean geometry.

Corollary 4.9. *If $a \neq b$, then $a/b \not\supset a, b$.*

Remark. By the corollary a proper ray (Section 4.1) never contains its endpoint and is analogous in this respect to an (open) ray in Euclidean geometry.

Theorem 4.10. *If $a \neq b$, then ab contains at least two points.*

PROOF. Suppose the theorem is false. Then since $ab \neq \varnothing$, $ab = c$ for some point c. Thus $ac = aab = ab = c$. Then $a \subset c/c = c$ and $a = c$. Similarly $b = c$. Thus $a = b$, contrary to hypothesis.   □

Theorem 4.11. $\mathcal{C}(a) = a$.

PROOF. $\mathcal{C}(a) \supset a$ (Theorem 2.19). To show $\mathcal{C}(a) \subset a$, let $p \subset \mathcal{C}(a)$. Then $pa \subset a$ (Theorem 2.20). Thus $p \subset a/a = a$ and $p = a$.   □

The result can also be proved using Theorem 4.10.

4.11 Formal Consequences Continued

J6 now comes into play in the derivation of two algebraic relations which involve both join and extension—they are comparable in a way to the Mixed Associative Law (Theorem 4.5).

Theorem 4.12. $a(b/c) \subset ab/c$.

PROOF. We show every point of the left member is in the right member. Let

$x \subset a(b/c)$. Then

$$x \approx a(b/c),$$

$$x/a \approx b/c \qquad \text{(Theorem 4.4)},$$

$$xc \approx ab \qquad \text{(J6)},$$

$$x \approx ab/c \qquad \text{(Theorem 4.4)},$$

and $x \subset ab/c$.   □

Corollary 4.12. $A(B/C) \subset AB/C$.

Theorem 4.13. $a/(b/c) \subset ac/b$.

PROOF. Let $x \subset a/(b/c)$. Then

$$x \approx a/(b/c),$$

$$x(b/c) \approx a \qquad \text{(Theorem 4.4)},$$

$$b/c \approx a/x \qquad \text{(Theorem 4.4)},$$

$$bx \approx ca = ac \qquad \text{(J6, J2)},$$

$$x \approx ac/b \qquad \text{(Theorem 4.4)},$$

and $x \subset ac/b$.   □

Corollary 4.13. $A/(B/C) \subset AC/B$.

By now we have encountered four laws of associative type, namely, Postulate J3 (the Associative Law), Theorem 4.5 (the Mixed Associative Law), and Theorems 4.12 and 4.13, which we have just proved. These principles are illustrated in Figures 4.18–4.21.

Postulate J3 is a familiar principle of elementary algebra. Theorem 4.5 is familiar as a principle for dividing a quotient (or a fraction) by a number. Theorems 4.12 and 4.13 correspond to familiar algebraic principles for multiplying and dividing by a quotient (or a fraction). Indeed, the right member in Theorem 4.13 is obtained by the familiar "invert and multiply" rule for division by a fraction.

Figure 4.18 Postulate J3. $(ab)c = a(bc)$

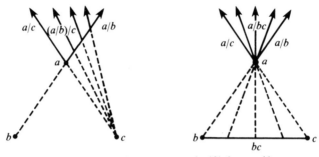

Figure 4.19   Theorem 4.5. $(a/b)/c = a/bc$

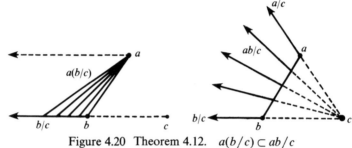

Figure 4.20   Theorem 4.12. $a(b/c) \subset ab/c$

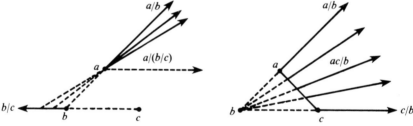

Figure 4.21   Theorem 4.13. $a/(b/c) \subset ac/b$

EXERCISES

1. Prove: If $p \subset a/b$, then $a/b \supset ap \cup p \cup p/b$. Interpret in Euclidean, triode and Cartesian models. Is it valid in all three?

2. Prove: $a(a/b) = a/b = a/ab$.

3. Prove: $a/(a/b) = ab/a$.

4. Prove that Theorem 4.9 is equivalent to J7 (in the face of the other postulates), so that Theorem 4.9 could have been chosen as a postulate instead of J7.

5. Prove that Theorem 4.12 is equivalent to J6. (See Exercise 4.) Similarly for Theorem 4.13.

6. Consider Theorem 4.12, $a(b/c) \subset ab/c$. Try to find when, if ever, equality holds by drawing diagrams for the two expressions in Euclidean geometry. Similarly for Theorem 4.13, $a/(b/c) \subset ac/b$.

7. (a) Prove: If $a/(b/c) = ac/b$, as in elementary algebra, then the basic set J contains at most one point. Similarly for $a(b/c) = ab/c$.
   (b) Describe the algebra of join and extension if J is a single point e.

8. The same as Exercise 7(a) if the converse of J6 is assumed: $ad \approx bc$ implies $a/b \approx c/d$.

9. (a) Prove: $\mathcal{C}(a/b) \supset a$, that is, the closure of a ray contains its endpoint. (Compare the Remark following Corollary 2.25.)
   (b) Prove: $\mathcal{C}(a/b) \supset b$ if and only if $a = b$.

4.12 Joins and Extensions of Rays

Some information is needed on the join of two rays a/b and c/d and their extension or quotient. It seems almost impossible to derive simple formulas, but we can obtain containment relations.

Theorem 4.14. $(a/b)(c/d) \subset ac/bd$.

PROOF.

$$
\begin{aligned}
(a/b)(c/d) &\subset (a/b)c/d && \text{(Corollary 4.12)} \\
&= c(a/b)/d \\
&\subset (ca/b)/d && \text{(Theorems 4.12, 4.1)} \\
&= ac/bd && \text{(Corollary 4.5).} \qquad \square
\end{aligned}
$$

Corollary 4.14. $(A/B)(C/D) \subset AC/BD$

Remarks. Theorem 4.14 and its corollary are quite useful, and a mnemonic hint may not be out of place. They can be remembered in terms of the familiar algebraic principle for multiplying two fractions, keeping in mind that the "product fraction" dominates the "product" of the given fractions.

The algebraic analogy is so striking that you may be led to think of the theorem as mere formalism. This would be a mistake. Theorem 4.14 expresses an important property of the join and extension operations in formal algebraic terminology; it is, nevertheless, well packed with concrete geometrical meaning. Figures 4.22 and 4.23 indicate its significance in Euclidean geometry when the four points are noncoplanar.

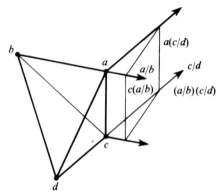

Figure 4.22

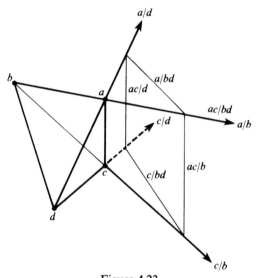

Figure 4.23

Next an analogue of Theorem 4.14 for the quotient of two rays.

Theorem 4.15. $(a/b)/(c/d) \subset ad/bc$.

PROOF.

$$(a/b)/(c/d) \subset (a/b)d/c \quad \text{(Corollary 4.13)}$$
$$\subset (ad/b)/c \quad \text{(Theorems 4.12, 4.1)}$$
$$= ad/bc \quad \text{(Corollary 4.5).} \qquad \square$$

Corollary 4.15. $(A/B)/(C/D) \subset AD/BC$.

A mnemonic can easily be formulated for Theorem 4.15 and its corollary similar to that for Theorem 4.14 and its corollary (Remarks following Corollary 4.14).

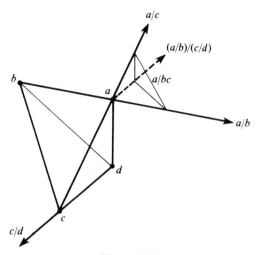

Figure 4.24

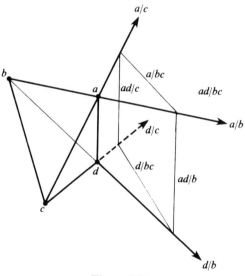

Figure 4.25

Theorem 4.15 is illustrated by Figures 4.24 and 4.25.

A consequence of Theorem 4.14 is the convexity of the extension of two convex sets.

Theorem 4.16. *Let A and B be convex. Then A / B is convex.*

(Compare Theorem 2.12.)

PROOF. $AA = A$ and $BB = B$ (Theorem 2.9). By Corollary 4.14

$$(A/B)(A/B) \subset AA/BB = A/B.$$

Thus A/B is convex (Theorem 2.9).   □

Corollary 4.16. *Any ray is convex.*

4.13 Solving Problems

We cannot give you any set of rules for solving all problems. We do not know any. However, we will make a few helpful suggestions in the course of solving the following problems.

EXAMPLE I. Prove: If $o \subset ab$ and $o \subset ac$, then $ob \approx oc$ (Figure 4.26).

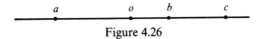

Figure 4.26

Solution. We apply a method of eliminating terms in containment or intersection relations. The hypothesis asserts

$$o \subset ab, \qquad o \subset ac. \tag{1}$$

Compare (1) with the desired conclusion

$$ob \approx oc. \tag{2}$$

What gross difference do you notice between (1) and (2)? The relation (1) contains o, a, b and c, but (2) contains o, b and c. So we eliminate a between the two components of (1). First "solve" each component statement in (1) for a to obtain

$$a \approx o/b, \qquad a \approx o/c. \tag{3}$$

The relation (3) implies

$$o/b \approx o/c \tag{4}$$

and a has been eliminated. Applying J6 to (4) yields $oc \approx ob$, and (2) is proved.   □

Remark. Sometimes it is more convenient to use another method of eliminating terms. Suppose, for example, we have

$$ab \approx cd, \qquad x \approx c/e \tag{1}$$

and wish to eliminate c. In (1) solve the second relation for c, getting $c \approx xe$ or $c \subset xe$. Then replace c in the first relation of (1) by xe, getting

$$ab \approx xed. \tag{2}$$

To justify (2) note that $c \subset xe$ implies

$$cd \subset xed \tag{3}$$

by the monotonic law for join (Theorem 2.1). Then the relations $ab \approx cd$ and (3) imply (2) by Theorem 4.3(c).

EXAMPLE II. Prove: If $x \subset ab$, $y \subset ac$ and $z \subset bc$, then $az \approx xy$ (Figure 4.27).

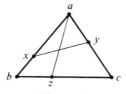

Figure 4.27

Solution. Observe that the hypothesis contains x, y, z, a, b, c; the conclusion, x, y, z, a. So we try to eliminate b and c from the hypothesis.

Solve $x \subset ab$ and $y \subset ac$ for b and c, getting

$$b \approx x/a, \qquad c \approx y/a.$$

Then "substitute" these relations in $z \subset bc$ to obtain

$$z \subset (x/a)(y/a). \tag{1}$$

The unwanted terms b and c have now been eliminated, but (1) does not resemble the conclusion

$$az \approx xy, \tag{2}$$

except in that they involve the same terms. Naturally we try to find some postulate or theorem which when applied to (1) will yield (2), or at least a result closer to (2). The principle for the product of two rays (Theorem 4.14) seems a good choice. It implies

$$(x/a)(y/a) \subset xy/a. \tag{3}$$

Then (1) yields $z \subset xy/a$. This implies $az \approx xy$, and the problem is solved.

□

Another Solution. The solution given depends for its success on the availability of Theorem 4.14. If this theorem had not been presented or not even discovered, the proof would fail.

Suppose, instead of eliminating b and c simultaneously, we eliminate b first and see what happens. We have

$$x \subset ab, \qquad y \subset ac, \qquad z \subset bc.$$

Then

$$b \approx x/a, \qquad b \approx z/c$$

so that

$$x/a \approx z/c$$

and

$$xc \approx az. \tag{4}$$

What can we do now? Note that c is common to (4) and the relation $y \subset ac$. The elimination of c will yield a relation involving a, z, x and y. So we eliminate c between $y \subset ac$ and (4). These yield

$$c \approx y/a, \qquad c \approx az/x,$$

so that

$$y/a \approx az/x$$

and

$$xy \approx aaz = az. \qquad \square$$

The advantage of the second method of solution is that it is more flexible: After one term is eliminated there is a chance to stop and assess the situation, to see whether a familiar elementary procedure will be effective. In the first method of solution, once both terms are eliminated there seems to be no recourse except the discovery of Theorem 4.14 or relation (3).

Finally let us underscore that the use of diagrams in the interpretation and analysis of problems, as in school geometry, is a helpful device which can lead to the discovery of solutions not readily obtained otherwise.

EXERCISES

1. (a) Prove: $(a/b)(a/c) \subset a/bc$.
   (b) Verify in Euclidean geometry: $(a/b)(a/c) = a/bc$.

2. Prove: If A and B are convex, then $\mathcal{S}(A \cap B) = \mathcal{S}(A) \cap \mathcal{S}(B)$, provided $\mathcal{S}(A) \approx \mathcal{S}(B)$. (Compare Exercise 4 at the end of Section 2.13.[5])

3. Prove: If $x \subset ab$, $y \subset ac$, $z \subset ad$ and $w \subset bcd$, then $aw \approx xyz$. Interpret in Euclidean 3-space and in a Euclidean plane. (Compare Example II above.)

4. (a) Prove: If $x \subset ab$ and $y \subset bc$, then $ac \approx bd$ implies $xy \approx bd$.
   (b) Can you prove a sort of converse of (a): If $x \subset ab$ and $y \subset bc$, then $xy \approx bd$ implies $ac \approx bd$? Justify your answer.

5. Prove: If ab, bc and ac have a common point, then $a = b = c$. (This indicates how badly a triangle will degenerate if we require its three sides to intersect.)

6. Prove: If ab, bc, cd have a common point o, then $da \supset o$.

7. Suppose $x \subset ab$ and $o \subset a$, bb'. Prove there exists x' such that $x' \subset a'b'$ and $o \subset xx'$. Interpret geometrically.

8. Generalize Exercise 7 to several points a, b, c or a, b, c, d. Interpret geometrically.

---

[5] Compare Rockafellar [1], p. 47, Theorem 6.5.

9. Suppose $p, p' \subset o/a$; $q, q' \subset o/b$; and $r \subset pq$. Prove that there is a ray o/c which contains r and intersects $p'q'$.

*10. Prove: If $ac \approx bd$ and if $x \approx ab$, $y \approx bc$, $z \approx cd$ and $w \approx da$, then $xz \approx yw$.

*11. Prove: If $ab \approx cd$, $ac \approx bd$ and $ad \approx bc$, then $a = b = c = d$.

Definition. Let point o be given. If $ao \approx bo$, we write $a \equiv b$ (mod o) and say a is *congruent* to b *modulo* o.

12. Prove that congruence modulo o is an equivalence relation, that is, (a) $a \equiv a$ (mod o), (b) $a \equiv b$ (mod o) implies $b \equiv a$ (mod o) and (c) $a \equiv b$ (mod o) and $b \equiv c$ (mod o) imply $a \equiv c$ (mod o). Interpret in Euclidean geometry.

Definition. Let Point o be given. If $ab \supset o$ we write $a|||b$ (mod o) and say a is *anticongruent* to b *modulo* o.

13. Prove and interpret in Euclidean geometry:
    (a) Ia $a \equiv b$ (mod o) and $b|||c$ (mod o), then $a|||c$ (mod o).
    (b) If $a|||b$ (mod o) and $b|||c$ (mod o), then $a \equiv c$ (mod o).
    (c) If $a|||b$ (mod o), $b|||c$ (mod o), $c|||d$ (mod o), then $a|||d$ (mod o).
    (d) If $a \equiv b$ (mod o) and $a|||b$ (mod o), then $a = b = o$.

*14. Prove: $a/(b/a) = a/b$.

15. Consider Corollary 4.16 and Exercise 4 at the end of Section 4.8. Is the triode model of J1–J4 also a model of J1–J7? Explain.

16. *Project.* Consider the following definition.
    A *fraction* (in $a_1, \ldots, a_n$) is a set of points which can be expressed by applying the operations $\cdot$ and $/$ a finite number of times to one or more of $a_1, \ldots, a_n$. Examples of fractions: ab/cd or $((abc/adf)pqr)/as$ or xyz or $(a/b)(c/d)$ or a.
    Try to develop a theory of fractions. Consider the following questions:
    (a) Are fractions convex?
    (b) What kind of figures can be obtained by interpreting fractions in a Euclidean plane or 3-space?
    (c) What can you say about the join and the extension of two fractions?
    (d) Given the family of fractions in $a_1, \ldots, a_n$. Does it have a least member; a greatest (Section 3.2)? What can you discover about a least or greatest member if it exists?

4.14 Extreme Points of Convex Sets

The formal theory, which has been augmented considerably by the adoption of postulates J5–J7, is applied in the remainder of the chapter to the study of specific geometrical topics.

The first of these is the notion of extreme point of a convex set, which is suggested by the intuitive idea of a "corner point" of a closed convex polygonal region (Figure 4.28).

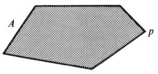

Figure 4.28

Definition. Let p be a point of convex set A. Then p is called an *extreme* point of A if p is never contained in the join of two distinct points of A.

First the defining property is cast in a slightly different form: Let p be a point of convex set A. Then p is an extreme point of A if and only if

$$p \subset ab, \qquad a, b \subset A$$

implies $p = a = b$.

Next, extreme points are characterized by a simple convexity condition.

Theorem 4.17. *Let A be convex. Then p is an extreme point of A if and only if $p \subset A$ and $A - p$ is convex.*

PROOF. Suppose p is an extreme point of A. Let $x, y \subset A - p$. We must show

$$xy \subset A - p. \tag{1}$$

Certainly $xy \subset A$. Suppose $p \subset xy$. Since p is an extreme point of A, $p = x = y$. Then $p \subset A - p$, which is impossible. Thus $p \not\subset xy$ and (1) holds.

Conversely let $p \subset A$ and $A - p$ be convex. Suppose p is not an extreme point of A. Then there exist x, y such that

$$p \subset xy, \qquad x \neq y, \qquad x, y \subset A. \tag{2}$$

Suppose $x = p$. By (2), $p \subset py$, and using J7, $y \approx p/p = p$. Thus $x = y$, contradicting (2), and we infer $x \neq p$. Similarly $y \neq p$. Hence $x, y \subset A - p$. Then $xy \subset A - p$ and (2) implies $p \subset A - p$, which is impossible. Hence our supposition is false, and p must be an extreme point of A. □

The defining property of extreme points is now generalized.

Theorem 4.18. *Let p be a point of convex set A. Then p is an extreme point of A if and only if p is never contained in the join of two or more distinct points of A.*

PROOF. Let p be an extreme point of A. Suppose

$$p \subset a_1 \ldots a_n, \tag{1}$$

where $a_1, \ldots, a_n \subset A$ and $n > 1$. We show the a's are identical. Rewrite (1) as

$$p \subset a_i(a_1 \ldots a_{i-1}a_{i+1} \ldots a_n), \qquad 1 \leqslant i \leqslant n.$$

This implies

$$p \subset a_i a, \qquad (2)$$

where

$$a \subset a_1 \ldots a_{i-1} a_{i+1} \ldots a_n \subset A.$$

Since p is an extreme point of A, (2) implies $p = a_i$. Thus

$$p = a_1 = \cdots = a_n,$$

and the forward implication is established. The converse is trivial and the theorem holds.   □

Corollary 4.18. *Let p be a point of convex set A. Then p is an extreme point of A if and only if*

$$p \subset a_1 \ldots a_n, \qquad a_1, \ldots, a_n \subset A$$

implies

$$p = a_1 = \cdots = a_n.$$

The next few theorems relate the ideas of extreme point and convex hull (Section 3.2).

Theorem 4.19. *An extreme point of convex set A belongs to every set of generators of A.*

PROOF. Let p be an extreme point of A, and S a set of generators of A. We must show $p \subset S$. Certainly

$$p \subset A = [S].$$

Hence p is in a join of points of S (Corollary 3.7.1):

$$p \subset s_1 \ldots s_n, \quad \text{where } s_1, \ldots, s_n \subset S \subset A.$$

Thus by the last corollary, $p = s_1 \subset S$, and the theorem holds.   □

Corollary 4.19. *Suppose set S generates the convex set A, and p is an extreme point of A. Then $S - p$ does not generate A.*

The theorem can be described by saying that an extreme point p of convex set A is an *essential* element of a generating set of A or, a bit loosely, that p is an *essential* generator of A.

The next theorem shows that a nonextreme point of convex set A is never an essential generator of A.

Theorem 4.20. *Suppose set S generates the convex set A, and p is a nonextreme point of A. Then $S - p$ also generates A.*

PROOF. We have to show $A = [S - p]$. Since $A = [S]$,

$$A \supset [S - p].$$

It remains to prove

$$[S - p] \supset A. \qquad (1)$$

Most of our work will be required to show

$$[S - p] \supset p, \qquad (2)$$

from which (1) will follow fairly easily. By hypothesis, p must be in the join of some pair of distinct points of A. Thus

$$p \subset ab, \quad a \neq b, \quad \text{where } a, b \subset A. \qquad (3)$$

Since $A = [S]$, a and b are in joins of points of S (Corollary 3.7.1) and we may write

$$a \subset x_1 \ldots x_m, \quad b \subset x_{m+1} \ldots x_n \quad \text{where } x_1, \ldots, x_n \subset S. \qquad (4)$$

We assert that at least two of the x's are distinct. For if all the x's are equal, the relation (4) implies $a = x_1$ and $b = x_1$, contrary to (3).

The relations (3) and (4) imply

$$p \subset x_1 \ldots x_n. \qquad (5)$$

By applying J4, the idempotent law, we can eliminate repetitions of terms in $x_1 \ldots x_n$, getting

$$x_1 \ldots x_n = y_1 \ldots y_r, \qquad (6)$$

where the y's are distinct and

$$\{y_1, \ldots, y_r\} = \{x_1, \ldots, x_n\}. \qquad (7)$$

The relations (5), (6), (7) and (4) yield

$$p \subset y_1 \ldots y_r, \quad \text{where } y_1, \ldots, y_r \subset S. \qquad (8)$$

Since in (7) two of the x's are distinct, so are two of the y's, and we have $r \geqslant 2$.

We show p is in a join of points of $S - p$. This is true by (8) if the y's are distinct from p. Suppose one of the y's, say y_1, equals p. Then

$$p \subset py_2 \ldots y_r,$$

and J7 implies

$$p = p/p \approx y_2 \ldots y_r.$$

Thus in any case p is in a join of points of $S - p$. Then (2) follows by Corollary 3.7.1.

In view of (2) we have $[S - p] \supset (S - p) \cup p = S$. Thus

$$[S - p] \supset [S] = A,$$

which verifies (1) and completes the proof.   □

Remark. The theorem asserts that if S is a set of generators of A, $p \subset S$, and p is a nonextreme point of A, then p is a redundant element of S.

Extreme points play an important role in the structure of polytopes.

Theorem 4.21. *The set of extreme points of a polytope P is finite and generates P.*

PROOF. By definition $P = [S]$, where S is finite. By Theorem 4.19, S contains all the extreme points of P. If S contains no other point, the theorem certainly holds.

Suppose S contains a nonextreme point of P. Let $\{p_1, \ldots, p_r\}$ be the set of those nonextreme points of P that are in S. By Theorem 4.20

$$P = [S - p_1].$$

Similarly if $r > 1$,

$$[S - p_1] = [(S - p_1) - p_2] = [S - \{p_1, p_2\}],$$

so that

$$P = [S - \{p_1, p_2\}].$$

Continuing in this way, we obtain finally

$$P = [S - \{p_1, \ldots, p_r\}].$$

Thus P is generated by its set of extreme points, which is finite, and the theorem is proved. $\qquad\square$

Corollary 4.21. *Each polytope P has an extreme point. If P is not a point, it has at least two extreme points.*

In view of the theorem, the extreme points of a polytope deserve a special name.

Definition. An extreme point of a polytope P is called a *vertex* of P.

It seems intuitively certain—and is true with a trivial exception—that extreme points are frontier points:

Theorem 4.22. *The frontier of a convex set A contains its extreme points, provided A is not a point.*

PROOF. Suppose A is not a point. Let p be an extreme point of A. Assume $p \subset \mathcal{I}(A)$. Choose x in A distinct from p. Then $p \subset xy$ for some y in A. This implies $p = x$, a contradiction. Thus $p \not\subset \mathcal{I}(A)$ and $p \subset \mathcal{F}(A)$ by definition (Section 2.11). $\qquad\square$

Corollary 4.22. *Let P be a polytope that is not a point. Then $\mathcal{F}(P) \neq \varnothing$.*

PROOF. Use the theorem in conjunction with Corollary 4.21. $\qquad\square$

EXERCISES

1. Prove: A polytope has a least set of generators.

2. Let $P = [a_1, \ldots, a_n]$, where $a_1, \ldots, a_n$ are distinct. Prove a_i is an extreme point of P if and only if

$$a_i \not\subset [a_1, \ldots, a_{i-1}, a_{i+1}, \ldots, a_n].$$

3. Suppose $a \neq b$. Prove:
   (a) $[a, b]$ has exactly two extreme points, namely, a and b.
   (b) ab has no extreme points.
   (c) a/b has no extreme points.

4. Prove the converse of Theorem 4.19: If p belongs to every set of generators of convex set A, then p is an extreme point of A.

5. Let p be a point of convex set A, and S a set of generators of A. Prove that p is not an extreme point of A if and only if (a) $p \subset [S - p]$; or (b) $[S - p] = A$.

6. Let $A = [S]$ and $p \subset A$. Prove that p is an extreme point of A if and only if p is never contained in the join of two or more distinct points of S. (Compare Theorem 4.18.)

7. (a) Must a convex set have extreme points?
   (b) Can a convex set consist solely of extreme points?
   (c) Can an extreme point of a convex set be an interior point?
   (d) Is every convex set generated by its set of extreme points?

8. Give examples in Euclidean geometry of an extreme point of a convex set which does not correspond to the intuitive idea of a "corner point" of a closed convex polygonal region.

9. (a) Find examples of a convex set A that is not a polytope and has a *minimal* set of generators, that is, a set of generators no proper subset of which generates A.
   (b) Find examples of convex sets that do not have minimal sets of generators.

10. Let S be a minimal set of generators of convex set A (see Exercise 9). Prove that S is a least set of generators of A (and so is uniquely determined); and that S is composed of extreme points of A.

11. Prove: If J has an extreme point p, then $J = p$.

12. Prove: If a convex set consists solely of extreme points, it contains at most one point.

13. Prove: If p is an extreme point of $[A_1, \ldots, A_n]$, then $p \subset A_i$ for some i, $1 \leqslant i \leqslant n$.

14. In a Euclidean plane find or construct a convex set that has infinitely many extreme points.

15. Prove: $[a, b] = [c, d]$ implies $\{a, b\} = \{c, d\}$. This is essentially equivalent to: Any proper closed segment has a unique pair of endpoints.

16. *Project.* Try to generalize the idea of extreme point from convex sets to arbitrary sets of points. When you have a definition that seems satisfactory, try to discover theorems and prove them.

4.15 Open Convex Sets

The formal theory is now applied to the study of a new family of convex sets—the open convex sets.

Many of the familiar convex sets in Euclidean geometry are naturally described—and intuitively conceived—as being open. Examples abound. One refers to an *open* segment, an *open* triangular region, an *open* spherical region. We proceed to make the idea precise and applicable uniformly to convex sets.

Figure 4.29

An examination of the three convex sets yields a simple common property: Each set contains none of its boundary points, or equivalently, every point of the set is an interior point of it. This enables us to define the idea of open convex set directly in terms of the interior operation on convex sets (Section 2.11).

Definition. A convex set A is said to be an *open* (convex) set or to be *geometrically open* if each point of A is an interior point of A, that is, if $A \subset \mathcal{I}(A)$.[6]

As was noted earlier (Section 2.9, the paragraph following the definition of convex sets), the basic set J, the set formed by a point a, and the empty set $\varnothing$ are convex. These sets can also be shown to be open: J by means of the Existence Postulate J5; a by the Idempotent Law J4; and $\varnothing$ (as usual) because the definition of openness imposes no restriction on it.

---

[6] It is important to distinguish this notion of geometrical openness of a convex set from the notion of open set in a metric space or a topological space. This is related to the distinction between geometrical interior and metrical interior (Section 2.25). Also compare Rockafellar [1], p. 44: relatively open convex set.

The study of open sets, which is introduced for its own sake, also throws light on the interior operation. It is involved in the derivation of several important interiority properties, including Theorem 4.30 (The Polytope Interior Theorem), $\mathscr{I}[a_1, \ldots, a_n] = a_1 \ldots a_n$; and Theorem 4.31 that $\mathscr{I}(AB) = \mathscr{I}(A)\mathscr{I}(B)$ provided the convex sets A and B have nonempty interiors.

First two simple characterizations of openness of convex sets.

Theorem 4.23. *Let A be convex. Then A is open if and only if* (a) $A = \mathscr{I}(A)$ *or* (b) $\mathscr{F}(A) = \varnothing$.

PROOF. (a) By definition A is open if and only if $A \subset \mathscr{I}(A)$. This holds if and only if $A = \mathscr{I}(A)$, since $\mathscr{I}(A) \subset A$ for every convex set A.

(b) For any convex set A, $\mathscr{F}(A) = A - \mathscr{I}(A)$ [Theorem 2.30(b)]. Note that $A \subset \mathscr{I}(A)$ if and only if $A - \mathscr{I}(A) = \varnothing$.   □

Corollary 4.23. *The interior of any convex set A is open.*

PROOF. Apply part (a) of the theorem to $\mathscr{I}(\mathscr{I}(A)) = \mathscr{I}(A)$ (Theorem 2.18).
                                                                                            □

Remark. Open sets can be manufactured at will from convex sets, merely by applying the interior operation. The operation, in effect, converts any convex set into an open set by deleting its frontier points. Every open set is obtainable by this process, since an open set is the interior of itself.

The next theorem gives a simple and convenient reformulation of the definition of interior point of a convex set.

Theorem 4.24. *Let A be convex and $p \subset A$. Then $p \subset \mathscr{I}(A)$ if and only if*

$$p \subset xA \quad \text{for each } x \subset A. \tag{1}$$

PROOF. Suppose $p \subset \mathscr{I}(A)$. By definition (Section 2.11), for each $x \subset A$ there exists $y \subset A$ such that $p \subset xy$. Thus $p \subset xA$ for each $x \subset A$, and (1) holds.

Conversely, suppose (1). Then for each $x \subset A$ there exists $y \subset A$ such that $p \subset xy$. By definition $p \subset \mathscr{I}(A)$.   □

Remark. The theorem shows that if A is nonempty, $\mathscr{I}(A)$ is the intersection of all sets xA for $x \subset A$.

Corollary 4.24. *The theorem holds if the condition $p \subset A$ is replaced by $A \neq \varnothing$.*

The theorem yields useful criteria for openness of a convex set.

Theorem 4.25. *Let A be convex. Then A is open if and only if* (a) $A \subset xA$ *for each* $x \subset A$ *or* (b) $A = xA$ *for each* $x \subset A$.

PROOF. (a) Suppose A open, and $x \subset A$. We show $A \subset xA$. Let $p \subset A$. Since A is open, $p \subset \mathcal{I}(A)$. By the last theorem, $p \subset xA$. Thus $A \subset xA$.

Conversely, suppose $A \subset xA$ for each $x \subset A$. Let $p \subset A$. Then $p \subset xA$ for each $x \subset A$. By the last theorem, $p \subset \mathcal{I}(A)$. Thus $A \subset \mathcal{I}(A)$ and is open by definition.

(b) We show that if $x \subset A$,

$$A = xA \tag{1}$$

is equivalent to

$$A \subset xA. \tag{2}$$

Certainly (1) implies (2). Since A is convex, $A \supset xA$, which with (2) implies (1). Thus (1) is equivalent to (2). Hence condition (b) of the theorem is equivalent to condition (a) and so to the condition that A is open. $\qquad\square$

Corollary 4.25.1. *A set A is open if and only if $A = xA$ for each $x \subset A$.*

PROOF. $A = xA$ for each $x \subset A$ implies A convex. $\qquad\square$

Restatement of Corollary 4.25.1. A set is open if and only if it absorbs properly each of its points.

Corollary 4.25.2. *Let A and B be open and $A \approx B$. Then $AB \supset A$ and $AB \supset B$.*

PROOF. Suppose $A, B \supset p$. Then $A \supset p$ implies, using the theorem, $AB \supset pB \supset B$. Similarly $AB \supset A$. $\qquad\square$

Corollary 4.25.3. *Let A be convex. Then A is open if and only if $p, q \subset A$ imply $p/q \approx A$.*

PROOF. Observe that $A \subset qA$ is equivalent to "$p \subset A$ implies $p \subset qA$". But $p \subset qA$ holds if and only if $p/q \approx A$. $\qquad\square$

EXERCISES

1. Prove: Any open polytope is a point.

2. Prove: The interior of a convex set A contains every open subset of A that it meets.

3. Prove: For a convex set A if $\mathcal{B}(A) = \varnothing$, then A is open. Does the converse hold?

4. If A is open and its complement $J - A$ is convex, is $J - A$ necessarily open? Can $J - A$ be open? Justify your answers.

5. Prove: If an open set A is not a point and $p \subset A$, then $A - p$ is not convex.

4.16 The Intersection of Open Sets

The intersection of two open sets is open.

Theorem 4.26 (The Open Set Intersection Theorem). *If A and B are open, then A ∩ B is open.*

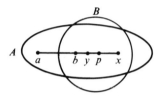

Figure 4.30

PROOF. Note that $A \cap B$ is convex by Theorem 2.13. We show each point of $A \cap B$ is an interior point of it. Suppose $p \subset A \cap B$ (Figure 4.30). Let $x \subset A \cap B$. Then $p \subset A$ and $x \subset A$. Since A is open, p is an interior point of A. Hence

$$p \subset xa \tag{1}$$

for some point $a \subset A$. By symmetry

$$p \subset xb \tag{2}$$

for some $b \subset B$. The relations (1) and (2) imply

$$p/a \approx x, \qquad p/b \approx x,$$

so that $p/a \approx p/b$ and $pb \approx pa$. Let y satisfy

$$y \subset pa, pb. \tag{3}$$

Note that $pa \subset A$ so that $y \subset A$. By symmetry $y \subset B$. Hence

$$y \subset A \cap B. \tag{4}$$

Now we want to relate p, x, y. By (1) and (3) $p \subset xa$, $y \subset pa$; so a is eliminated between these relations. We have

$$p/x \approx a, \qquad y/p \approx a,$$

so that $p/x \approx y/p$ and $p \approx xy$. This with (4) asserts p is an interior point of $A \cap B$. By definition $A \cap B$ is open, and the theorem is proved. ☐

Corollary 4.26. *Let $A_1, \ldots, A_n$ be open. Then $A_1 \cap \cdots \cap A_n$ is open.*

4.17 The Join of Open Sets

We come now to one of the key theorems in the theory of open sets: The family of open sets is closed under the operation join.

Theorem 4.27 (The Open Set Join Theorem). *Let A and B be open. Then AB is open.*

PROOF.

Case I. A = a and B = b. We have to prove that ab, the join of two points, is open, and employ Theorem 4.25(a) for this purpose. First note that ab is convex (Theorem 2.10). Then suppose $x \subset ab$. We show

$$ab \subset xab. \tag{1}$$

Let $p \subset ab$. Then

$$a \subset p/b, \qquad b \subset p/a. \tag{2}$$

Using (2) and the theorem on the product of two rays (Theorem 4.14), we have

$$x \subset ab \subset (p/b)(p/a) \subset p/ab.$$

Thus $xab \approx p$ and $p \subset xab$. Hence (1) holds, and ab is open [Theorem 4.25(a)].

Case II. A and B are arbitrary. Let $x \subset AB$ (Figure 4.31). We show

$$AB = xAB. \tag{3}$$

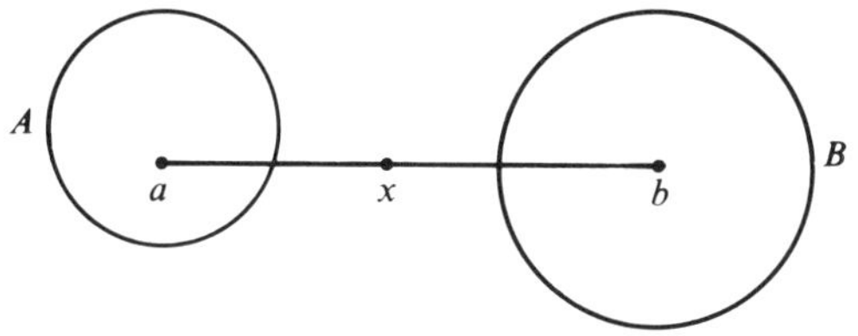

Figure 4.31

Since $x \subset AB$, we have $x \subset ab$, where $a \subset A$, $b \subset B$. By hypothesis and Case I, A, B and ab are open. Hence Corollary 4.25.1 implies

$$A = aA, \qquad B = bB, \qquad ab = xab.$$

Then

$$AB = aAbB = abAB = xabAB = xaAbB = xAB,$$

and (3) holds. Thus AB is open (Corollary 4.25.1) and the proof is complete. □

Corollary 4.27. *Let $A_1, \ldots, A_n$ be open. Then $A_1 \ldots A_n$ is open.*

The special case of the corollary where $A_1, \ldots, A_n$ are individual points $a_1, \ldots, a_n$ is—to signalize its importance—stated as a theorem.

Theorem 4.28. *Any join of points $a_1 \ldots a_n$ is open.*

Remark. Joins of points have already appeared in two nontrivial roles: first as convex sets (Theorem 2.10), second as building blocks of a sort for the construction of any convex set from a generating set (Theorem 3.7). Now in a significant sharpening of the convexity property they appear as open sets—indeed, they may be considered to form the simplest family of open sets, since, in a sense, they are finitely determined.

Theorem 4.28, applied to Euclidean geometry, assures us that joins of distinct points such as $ab, abc, abcd, \ldots$, which appeared as interiors of segments, triangles, "convex" quadrilaterals, tetrahedrons, $\ldots$, truly are open sets in a precise and natural sense.

Finally the theorem shows that any segment ab in the abstract theory is an open set, like its prototype the (open) segment of Euclidean geometry.

Corollary 4.28. $\mathcal{I}(a_1 \ldots a_n) = a_1 \ldots a_n.$

PROOF. Apply Theorem 4.23. ☐

4.18 Segments Are Infinite

Of course, we mean proper segments are infinite sets.

Using results on extreme points and open sets obtained above, we can give a brief but sophisticated proof of this familiar Euclidean property.

Theorem 4.29. *If $a \neq b$, then ab is an infinite set.*

PROOF. Assume ab is finite; then $ab = \{p_1, \ldots, p_n\}$, since $ab \neq \varnothing$. Then

$$ab = [ab] = [p_1, \ldots, p_n]$$

and ab is a polytope. Hence ab has an extreme point p (Corollary 4.21). But ab contains two distinct points (Theorem 4.10), and so $\mathcal{F}(ab) \supset p$ (Theorem 4.22). Since ab is open (Theorem 4.28), $\mathcal{F}(ab) = \varnothing$ (Theorem 4.23), which is impossible. Thus our assumption is false and ab is infinite. ☐

Corollary 4.29.1. *If a convex set contains two distinct points, it is infinite.*

Proper rays are infinite:

Corollary 4.29.2. *If $a \neq b$, then a/b is an infinite set.*

PROOF. Show a/b has two distinct points.   □

Remark on the Proof of the Theorem. The proof of Theorem 4.29 may strike you as somewhat artificial. It may seem more natural to prove the theorem by showing that a set ab, $a \neq b$, contains a point p_1, that it contains a point p_2, $p_2 \neq p_1$, and by some kind of inductive argument to show that ab contains as many points as we please. This is a "natural" way to try to prove the theorem, though not necessarily easy to carry out. Let us remark, however, that the naturalness of a proof depends to some extent on the context. In relation to the present involvement with open sets and extreme points the proof—although unexpected—may not seem very unnatural. In any case, the result was not needed earlier, and the proof seems to us an attractive application of current concepts.

EXERCISES

1. Prove Theorem 4.29 by induction.

2. Complete the proof of Corollary 4.29.2.

3. Show that the intersection of a family of open sets need not be open.

4. Suppose A and B are open sets, $A \approx B$ and $A \cup B$ is convex. Prove $A \cup B$ is open. How are $A \cup B$ and AB related? Will $A \cup B$ be open if $A \cap B = \varnothing$?

5. Prove: Any ray is open.

6. Prove: $a_1 \ldots a_n$ contains every join of points of $\{a_1, \ldots, a_n\}$ that it meets.

4.19 The Polytope Interior Theorem

Our treatment of open sets in Sections 4.15 and 4.17 leads to the resolution of the problem, posed in Chapter 3, of a formula for the interior of a polytope in terms of a finite set of generators (Section 3.7, the first five paragraphs).

Theorem 4.30 (The Polytope Interior Theorem).
$$\mathcal{I}[a_1, \ldots, a_n] = a_1 \ldots a_n.$$

PROOF.
$$\mathcal{I}[a_1, \ldots, a_n] = \mathcal{I}(a_1 \ldots a_n) \qquad \text{(Theorem 3.5),}$$

$$\mathcal{I}(a_1 \ldots a_n) = a_1 \ldots a_n \qquad \text{(Corollary 4.28),}$$

and the theorem holds.   □

Restatement of the Theorem. The interior of a polytope P is the join of the points of any finite set of generators of P.

Since a polytope is generated by its set of vertices, which is finite (Theorem 4.21), we have

Corollary 4.30. *The interior of a polytope is the join of its vertices.*

4.20 The Interior of a Join of Convex Sets

Now we proceed to extend the polytope interior theorem to a generalized polytope (Section 3.13). For this purpose we employ the following interiority property.

Theorem 4.31. *Let A and B be convex. Then $\mathcal{I}(AB) = \mathcal{I}(A)\mathcal{I}(B)$, provided $\mathcal{I}(A)$ and $\mathcal{I}(B)$ are nonempty.*

PROOF. Note that $\mathcal{I}(AB)$ is defined, since AB is convex. First we prove

$$\mathcal{I}(AB) \subset \mathcal{I}(A)\mathcal{I}(B). \tag{1}$$

Let $p \subset \mathcal{I}(AB)$. By Theorem 4.24, if $x \subset AB$, then

$$p \subset xAB. \tag{2}$$

Note that $\mathcal{I}(A)\mathcal{I}(B) \neq \varnothing$. In (2) choose x so that $x \subset \mathcal{I}(A)\mathcal{I}(B)$. Then using Corollary 2.16,

$$p \subset xAB \subset \mathcal{I}(A)\mathcal{I}(B)AB = \mathcal{I}(A)A \cdot \mathcal{I}(B)B = \mathcal{I}(A)\mathcal{I}(B),$$

and (1) holds.

Conversely we show

$$\mathcal{I}(A)\mathcal{I}(B) \subset \mathcal{I}(AB). \tag{3}$$

Let $p \subset \mathcal{I}(A)\mathcal{I}(B)$ (Figure 4.32). Note $p \subset AB$. Then

$$p \subset ab, \quad \text{where} \quad a \subset \mathcal{I}(A), \quad b \subset \mathcal{I}(B). \tag{4}$$

Let $x \subset AB$. We show $p \subset xAB$. We have

$$x \subset a'b', \quad \text{where} \quad a' \subset A, \quad b' \subset B. \tag{5}$$

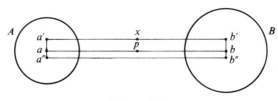

Figure 4.32

The relation (4) implies

$$a \subset a'a'', \quad b \subset b'b'', \quad \text{where} \quad a'' \subset A, \quad b'' \subset B. \tag{6}$$

By (4) and (6)

$$p \subset a'a''b'b''. \tag{7}$$

By (5) and Theorems 4.28, 4.25

$$a'b' = xa'b'. \tag{8}$$

Hence (7) and (8) imply

$$p \subset a'b'a''b'' = xa'b'a''b'' = xa'a''b'b'' \subset xAB.$$

By Theorem 4.24, $p \subset \mathcal{I}(AB)$ so that (3) holds and the theorem is proved. $\qquad\square$

It is easier to generalize the theorem to n sets than to prove it.

Theorem 4.32. *If* $A_1, \ldots, A_n$ *are convex and* $\mathcal{I}(A_1), \ldots, \mathcal{I}(A_n)$ *are non-empty, then*

$$\mathcal{I}(A_1 \ldots A_n) = \mathcal{I}(A_1) \ldots \mathcal{I}(A_n).$$

PROOF. We employ a step by step argument. The theorem holds for $n = 1$. Suppose $n > 1$. We have

$$\mathcal{I}(A_1 A_2) = \mathcal{I}(A_1)\mathcal{I}(A_2) \neq \varnothing \qquad \text{(Theorems 4.31, 2.2)}.$$

Similarly if $n > 2$,

$$\mathcal{I}(A_1 A_2 A_3) = \mathcal{I}(A_1 A_2)\mathcal{I}(A_3) \qquad \text{(Theorem 4.31)}$$

$$= \mathcal{I}(A_1)\mathcal{I}(A_2)\mathcal{I}(A_3) \neq \varnothing \qquad \text{(Theorems 4.31, 2.6)}.$$

Continuing in this way, we obtain finally

$$\mathcal{I}(A_1 \ldots A_n) = \mathcal{I}(A_1) \ldots \mathcal{I}(A_n). \qquad\square$$

If you prefer you can cast the proof in inductive form—it is especially simple if the conclusion of the theorem is changed to

$$\mathcal{I}(A_1 \ldots A_n) = \mathcal{I}(A_1) \ldots \mathcal{I}(A_n) \neq \varnothing.$$

4.21 The Generalized Polytope Interior Theorem

The polytope interior theorem (Theorem 4.30) can now be extended to a result on the interior of a generalized polytope (Section 3.13).

Theorem 4.33 (The Generalized Polytope Interior Theorem). *Let* $A_1, \ldots, A_n$ *be convex, and* $\mathcal{I}(A_1), \ldots, \mathcal{I}(A_n)$ *be nonempty. Then*

$$\mathcal{I}[A_1, \ldots, A_n] = \mathcal{I}(A_1) \ldots \mathcal{I}(A_n). \tag{1}$$

(Compare Theorem 4.30.)

PROOF. The proof is almost the same as that for Theorem 4.30, the polytope interior theorem. Since $\mathcal{I}(A_1), \ldots, \mathcal{I}(A_n)$ are nonempty, so are $A_1, \ldots, A_n$. Then

$$\mathcal{I}[A_1, \ldots, A_n] = \mathcal{I}(A_1 \ldots A_n) \qquad \text{(Theorem 3.12),}$$

$$\mathcal{I}(A_1 \ldots A_n) = \mathcal{I}(A_1) \ldots \mathcal{I}(A_n) \qquad \text{(Theorem 4.32),}$$

and (1) *holds.*   □

Two corollaries follow, one with weaker and one with stronger hypothesis than the theorem.

Corollary 4.33.1. $\mathcal{I}[S_1, \ldots, S_n] = \mathcal{I}[S_1] \ldots \mathcal{I}[S_n]$, *provided* $\mathcal{I}[S_1], \ldots,$ $\mathcal{I}[S_n]$ *are not empty.*

Corollary 4.33.2. *Let* $A_1, \ldots, A_n$ *be open and not empty. Then*

$$\mathcal{I}[A_1, \ldots, A_n] = A_1 \ldots A_n.$$

(Compare Theorem 4.30.)

The theorem is a strong generalization of the polytope interior theorem. It can be verified in the Euclidean examples of generalized polytopes of Section 3.15.

4.22 Closed Convex Sets

In introducing the idea of an open convex set (Section 4.15) we pointed out that the notion was grounded in experience with Euclidean convex sets which were naturally described as open. Equally familiar are Euclidean convex sets which are naturally described as closed, for example, a *closed* segment, a *closed* triangular region and a *closed* spherical region (Figure 4.33). The three convex sets have a common property: Each boundary point of the set is a point of the set, or equivalently, each contact point of the set is a point of it. This enables us to define the idea of closed convex set directly in terms of the closure operation on convex sets (Section 2.16).

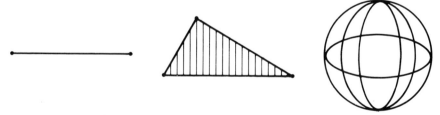

Figure 4.33

Definition. A convex set A is said to be a *closed* (convex) set or to be *geometrically closed* if each contact point of A is a point of A, that is, if $\mathcal{C}(A) \subset A$.

Observe the analogy between open and closed convex sets: A is characterized as open by the condition $A \subset \mathcal{I}(A)$, as closed by $A \supset \mathcal{C}(A)$.

Note that J and $\varnothing$ are closed. Each point a is closed, since $\mathcal{C}(a) = a$ (Theorem 4.11).

A few elementary properties of closed sets are proved.

Theorem 4.34. *Let A be convex. Then A is closed if and only if* (a) $\mathcal{C}(A) = A$ *or* (b) $\mathcal{B}(A) \subset A$.

(Compare Theorem 4.23.)

PROOF. (a) Use $A \subset \mathcal{C}(A)$ (Theorem 2.19).

(b) Use $\mathcal{C}(A) = \mathcal{I}(A) \cup \mathcal{B}(A)$ [Theorem 2.30(c)]. □

Corollary 4.34.1. *The closure of convex set A is closed, provided the interior of A is nonempty.*

(Compare Corollary 4.23.)

PROOF. By Theorem 2.26, $\mathcal{C}(\mathcal{C}(A)) = \mathcal{C}(A)$. □

If you are interested in the question of the existence of a convex set whose closure is not closed, see the Discussion following the proof of Theorem 2.26.

Corollary 4.34.2. *If A is open, then $\mathcal{C}(A)$ is closed.*

Theorem 4.35 (The Closed Set Intersection Theorem). *Let A and B be closed. Then $A \cap B$ is closed.*

(Compare Theorem 4.26.)

PROOF. Note that $A \cap B$ is convex. $A \cap B \subset A$ implies $\mathcal{C}(A \cap B) \subset \mathcal{C}(A)$ (Theorem 2.21). $\mathcal{C}(A) \subset A$, since A is closed. Thus $\mathcal{C}(A \cap B) \subset A$. By symmetry $\mathcal{C}(A \cap B) \subset B$. Thus $\mathcal{C}(A \cap B) \subset A \cap B$ and $A \cap B$ is closed by definition. □

Using a similar method we can prove the following generalization of the theorem.

Theorem 4.36. *Let F be a family of closed sets. Then the intersection of the sets of F is also closed.*

(Compare Theorem 2.14.)

Remark. In Euclidean geometry a closed segment $[a, b]$ is a closed set. But this need not hold in the theory. [See Exercise 4(a) at the end of Section 5.3.]

EXERCISES

1. Prove: A convex set A is closed if and only if $A \supset pq$ implies $A \supset p, q$.

2. Prove: A convex set A is closed if and only if $\mathfrak{F}(A) = \mathfrak{B}(A)$.

3. Suppose A and B are closed. Must AB be closed; must $[A, B]$? Can AB be open; can $[A, B]$?

4. Prove: For any convex set A there exists a unique least closed convex set that contains A.

5. Prove: For convex sets A and B, if $A \subset B$ and B is closed, then $\mathcal{C}(A) \subset B$.

*6. Prove: If A and B are closed and $A \cup B$ is convex, then $A \cup B$ is closed.

7. Prove: If A, B and $A \cup B$ are convex, then $\mathcal{C}(A \cup B) = \mathcal{C}(A) \cup \mathcal{C}(B)$. [Compare Exercise 4(c) at the end of Section 2.15.]

8. Prove: If A, B and $A \cup B$ are convex and $\mathfrak{I}(A) \approx \mathfrak{I}(B)$, then $\mathfrak{I}(A \cup B) = \mathfrak{I}(A) \cup \mathfrak{I}(B)$. [Compare Exercise 5(c) at the end of Section 2.12.]

9. Prove: If A is open, B is closed and $A - B$ is convex, then $A - B$ is open.

10. Prove: If A is closed, B is open, $B \approx \mathfrak{I}(A)$ and $A - B$ is convex, then $A - B$ is closed.

11. Suppose $J - A$, the complement of A, is convex. What can you say about $J - A$ if A is open; is closed? Justify your answers.

4.23 The Theory of Order

The major results of the chapter have now been derived. The remainder is devoted to three related lower key ideas: order of points, in terms of betweenness; separation of points; and perspectivity of points.

The study of order gives a new perspective on our subject by affording a new mode of comparison with Euclidean geometry, which is pervaded by order relations of points.

In Euclidean geometry the join ab of the distinct points a and b is defined to be the segment ab—and segment ab is characterized as the set of points between a and b (Section 2.1). So we may assert: In Euclidean geometry a point is in the join of a and b, $a \neq b$, if and only if it is *between* a and b. This property is easily converted into a definition of betweenness in the abstract theory.

Definition. Suppose $x \subset ab$ and $a \neq b$. Then we say x is *between* a and b and write (axb).

Many of the familiar properties of betweenness in Euclidean geometry are covered in the following theorems, which are included to shed light on the order relations of points.[7]

Theorem 4.37. (abc) *implies the distinctness of* a, b *and* c.

PROOF. By definition $b \subset ac$ and $a \neq c$. Then Theorem 4.9 implies $a \neq b$ and $c \neq b$. ☐

The following result is easily proved.

Theorem 4.38. (abc) *implies* (cba).

Theorem 4.39. (abc) *implies that* (bca) *is false.*

PROOF. Suppose (abc) and (bca). By definition

$$b \approx ac \quad \text{and} \quad c \approx ba. \tag{1}$$

Eliminating b between the two relations in (1), we have $c \approx aca = ac$, so that $a \approx c/c = c$. This contradicts (abc), and the theorem follows. ☐

Corollary 4.39. (abc) *implies that* (bac), (cab), (bca) *and* (acb) *are false.*

Next an existence theorem.

Theorem 4.40. *If* $a \neq b$, *there exist* x, y *and* z *such that* (axb), (yab) *and* (abz).

PROOF. J1 implies $x \subset ab$ for some x. By definition (axb).

J5 implies $y \subset a/b$ for some y. Then $a \subset by = yb$. If $y = b$ then $a = b$, contrary to hypothesis. Thus $y \neq b$, and by definition (yab). Similarly for the third conclusion. ☐

Theorems 4.37–4.40 fairly well exhaust the order properties of three points. We continue with properties of four points which satisfy two order relations.

Theorem 4.41. (abc), (bcd) *imply* (abd), (acd).

PROOF. By hypothesis

$$b \approx ac, \quad c \approx bd. \tag{1}$$

Eliminating c in (1), we get $b \approx abd$, so that $b/b \approx ad$ and $b \approx ad$. Suppose

---

[7] See Prenowitz and Jordan [1], Chapter 10, Sections 1–3, 15.

$a = d$. Then $b \approx ad$ implies $b \approx dd = d$, contrary to (bcd). Thus $a \neq d$ and by definition (abd). The same method works for (acd). □

The following theorem can be proved somewhat similarly.

Theorem 4.42. (abc), (acd) *imply* (abd), (bcd).

Theorem 4.43. (abx), (aby) *imply* (xay) *is false and* (xby) *is false.*

PROOF. Suppose (xay). By the last theorem, (xba), (xay) imply (bay). By Corollary 4.39, (bay) implies (aby) is false, contrary to the hypothesis. Thus (xay) is false.

Suppose (xby). The hypothesis and (xby) imply

$$b \approx ax, \qquad b \approx ay, \qquad b \approx xy. \tag{1}$$

Now we eliminate a between the first two relations in (1). We have

$$b/x \approx a, \qquad b/y \approx a,$$

so that $b/x \approx b/y$ and J6 yields

$$by \approx bx. \tag{2}$$

Next we eliminate x between (2) and $b \approx xy$. We have

$$by/b \approx x, \qquad b/y \approx x,$$

so that $by/b \approx b/y$, which implies

$$by \approx b.$$

Thus $y \approx b/b = b$. But (aby) implies $y \neq b$ (Theorem 4.37). We infer from the contradiction the falsity of (xby). □

The following theorem can be proved in a similar manner.

Theorem 4.44. (axb), (ayb) *imply* (xay) *is false and* (xby) *is false.*

EXERCISE

Prove Theorems 4.42 and 4.44, and complete the proof of Theorem 4.41.

4.24 Ordered Sets of Points

Not all of the familiar properties of linear order in Euclidean geometry hold in our theory. In order to express the basic difference we introduce the following definition.

Definition. A set of points S is *ordered* if for any three distinct points of S, one is contained in the join of the other two, or equivalently, one is between the other two.

In Euclidean geometry every line is an ordered set. This property is not deducible from J1–J7. (See Exercise 1 at the end of Section 6.10.)

EXERCISE

In the triode model and the Cartesian model (Sections 2.22, 2.24) is every segment ordered? Every ray? Explain.

4.25 Separation of Points by a Point

The notion of separation of figures is one of the most basic and familiar ideas in geometry and has its roots in experience as universal as crossing a road. The simplest case of the concept is now introduced.

Definition. If $p \approx ab$ we say p *separates* a and b. If in addition $p \neq a, b$ we say p *strictly* separates a and b.

Note the following properties:

(1) p strictly separates a and b if and only if $p \approx ab$ and p, a, b are distinct.
(2) p separates a and b, but not strictly, if and only if $p = a = b$.

Observe that an interior point of a convex set A separates each point of A from some point of A; an extreme point of A strictly separates no two points of A.

An Objection. It may seem a misuse of the term separation to permit nonstrict separation—to say p separates p and p. Certainly the important case is that of strict separation. However, nonstrict separation may be considered a degenerate case comparable to and corresponding to the terminology that permits a point to be a degenerate segment. The inclusion of the degenerate case causes no trouble in the application of the separation concept, but in fact facilitates it.

EXERCISES (SEPARATION OF POINTS)

1. Prove: If o separates the pairs a, b and b, c, then o does not separate a, c provided $a, b, c \neq o$.

2. Prove: If o separates the pairs a, b and b, c and c, d, then o separates a, d.

3. Prove: If o separates the pairs a, a' and b, b', then o separates each point of ab from some point of $a'b'$ and vice versa.

4. (a) Let points a and b be given. Prove that the set of points x that are separated from b by a is the ray a/b.
   (b) Prove: If a separates b and c, then a separates each point of segment ab from c.

(c) Prove: If a separates b and c, then a separates each point of ab from each point of ac.

(d) Prove: If a separates b and c, then a separates each point of a/b from each point of a/c.

4.26 Perspectivity and Precedence of Points

A third way of viewing the relation $x \subset ab$ is now given, which leads to geometric relations quite different from betweenness and separation.

The notion of perspective points (or perspective figures) is very familiar in projective geometry and has its origin, as the term perspective suggests, in visual experience. In a projective geometry points p and q are said to be *perspective from* point a if p and q are collinear with and distinct from a (Figure 4.34). The idea is easily adapted to a geometry in which segment or join is the central idea. If $a \approx ob$, we may say a is perspective to b from o (Figure 4.35). Suppose $o \neq b$. Then the relation of a and b to o can be expressed in visual terms: sighting from o we can see through a to b; or in kinematic terms: moving directly from o to b we encounter a before b.

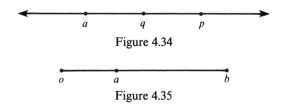

Figure 4.34

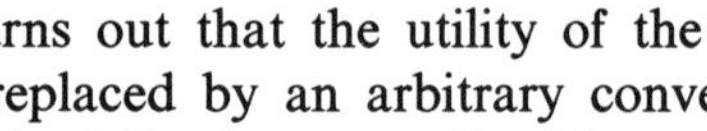

Figure 4.35

These interpretations—indeed, the basic geometric situation itself—suggest describing the relation $a \approx ob$ in terms of "precedence". Thus if $a \approx ob$, we say a *precedes b with respect to o* and write $a < b(o)$.

In the sequel it turns out that the utility of the concept is amplified considerably if o is replaced by an arbitrary convex set. So the formal definition is stated in the following generalized form (Figure 4.36).

Definition. Let the convex set K be given. If $a \subset bK$, we say *a precedes b with respect to K* or *b succeeds a with respect to K* and write $a < b(K)$ or $b > a(K)$.

Suppose $a < b(K)$, $a \neq b$ (Figure 4.37). Then $a \subset bK$ so that $a \subset bk$ where $k \subset K$. Thus the relation can be expressed in this way: from some point of K we can see through a to b; or reversely, from b we can see through a to a point of K.

The situation is suggestive of order properties, not in the sense of betweenness but in the sense of a 2-term relation of order such as "less than" for real numbers or "to the left of" for points of a line.

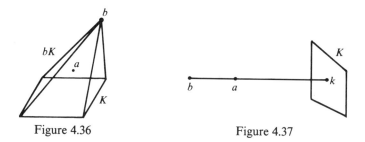

Figure 4.36   Figure 4.37

The relation of precedence with respect to K does have two familiar order properties: it is transitive for every triple of points, and irreflexive for every point *not* in K.

Theorem 4.45.

(a) (Transitive Property) $a < b(K)$ *and* $b < c(K)$ *imply* $a < c(K)$, *for all* a, b, c.
(b) (Irreflexive Property) $a \not< a(K)$ *if and only if* $a \not\subset K$.

PROOF. (a) Given $a \subset bK$, $b \subset cK$, then $a \subset cKK = cK$ and $a < c(K)$ by definition.

(b) We show $a < a(K)$ is equivalent to $a \subset K$. Suppose $a < a(K)$. Then $a \subset aK$, and by J7 $a = a/a \approx K$. Conversely suppose $a \subset K$. Then $a = aa \subset aK$ and $a < a(K)$. □

Corollary 4.45. $a < a(K)$ *if and only if* $a \subset K$.

Next a sort of double perspectivity relation is introduced.

Definition. Let K be a convex set (Figure 4.38). Suppose $a, b \subset K$, $a < b(K)$ and $b < a(K)$. Then we say a and b are (*mutually*) *perspective in* K and write $a \equiv b(K)$.

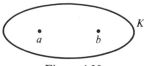

Figure 4.38

Observe that $a \equiv b(K)$, $a \neq b$, may be described so: In K one can see from a through b and from b through a.

It is not hard to prove that the relation of perspectivity in K is an "equivalence in K" in the sense of the following definition.

Definition. Let S be a set, not necessarily a set of points, and R a relation that relates pairs of elements of S. If a has the relation R to b, we write $a \, R \, b$. Suppose R satisfies the following conditions for a, b, c in S.

(1) $a R a$ (reflexive property)
(2) $a R b$ implies $b R a$ (symmetric property)
(3) $a R b$ and $b R c$ imply $a R c$ (transitive property).

Then we call R an *equivalence relation* (in S) or simply an *equivalence*.

Theorem 4.46. *The relation of perspectivity in a convex set K is an equivalence relation in the set K.*

PROOF. Apply the last theorem and corollary.   □

Finally the relation of perspectivity is linked to the notion of open set.

Theorem 4.47. *Let K be a convex set. Then the following conditions are equivalent:*

(i) $a \equiv b(K)$;
(ii) *a and b are in the join of two points of K;*
(iii) *a and b are in an open subset of K.*[8]

PROOF. We show that (i), (ii), (iii) imply one another cyclically.
    Suppose (i). By definition $a \subset bK$, $b \subset aK$. Then $a \subset bp$ and $b \subset aq$, where $p, q \subset K$ (Figure 4.39). Thus $a \subset aqp$, and $a \subset pq$ follows. By symmetry $b \subset pq$ and condition (ii) holds.
    Suppose (ii). Then (iii) follows, since any join of points is open (Theorem 4.28).

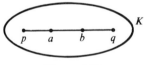

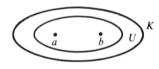

Figure 4.39   Figure 4.40

Suppose (iii). Say $a, b \subset U \subset K$ and U is open (Figure 4.40). Using Theorem 4.25 we have

$$U \subset aU \subset aK.$$

Thus $b \subset aK$ and $b < a(K)$ by definition. Similarly $a < b(K)$, and (i) follows.   □

Corollary 4.47. *Let K be an open set. Then any two points of K are perspective in K.*

EXERCISES (PRECEDENCE AND PERSPECTIVITY OF POINTS)

1. Prove: If K is convex, $a < b(K)$ and $b < a(K)$, then $a, b \subset K$.

2. Let K be convex and nonempty. Prove that x is an interior point of K if and only if x precedes each point of K with respect to K.

[8] Compare Rockafellar [1], p. 164, last paragraph.

3. Let K be convex. Prove that x is an interior point of K if and only if (a) x precedes some interior point of K with respect to K, or (b) x is perspective in K to some interior point of K.

4. Let K be convex. Prove that x is an extreme point of K if and only if x precedes no point of K with respect to K, except itself.

5. Choose a convex set K in Euclidean geometry. Choose p in K. Find all points x in K such that (a) $x < p(K)$; (b) $x > p(K)$; (c) $x \equiv p(K)$. Make different choices of p. Do you always get the same set for (a)? If not, what changes do you observe? The same for (b); for (c).

6. Prove: If K is convex and any two points of K are mutually perspective in K, then K is open. (Compare Corollary 4.47.)

5 Join Geometries

Having completed the formulation of our basic set of postulates J1–J7 in Chapter 4, we now characterize our basic object of study, the idea of a join geometry. A set of examples of join geometries is presented to give body to the theory and indicate its range of application. Many of the examples have properties that are not valid in Euclidean geometry and can be used to settle questions of deducibility—specifically to show that certain statements are not deducible from J1–J7. One of the examples is essentially n-dimensional Euclidean geometry presented in algebraic form—a subject of considerable importance in modern mathematics. Another example is an infinite dimensional geometry which contains nonempty convex sets that have empty interiors. The chapter contains a discussion of isomorphism of join geometries.

5.1 The Concept of a Join Geometry

Our postulates have been presented—for simplicity of exposition—in two installments: J1–J4 in Chapter 2 and J5–J7 in Chapter 4. However, from the beginning our intention has been to develop a geometrical theory based on J1–J7. Here is a list of the postulates.

J1 (Existence Law). $ab \neq \emptyset$.
J2 (Commutative Law). $ab = ba$.
J3 (Associative Law). $(ab)c = a(bc)$.
J4 (Idempotent Law). $aa = a$.
J5 (Existence Law). $a/b \neq \emptyset$.

J6 (Four Term Transposition Law). If $a/b \approx c/d$, then $ad \approx bc$.
J7 (Idempotent Law). $a/a = a$.

Now our basic object of study is introduced.

Definition. A *join geometry* (or *join space*) is a model or interpretation of postulates J1–J7.

These postulates, as we know (Section 2.4), involve two basic (or primitive) terms J and $\cdot$, which denote a set and a join operation in the set. A *join geometry*, then, is a specific pair $(J, \cdot)$, consisting of a set J and a join operation $\cdot$ in J, which satisfy postulates J1–J7. Sometimes it is convenient to say, a bit imprecisely, that J is a join geometry if the context makes clear which operation $\cdot$ is being considered.

J1–J7 can be called the postulates for the theory of join geometries or simply the postulates for a join geometry.

An Example of a Join Geometry

Our first example of a join geometry may be unfamiliar but is not complicated. Let C be a Euclidean semicircle (Figure 5.1) which is *open*, that is, C does not contain its endpoints. Let a and b be distinct points of C. Then the *open arc ab* of C consists of all points of C that are between a and b; a and b, the *endpoints* of the open arc, are not contained in it.

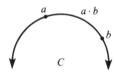

Figure 5.1

Take J to be the set C. Define the operation $\cdot$ in J by taking $a \cdot b$ to be the open arc ab of C, provided $a \neq b$, and taking $a \cdot a$ to consist of a. Observe that $\cdot$ is a join operation in J (Section 2.2).

The join system $(J, \cdot)$ is a join geometry and will be designated by JG1. To show JG1 is a join geometry we must verify that it satisfies postulates J1–J7. Thus in J1–J7 the abstract terms J and $\cdot$ must be assigned the specific meanings given in the definition of JG1, and the resulting statements shown to be true.

J1 is satisfied, since the join $a \cdot b$ as defined in JG1 always contains at least one point. J2 is easily seen to be satisfied, since the open arc ab and the open arc ba are identical sets.

Consider now J3: $(ab)c = a(bc)$. To verify J3 we must apply the definition of join of two sets to $(ab)c$ and $a(bc)$, keeping in mind that "element of J" and the "join of two elements of J" are specified in the definition of JG1. When this is done it turns out that $(ab)c$ and $a(bc)$ are identical sets for all choices of a, b, c in C.[1] For example, if b is between a and c (Figure 5.2), then $(ab)c$ and $a(bc)$ are equal to the open arc ac. Thus JG1 satisfies J3.

It is trivial that JG1 satisfies J4.

To verify J5, J6 and J7, the postulates that involve extension, we must interpret this defined term in JG1. First consider the case $a \neq b$ (Figure 5.3). Then a/b is seen to be the set of points x of C such that the open arc bx contains a. Let p be the endpoint of C which satisfies the condition that a is between b and p. Then a/b can be identified as the set of points of C that are between a and p. Now consider a/a. Suppose $x \neq a$. Then $a \cdot x \not\ni a$, since $a \cdot x$ is an open arc. Thus $a/a \not\ni x$, and $a/a = a$ easily follows.

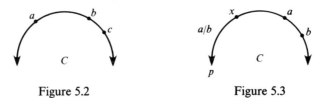

Figure 5.2   Figure 5.3

In any case $a/b \neq \emptyset$ and JG1 satisfies J5. Incidentally we have shown that JG1 satisfies J7.

Finally, by patiently considering cases (a few of which are illustrated in Figure 5.4), it can be verified that JG1 satisfies J6:

$$a/b \approx c/d \quad \text{implies} \quad ad \approx bc.[2]$$

Conclusion: JG1 is a join geometry.

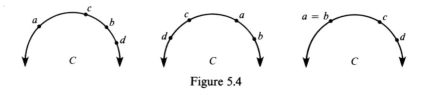

Figure 5.4

[1] The verification is analogous to the verification of EJ3 for collinear points (Section 2.1).

[2] The verification is analogous to that of J6 for collinear points in a Euclidean geometry (Exercise 1 at the end of Section 4.9).

5.2 A List of Join Geometries

Figure 5.5 gives, in tabular form, a list of join geometries. A discussion of the examples follows.

Figure 5.5

Join Geometry	J	$a \cdot b$	Diagram
JG1	An open Euclidean semicircle C.	The open arc ab of C if $a \neq b$; a, if $a = b$.	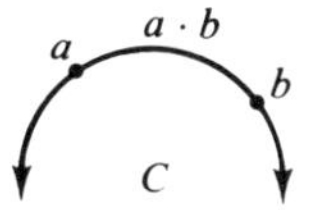
JG2	An open Euclidean hemisphere H.	The great circle open arc ab of H if $a \neq b$; a, if $a = b$.	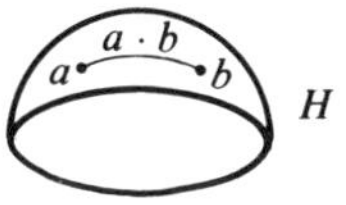
JG3	A Euclidean line.	Segment ab if $a \neq b$; a, if $a = b$.	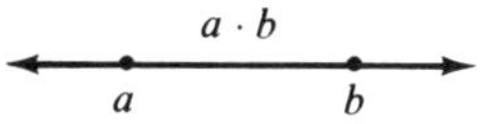
JG4	A Euclidean plane.	Segment ab if $a \neq b$; a, if $a = b$.	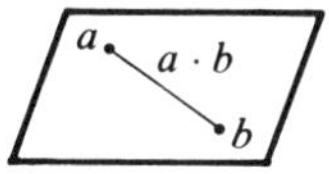
JG5	A Euclidean 3-space.	Segment ab if $a \neq b$; a, if $a = b$.	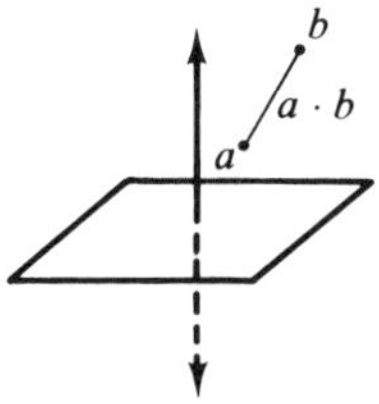
JG6	The interior of a Euclidean triangle.	Segment ab if $a \neq b$; a, if $a = b$.	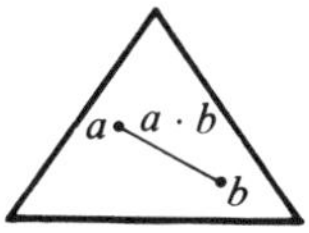
JG7	The interior of a Euclidean tetra-hedron.	Segment ab if $a \neq b$; a, if $a = b$.	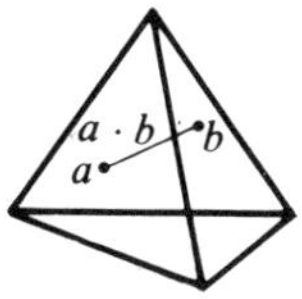
JG8	The interior of a Euclidean circle.	Segment ab if $a \neq b$; a, if $a = b$.	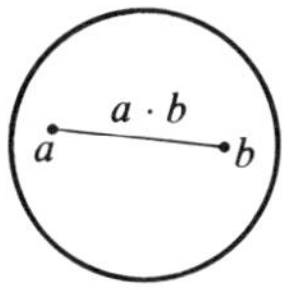

Figure 5.5 continued

Join Geometry	J	$a \cdot b$	Diagram
JG9	The interior of a Euclidean sphere.	Segment ab if $a \neq b$; a, if $a = b$.	
JG10	The union of three distinct open rays R_1, R_2, R_3 with endpoint o in a Euclidean plane.	Point a, if $a = b$; segment ab if $a \neq b$ and a, b are in the same ray R_1, R_2 or R_3; otherwise the union of segments oa and ob.	
JG11	In a Euclidean plane the family F of open rays with endpoint o, which lie on a given side of a given line L containing o. (F is an *open halfpencil* of rays.)	The set of rays of family F which lie between ray a and ray b, if $a \neq b$; a, if $a = b$.	
JG12	In a Euclidean 3-space the family F of open rays with endpoint o, which lie on a given side of a given plane P containing o. (F is an *open halfbundle* of rays.)	The set of rays of family F which lie between ray a and ray b, if $a \neq b$; a, if $a = b$.	
JG13	The real line as the set $\mathbb{R}$ of all real numbers.	The open interval or segment ab of $\mathbb{R}$ if $a \neq b$, that is, the set of real numbers between a and b; a, if $a = b$.	
JG14	The "punctured" real line as the set $\mathbb{R}'$ of real numbers distinct from 0.	Number a, if $a = b$; segment ab of $\mathbb{R}'$ if $a \neq b$ and 0 is not between a and b; otherwise the segment ab with 0 deleted.	

Figure 5.5 continued

Join Geometry	J	$a \cdot b$	Diagram

| JG15 | The rational line as the set $\mathbb{Q}$ of all rational numbers. | The open interval or segment ab of $\mathbb{Q}$ if $a \neq b$, that is, the set of rational numbers between a and b; a, if $a = b$. | |
| JG16 | The Cartesian plane as the set of all ordered pairs (x_1, x_2) of real numbers. Notation: $a = (a_1, a_2)$, $b = (b_1, b_2)$ and so on. | The set of all $x = (x_1, x_2)$ such that $x_1 \subset a_1 \cdot b_1$, $x_2 \subset a_2 \cdot b_2$; $a_1 \cdot b_1$ and $a_2 \cdot b_2$ as in JG13. | |

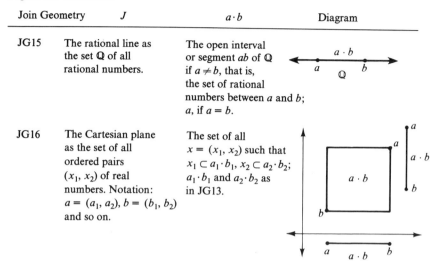

Discussion of the Join Geometries

JG1 was discussed in Section 5.1. In JG2 any arc of a great circle lies in a plane that contains the center of H. But such a plane intersects H in an open semicircle. Hence each join $a \cdot b$ in H is contained in an open semicircle of H. Now suppose C in JG1 is chosen as an open semicircle of H. Let a, $b \subset C$. Then the join of a and b as determined in JG1 is precisely its join as determined in JG2. In other words the join operation in JG1 is the same as that in JG2 but restricted to the points of C. In a certain sense, then, JG1 is a *subgeometry* of JG2.

JG3, JG4 and JG5 will naturally be called *Euclidean join geometries*. Observe that JG4 is a subgeometry of JG5 in the sense indicated above. Similarly JG3 is a subgeometry of JG4 and so of JG5.

Now consider JG6–JG9. JG6 may be considered to arise from JG4 by taking J to be a certain open convex subset of JG4 and the join operation to be that of JG4 restricted in its application to the set J. JG8 arises similarly from JG4, and the pair JG6, JG8 may be considered subgeometries of JG4. JG7 and JG9 are related to JG5 as JG6 and JG8 are to JG4. In a sense the triple JG6, JG8, JG4 gives rise to the triple JG7, JG9, JG5 by "increasing dimensions by 1".

Query. Does JG3 have subgeometries?

JG10 and the triode model (Section 2.22) are similar. However, they differ in that the rays in the triode model are closed, while those in JG10 are open. The difference is not trivial: the triode model is not a join geometry. [See Exercise 11(c) below.]

JG11 and JG12 may be called *ray geometries*, since the term point (element of J) is interpreted to mean ray. Note in JG11 and JG12 that no ray in line L or plane P is in J. JG11 bears a close analogy to JG1. This is seen quite vividly by choosing semicircle C of JG1 to have center o and to lie on the side of L that contains the rays of F (Figure 5.6). Then let each point p of C correspond to the ray of F that contains p. Now let points a, b of C correspond to rays R, S of F. Observe that the rays that correspond to the points of $a \cdot b$ form $R \cdot S$. Next note the similar analogy JG12 bears to JG2. A study of ray geometries as generalizations of Euclidean spherical geometry will be made in Chapter 10.

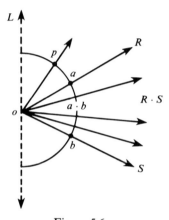

Figure 5.6

JG14 is a sort of subgeometry of JG13—but of a new type, since it arises from a *nonconvex* subset of JG13. It also may be considered an arithmetic version of JG10 where, instead of three, only two rays are used.

JG15 is an analogue of JG13 gotten by changing the number field.

JG16 is the example most remote from our elementary geometrical experience. It is however a purely algebraic formulation of the Cartesian model (Section 2.24).

The verification that JG1–JG16 are join geometries is a tedious though in most cases a routine chore and is left to the exercises below. J1, J2, J4, J5 and J7 are quite immediate in each example, but J3 and J6 usually require arguments by cases. That JG1 is a join geometry was indicated in Section 5.1, although not all of the details were presented.

EXERCISES

1. (a) Fill in the details of the verification that JG1 is a join geometry.
   (b) Verify that JG11 is a join geometry.

2. (a) Verify by intuitive geometric reasoning that JG3–JG5 are join geometries.
   (b) Verify that JG6–JG9 are join geometries. How can part (a) be used here?

3. Verify that JG13–JG15 are join geometries.

4. In JG2, describe and illustrate
   (a) if a and b are distinct: $[a, b]$, $\mathcal{I}[a, b]$, $\mathcal{F}[a, b]$, $\mathcal{C}[a, b]$, $\mathcal{B}[a, b]$, a/b, ab/ab;
   (b) if a, b, c are distinct and no one is in the join of the other two: abc, $[a, b, c]$, $\mathcal{I}[a, b, c]$, $\mathcal{F}[a, b, c]$, $\mathcal{C}[a, b, c]$, $\mathcal{B}[a, b, c]$, a/bc, $a/(b/c)$, abc/a, abc/ab, abc/abc.

5. (a) Verify that JG2 is a join geometry.
   (b) The same for JG12.

6. The same as Exercise 4 for JG8.

7. The same as Exercise 4 for JG10.

8. (a) Verify that JG10 is a join geometry.
   (b) Could a join geometry similar to JG10 be constructed from more than three open rays emanating from the same point; less than three such rays? Explain.

9. The same as Exercise 4 for JG16. Try not to miss any cases.

10. (a) Verify by intuitive geometric reasoning that JG16 is a join geometry. (See Section 2.24.)
    (b) Try to prove that JG16 is a join geometry by using the fact that JG13 is a join geometry and that join in JG16 is defined in terms of join in JG13.

11. In each part show that the given join system (Section 2.2) satisfies J1–J4 but is *not* a join geometry.
    (a) J is a Euclidean plane; $a \cdot b$ is the closed segment ab if $a \neq b$; a if $a = b$. Does the conclusion hold if plane is replaced by 3-space; by line?
    (b) J is a closed Euclidean circular region; $a \cdot b$ is the segment ab if $a \neq b$; a if $a = b$. Does the conclusion hold if circular is replaced by triangular; spherical; tetrahedral?
    (c) *The triode model* (Section 2.22). J is the union of three distinct closed rays with endpoint o in a Euclidean plane or 3-space; $a \cdot b$ is a if $a = b$, segment ab if $a \neq b$ and a, b are in the same closed ray; otherwise it is the union of point o and segments oa and ob.

12. In each of the following parts show that the given join system constructed from the set of natural numbers $\{1, 2, 3, \ldots\}$ is not a join geometry. Determine which of J1–J7 are satisfied.
    (a) J is the set of natural numbers; $a \cdot b = \{a, b\}$.
    (b) J is the set of natural numbers; $a \cdot b = $ GCD (a, b), the greatest common divisor of a and b, that is, the common divisor of a and b which is divisible by every common divisor of a and b.
    (c) J is the set of natural numbers; $a \cdot b$ is the set of common divisors of a and b.
    (d) J is the family of nonempty subsets of the set of natural numbers; $a \cdot b = a \cup b$, the union of sets a and b.

13. Consider a punctured Euclidean plane which is JG4 modified as follows. Delete a given point o from a Euclidean plane to obtain J. To obtain $a \cdot b$, join

a and *b* as in JG4 but delete *o* if *o* is a point of this join. Show that a punctured Euclidean plane is not a join geometry. Determine which of J1–J7 hold for a punctured Euclidean plane.

14. Suppose $(J, \cdot)$ is a finite join geometry, that is, J is a finite set. Prove that J contains at most one point.

5.3 Deducibility and Counterexamples

In the development of a mathematical theory the question often arises whether a certain statement is or is not deducible from the postulates. We may attempt to deduce it. If we succeed, there is no problem—but suppose our efforts to deduce it fail. We may begin to suspect that the statement is just not deducible from the postulates. There is an important procedure— though not an automatic one—for confirming that a statement is not deducible from a given set of postulates. We illustrate the procedure by means of an example.

In Euclidean geometry a point of a segment "divides the segment into two segments" (Figure 5.7). This somewhat vague principle can be formulated in our theory as follows:

$$\text{If } p \subset ab, \text{ then } ab = ap \cup p \cup pb. \qquad \text{(A)}$$

The question is this: Can principle (A) be deduced from J1–J7? Suppose the answer is yes. Then any join geometry $(J, \cdot)$ must satisfy (A), since it already satisfies J1–J7. Suppose we can find a join geometry J which falsifies (A). Then we infer that (A) cannot be deduced from J1–J7, for otherwise a contradiction results in J.

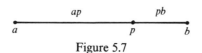

Figure 5.7

Let us then search the list of join geometries in Section 5.2 for one which falsifies (A). We need an example containing a segment that is quite unlike Euclidean segments. JG16 furnishes such a segment. In JG16 let $a = (1, 1)$, $b = (-1, -1)$ and $p = (0, 0)$. Then $p \subset ab$, but ab contains points other than those of $ap \cup p \cup pb$, namely, the points indicated by the shading in Figure 5.8.

We conclude that JG16 does not satisfy (A) and infer that (A) is not deducible from J1–J7. The join geometry JG16 is called, since it falsifies (A), a *counterexample* for (A).

Providing counterexamples is an important function of the list JG1– JG16. It also indicates the richness and variety of application that is made

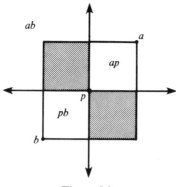

Figure 5.8

possible by an abstract postulational treatment of a mathematical discipline.

Remarks.

(1) Our example above of a proof of nondeducibility may foster a false impression. It may suggest that counterexamples are easily found to settle questions of deducibility. Actually we chose principle (A) as an illustration because we knew the list JG1–JG16 *contained* the required counterexample. The work involved in constructing or discovering a counterexample in a mathematical theory may be very difficult and time-consuming.

(2) The method of proving a statement is not deducible from a given set of postulates is not an absolute one. To see this let us analyze the example above. We assumed (A) *was* deducible from J1–J7 and showed that JG16 would have two contradictory properties. But JG16 was constructed from the real number system. Thus the deducibility of (A) would imply the inconsistency of the real number system. Hence our conclusion that (A) was *not* deducible from J1–J7 was based on the implicit assumption that the real number system is consistent. This indicates that the method of establishing nondeducibility by means of a counterexample is a relative, not an absolute one. It is nevertheless very important and very useful.

EXERCISES

1. Show that each of the postulates J5, J6 and J7 is not deducible from postulates J1–J4. (*Hint.* Examine the join systems in Exercise 11 at the end of Section 5.2.)

2. Show that each of the following statements is not deducible from J1–J7.
   (a) Every segment is ordered (Section 4.24).
   (b) Every ray is ordered.
   (c) If some nondegenerate segment is ordered, then every segment is ordered.

(d) If some nondegenerate segment is ordered, then some segments are not ordered.

(e) If every segment is ordered, then every ray is ordered.

3. Show that the formula

$$(a/b)(a/c) = a/bc$$

is not deducible from J1–J7. Compare Exercise 1 at the end of Section 4.13. How does this exercise pertain to the notion of interior of an angle?

4. Show that the following statements are not deducible from J1–J7.
   (a) The closed segment $[a, b]$ is a closed convex set (Section 4.22).
   (b) A segment ab has a unique pair of endpoints.
   (c) If $ab \approx cd$, there exist p, q, r such that $a, b, c, d \subset pqr$.

5. Is the following statement deducible from J1–J7: If A and B are convex, $A \cup B = J$ and $A \cap B = \varnothing$, then A and B have a common boundary point? Justify your answer.

6. Try to determine whether each formula below is deducible from J1–J7. Justify your assertions.
   (a) $a(b/a) = ab/a$;
   (b) $a/(b/a) = a/b$.

7. Show that J1 is deducible from J2–J7. Which of J2–J7 did you use?[3]

5.4 The Existence of Points

Do J1–J7 imply the existence of points, or equivalently, that $J \neq \varnothing$? This may appear trivial when first encountered, for the postulates seem to involve that points do exist. Consider J1. In the form $ab \neq \varnothing$, it asserts ab contains at least one point and so seems to imply $J \neq \varnothing$. However, the symbols a, b in J1 denote points—elements of J. Thus J1 asserts:

If $a, b \subset J$, then $ab \neq \varnothing$.

It says: If you can find points a and b, then you are assured of the existence of a point in ab. Thus J1 involves existence only hypothetically. The same is true for J2–J7. None of the postulates asserts categorically the existence of points. So it is hard to see how to frame a proof that $J \neq \varnothing$.

Can we find a counterexample for the statement $J \neq \varnothing$? Can we find a join geometry with no points? Yes. But the answer may seem strange at first sight. Choose J to be $\varnothing$. Define $\cdot$ so: If $a, b \subset J$, then $a \cdot b = \varnothing$. The operation $\cdot$ may be called the *null* join operation in $\varnothing$. We assert that $(J, \cdot)$ satisfies postulates J1–J7. Consider for instance J3. It asserts: If $a, b, c \subset J$, then $(ab)c = a(bc)$. Since $J = \varnothing$, postulate J3 imposes no restriction on $(J, \cdot)$—we say J3 is satisfied *vacuously*. The same argument

---

[3] Observe that the standard set of postulates for a field is not independent.

applies to the other postulates, and we conclude $(J, \cdot)$ is a join geometry. Thus no absolute statement on the existence of points is deducible from J1–J7.

5.5 Isomorphism of Join Systems

In this section the idea that two join geometries have the same structure or are isomorphic is introduced. Actually the idea will be defined more broadly for any two join systems and will be applied to systems that are not join geometries (see Section 10.17).

In a Euclidean line consider points $p_1, \ldots, p_5$ related as indicated in Figure 5.9. The points are used to construct a finite join system (Section 2.2). Let $J = \{p_1, p_2, p_3, p_4, p_5\}$. Define $\cdot$ in J as follows: $a \cdot b$ is the set of points of J that are between a and b if $a \neq b$; $a \cdot a = a$. Thus $p_1 \cdot p_5 = \{p_2, p_3, p_4\}$, $p_4 \cdot p_2 = p_3$ and $p_1 \cdot p_2 = \varnothing$. The pair $(J, \cdot)$ is a join system. In effect the closed segment $p_1 p_5$ has been reduced to five points, and the Euclidean join operation has been reduced correspondingly to yield a join operation in J.

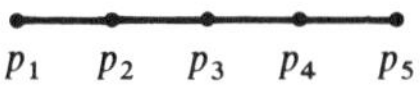

$$\bullet\!\!-\!\!\!-\!\!\!\bullet\!\!-\!\!\!-\!\!\!\bullet\!\!-\!\!\!-\!\!\!\bullet\!\!-\!\!\!-\!\!\!\bullet$$
$$p_1 \quad p_2 \quad p_3 \quad p_4 \quad p_5$$

Figure 5.9

In a Euclidean plane consider, as indicated in Figure 5.10, open rays $R_1, \ldots, R_5$ with endpoint o lying on a given side of a line L containing o. (Compare JG11.) Let $J' = \{R_1, R_2, R_3, R_4, R_5\}$. Define $\circ$ in J' in this way: $A \circ B$ is the collection of rays of J' that lie between A and B, if $A \neq B$; $A \circ A = A$. Then $(J', \circ)$ is a join system.

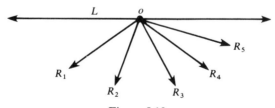

Figure 5.10

$(J, \cdot)$ and $(J', \circ)$ are similar in some sense—they seem to have the same structure although composed of different material. If this is so, each element of J should have a corresponding element in J', and the two elements should play similar roles in their respective systems.

To illustrate this consider p_1. It plays a special role in J as a sort of extreme point. Thus it should correspond to R_1 or R_5, which are "extreme rays" in J'. Suppose we make p_1 correspond to R_1, writing $p_1 \rightarrow R_1$. Similarly we require $p_5 \rightarrow R_5$. Since p_2 is "next to" p_1, we require $p_2 \rightarrow R_2$; similarly $p_4 \rightarrow R_4$. Finally we must have $p_3 \rightarrow R_3$. Thus we have defined a one-to-one correspondence between J and J' symbolized by

$$p_i \rightarrow R_i, \qquad i = 1, \ldots, 5. \tag{1}$$

The correspondence (1) relates the join of two points in J to the join of their corresponding rays in J'. To indicate the nature of the relation, consider p_1 and p_4 in J and the corresponding rays R_1 and R_4. By definition $p_1 \cdot p_4 = \{p_2, p_3\}$ and $R_1 \circ R_4 = \{R_2, R_3\}$. Since $p_2 \rightarrow R_2, p_3 \rightarrow R_3$, we see that (1) effects a correspondence between the points of $p_1 \cdot p_4$ and the rays of $R_1 \circ R_4$. It isn't hard to verify this property for any pair of points p_i, p_j of J and the corresponding rays R_i, R_j. The result can be expressed in this way:

Let p denote any element of J and R its correspondent in J' as determined by (1). Then

$$p \subset p_i \cdot p_j \quad \text{if and only if} \quad R \subset R_i \circ R_j.$$

The relation between the two join systems becomes particularly simple if J is chosen so that its points lie on their corresponding rays, as in Figure 5.11.

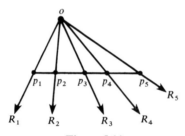

Figure 5.11

Our discussion suggests the following

Definition. Let $(J, \cdot)$ and $(J', \circ)$ be join systems. Let there exist a one-to-one correspondence $x \rightarrow x'$ between J and J' which satisfies the following condition:

$$a \rightarrow a', \qquad b \rightarrow b'$$

implies

$$x \subset a \cdot b \quad \text{if and only if} \quad x' \subset a' \circ b'. \tag{2}$$

Then we say the join systems $(J, \cdot)$ and $(J', \circ)$ are *isomorphic* (or have the *same structure*) or that $(J, \cdot)$ is *isomorphic* to $(J', \circ)$. Sometimes it is convenient to say $(J, \cdot)$ and $(J', \circ)$ are *isomorphic under* the correspondence $x \rightarrow x'$.

Note that the two join systems discussed above are isomorphic under the correspondence $p_i \to R_i$.

A Euclidean Application of the Isomorphism Concept

Most of the join geometries in the list JG1–JG16 are not individual mathematical systems but families of similar ones. Consider JG3. The Euclidean line in its description is not specified: It could be any line in a Euclidean plane or 3-space. So there are an infinitude of geometries of the type JG3. They seem to be equivalent in some sense—their join operations behave in the same way. We assert that any two examples of JG3 are isomorphic—and proceed to justify this for a particular case.

Let L and L' be parallel lines in a Euclidean plane P (Figure 5.12). Let $\cdot$ and $\circ$ be the Euclidean join operations in L and L'. Choose o in P not on L or L'. Let x be any point of L. Define x' to be the intersection of line ox and L'. Then $x \to x'$ is a one-to-one correspondence between the lines L and L'. Let $a, b \subset L$. Suppose $x \subset a \cdot b$. Then $x' \subset a' \circ b'$. Conversely, if $x' \subset a' \circ b'$, then $x \subset a \cdot b$. Thus condition (2) above holds and $(L, \cdot)$ is isomorphic to $(L', \circ)$ by definition.

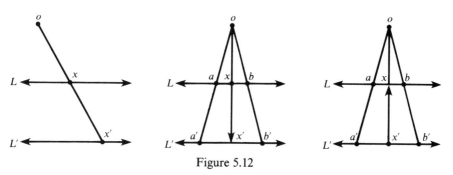

Figure 5.12

Remark. The correspondence $x \to x'$ between L and L' in the example above can be described in simple terms without referring to the join operations: Let a, b, c be points of L, so that a', b', c' are their correspondents in L' (Figure 5.13). Then c is between a and b if and only if c' is

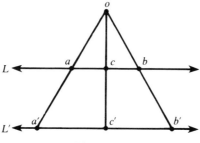

Figure 5.13

between a' and b'. In view of this the correspondence is said to *preserve betweenness*.

Isomorphic and Automorphic Correspondences

The correspondences which establish that two join systems are isomorphic deserve a name.

Definition. Suppose $(J, \cdot)$ and $(J', \circ)$ are join systems and $x \to x'$ is a one-to-one correspondence between J and J' such that

$$a \to a', \qquad b \to b'$$

implies

$$x \subset a \cdot b \quad \text{if and only if} \quad x' \subset a' \circ b'.$$

Then $x \to x'$ is called an *isomorphic correspondence* or an *isomorphism* between $(J, \cdot)$ and $(J', \circ)$. An isomorphism between $(J, \cdot)$ and itself is an *automorphic correspondence* or *automorphism* of $(J, \cdot)$.

The simplest example of an automorphism of $(J, \cdot)$ is the *identity correspondence* in J defined by $x \to x$.

Query. In the example above can you find another isomorphism between $(L, \cdot)$ and $(L', \circ)$?

Notation. For the sake of convenience we often employ the symbol $\cdot$ to refer to two or more join operations. For example, in studying isomorphic systems we may say $(J, \cdot)$ is isomorphic to $(J', \cdot)$. If the context indicates the operation being studied we may dispense with the symbol $\cdot$ altogether, merely saying J is isomorphic to J' and denoting joins in J and J' by ab and $a'b'$ respectively. Similarly the symbol $/$ may be used to denote two or more extension operations.

Isomorphisms and the Extension Operation

An isomorphism $x \to x'$ between join systems J and J' affects extensions as well as joins. (Note that the definition of extension in Section 4.1 is applicable to any join system.) Suppose $a, b \subset J$ and $x \subset a/b$. Then $a \subset bx$. By definition of isomorphism $a' \subset b'x'$, so that $x' \subset a'/b'$.

Conversely suppose $x \subset J$ and $x' \subset a'/b'$. Then $a' \subset b'x'$, and by definition of isomorphism $a \subset bx$. Thus $x \subset a/b$.

Summary Statement (A). Let $x \to x'$ be an isomorphism between join systems J and J'. Then

$$a \to a', \qquad b \to b'$$

implies

$$x \subset a/b \quad \text{if and only if} \quad x' \subset a'/b'.$$

Correspondences Operate on Sets

A one-to-one correspondence C, $x \to x'$, between two sets S and S' is easily extended to apply to subsets of S. Let $A \subset S$. Then each x in A has a correspondent x' in S'. The set of all such x' is called the *correspondent* of *set A* (*under C*) and is denoted by A'. (In effect if we limit the application of the correspondence to the elements of A, we get a "restricted" correspondence between A and A'.)

The idea of correspondent of a set is useful in dealing with isomorphisms.

Suppose $x \to x'$ is an isomorphism between J and J'. Let $a, b \subset J$. Then $a \to a'$, $b \to b'$. We are interested in how the correspondence $x \to x'$ relates the sets ab and $a'b'$.

First suppose $x \subset ab$. Then by definition of isomorphism $x' \subset a'b'$. Thus the correspondent of each element of ab is in $a'b'$.

Now let $y \subset a'b'$. Then $y \subset J'$ and therefore is the correspondent of some element $z \subset J$, that is, $y = z'$. Then $z' \subset a'b'$, and by definition of isomorphism $z \subset ab$. Thus each element of $a'b'$ is the correspondent of some element of ab. Hence $a'b'$ is identified as the set of correspondents of the elements of ab, that is, $a'b' = (ab)'$.

Summary Statement (B). Let $x \to x'$ be an isomorphism between the join systems J and J'. Then $a, b \subset J$ implies

$$(ab)' = a'b'. \tag{3}$$

In view of (3) we may say the isomorphism $x \to x'$ "maps joins on joins" or "preserves joins". The condition (3) can be used to characterize isomorphisms.

A similar result for extension can be derived from Summary Statement (A) above.

Let $x \to x'$ be an isomorphism between the join systems J and J'. Then $a, b \subset J$ implies

$$(a/b)' = a'/b'. \tag{4}$$

In view of (4) we say the isomorphism "maps extensions on extensions" or "preserves extensions".

Finally we show that an "isomorph" of a join geometry must be a join geometry.

Theorem 5.1. *Let J and J' be isomorphic join systems. Then if J is a join geometry, so is J'.*

PROOF. Given that J satisfies postulates J1–J7, we must verify that J' satisfies J1–J7. Since J and J' are isomorphic, there exists an isomorphism $x \to x'$ between J and J'.

To verify that J' satisfies J1, we suppose $p, q \subset J'$ and show $pq \neq \emptyset$. Since $p, q \subset J'$, p and q are the correspondents of elements, say $a, b \subset J$, that is, $p = a'$, $q = b'$. We know $ab \neq \emptyset$. Let $x \subset ab$. Then by definition of isomorphism $x' \subset a'b' = pq$, and $pq \neq \emptyset$.

Verification of J2: As was seen above, we can take a', b' ($a, b \subset J$) as arbitrary elements of J'. Then by (3) and postulate J2 for J we have

$$a'b' = (ab)' = (ba)' = b'a'.$$

Verification of J3: We show

$$(a'b')c' = a'(b'c') \qquad\qquad\qquad\qquad \text{(i)}$$

for $a, b, c \subset J$. Suppose

$$x' \subset (a'b')c'. \qquad\qquad\qquad\qquad\qquad \text{(ii)}$$

We show

$$x' \subset a'(b'c'). \qquad\qquad\qquad\qquad\qquad \text{(iii)}$$

The relation (ii) implies

$$x' \subset y'c', \quad \text{where } y' \subset a'b'.$$

By definition of isomorphism

$$x \subset yc, \qquad y \subset ab,$$

so that $x \subset (ab)c = a(bc)$. Now $x \subset az$, $z \subset bc$, so that $x' \subset a'z'$, $z' \subset b'c'$, and (iii) follows. In the same way it can be shown that (iii) implies (ii), and we infer (i).

To verify J4 observe that

$$a'a' = (aa)' = a'.$$

J5 can be verified by the method used for J1.

Verification of J6: Suppose $a'/b' \approx c'/d'$. Then

$$a/b \approx c/d \qquad [\text{Summary Statement (A)}],$$

$$ad \approx bc \qquad [\text{J6 for } J],$$

$$a'd' \approx b'c' \qquad [(2)].$$

To verify J7 we have

$$a'/a' = (a/a)' = a'. \qquad\qquad\qquad \square$$

Remark on the Theorem. Note in the proof that the validity of a postulate for J' was deduced from its validity for J. The type of argument employed in the proof can be used to show that if J and J' are isomorphic join systems, then any property of J as a *join system* will hold for J'. To illustrate this let $x \to x'$ be an isomorphism between J and J'. Suppose that in J, a and b satisfy $ab = \varnothing$. Then their correspondents a' and b' satisfy $a'b' = \varnothing$. If A is a convex set in J, then A' will be a convex set in J'. If in J, $ab = ap \cup p \cup pb$, then in J', $a'b' = a'p' \cup p' \cup p'b'$.

EXERCISES (ISOMORPHISM)

1. Show informally that the relation "J is isomorphic to J'" is an equivalence in the collection of all join systems (Section 4.26).

2. Show that the given pair of join geometries are isomorphic.
   (a) JG1 and JG11, choosing semicircle C of JG1 in the plane of JG11 with center o and lying on the side of L that contains the rays of F.
   (b) JG2 and JG12, choosing H of JG2 in the 3-space of JG12 with center o and lying on the side of P that contains the rays of F.
   (c) JG3 and JG13.

3. Show in each part that J is isomorphic to J', where the operation in J and J' is Euclidean join.
   (a) J and J' are coplanar lines.
   (b) J and J' are noncoplanar (skew) lines.
   (c) J and J' are noncollinear segments with a common endpoint.
   (d) J and J' are segments lying on parallel lines.
   (e) J and J' are any segments in 3-space.
   (f) J is a segment and J' a ray that have a common endpoint and are noncollinear.

4. In a Euclidean plane find a segment J and a line J' such that, if $\cdot$ is Euclidean join in each, you can prove J is isomorphic to J'. [*Suggestion.* Try to apply Exercise 3(f).]

5. Let a, b, o be noncollinear points in a Euclidean plane. Let $J = ab$ and $\cdot$ be the Euclidean join in J. Let J' be the set of rays with endpoint o that lie inside $\angle aob$. Define $\cdot$ in J' so: $R \cdot S$ is the set of rays between R and S if $R \neq S$; $R \cdot R = R$. Show J and J' are isomorphic.

6. Try to generalize Exercise 5.

7. (a) Find some automorphisms of JG13 in which 0 is fixed, that is, corresponds to itself.
   (b) Find an automorphism of JG13 in which 0 corresponds to 1; 0 corresponds to a given number a; a corresponds to b, where a and b are given numbers.

8. (a) Let $J = \mathbb{N}$, the set of natural numbers $1, 2, 3, \ldots$. Define $\cdot$ in J so: $a \cdot b$ is the set of numbers of $\mathbb{N}$ that are between a and b if $a \neq b$; $a \cdot a = a$. Let $J' = \mathbb{N} - \{1\}$, and define $\cdot$ as in J. Show J is isomorphic to J'.
   (b) How many automorphisms does J have? Justify your answer.

9. Suppose J and J' are isomorphic join geometries. Prove if J satisfies the given condition, then J' does also.
   (a) If $p \subset ab$, then $ab = ap \cup p \cup bp$.
   (b) If a, b, c are distinct, one of $a \subset bc$, $b \subset ac$ or $c \subset ab$ holds.
   (c) If $ab = cd$, then $\{a, b\} = \{c, d\}$.

10. Suppose the correspondence $x \to x'$ is an isomorphism between join geometries J and J'. Prove:
    (a) If A is convex in J, then A' is convex in J'.
    (b) If A is convex and $p \subset \mathcal{I}(A)$, then $p' \subset \mathcal{I}(A')$.
    (c) If A is convex and $p \subset \mathcal{C}(A)$, then $p' \subset \mathcal{C}(A')$.

11. Show that a one-to-one correspondence $x \to x'$ between two join systems J and J' which satisfies the condition

$$a, b \subset J \quad \text{implies} \quad (ab)' \subset a'b'$$

need not be an isomorphism between J and J'.

5.6 A Class of Join Geometries of Arbitrary Dimension

Our object now is to construct a class of join geometries which will turn out to be algebraic counterparts of Euclidean geometries of dimensions $1, 2, 3, \ldots$. Our approach is suggested by analytic geometry. JG3, JG4, JG5 correspond to Euclidean geometries of dimensions 1, 2, 3. The latter can be treated algebraically by means of coordinate systems in which points are represented by real numbers x_1, ordered pairs of real numbers (x_1, x_2), ordered triples of real numbers (x_1, x_2, x_3).

It is natural then to choose the basic set for an arbitrary join geometry of our class to be the set of all ordered n-tuples $(x_1, \ldots, x_n)$ of real numbers. This is denoted by $\mathbb{R}^n$, where $\mathbb{R}$ stands for the real number system. We take $\mathbb{R}^1$ to be $\mathbb{R}$.

How shall we define join in $\mathbb{R}^n$? An easier question first: How can join be defined in $\mathbb{R}^1$ and $\mathbb{R}^2$? This is our approach: $\mathbb{R}^1$ and $\mathbb{R}^2$ are algebraic counterparts of a line L and a plane P in Euclidean geometry. In L and P join is defined in terms of segment or betweenness. It should be possible using coordinate representation in L and P to obtain algebraic counterparts of segment or betweenness.

The definition of join in $\mathbb{R}^n$ is motivated by considering the problem in $\mathbb{R}^1$ and $\mathbb{R}^2$. The reader may, if he wishes, omit the discussion and proceed directly to the definition which appears in the final paragraph of this section.

Join in $\mathbb{R}^1$

The definition of join in $\mathbb{R}^1$ is easily obtained. Let p, q, r be points of line L and a, b, c their coordinates (Figure 5.14). Then r is geometrically between p and q if and only if c is algebraically between a and b, that is,

$$a < c < b \quad \text{or} \quad b < c < a.$$

Thus we have the desired definition of join.

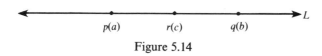

Figure 5.14

Definition (Initial Formulation). In $\mathbb{R}^1$ the operation $\cdot$ is defined as follows: $a \cdot b$ is the set of real numbers between a and b if $a \neq b$; $a \cdot a = a$. [Note then that $(\mathbb{R}^1, \cdot)$ is JG13.]

There is no question that this is the correct definition—but it is not useful for our purposes as stated. It will not generalize to $\mathbb{R}^2$. It has the disadvantage that it does not distinguish one number between a and b from any other. Can we find an algebraic formulation of betweenness which turns out numbers between a and b at will?

The answer we seek is found in the process of forming weighted averages. Suppose in a course you have grades of 70 and 90, but do not know the *weights* to be assigned to them. What can your average be? The answer is: Any number between 70 and 90. For example, if the weights are 2 and 3, the average is $(2 \times 70 + 3 \times 90)/5 = 82$. The numbers between 70 and 90 are precisely all the weighted averages of 70 and 90. (Note that if in the illustration the weights $\frac{2}{5}$ and $\frac{3}{5}$ are used, the same result is obtained.)

Proposition I. *Let a, b, x be real numbers, $a \neq b$. Then x is between a and b if and only if there exist real numbers λ, μ such that*

$$x = \lambda a + \mu b, \quad 0 < \lambda, \mu, \quad \lambda + \mu = 1. \tag{1}$$

PROOF. Suppose (1). Assume $a < b$. Then

$$x = \lambda a + (1 - \lambda)b = b - \lambda(b - a) < b.$$

Similarly

$$x = (1 - \mu)a + \mu b = a + \mu(b - a) > a.$$

Thus $a < x < b$, and x is between a and b. Similarly for the case $b < a$.

Conversely suppose x is between a and b. Solving the equations

$$x = \lambda a + \mu b, \quad \lambda + \mu = 1$$

for λ, μ, we get

$$\lambda = \frac{x - b}{a - b}, \qquad \mu = \frac{a - x}{a - b}.$$

These values for λ, μ satisfy the equations in (1). The condition $0 < \lambda$, μ follow easily from the hypothesis. Thus (1) is verified and the proof is complete.   □

Now the effective definition of join in $\mathbb{R}^1$ can be stated.

Definition (Final Formulation). For a, b in $\mathbb{R}^1$, $a \cdot b$ is the set of all real numbers x expressible in the form

$$x = \lambda a + \mu b,$$

where λ, μ are real numbers that satisfy

$$0 < \lambda, \mu, \qquad \lambda + \mu = 1.$$

Note that the definition applies uniformly for all $a \cdot b$ and yields $a, a = a$.

Join in $\mathbb{R}^2$

We present a purely algebraic criterion for betweenness of three points in a plane.

Proposition II. *Let P be a Euclidean plane in which a Cartesian coordinate system has been introduced. Let a, b, p be points of P ($a \neq b$) with respective coordinate pairs (a_1, a_2), (b_1, b_2), (p_1, p_2). Then p is between a and b if and only if there exist real numbers λ, μ such that*

$$p_1 = \lambda a_1 + \mu b_1, \tag{1}$$

$$p_2 = \lambda a_2 + \mu b_2 \tag{2}$$

and

$$0 < \lambda, \mu, \qquad \lambda + \mu = 1. \tag{3}$$

PROOF. Let L be the line determined by a and b (Figure 5.15).
    Case I. $a_1 \neq b_1$. Then L is not vertical—let its equation be

$$x_2 = mx_1 + n. \tag{4}$$

Suppose p is between a and b. Then p_1 is between a_1 and b_1. By Proposition I there exist real numbers λ, μ such that (1) and (3) hold.

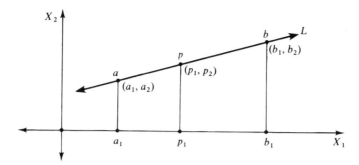

Figure 5.15

Employing (4), (1) and (3), we have

$$p_2 = mp_1 + n, \tag{5}$$

$$p_2 = m(\lambda a_1 + \mu b_1) + n,$$

$$p_2 = \lambda(ma_1 + n) + \mu(mb_1 + n),$$

$$p_2 = \lambda a_2 + \mu b_2. \tag{6}$$

Thus (2) holds and the forward implication is established.

Conversely, suppose there exist numbers λ, μ satisfying (1), (2), (3). Then by retracing steps starting from (6) we reach (5), so that p is on line ab. By Proposition I, (1) and (3) imply p_1 is between a_1 and b_1. Thus p is between a and b.

Case II. $a_1 = b_1$. Then L is vertical and is representable by $x_1 = n$. Since this can be written $x_1 = mx_2 + n$ (where $m = 0$), the argument of Case I applies. $\qquad\qquad\square$

Proposition II suggests the following

Definition of Join in $\mathbb{R}^2$. Let (a_1, a_2), (b_1, b_2) be elements of $\mathbb{R}^2$. Then their *join* $(a_1, a_2) \cdot (b_1, b_2)$ is the set of all elements (x_1, x_2) of $\mathbb{R}^2$ that satisfy

$$x_1 = \lambda a_1 + \mu b_1, \qquad x_2 = \lambda a_2 + \mu b_2,$$

where λ, μ are real numbers such that

$$0 < \lambda, \mu, \qquad \lambda + \mu = 1.$$

The definition is immediately extendible from ordered pairs to ordered n-tuples and yields the

Definition of Join in $\mathbb{R}^n$. Let $(a_1, \ldots, a_n)$, $(b_1, \ldots, b_n)$ be elements of $\mathbb{R}^n$. Then their *join* $(a_1, \ldots, a_n) \cdot (b_1, \ldots, b_n)$ is the set of all elements $(x_1, \ldots, x_n)$ of $\mathbb{R}^n$ that satisfy

$$x_1 = \lambda a_1 + \mu b_1, \ldots, x_n = \lambda a_n + \mu b_n,$$

where λ, μ are real numbers such that

$$0 < \lambda, \mu, \qquad \lambda + \mu = 1.$$

5.7 $\mathbb{R}^n$ Is Converted into a Vector Space

$\mathbb{R}^n$ is a set of elements which may be called *points* or *(arithmetic) vectors*. In itself it has no structure. By defining join in $\mathbb{R}^n$ we have imposed a structure on it—we have formed a join system. The join of two points $(a_1, \ldots, a_n)$ and $(b_1, \ldots, b_n)$ is defined by operations of multiplication and addition applied to the individual components of the points. In order to study the properties of $\cdot$ in $\mathbb{R}^n$, in particular to prove that $(\mathbb{R}^n, \cdot)$ is a join geometry, it becomes very convenient to define elementary algebraic operations on *points* of $\mathbb{R}^n$ and to study how they act on the points without explicitly considering components of the points.

In technical terms we are going to convert $\mathbb{R}^n$ into a vector space over the real number system—but no knowledge of the concept of vector space will be assumed.

First a convenient notation which facilitates consideration of n-tuples as individual entities. An element of $\mathbb{R}^n$ is represented by one of the letters $a, b, c, \ldots$, and components of the element by the same letter with appropriate subscript $1, \ldots, n$. Thus the real number a_i is the ith component of a—so that $a = (a_1, \ldots, a_n)$, $b = (b_1, \ldots, b_n)$ and so on.

In $\mathbb{R}^n$ we take equality naturally to mean equality of corresponding components, that is, $a = b$ or $(a_1, \ldots, a_n) = (b_1, \ldots, b_n)$ means $a_1 = b_1, \ldots, a_n = b_n$.

Two algebraic operations are defined in $\mathbb{R}^n$.

Definition. If a and b are elements of $\mathbb{R}^n$, their *sum* $a + b$ or $(a_1, \ldots, a_n) + (b_1, \ldots, b_n)$ stands for $(a_1 + b_1, \ldots, a_n + b_n)$.

Definition. If λ is a real number and a is an element of $\mathbb{R}^n$, their *product* λa or $\lambda(a_1, \ldots, a_n)$ stands for $(\lambda a_1, \ldots, \lambda a_n)$.

$\mathbb{R}^n$ is closed under the operation of addition, since the sum of two elements of $\mathbb{R}^n$ is a uniquely determined element of $\mathbb{R}^n$. Similarly $\mathbb{R}^n$ is closed under the operation of multiplying by a real number.

Addition in $\mathbb{R}^n$ satisfies the familiar commutative and associative laws of school algebra:

$$a + b = b + a, \tag{1}$$

$$(a + b) + c = a + (b + c). \tag{2}$$

The element $0 = (0, \ldots, 0)$ is called the *zero element* (or *zero vector*) of $\mathbb{R}^n$ and has the familiar property

$$a + 0 = 0 + a = a. \tag{3}$$

Subtraction can be treated in several different but equivalent ways. We choose the following. Define the *negative* of a by

$$-a \text{ or } -(a_1, \ldots, a_n) \quad \text{stands for} \quad (-a_1, \ldots, -a_n).$$

Observe the familiar property

$$a + (-a) = -a + a = 0. \tag{4}$$

Then we define the *difference* of a and b by $a - b = a + (-b)$, and have

$$a - b = c \quad \text{if and only if} \quad a = b + c. \tag{5}$$

In effect the properties (1)–(5) enable us to apply the familiar theory of addition and subtraction in school algebra to the elements of $\mathbb{R}^n$, as if they were real numbers rather than n-tuples of real numbers.

Consider now the multiplication operation. The following properties hold, in which λ, μ denote real numbers:

$$1a = a, \quad (-1)a = -a, \quad 0a = 0, \tag{6}$$

$$\lambda(\mu a) = (\lambda \mu)a, \tag{7}$$

$$\lambda(a + b) = \lambda a + \lambda b, \tag{8}$$

$$(\lambda + \mu)a = \lambda a + \mu a, \tag{9}$$

$$\lambda a = 0 \quad \text{if and only if} \quad \lambda = 0 \text{ or } a = 0. \tag{10}$$

In effect properties (6)–(10) enable us to apply the familiar theory of the multiplicative algebra of real numbers. But it is important to keep in mind that we are multiplying neither two numbers nor two vectors but a vector by a number of produce a vector. Also note that two zeros have been denoted by the same symbol: the zero *number* 0 and the zero *vector* 0. No confusion should arise, since the context will indicate the meaning intended.

A Word on Notation. Elements of $\mathbb{R}^n$ will be denoted by $a, b, c, \ldots$. Greek letters $\alpha, \beta, \gamma, \ldots$ will denote real numbers. Sometimes quotients of real numbers are indicated by the stroke /, but no essential ambiguity is involved, since the context will indicate whether quotient of real numbers or extension of points is intended.

With the above qualifications it will be observed that the familiar formal manipulative properties of addition, subtraction and multiplication in school algebra are valid. Thus we can apply the theory of linear equations in the usual way. As an illustration we can solve

$$\alpha x + \beta y + \gamma z = 0$$

for x, assuming of course $\alpha \neq 0$, and get

$$x = \left(-\frac{\beta}{\alpha}\right)y + \left(-\frac{\gamma}{\alpha}\right)z.$$

5.8 Restatement of the Definition of Join in $\mathbb{R}^n$

The definition of join in $\mathbb{R}^n$ is now restated in terms of the operations of addition of vectors and multiplication of vectors by real numbers studied in the last section.

The definition (Section 5.6, last paragraph) asserts that the join

$$(a_1, \ldots, a_n) \cdot (b_1, \ldots, b_n)$$

is the set of all $(x_1, \ldots, x_n)$ of $\mathbb{R}^n$ that satisfy

$$x_1 = \lambda a_1 + \mu b_1, \ldots, x_n = \lambda a_n + \mu b_n, \tag{1}$$

where λ, μ are real numbers such that

$$0 < \lambda, \mu, \qquad \lambda + \mu = 1.$$

The condition (1) implies

$$x = (x_1, \ldots, x_n) = (\lambda a_1 + \mu b_1, \ldots, \lambda a_n + \mu b_n)$$

$$= (\lambda a_1, \ldots, \lambda a_n) + (\mu b_1, \ldots, \mu b_n)$$

$$= \lambda(a_1, \ldots, a_n) + \mu(b_1, \ldots, b_n)$$

$$= \lambda a + \mu b,$$

so that

$$x = \lambda a + \mu b. \tag{2}$$

It is easily seen, by retracing steps, that (2) also implies (1). Thus join in $\mathbb{R}^n$ may be characterized as follows.

Restatement of Definition. Let a and b be points of $\mathbb{R}^n$. Then their *join* $a \cdot b$ is the set of all points x of $\mathbb{R}^n$ expressible in the form

$$x = \lambda a + \mu b,$$

where λ, μ are real numbers which satisfy

$$0 < \lambda, \mu, \qquad \lambda + \mu = 1.$$

In this characterization of the operation, no mention is made of the fact that elements of $\mathbb{R}^n$ are sequences of real numbers, or even of the index n itself. This will make possible a formal algebraic study of join in $\mathbb{R}^n$ based on the vector algebraic properties (1)–(10) of Section 5.7. This is not a trivial advantage. It signifies, for example, that $\mathbb{R}^{27}$ is no more difficult to study than $\mathbb{R}^2$.

5.9 Proof That ($\mathbb{R}^n$, ·) Is a Join Geometry

($\mathbb{R}^n$, ·), where · is the join operation defined in $\mathbb{R}^n$, is a join geometry, as we now proceed to prove. It is worth noting that the only properties of $\mathbb{R}^n$ employed are the "vector algebraic" principles (1)–(10) of Section 5.7. Of course properties of real numbers are used.

First two lemmas are needed.

Lemma 5.1. *In $\mathbb{R}^n$, $x \subset (ab)c$ if and only if x is expressible in the form*

$$x = \lambda a + \mu b + \nu c, \qquad 0 < \lambda, \mu, \nu, \quad \lambda + \mu + \nu = 1. \tag{1}$$

PROOF. Suppose

$$x \subset (ab)c. \tag{2}$$

Then

$$x \subset yc, \qquad y \subset ab \tag{3}$$

holds for some y. By definition of join in $\mathbb{R}^n$, (3) yields

$$x = \alpha y + \beta c, \qquad 0 < \alpha, \beta, \quad \alpha + \beta = 1; \tag{4}$$

$$y = \gamma a + \delta b, \qquad 0 < \gamma, \delta, \quad \gamma + \delta = 1. \tag{5}$$

Eliminating y between (4), (5), we have

$$x = (\alpha\gamma)a + (\alpha\delta)b + \beta c. \tag{6}$$

Let λ, μ, ν be determined by

$$\lambda = \alpha\gamma, \qquad \mu = \alpha\delta, \qquad \nu = \beta. \tag{7}$$

Substituting in (6) the values of λ, μ, ν given by (7), we have

$$x = \lambda a + \mu b + \nu c. \tag{8}$$

The relations (7), (4) and (5) imply

$$0 < \lambda, \mu, \nu, \tag{9}$$

since the product of positive numbers is positive. Furthermore we have

$$\lambda + \mu + \nu = \alpha\gamma + \alpha\delta + \beta$$

$$= \alpha(\gamma + \delta) + \beta = \alpha + \beta = 1. \tag{10}$$

The relations (8), (9) and (10) yield (1), so that (2) implies (1).

Conversely suppose x satisfies (1). In essence we retrace our steps to (2). First we determine $\alpha, \beta, \gamma, \delta$ to satisfy (7) and the relation $\alpha + \beta = 1$; we get

$$\alpha = 1 - \nu = \lambda + \mu, \qquad \beta = \nu,$$

$$\gamma = \frac{\lambda}{\lambda + \mu}, \qquad \delta = \frac{\mu}{\lambda + \mu}. \tag{11}$$

The relations (11) and (1) imply

$$0 < \alpha, \beta, \gamma, \delta \quad \text{and} \quad \gamma + \delta = 1. \tag{12}$$

Substituting (7) into (1), we have

$$x = \alpha\gamma a + \alpha\delta b + \beta c = \alpha(\gamma a + \delta b) + \beta c. \tag{13}$$

Let $y = \gamma a + \delta b$, so that (5) holds. Substituting y for $\gamma a + \delta b$ in (13) we get

$$x = \alpha y + \beta c. \tag{14}$$

By (12), $0 < \alpha, \beta$, and (11) implies $\alpha + \beta = 1$. These relations with (14) yield (4). The relations (4) and (5) yield (3), from which (2) follows, and the proof is complete.   □

Similarly we can prove the following result.

Lemma 5.2. *In* $\mathbb{R}^n$, $x \subset a(bc)$ *if and only if* x *is expressible in the form* (1).

Theorem 5.2. *Let* · *be the join operation defined in* $\mathbb{R}^n$. *Then* $(\mathbb{R}^n, \cdot)$ *is a join geometry.*

PROOF. We verify the postulates in order.

Verification of J1, $ab \neq \emptyset$. $\frac{1}{2}a + \frac{1}{2}b$ is an element of ab.

Verification of J2, $ab = ba$. This follows directly from the definition of join.

Verification of J3, $(ab)c = a(bc)$. By Lemma 5.1, $x \subset (ab)c$ is equivalent to

$$x = \lambda a + \mu b + \nu c, \quad 0 < \lambda, \mu, \nu, \quad \lambda + \mu + \nu = 1. \tag{1}$$

By Lemma 5.2, $x \subset a(bc)$ is equivalent to (1). Thus $x \subset (ab)c$ if and only if $x \subset a(bc)$, and J3 holds.

Verification of J4, $aa = a$. aa is the set of x expressible in the form $\lambda a + \mu a$, where $0 < \lambda, \mu$ and $\lambda + \mu = 1$. $\lambda a + \mu a = (\lambda + \mu)a = a$. Hence $aa = a$.

Verification of J5, $a/b \neq \emptyset$. $2a - b$ is an element of a/b. For

$$a = \tfrac{1}{2}(2a - b) + \tfrac{1}{2}b.$$

Verification of J6, $a/b \approx c/d$ *implies* $ad \approx bc$. Suppose $a/b \approx c/d$. Then $x \approx a/b$ and $x \approx c/d$ for some x. Hence

$$a \approx bx, \quad c \approx dx,$$

which imply

$$a = \lambda b + \mu x, \quad 0 < \lambda, \mu, \quad \lambda + \mu = 1, \tag{2}$$

$$c = \lambda' d + \mu' x, \quad 0 < \lambda', \mu', \quad \lambda' + \mu' = 1. \tag{3}$$

Eliminating x between (2) and (3), we have

$$\mu' a + \mu\lambda' d = \mu'\lambda b + \mu c. \tag{4}$$

Furthermore

$$\mu' + \mu\lambda' = \mu' + \mu(1 - \mu') = \mu'(1 - \mu) + \mu = \mu'\lambda + \mu. \qquad (5)$$

Let ρ be the common value of the members of (5). Note ρ is positive, so that (4) implies

$$\frac{\mu'}{\rho}a + \frac{\mu\lambda'}{\rho}d = \frac{\mu'\lambda}{\rho}b + \frac{\mu}{\rho}c. \qquad (6)$$

Let y denote either member of (6). Then one easily checks that

$$y \subset ad, \qquad y \subset bc.$$

Thus $ad \approx bc$, and J6 holds.

Verification of J7, $a/a = a$. Suppose $x \subset a/a$. Then $ax \supset a$ and

$$a = \lambda a + \mu x, \qquad 0 < \lambda, \mu, \quad \lambda + \mu = 1.$$

Solving for x, we have

$$x = \mu^{-1}(a - \lambda a) = \mu^{-1}(1 - \lambda)a = \mu^{-1}\mu a = a.$$

Thus a is the only possible member of a/a, and $a/a = a$ by J5.

This completes the proof that ($\mathbb{R}^n$, ·) is a join geometry. □

The join geometry ($\mathbb{R}^n$, ·) may be called the *arithmetic* or *vector* join geometry of *dimension n*, or simply the join geometry $\mathbb{R}^n$. It is constantly used as a model to interpret and test the ideas and principles of the theory. Its principal advantage is that it provides in concrete numerical terms a representation of the linear geometry of Euclidean *n*-space.

EXERCISES (ON THE JOIN GEOMETRY $\mathbb{R}^n$)

1. Prove directly from the definitions of convexity and join in $\mathbb{R}^n$ that ab is convex.

2. Prove Lemma 5.2.

3. Prove: abc is the set of all x expressible in the form
$$x = \lambda a + \mu b + \nu c, \qquad 0 < \lambda, \mu, \nu, \quad \lambda + \mu + \nu = 1.$$

4. Prove:
   (a) $[a, b]$ is the set of all x expressible in the form
   $$x = \lambda a + \mu b, \qquad 0 \leqslant \lambda, \mu, \quad \lambda + \mu = 1.$$
   (b) $[a, b, c]$ is the set of all x expressible in the form
   $$x = \lambda a + \mu b + \nu c, \qquad 0 \leqslant \lambda, \mu, \nu, \quad \lambda + \mu + \nu = 1.$$

5. Prove:
   (a) a/b is the set of all x expressible in the form
   $$x = \lambda a + \mu b, \qquad \mu < 0, \quad \lambda + \mu = 1.$$
   (b) $[a, b]/[a, b]$ is the set of all x expressible in the form
   $$x = \lambda a + \mu b, \qquad \lambda + \mu = 1.$$

6. Prove:
   (a) $\mathcal{C}(ab) = [a, b]$.
   (b) $\mathcal{C}[a, b] = [a, b]$. [Compare Exercise 4(a) at the end of Section 5.3.]

7. Prove: If $ab = cd$, then $\{a, b\} = \{c, d\}$. [Compare Exercise 4(b) at the end of Section 5.3.]

Notation. In Exercises 8 through 11 $a_1, \ldots, a_m$ denote m points of $\mathbb{R}^h$, not the components of a point (vector) of $\mathbb{R}^m$.

8. Prove: $a_1 \ldots a_m$ is the set of all points x expressible in the form
   $$x = \lambda_1 a_1 + \cdots + \lambda_m a_m, \quad 0 < \lambda_1, \ldots, \lambda_m, \quad \lambda_1 + \cdots + \lambda_m = 1.$$
   (*Hint.* Use induction.)

9. Prove: $[a_1, \ldots, a_m]$ is the set of all points x expressible in the form
   $$x = \lambda_1 a_1 + \cdots + \lambda_m a_m, \quad 0 \leqslant \lambda_1, \ldots, \lambda_m, \quad \lambda_1 + \cdots + \lambda_m = 1.$$

10. Prove: $[a_1, \ldots, a_m]/[a_1, \ldots, a_m]$ is the set of all x expressible in the form
    $$x = \lambda_1 a_1 + \cdots + \lambda_m a_m, \quad \lambda_1 + \cdots + \lambda_m = 1.$$

11. Prove:
    (a) $\mathcal{C}(a_1 \ldots a_m) = [a_1, \ldots, a_m]$.
    (b) $\mathcal{C}[a_1, \ldots, a_m] = [a_1, \ldots, a_m]$.
    (c) $\mathcal{B}[a_1, \ldots, a_m] = \mathcal{F}[a_1, \ldots, a_m]$.

12. In $\mathbb{R}^n$ let α be a fixed nonzero real number and a a fixed vector. Prove that each of the following correspondences is an automorphism of $\mathbb{R}^n$:
    $$\text{(a) } x \to \alpha x; \quad \text{(b) } x \to x + a.$$

13. Suppose a one-to-one correspondence $x \to x'$ between $\mathbb{R}^n$ and itself has the properties:
    (i) $(x + y)' = x' + y'$ for x, y in $\mathbb{R}^n$;
    (ii) $(\alpha x)' = \alpha x'$ for each x in $\mathbb{R}^n$ and each real number α.

    Prove that $x \to x'$ is an automorphism of $\mathbb{R}^n$.

14. *Project.* In the discussion in Sections 5.6–5.9 replace the system of real numbers $\mathbb{R}$ by the system of rational numbers $\mathbb{Q}$. Develop a theory for the set $\mathbb{Q}^n$ of n-tuples of rational numbers analogous to that for $\mathbb{R}^n$ and obtain a join geometry $(\mathbb{Q}^n, \cdot)$ analogous to $(\mathbb{R}^n, \cdot)$.

5.10 Linear Inequalities and Halfspaces

In $\mathbb{R}^2$, let $a = (0, 0)$, $b = (1, 0)$ and $c = (0, 1)$. Then by Lemmas 5.1 and 5.2, abc is the set of all $x = (x_1, x_2)$ expressible in the form
$$x = \lambda(0, 0) + \mu(1, 0) + \nu(0, 1) = (\mu, \nu)$$

where

$$0 < \lambda, \mu, \nu, \qquad \lambda + \mu + \nu = 1.$$

Thus $x_1 = \mu$, $x_2 = \nu$ and it follows quickly that abc is the set of all $x = (x_1, x_2)$ satisfying

$$x_1 > 0, \qquad x_2 > 0, \qquad x_1 + x_2 < 1. \tag{1}$$

Let us interpret these results in a plane with given coordinate system. Represent the element $x = (x_1, x_2)$ of $\mathbb{R}^2$ by the point x' with coordinates (x_1, x_2). First plot the elements a, b, c (Figure 5.16). To plot the *linear* inequalities in (1) consider the associated *linear* equalities

$$x_1 = 0, \qquad x_2 = 0, \qquad x_1 + x_2 = 1. \tag{2}$$

These are represented by *lines* $a'c'$, $a'b'$, $b'c'$. Then the inequality $x_1 > 0$ is represented by the open halfplane that lies on the right of the line $a'c'$—that is, a *point* x' is in this open halfplane if and only if $x_1 > 0$. Similarly $x_2 > 0$ is represented by the open halfplane that lies above the line $a'b'$; and $x_1 + x_2 < 1$ is represented by the open halfplane that lies below and to the left of the line $b'c'$. Thus an element $x = (x_1, x_2)$ satisfies the system of inequalities (1) if and only if its corresponding point x' lies in the three indicated halfplanes. Hence the join abc is represented by the intersection of the three halfplanes: this is the interior of triangle $a'b'c'$, namely, the join $a'b'c'$.

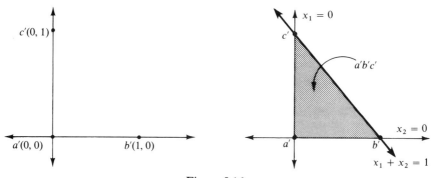

Figure 5.16

Now combine the inequalities (1) with the equalities (2) to obtain

$$x_1 \geqslant 0, \qquad x_2 \geqslant 0, \qquad x_1 + x_2 \leqslant 1. \tag{3}$$

These inequalities are represented by *closed* halfplanes. The set of elements which satisfy the system of inequalities (3) can be shown [by means of Exercise 4(b) at the end of Section of 5.9] to be precisely the polytope $[a, b, c]$. While the intersection of the closed halfplanes is precisely the closed triangular region $a'b'c'$, that is, the convex hull $[a', b', c']$.

EXERCISES

1. In each part express the join and the convex hull of the given points of $\mathbb{R}^2$ as the solution set of a system of linear inequalities. Employ analytic geometry—no proof is required.
   (a) $(1, 0)$, $(0, 1)$, $(1, 1)$.
   (b) $(1, 1)$, $(2, 4)$, $(3, 0)$.
   (c) $(0, 0)$, $(1, 0)$, $(0, 1)$, $(1, 1)$.
   (d) $(0, 0)$, $(2, 0)$, $(1, 1)$, $(3, 1)$.
   (e) $(1, 2)$, $(2, 4)$, $(3, 0)$, $(2, -2)$.

2. In each part describe and illustrate geometrically the solution set of the given system of inequalities in $\mathbb{R}^2$.
   (a) $x_2 > 0$, $x_1 + x_2 < 1$, $x_1 - x_2 > -1$.
   (b) $x_1 < 1$, $x_1 > -1$, $x_2 > 0$.
   (c) $x_2 \leqslant 3$, $2x_1 - x_2 \geqslant 0$, $4x_1 - x_2 \leqslant 0$.
   (d) $x_2 \geqslant 0$, $x_2 \leqslant 2$, $x_1 + x_2 \leqslant 4$, $x_1 - x_2 \geqslant -4$.

In $\mathbb{R}^2$ an equation of the form

$$a_1x_1 + a_2x_2 = k, \qquad (a_1, a_2) \neq (0, 0),$$

is represented geometrically by a line L. The associated *strict* linear inequalities

$$a_1x_1 + a_2x_2 > k, \qquad a_1x_1 + a_2x_2 < k$$

are represented by the open halfplanes or sides of L, and the associated linear inequalities

$$a_1x_1 + a_2x_2 \geqslant k, \qquad a_1x_1 + a_2x_2 \leqslant k$$

are represented by the closed halfplanes of L.

Analogous results hold for $\mathbb{R}^1$ and $\mathbb{R}^3$. In $\mathbb{R}^1$ a linear equation $a_1x_1 = k$, $a_1 \neq 0$, is represented by a point on a linear scale (Figure 5.17), namely, the point $p(k/a_1)$. The linear inequalities $a_1x_1 > k$, $a_1x_1 < k$ are represented by the open halflines with endpoint p (or sides of p), and $a_1x_1 \geqslant k$, $a_1x_1 \leqslant k$ by the associated *closed* halflines. Finite joins of points and polytopes in $\mathbb{R}^1$ can be represented as intersections of open and closed halflines respectively. For example, if $a = 0$, $b = 1$, then ab is represented by the intersection of the open halflines given by $x_1 > 0$ and $x_1 < 1$, and $[a, b]$ by the intersection of the corresponding closed halflines.

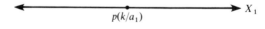

$$p(k/a_1)$$

$$X_1$$

Figure 5.17

In $\mathbb{R}^3$, a linear equation has the form

$$a_1x_1 + a_2x_2 + a_3x_3 = k, \qquad (a_1, a_2, a_3) \neq (0, 0, 0),$$

and is represented by a plane P in a given 3-space. The associated strict and nonstrict linear inequalities are represented respectively by the open and closed halfspaces of P. Once again finite joins of points and polytopes can be represented by intersections of halfspaces. For example, if $a = (0, 0, 0)$, $b = (1, 0, 0)$, $c = (0, 1, 0)$, $d = (0, 0, 1)$, then $abcd$ is the set of solutions of the system

$$x_1 > 0, \qquad x_2 > 0, \qquad x_3 > 0, \qquad x_1 + x_2 + x_3 < 1.$$

Each inequality is represented by an open halfspace, and $abcd$ by their intersection, which is the join $a'b'c'd'$ of the points a', b', c', d' that represent a, b, c, d. Similarly the polytope $[a, b, c, d]$ is the solution set of the system

$$x_1 \geqslant 0, \qquad x_2 \geqslant 0, \qquad x_3 \geqslant 0, \qquad x_1 + x_2 + x_3 \leqslant 1,$$

and turns out to be represented by $[a', b', c', d']$ (Figure 5.18).

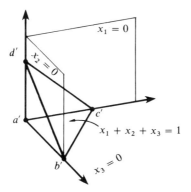

Figure 5.18

EXERCISES

1. In each part express the join and the convex hull of the given points of $\mathbb{R}^3$ as the solution set of a system of linear inequalities. Employ analytic geometry—no proof is required.
   (a) $(1, 0, 0)$, $(0, 1, 0)$, $(0, 0, 1)$, $(1, 1, 1)$.
   (b) $(0, 0, 0)$, $(1, 0, 0)$, $(0, 1, 0)$, $(1, 1, 0)$, $(0, 0, 1)$.
   (c) $(0, 0, 0)$, $(1, 0, 0)$, $(0, 1, 0)$, $(0, 0, 1)$, $(1, 0, 1)$, $(0, 1, 1)$.

2. In each part describe and illustrate the solution set of the given system of inequalities in $\mathbb{R}^3$.
   (a) $x_2 > 0$, $x_1 + x_2 > 1$, $x_1 - x_2 > -1$, $x_3 > 0$, $x_3 < 1$.
   (b) $x_1 < 1$, $x_1 > -1$, $x_2 > 0$, $x_1 - x_2 > -2$, $x_3 > 0$, $x_3 < 2$.
   (c) $x_2 \geqslant 0$, $x_1 + x_2 \leqslant 1$, $x_1 - x_2 \geqslant -1$, $x_3 \geqslant 0$, $x_2 + x_3 \leqslant 1$.

3. In $\mathbb{R}^4$ express the join and convex hull of the points $(0, 0, 0, 0)$, $(1, 0, 0, 0)$, $(0, 1, 0, 0)$, $(0, 0, 1, 0)$, $(0, 0, 0, 1)$ as the solution set of a system of linear inequalities. No proof is required.

We now make a formal definition in $\mathbb{R}^n$.

Definition. Consider the linear equation

$$a_1 x_1 + \cdots + a_n x_n = k, \qquad (a_1, \ldots, a_n) \neq (0, \ldots, 0). \qquad (1)$$

The set L of solutions $x = (x_1, \ldots, x_n)$ of (1) is called a *hyperplane* of $\mathbb{R}^n$. Consider the following linear inequalities associated with (1):

$$a_1 x_1 + \cdots + a_n x_n > k, \qquad a_1 x_1 + \cdots + a_n x_n < k \qquad (2)$$

and

$$a_1 x_1 + \cdots + a_n x_n \geqslant k, \qquad a_1 x_1 + \cdots + a_n x_n \leqslant k. \qquad (3)$$

The solution sets of the inequalities in (2) are called *sides* of L or the (*open*) *halfspaces* determined by L. The solution sets of the inequalities in (3) are called the *closed halfspaces* determined by L.

As the examples have indicated, finite joins of points and polytopes in $\mathbb{R}^n$ can be represented as intersections of open and closed halfspaces, respectively. In Chapter 13, after the assumption of an additional postulate, similar results for a join geometry will be proved.

5.11 Pathological Convex Sets

In Chapter 2 several results were encountered whose proof required the apparently unnecessary assumption that a nonempty convex set has a nonempty interior (see Theorem 2.26 and the discussion following its proof). So far the assumption is verified by all our experience with convex sets and would seem to be deducible from postulates J1–J7. Nevertheless this assumption is not deducible from J1–J7, and we are almost ready to construct the necessary counterexample. The strange kind of convex set we seek deserves a name.

Definition. Suppose K is convex, $K \neq \varnothing$, but $\mathcal{I}(K) = \varnothing$. Then we say K is a *pathological* (convex) set or is *pathological*.

5.12 An Infinite Dimensional Join Geometry

In order to exhibit a pathological convex set, we find it necessary to construct an infinite dimensional join geometry. Such geometrical systems are interesting in themselves and for the light they throw on finite dimensional join geometries such as $\mathbb{R}^n$.

To construct the desired infinite dimensional join geometry we proceed as for the join geometry $(\mathbb{R}^n, \cdot)$, but start with infinite sequences rather than n-tuples. Let $\mathbb{R}^*$ be the set of all infinite sequences $(x_1, \ldots, x_n, \ldots)$ of real numbers. The elements of $\mathbb{R}^*$ are called *points* or *(arithmetic) vectors*. An element of $\mathbb{R}^*$ is represented by one of the letters $a, b, c, \ldots$, and components of the element by the same letter with appropriate subscript. Thus a_i is the ith component or ith term of the infinite sequence a, and $a = (a_1, \ldots, a_n, \ldots)$.

Equality in $\mathbb{R}^*$ means term by term equality. Thus in $\mathbb{R}^*$, $a = b$ or $(a_1, \ldots, a_n, \ldots) = (b_1, \ldots, b_n, \ldots)$ means $a_n = b_n$ for $n \geq 1$.

Addition in $\mathbb{R}^*$ denotes term by term addition. Thus in $\mathbb{R}^*$, $a + b$ stands for the infinite sequence

$$(a_1 + b_1, \ldots, a_n + b_n, \ldots).$$

Similarly if λ is a real number and a is an element of $\mathbb{R}^*$, their product λa stands for

$$(\lambda a_1, \ldots, \lambda a_n, \ldots).$$

$\mathbb{R}^*$ is closed under the operation of addition and of multiplication by a real number.

The vector algebraic additive concepts of zero element $[0 = (0, \ldots, 0, \ldots)]$, negative and subtraction immediately generalize to $\mathbb{R}^*$, and the vector algebraic properties (1)–(10) of $\mathbb{R}^n$ (Section 5.7) are easily seen to be valid for $\mathbb{R}^*$.

A join operation can be defined in $\mathbb{R}^*$ exactly as in $\mathbb{R}^n$ (see Restatement of Definition in Section 5.8).

Definition. Let a and b be points of $\mathbb{R}^*$. Then their join $a \cdot b$ is the set of all points x of $\mathbb{R}^*$ expressible in the form

$$x = \lambda a + \mu b,$$

where λ, μ are real numbers which satisfy

$$0 < \lambda, \mu, \qquad \lambda + \mu = 1.$$

This definition of join converts $\mathbb{R}^*$ into a join geometry—that is, $(\mathbb{R}^*, \cdot)$ is a join geometry. The proof (Section 5.9) that $(\mathbb{R}^n, \cdot)$ is a join geometry is "vector algebraic"—it rests on properties (1)–(10) of Section 5.7—and holds without change for $(\mathbb{R}^*, \cdot)$.

5.13 Three Pathological Convex Sets

Pathological convex sets can be defined in the join geometry $\mathbb{R}^*$ without too much trouble.

EXAMPLE I. Let K be the set of points of $\mathbb{R}^*$ in which there are no negative terms and only a finite number of postive terms. Thus $a \subset K$ if and only if $a_n \geq 0$ for all n and there exists an r such that $a_n = 0$ for $n > r$.

To show K is convex, let $a, b \subset K$. Let $c \subset ab$. Then by definition of join,

$$c = \lambda a + \mu b$$

for some real numbers λ, μ satisfying

$$0 < \lambda, \mu, \qquad \lambda + \mu = 1.$$

Then each component c_n of c is expressible as

$$c_n = \lambda a_n + \mu b_n.$$

It follows that $c_n \geqslant 0$ for all n and $c_n > 0$ only for finitely many n, since the same is true of a_n and b_n. Thus $c \subset K$ and K is convex.

Now we show $\mathcal{G}(K) = \varnothing$. Let p be any point of K (Figure 5.19). We show

$$p \not\subset \mathcal{G}(K). \tag{1}$$

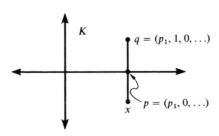

Figure 5.19

Now p can be represented in the form

$$p = (p_1, \ldots, p_r, 0, \ldots).$$

Let

$$q = (p_1, \ldots, p_r, 1, 0, \ldots).$$

Note q is a point of K. Consider the relation

$$p \subset qx \tag{2}$$

where $x \subset \mathbb{R}^*$. The relation (2) implies the existence of λ, μ for which

$$p_n = \lambda q_n + \mu x_n, \qquad 0 < \lambda, \mu, \quad \lambda + \mu = 1. \tag{3}$$

In (3) let $n = r + 1$. Then (3) becomes

$$0 = \lambda + \mu x_{r+1}.$$

Hence $x_{r+1} = -\lambda/\mu < 0$, and $x \not\subset K$. Thus (2) has no solution x in K, so that (1) holds and $\mathcal{G}(K) = \varnothing$.

EXAMPLE II. Let K be the set of points x of $\mathbb{R}^*$ for which there exists an r such that $x_n > 0$, $1 \leqslant n \leqslant r$; $x_n = 0$, $n > r$; and $x_1 > 1/r$.

We show K is convex. Let $a, b \subset K$,

$$a = (a_1, \ldots, a_r, 0, \ldots), \qquad a_r > 0,$$

$$b = (b_1, \ldots, b_s, 0, \ldots), \qquad b_s > 0.$$

Let $c \subset ab$. Then for some λ, μ,

$$c_n = \lambda a_n + \mu b_n, \qquad 0 < \lambda, \mu, \quad \lambda + \mu = 1.$$

Let us suppose $r \geqslant s$, which is not restrictive. Then

$$c_n > 0 \quad \text{for } n \leqslant r, \qquad c_n = 0 \quad \text{for } n > r,$$

$$c_1 = \lambda a_1 + \mu b_1 > \lambda(1/r) + \mu(1/s)$$

$$\geqslant \lambda(1/r) + \mu(1/r) = (\lambda + \mu)(1/r) = 1/r.$$

Thus $c \subset K$, and K is convex.

The convex set K has the interesting property that

$$\mathcal{C}(\mathcal{C}(K)) \neq \mathcal{C}(K). \tag{1}$$

We outline a proof of this. Let

$$e^{(t)} = (1/t, 0, \ldots), \qquad t = 1, 2, \ldots.$$

It can be shown that each point $e^{(t)}$ is in $\mathcal{C}(K)$. Since $\mathcal{C}(K)$ is convex (Theorem 2.22), $\mathcal{C}(K)$ must contain the join of any two points $e^{(u)}, e^{(v)}$. It follows then that $\mathcal{C}(K)$ contains the join of $0 = (0, \ldots, 0, \ldots)$ and $e^{(1)}$. Thus $0 \subset \mathcal{C}(\mathcal{C}(K))$. But it can be shown that $0 \not\subset \mathcal{C}(K)$. Hence (1) holds, and we infer by Theorem 2.26 that $\mathcal{I}(K) = \varnothing$. Thus K is pathological.

EXAMPLE III. Let K be the set of all points x of $\mathbb{R}^*$ for which there exists an r such that $x_r > 0$ and $x_n = 0$ for $n > r$. The verification that K is convex and that $\mathcal{I}(K) = \varnothing$ is left to the reader (Exercise 3 below).

5.14 Is There a Simple Way to Construct Pathological Convex Sets?

The discovery of pathological convex sets is both astonishing and rewarding. But one wonders if there is a simpler way to find them. The answer is no: If a join geometry is finite dimensional in an appropriate sense, it has no pathological convex subsets. (See Section 6.22.) This should be reassuring to your intuition for convex sets, which may have been traumatized by the very existence of pathological sets. It says that pathological sets—if they must exist—are confined to a transfinite realm and can not insinuate themselves into one of our finite dimensional domains.

Remark. The terms finite and infinite dimensional as applied to the join geometries $\mathbb{R}^n$ and $\mathbb{R}^*$ have been used purely descriptively. In the next chapter we begin to clarify them (see Section 6.19), and in Chapter 11, with the assumption of an additional postulate, develop a satisfactory theory of dimension. The theory, incidentally, assigns a dimension to $\mathbb{R}^n$, which turns out to be—as you may have expected—n (Exercise 4 at the end of Section 11.6).

EXERCISES

1. In Example I of Section 5.13, prove K is a closed convex set.

2. Consider Example II of Section 5.13.
   (a) Fill in the details of the proof that $\mathcal{C}(\mathcal{C}(K)) \neq \mathcal{C}(K)$.
   (b) Prove that $\mathcal{C}(K)$ is the set of points x of $\mathbb{R}^*$ for which $x_1 > 0$, $x_n \geqslant 0$ for all n and there exists an r such that $x_n = 0$ for $n > r$.
   (c) Find $\mathcal{C}(\mathcal{C}(K))$. Does $\mathcal{C}(\mathcal{C}(\mathcal{C}(K))) = \mathcal{C}(\mathcal{C}(K))$?

3. Consider Example III of Section 5.13.
   (a) Prove: K is convex.
   (b) Prove: $\mathcal{I}(K) = \varnothing$.
   (c) Prove: $\mathcal{C}(K)$ is the set of all points of $\mathbb{R}^*$ in which there are only a finite number of nonzero terms.
   (d) Prove: $\mathcal{C}(\mathcal{C}(K)) = \mathcal{C}(K)$.
   (e) Prove: $\mathcal{I}(\mathcal{C}(K)) = \mathcal{C}(K)$. Is $\mathcal{I}(\mathcal{C}(K)) = \mathcal{I}(K)$?
   (f) Prove: If $a, b \subset \mathcal{C}(K)$, then $a/b \subset \mathcal{C}(K)$.

Linear Sets

6

This chapter is devoted to a very prominent subfamily of the family of convex sets—the one formed by the convex sets which are "linear" or linelike in the sense of extending endlessly. The concept is suggested, of course, by the familiar linear sets or linear spaces of Euclidean geometry: point, line, plane, 3-space. The notion of linear set is defined without referring to the concept of line; then line is defined as a special kind of linear set. Emphasis is put on *general* properties of linear sets, that is, properties common to all linear sets, and specifically on dimension-free properties which make no reference to "dimension"—to what distinguishes a plane from a line or a 4-space. We study the generation of linear sets as we did that of convex sets in Chapter 3, and introduce the notion of *linear hull*, the analogue of convex hull for the theory of linear sets. Among the ideas introduced are: linear independence of points, simplex (a generalization of closed triangular or tetrahedral region), and hyperplane in a linear set (a generalization of plane in a 3-space). An approach is made toward characterizing finite dimensional linear sets. Finally we note that new results are obtained on three familiar topics: the interior operation, open sets and pathological sets.

As in Chapter 4 the treatment is based on Postulates J1–J7. In the language of Chapter 5, we are studying an arbitrary join geometry $(J, \cdot)$, although ordinarily no explicit reference to $(J, \cdot)$ will be made.

6.1 The Notion of Linear Set

Linear sets may be described as flat or uncurved subspaces of a given space. Why should the idea be introduced? In Euclidean geometry there is no great need to introduce it, since the only nontrivial examples of linear sets are lines, planes and 3-space—and they are not studied in terms of a common characteristic.

In our theory, as we saw in Chapter 5 (Sections 5.6–5.9, 5.12) there exist join geometries $\mathbb{R}^n$ of arbitrary finite dimension n as well as the join geometry $\mathbb{R}^*$ of infinite dimension. In such a join geometry of higher dimension, say $\mathbb{R}^4$ or $\mathbb{R}^{19}$ or $\mathbb{R}^*$, we may expect to find higher dimensional analogues of lines, planes and 3-space. But we cannot really expect to say anything about these geometric creatures that may inhabit a higher dimensional join geometry—we cannot even say whether or not they exist—until we decide on how to define them. What we need then is a common property of Euclidean line, plane, and 3-space which can be used to define a linear set or flat manifold in a join geometry.

In a *Euclidean* geometry it is not at all difficult to find such a property and formulate a suitable definition:

In a Euclidean geometry a set of points S is a *linear set* if

$$S \supset a, b \, (a \neq b) \quad \text{implies} \quad S \supset \text{line } ab.$$

That this property holds for a plane is a postulate in the usual treatment of Euclidean geometry. It is seen to hold for a line and a 3-space; it holds trivially for a point and $\varnothing$. The definition is simple and attractive. Can it be employed in our theory? To answer this we must face another question: How in abstract join theory can the idea of line be introduced? The join ab corresponds to Euclidean segment and the extension a/b to Euclidean ray —but it is hard to see how to give a simple and natural construction for line ab which is appropriate in the theory.

In Euclidean geometry (see Section 4.1) line ab can be characterized in terms of join and extension in this way (see Figure 6.1)

$$\text{line } ab = ab \cup a/b \cup b/a \cup a \cup b \quad (a \neq b). \tag{1}$$

This property of line is intimately related to and is usually derived from the Euclidean proposition that a line is an ordered set of points: For any three distinct points of a line, one is between the other two (Section 4.24). As a characterization of line in join theory it does not seem quite suitable —but there is no obvious alternative.

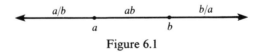

Figure 6.1

To sum up: We have an apparently satisfactory definition of linear set which requires us to define line. But we do not know how to define line in a join geometry.

So we take a new tack. Instead of adhering to the Euclidean approach, we try to follow a natural development for a join-centered treatment of geometry. We lay aside the question of line and try first to define linear set independently of the idea of line. Then line will be characterized in a sense as the simplest nontrivial type of linear set.

Considered intuitively, a linear set is a "flat" space or manifold as distinguished from a curved one such as a cone or a sphere. Consequently we require a linear set to be *convex*. This is quite natural from another viewpoint: Every Euclidean linear set S must be convex. For if $S \supset a, b$ ($a \neq b$), then $S \supset$ line ab and certainly $S \supset ab$. However, the requirement of convexity is not sufficient to characterize linearity—for in Euclidean geometry a convex set need not be a linear set, as a segment or circular region attests. A linear set must, so to speak, extend out endlessly in all directions possible. This suggests a second characteristic:

A linear set must contain the extension of a and b whenever it contains a and b.

Now we have a basis for defining linear set without using the concept of line.

6.2 The Definition of Linear Set

Thus we are led to the following definition.

Definition. S is a *linear* set or a *linear space* or a *linear subspace* of J if (a) S is convex and (b) $S \supset a, b$ implies $S \supset a/b$. A set which satisfies condition (b) is said to be *closed under* the operation *extension*.

Note the defining conditions for linearity of S may be expressed as (1) $S \supset a, b$ implies $S \supset ab$ and $S \supset a/b$ or (2) S is closed under the operations join and extension.

A linear set is a convex set which is, so to speak, "fully extended". Observe that each point a is linear, since $a = aa = a/a$ by postulates J4 and J7. Also J and $\varnothing$ are linear.

The Notion of Linear Set in Euclidean Geometry

In a Euclidean geometry (considered as a join geometry) our definition of linear set is equivalent to the familiar one (Section 6.1 above).

For suppose a Euclidean set S is closed under join and extension. Then $S \supset a, b$ ($a \neq b$) implies

$$S \supset ab \cup a/b \cup b/a \cup a \cup b$$

and so S contains the Euclidean line ab [Section 6.1, (1)].

Conversely, suppose a Euclidean set S has the property that $S \supset a, b$ ($a \neq b$) implies $S \supset$ line ab. Let $S \supset a, b$. Assume $a \neq b$. Then

$$S \supset \text{line } ab = ab \cup a/b \cup b/a \cup a \cup b.$$

Thus $S \supset ab$ and $S \supset a/b$. These relations certainly hold if $a = b$. Hence S is closed under join and extension, and our assertion is justified.

Thus the concept of linear set in the theory may be considered a generalization of the Euclidean notion of linear set. It will enable us to formulate a definition of line (Section 6.10 below) which is sufficiently weak to permit the existence of "partially ordered" join geometries, in which a line need not be an ordered set of points (see Exercise 1 at the end of Section 6.10 below).

EXERCISES

1. (a) In JG16 (Section 5.2) consider the set

   $$S = ab \cup a/b \cup b/a \cup a \cup b,$$

   where a and b are not on the same vertical or horizontal line. Show that S is not a convex set in JG16. This indicates that (1) of Section 6.1 above is not suitable as a definition of line in JG16.
   (b) In JG10 show that $ab \cup a/b \cup b/a \cup a \cup b$ is convex but not linear when a and b are not both in the same ray R_1, R_2 or R_3.

2. Find all linear sets in each of the following join geometries: JG2, JG5, JG6, JG8, JG9, JG10, JG16.

3. Prove: Any linear set is both open and closed. Show this property is not sufficient to characterize linearity: that is, find a join geometry that contains a convex set that is both open and closed but not linear.

4. Prove: If A is linear and $p \subset A$, then $A = pA$.

6.3 Conditions for Linearity

Theorem 6.1. *A is linear if and only if it is convex and satisfies one of the equivalent conditions*: (a) $A \supset A/A$; (b) $A = A/A$.

(Compare Theorem 2.9.)

PROOF. The equivalence of (a) and (b) follows from $A/A \supset A$ (Corollary 4.1.2).

Suppose A linear. Then A is convex and closed under extension. Let $p \subset A/A$. Then $p \subset x/y$, where $x, y \subset A$. Thus $x/y \subset A$, so that $p \subset A$ and (a) holds. The converse is easily proved, since (a) implies that A is closed under extension. $\square$

Corollary 6.1. *Let A be linear. Then $A \supset X, Y$ implies $A \supset XY$ and $A \supset X/Y$.*

(Compare Theorem 2.11.)

PROOF. $A \supset XY$, since A is convex. Note $A \supset X$, $A \supset Y$ imply $A/A \supset X/Y$ (Corollary 4.1.1). By the theorem, $A \supset A/A$, and $A \supset X/Y$ follows. $\square$

The convexity requirement in the definition of linear set is redundant.

Theorem 6.2. *A is linear if and only if A is closed under extension.*

PROOF. Suppose A closed under extension. We show A convex. Let $A \supset a, b$. Then $A \supset a/b$. Since $A \supset a$ and $A \supset a/b$, the closure of A under extension implies

$$A \supset a/(a/b). \qquad (1)$$

By Corollary 4.7(c)

$$b \subset a/(a/b).$$

Multiplying both members by a and applying Corollary 4.12, we have

$$ab \subset a(a/(a/b)) \subset aa/(a/b) = a/(a/b). \qquad (2)$$

By (1) and (2), $A \supset ab$ which implies A is convex and so linear. The converse is trivial. $\square$

Corollary 6.2.1. *A is linear if and only if $A \supset A/A$, or equivalently $A = A/A$.*

Corollary 6.2.2. *Any linear set is open.*

PROOF. A convex set A is open if $A \supset p, q$ implies $A \approx p/q$ (Corollary 4.25.3). $\square$

6.4 Constructing Linear Sets from Linear Sets

The simplest and most obvious way to obtain linear sets—if not the most useful—is to take intersections of linear sets.

Theorem 6.3. *The intersection of two linear sets is also linear.*

(Compare Theorem 2.13.)

PROOF. Let A and B be linear. Suppose $A \cap B \supset x, y$. Then $A \supset x, y$, so that $A \supset x/y$. Similarly $B \supset x/y$. Thus $A \cap B \supset x/y$, and $A \cap B$ is linear (Theorem 6.2). $\square$

Theorem 6.4. *Let F be a family of linear sets. Then the intersection of the sets of F is also linear.*

(Compare Theorem 2.14.)

PROOF. Use the method of Theorem 2.14, the analogous theorem for a family of convex sets.   □

The results that the join and the extension of two convex sets are convex (Theorems 2.12, 4.16) do not go over to linear sets. It is not hard to find counterexamples. In the Euclidean join geometries JG4 and JG5 (Section 5.2), if A and B are parallel lines, neither AB nor A/B is linear. Linear sets cannot be constructed from linear sets quite so easily as convex sets from convex ones.

If however two linear sets intersect, their extension *is* linear. The following lemma is employed in the proof.

Lemma 6.5. *Let A be linear and $p \subset A$. Then*

$$A = A/p = p/A.$$

PROOF. Since A is linear, $A \supset A/p$. For the reverse inclusion suppose $a \subset A$. Then $A \supset ap$ and $A/p \supset a$. Thus $A/p \supset A$ and $A = A/p$.

Similarly for $A = p/A$. We have $A \supset p/A$. Suppose $a \subset A$. Then $A \supset p/a$, which yields $a \subset p/A$. Thus $p/A \supset A$ and $A = p/A$.   □

Theorem 6.5. *Let A and B be linear and $A \approx B$. Then A/B is linear.*

PROOF. Let $p \subset A, p \subset B$. By Lemma 6.5, $A = p/A$ and $B = p/B$. Then

$$(A/B)/(A/B) = (A/B)/((p/A)/(p/B))$$

$$\subset (A/B)/(pB/pA) \qquad \text{(Corollary 4.15)}$$

$$\subset (A/B)/(B/A)$$

$$\subset AA/BB \qquad \text{(Corollary 4.15)}$$

$$= A/B.$$

Thus $(A/B)/(A/B) \subset A/B$, and A/B is linear (Corollary 6.2.1).   □

The result is not trivial, since it provides a basic method for constructing more complicated linear sets from simpler ones. Despite its generality (it involves no dimensional restriction), it has strong intuitive geometric content. For example, in JG5 (Section 5.2) let line A meet plane B in a single point (Figure 6.2). Observe how the extensions a/b ($a \subset A$, $b \subset B$) cover the Euclidean 3-space.

Remark. One might expect the theorem to hold for the operation join as well as extension. This is so in Euclidean geometry but not in general. (See Exercise 14 in the first group below.)

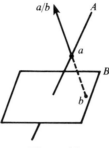

Figure 6.2

EXERCISES

1. Prove: If L is linear and L meets ab, bc, ca, then $L \supset a, b, c$. What does this tell you about a Euclidean triangle?

2. Prove: If L is linear and L meets ab, bc, cd, then L meets ad. Interpret in Euclidean 3-space.

3. Prove: If a nonempty linear set A absorbs X (Section 2.14), then $A \supset X$.

4. (a) Is A linear if $A = p/A$ for some p? Explain.
   (b) Is A linear if A is convex and $A = p/A$ for some p? Explain.

5. Prove: If A and B are convex, then $(A/B) \cap (B/A)$ is linear. It is nonempty if and only if $A \approx B$.

6. Prove: If L is linear and nonempty, then $La \approx Lb$, $L/a \approx L/b$ and $L/a = L/b$ are equivalent. Interpret in JG4, JG5, JG16.

7. Prove: If L is linear, then $L/(L/(L/a)) = L/a$. Interpret in JG4, JG5. (Compare Exercise 6.)

8. Prove: If L is linear, then $L/(L/a) = La/L$. Interpret in JG4, JG5. (Compare Exercises 6, 7.)

9. Prove: If A and B are linear and $A \cup B$ is convex, then $A \subset B$ or $B \subset A$.

10. Prove: If A and B are linear and $A \approx B$, then $A/B = B/A$.

11. (a) If $A/B = B/A$ and $A, B \neq \emptyset$, must $A \approx B$? Explain.
    (b) If A/B is linear and $A, B \neq \emptyset$, must $A \approx B$? Explain.
    (c) If $A/B = B/A$, A/B is linear and $A, B \neq \emptyset$, must $A \approx B$? Explain.
    (d) If $A/B = B/A$ and A/B is linear, must A and B be linear? Explain.

12. Prove: If A and B are linear and contain p, then AB/p is linear. (Compare Theorem 6.5.)

13. Prove: If A and B are linear and contain p, then $AB/p = p/AB = A/B$.

14. (a) Verify in the Euclidean join geometries JG4, JG5 the following proposition: If A and B are linear and $A \approx B$, then AB is linear.
    (b) Prove that the proposition in (a) is not deducible from J1–J7 (see Section 5.3).

EXERCISES (CONGRUENCE AND ANTICONGRUENCE RELATIONS)

Definition. Let the nonempty linear set L be given. Then $a \equiv b$ (mod L) (read a is *congruent* to b *modulo* L) means $aL \approx bL$.

1. Prove that congruence modulo L is an equivalence relation (see Section 4.26). Interpret in JG4, JG5. (Compare Exercise 12 at the end of Section 4.13.)

Definition. Let the nonempty linear set L be given. Then $a\|\|b$ (mod L) (read a is *anticongruent* to b *modulo* L or a and b are *separated* by L) means $ab \approx L$.

2. Prove: If $a \equiv b$ (mod L), $b\|\|c$ (mod L) then $a\|\|c$ (mod L). [Compare Exercise 13(a) at the end of Section 4.13.]

3. Prove: If $a\|\|b$ (mod L), $b\|\|c$ (mod L) then $a \equiv c$ (mod L). [Compare Exercise 13(b) at the end of Section 4.13.]

4. Prove: If $a\|\|b$ (mod L), $b\|\|c$ (mod L), $c\|\|d$ (mod L) then $a\|\|d$ (mod L). [Compare Exercise 13(c) at the end of Section 4.13.]

5. Can $a \equiv b$ (mod L) and $a\|\|b$ (mod L) both hold? Explain. [Compare Exercise 13(d) at the end of Section 4.13.]

6. Let L be linear and nonempty.
   (a) Prove: $x \equiv a$ (mod L) if $x \subset aL$ or $x \subset a/L$.
   (b) Is the converse valid? Explain.

7. (a) Given point a and a nonempty linear set L, find in terms of a and L the set of all points x that satisfy the relation $x \equiv a$ (mod L). Interpret your result in JG5, letting L be a point; a line; a plane.
   (b) Apply the result of (a) to Exercise 6(a). What do you discover? Interpret geometrically.

8. The same as Exercise 7(a) for the relation $x\|\|a$ (mod L).

9. Prove: $a \equiv b$ (mod L) if and only if $a\|\|p$ (mod L) and $b\|\|p$ (mod L) for some point p.

10. (a) Given point a and a nonempty linear set L, make use of Exercise 9 to prove the set of all points x that satisfy $x \equiv a$ (mod L) is $L/(L/a)$.
    (b) Compare your result with that in Exercise 7(a). What do you discover? Interpret geometrically.

11. Suppose $a \equiv b$ (mod L) and c is any point. Prove that ac and bc are related so: Each point of ac is congruent modulo L to some point of bc, and each point of bc is congruent modulo L to some point of ac. Interpret geometrically.

Definition. Let the nonempty linear set L be given. Suppose sets A and B have the property that every point of each is congruent modulo L to some point of the other. Then we say A is *congruent* to B *modulo* L and write $A \equiv B$ (mod L).

12. Prove that the relation of congruence modulo L for sets is an equivalence relation (see Section 4.26). (Compare Exercise 1.)

13. Prove: $a \equiv b$ (mod L) and $c \equiv d$ (mod L) imply $ac \equiv bd$ (mod L). (Recall Exercise 11.) Give several interpretations in Euclidean geometry, including one in which L is a line and L, a, b, c, d are not coplanar. Observe in this case that each point of ac is congruent modulo L to a unique point of bd, and similarly each point of bd is congruent modulo L to a unique point of ac. Hence in this case the relation $ac \equiv bd$ (mod L) effects a one to one correspondence between the sets ac and bd.

Definition. Let the nonempty linear set L be given. Suppose sets A and B have the property that every point of each is anticongruent modulo L to some point of the other. Then we say A is *anticongruent* to B *modulo* L and write $A \||| B$ (mod L).

14. The same as Exercise 2 for sets A, B, C.

15. The same as Exercise 3 for sets A, B, C.

16. Prove: If $a\|||b$ (mod L) and $c\|||d$ (mod L) then $ac\|||bd$ (mod L). Give several interpretations in Euclidean geometry, including one in which L is a line and L, a, b, c, d are not coplanar. Observe in this case the relation $ac\|||bd$ (mod L) effects a one to one correspondence between ac and bd as in Exercise 13.

6.5 The Construction of a Linear Set from a Convex Set

Linear sets are easily obtained from convex sets.

Theorem 6.6. *Let A be convex. Then A/A is linear.*

PROOF. Using the principle on the quotient of two extensions (Corollary 4.15), we have

$$(A/A)/(A/A) \subset AA/AA = A/A.$$

Thus A/A is linear (Corollary 6.2.1). ▫

The importance and the utility of the theorem are in inverse relation to the length of its proof. As an illustration, let A be the interior of a Euclidean sphere. Then A/A is the union of all rays a/b for all a, b in A, and therefore is the 3-space containing A (Figure 6.3).

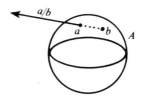

Figure 6.3

A very important special case of the theorem occurs when A is a join of points $a_1 \ldots a_n$, which is known to be convex (Theorem 2.10). Thus we have the following result.

Corollary 6.6.1. $a_1 \ldots a_n / a_1 \ldots a_n$ *is linear.*

This principle enables us to construct a linear set from any finite set of points. For example, if a and b are distinct points in a Euclidean geometry, ab/ab is the line determined by a and b. Similarly if a, b, c, d are noncoplanar points in a Euclidean 3-space, $abcd/abcd$ is that 3-space. Finally, in higher dimensional Euclidean geometry, if a, b, c, d, e are not in a 3-space, then $abcde/abcde$ will be a 4-space.

The theorem yields a construction for a linear set which contains a given set.

Corollary 6.6.2. $[S]/[S]$ *is a linear set that contains S.*

PROOF. $[S]$ is convex, and so $[S]/[S]$ is linear by the theorem. Moreover, using Corollary 4.1.2, we have

$$[S]/[S] \supset [S] \supset S. \qquad \square$$

6.6 Linear Sets Give Rise to Join Geometries

Now we take up an important theoretical point involving linear sets. Suppose S is a linear set of join geometry $(J, \cdot)$. Since S is convex, it is natural to study the application of the operation $\cdot$ to the points of S. Stated more formally, we study the pair $(S, \cdot)$, where $\cdot$ denotes the original operation restricted in its application to the set S. (No essential ambiguity is involved in this double use of the symbol $\cdot$, since S is a subset of J.)

Is $(S, \cdot)$ a join geometry? First note that $\cdot$ is a join operation in S, since S is convex. Next observe that postulates J1–J4 hold in $(S, \cdot)$, since they hold in $(J, \cdot)$. To consider the remaining postulates we must define the extension operation *in* $(S, \cdot)$. Given a, b in S, consider the set of x *in S* for which $bx \supset a$. But S is closed under $/$, the extension operation in $(J, \cdot)$. Thus any solution x in J of $bx \supset a$ must be in S. Hence the extension operation in $(S, \cdot)$ must be $/$, merely restricted in its application to the points of S. It is now not hard to see that J5, J6 and J7 are valid in $(S, \cdot)$. Thus $(S, \cdot)$ is indeed a join geometry. The result suggests the following

Definition. Let $(J, \cdot)$ be a join geometry and S a subset of J. Consider $(S, \cdot)$, where the operation is restricted in its application to S. If $(S, \cdot)$ is a join geometry, it is called a *sub-join-geometry* of $(J, \cdot)$, or a *subgeometry* of

$(J, \cdot)$. Sometimes we say simply S is a *subgeometry* of J if the context makes clear which join operation is involved.

The notion of subgeometry is analogous to the subsystem concept in modern algebra, as exemplified by subgroup of a group, subfields of a field, and so on.

The discussion above yields the following theorem.

Theorem 6.7. *Let L be a linear set of a join geometry $(J, \cdot)$. Then $(L, \cdot)$ is a subgeometry of $(J, \cdot)$.*

Remark. The theorem says in effect that the join operation of J when applied to L converts L into a join geometry. Thus any linear set of a join geometry may be considered a join geometry in its own right to which the theory is applicable.

EXERCISE

(a) Let $(J, \cdot)$ be a join geometry and $S \subset J$. Prove that $(S, \cdot)$ is a subgeometry of $(J, \cdot)$ if and only if S is an open convex subset of J. (Be careful to distinguish between the extension operation in J and that in S.)
(b) Use (a) to prove that JG6–JG9 (Section 5.2) are join geometries.
(c) Construct some new join geometries by taking open sets in various join geometries.

6.7 The Generation of Linear Sets: Two Euclidean Examples

Suppose a set S is given. It is natural to try to *linearize* it—that is, to convert S into a linear set containing S in the simplest possible way.

In Euclidean geometry, when we form the line determined by two distinct points or the plane determined by three noncollinear points, we are in effect linearizing certain pairs or triples of points. Let us examine these examples more closely.

EXAMPLE I. Let $S = \{a, b\}$, $a \ne b$ (Figure 6.4). If a linear set L contains S, then L must contain ab, a/b and b/a. Thus

$$L \supset S^* = a \cup b \cup ab \cup a/b \cup b/a,$$

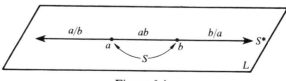

Figure 6.4

which is the Euclidean line ab [Section 6.1, (1)]. Thus S^* is the simplest or least linear set that contains $\{a, b\}$ and in forming S^*, we have linearized $\{a, b\}$.

EXAMPLE II. Similarly, let $S = \{a, b, c\}$, where a, b and c are noncollinear points (Figure 6.5). Any linear set L that contains S must contain the joins ab, bc, ac, abc, the extensions $a/b, b/a, b/c, c/b, a/c, c/a$ and the sets denoted by the mixed expressions $a/bc, b/ac, c/ab, ab/c, bc/a, ac/b$. Thus

$$L \supset S^* = a \cup b \cup c \cup ab \cup bc \cup ac \cup abc$$

$$\cup a/b \cup b/a \cup b/c \cup c/b \cup a/c \cup c/a$$

$$\cup a/bc \cup b/ac \cup c/ab \cup ab/c \cup bc/a \cup ac/b,$$

which is precisely the Euclidean plane abc (see Figure 6.5). In constructing S^* we have linearized $\{a, b, c\}$. (Compare Exercise 7 in the second set at the end of Section 4.2.)

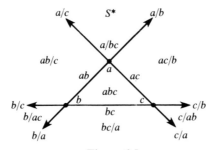

Figure 6.5

This process, which may be described as "linearization by adjunction," is valid in a Euclidean join geometry (or a Euclidean geometry) of arbitrary dimension. It holds, more broadly, in any ordered join geometry (see Section 12.1 and Theorem 12.20)—but is not valid in general (see Exercise 1 at the end of Section 6.2).

6.8 The Generation of Linear Sets: General Case

However, we need not be dismayed. The convexification process employed in Section 3.2 which led to the convex hull concept does have an exact analogue for linearization of sets. We state the corresponding definitions and theorem.

Definition. A linear set which contains a given set S is called a *linear extension* of S.

Theorem 6.8 (Existence). *Let set S be given. Then the family of linear extensions of S has one and only one least member.*

(Compare Theorem 3.1.)

PROOF. Use the method of Theorem 3.1.   □

Restatement of Theorem 6.8. For every set there exists a unique least linear extension.

Definition. Let S be any set. The least linear extension of S, denoted by $\langle S \rangle$, is called the *linear hull* of S or the linear set *generated* (or *spanned* or *determined*) by S. If L is a given linear set and S a set for which $\langle S \rangle = L$, we call S a *set of generators* of L (in the sense of *linearity*), and say S generates L (*linearly*) or L is generated by S (*linearly*). If a linear set has a finite set of generators, it is said to be *finitely generated* (*linearly*).

Note that $\langle S \rangle \supset [S]$. Also $\langle S \rangle = S$ if and only if S is linear. Finally observe that the familiar linear sets of Euclidean geometry ($\varnothing$, point, line, plane and 3-space) are finitely generated; for instance $\varnothing = \langle \varnothing \rangle$ and plane $abc = \langle a, b, c \rangle$.

In summary, any set of points has a linear hull—any geometric figure can be linearized.

6.9 The Linear Hull of a Finite Set: Finitely Generated Linear Sets

As in the study of convex hulls, we seek ways of expressing the linear hull of a set S in terms of geometric operations on the points of S. The case of nonempty finite S is covered first.

Notation. $\langle \{a_1, \ldots, a_n\} \rangle$ is conveniently written $\langle a_1, \ldots, a_n \rangle$.

Theorem 6.9. $\langle a_1, \ldots, a_n \rangle = a_1 \ldots a_n / a_1 \ldots a_n.$

PROOF. We have to show that $a_1 \ldots a_n / a_1 \ldots a_n$ is the least linear extension of $\{a_1, \ldots, a_n\}$. We know $a_1 \ldots a_n / a_1 \ldots a_n$ is linear (Corollary 6.6.1). To show it is an extension of $\{a_1, \ldots, a_n\}$, we must show

$$a_i \approx a_1 \ldots a_n / a_1 \ldots a_n, \qquad 1 \leqslant i \leqslant n. \qquad (1)$$

The relation (1) is equivalent to $a_i a_1 \ldots a_n \approx a_1 \ldots a_n$, and so to $a_1 \ldots a_n$

$\approx a_1 \ldots a_n$, which clearly holds. Thus $a_1 \ldots a_n / a_1 \ldots a_n$ *is* a linear extension of $\{a_1, \ldots, a_n\}$. It is the least such, for if L is any linear extension of $\{a_1, \ldots, a_n\}$, then $L \supset a_1 \ldots a_n$ and so $L \supset a_1 \ldots a_n / a_1 \ldots a_n$. □

Remark. Any bounded set S (Section 3.11) is contained in a finitely generated linear set. For by definition $S \subset [a_1, \ldots, a_n]$. Then certainly $S \subset \langle a_1, \ldots, a_n \rangle$. It will be proved in Chapter 11, for a join geometry satisfying a postulate equivalent to "two points determine a line", that a finitely generated linear set is finite dimensional. (See Corollary 11.16.) Thus in a broad class of join geometries any bounded set must be contained in a finite dimensional linear subspace.

EXERCISES (LINEAR HULLS)

1. In Euclidean geometry what is $\langle A \rangle$ if A is a segment? A closed triangular region? The interior of a sphere? The join of three points? A circle? A right angle?

2. (a) In Euclidean geometry suppose $\langle A \rangle = L$, where L is a plane. Give several choices for A which satisfy the equation. Is there a greatest solution A? A least one? A minimal one—that is, one no proper subset of which is a solution?
   (b) The same if L is a 3-space.

3. (a) Prove: If L is linear and L meets ab, bc, ca, then $L \supset \langle a, b, c \rangle$. (Compare Exercise 1 in the first set at the end of Section 6.4.)
   (b) Infer that a linear set L meets all three of ab, bc, ca if and only if $L \supset \langle a, b, c \rangle$.

4. Prove: $A \supset B$ implies $\langle A \rangle \supset \langle B \rangle$.

5. Prove: $\langle \langle A \rangle \rangle = \langle A \rangle$.

6. Prove: $\langle a, b \rangle = \langle ab \rangle$.

7. Prove: $\langle a, b \rangle = \langle a/b \rangle$.

6.10 The Definition of Line

Now we are prepared to give an answer to the question: What should a line be? In Section 6.1 it was indicated that the conventional characterization in Euclidean geometry,

$$\text{line } ab = ab \cup a/b \cup b/a \cup a \cup b \qquad (a \neq b), \qquad (1)$$

does not seem a satisfactory answer. (In this regard consider Exercise 1 at the end of Section 6.2.) So we seek an alternative. We should not expect that a line in our theory will *necessarily* enjoy all the properties of a Euclidean line. A choice then must be made of which properties of a Euclidean line are to be considered fundamental. Here is our choice.

If a and b are distinct points, line ab has the following properties:

(a) Line ab contains a and b.
(b) Line ab is a linear set.
(c) If any linear set contains a and b, it must contain line ab.

These properties characterize line ab in Euclidean geometry as the least linear set that contains a and b. So we adopt the following

Definition. The linear hull of two distinct points is called a *line*. The linear hull $\langle a, b \rangle$, if $a \neq b$, is called *line ab*.

The last theorem yields the formula
$$\text{line } ab = ab/ab \qquad (a \neq b).$$
Observe that by definition $\langle a, b \rangle$ contains a, b and so ab, a/b, b/a. This gives
$$\langle a, b \rangle \supset ab \cup a/b \cup b/a \cup a \cup b, \tag{2}$$
a simple but important relation which connects, when $a \neq b$, the conventional Euclidean characterization of line [(1) above] and the one we have adopted. The relation (2) implies the equivalence of the two characterizations in Euclidean geometry, since the right member of (2) is a linear set in Euclidean geometry and (2) becomes an equality.

As was indicated above, lines will not have all the familiar Euclidean properties. Postulates J1–J7 form a rather weak basis for a geometrical theory and omit many familiar properties of lines, for example, the existence of a unique line containing two distinct points and that a line is an ordered set of points (see Exercises 1 and 2 below). Indeed, we are assuming so little that you may feel our "lines" are not really lines—and that the concept should not be defined at all. The choice is to define line or dispense with the concept. If we do the latter, our theory cannot be compared with or applied to the Euclidean theory of lines. This is a severe restriction, since Euclidean geometry (and $\mathbb{R}^n$, its algebraic counterpart) *is* a model of our theory and in it lines play a central role. So we have made the decision to define line and adopted a generalization of the Euclidean line concept that seems suitable. It should be noted that as the theory is developed and additional assumptions are introduced (Chapters 11 and 12), our "lines" will take on more of the properties of the "real" lines you studied in high school geometry.

EXERCISES (LINES)

1. (a) Determine the significance of line in each of the join geometries: JG2, JG5, JG6, JG8, JG9, JG10, JG16.
   (b) Which of the join geometries in (a) have the property that every line is an ordered set (Section 4.24)?
   (c) Show that the property in (b) is not deducible from Postulates J1–J7.

2. (a) If $a \neq b$, is there a line containing a and b? Justify your answer.
   (b) If $a \neq b$, is there a unique line containing a and b? Justify your answer.

3. In Euclidean geometry any nonempty linear subset of a line must be a point. Is this true in any join geometry? Justify your answer.

*4. Prove: If $p \subset ab$, then $\langle a, b \rangle = ab/p = p/ab$.

6.11 The Linear Hull of an Arbitrary Set

Having disposed of the notion of line (the linear hull of a pair of distinct points), we return to the problem of expressing $\langle S \rangle$ in terms of S and obtain two results for arbitrary S.

Theorem 6.10 (Linear Hull Representation Theorem). $\langle S \rangle$ *is the union of all quotients of the form*

$$a_1 \ldots a_n/a_1 \ldots a_n, \tag{1}$$

where the a's are points of S.

(Compare Theorems 6.9, 3.7.)

PROOF. Let S^* be the union of all sets (1) where the a's are in S. The theorem asserts $\langle S \rangle = S^*$. We first show $\langle S \rangle \supset S^*$. Let $x \subset S^*$. Then $x \subset a_1 \ldots a_n/a_1 \ldots a_n$, where the a's are in S. Thus

$$\langle S \rangle \supset S \supset a_1, \ldots, a_n.$$

Since $\langle S \rangle$ is linear, $\langle S \rangle \supset a_1 \ldots a_n$ and so $\langle S \rangle \supset a_1 \ldots a_n/a_1 \ldots a_n$. Thus $\langle S \rangle \supset x$ and $\langle S \rangle \supset S^*$.

Now we prove the reverse inclusion $S^* \supset \langle S \rangle$. First we show $S^* \supset S$. Suppose $a \subset S$. Then $a = a/a$, which is an expression of the form (1) for $n = 1$. Thus $S^* \supset a$, so that $S^* \supset S$. Next we show S^* linear. Let $x, y \subset S^*$. By definition of S^*

$$x \subset a_1 \ldots a_r/a_1 \ldots a_r, \qquad y \subset b_1 \ldots b_s/b_1 \ldots b_s,$$

where the a's and b's are in S. Thus using the principle on the quotient of two quotients (Corollary 4.15),

$$x/y \subset (a_1 \ldots a_r/a_1 \ldots a_r)/(b_1 \ldots b_s/b_1 \ldots b_s)$$

$$\subset a_1 \ldots a_r b_1 \ldots b_s/a_1 \ldots a_r b_1 \ldots b_s \subset S^*.$$

Hence S^* is closed under extension and is linear (Theorem 6.2). Thus S^* is a linear extension of S, and so $S^* \supset \langle S \rangle$, the least linear extension of S. This completes the proof. □

Corollary 6.10.1. $x \subset \langle S \rangle$ *if and only if there exist finitely many points* $a_1, \ldots, a_n$ *of* S *such that* $x \subset a_1 \ldots a_n / a_1 \ldots a_n$.

(Compare Corollary 3.7.1.)

Corollary 6.10.2. $\langle S \rangle$ *is the union of all linear sets generated by finite subsets of* S.

(Compare Corollary 3.7.2.)

PROOF. $a_1 \ldots a_n / a_1 \ldots a_n = \langle a_1, \ldots, a_n \rangle$ by Theorem 6.9. □

Corollary 6.10.3 (Finiteness of Generation). $x \subset \langle S \rangle$ *if and only if there exist finitely many points* $a_1, \ldots, a_n$ *of* S *such that* $x \subset \langle a_1, \ldots, a_n \rangle$.

(Compare Corollary 3.7.3.)

Now a compact formula for $\langle S \rangle$.

Theorem 6.11. $\langle S \rangle = [S]/[S]$.

PROOF. $[S]/[S]$ is a linear extension of S (Corollary 6.6.2). It is the least linear extension of S, since any linear set that contains S must contain $[S]$ and so $[S]/[S]$. □

One hardly could have expected so simple a result: To linearize S, convexify it and "divide" the result by itself. Note that in a sense most of the work lies in convexifying S—when this is done, one extension operation completes the job.

Corollary 6.11.1. *Let* A *be convex. Then* $\langle A \rangle = A/A$.

Corollary 6.11.2. $\langle a_1, \ldots, a_n \rangle = \langle a_1 \ldots a_n \rangle$.

PROOF. By Theorem 6.9 and the first corollary,
$$\langle a_1, \ldots, a_n \rangle = a_1 \ldots a_n / a_1 \ldots a_n = \langle a_1 \ldots a_n \rangle.$$ □

EXERCISES

1. Prove: $\langle S \rangle = S/S \cup S^2/S^2 \cup \cdots \cup S^n/S^n \cup \cdots$.

2. If S is a finite set, will the terms in the series in Exercise 1 become constant? Explain.

3. Can you find examples where S is infinite and the terms in the series in Exercise 1 become constant? Explain.

4. Prove: $[a_1, \ldots, a_n]/[a_1, \ldots, a_n] = a_1 \ldots a_n / a_1 \ldots a_n$.

5. Does $\langle S \rangle = [S/S]$? Explain.

6. Prove: If A and B are nonempty convex sets, $A \not\approx B$ and $A \cup B$ is linear, then $A \cup B = \langle A \rangle = \langle B \rangle$.

6.12 The Linear Hull of a Finite Family of Sets

The notion of linear hull can easily be generalized to several sets or to an arbitrary family of sets. The desirability of doing this is indicated by the familiar use in high school geometry of phrases such as "the plane determined by a line and a point not on the line" or "the plane determined by two distinct intersecting lines". It is desirable to extend this usage to be able to refer in the first example to the case where the point may be on the line, or in the second to the case where the lines may coincide or not intersect. These examples indicate that a line and a point may "determine" a line rather than a plane, and a line and a line, if they are noncoplanar or skew, a 3-space.

It suffices for our purposes to generalize the concept of linear hull to a finite nonempty family of sets, although the treatment applies without essential change to an arbitrary family of sets. The treatment is exactly analogous to that for the linear hull of a single set (Section 6.8) and is presented in outline form. (You may want to compare it with that for convex hull of a finite nonempty family of sets in Section 3.13.)

Definition. M is a *linear extension* of the sets $S_1, \ldots, S_n$ if M is linear and contains each of them.

Theorem 6.12. *Let sets $S_1, \ldots, S_n$ be given. Then the family of linear extensions of $S_1, \ldots, S_n$ has a unique least member.*

(Compare Theorems 6.8 and 3.9.)

Definition. Let $S_1, \ldots, S_n$ form a family of sets. The least linear extension of $S_1, \ldots, S_n$, denoted by $\langle S_1, \ldots, S_n \rangle$, is called the *linear hull* of (the family) $S_1, \ldots, S_n$ or the linear set *generated* (or *spanned* or *determined*) by $S_1, \ldots, S_n$. If L is a given linear set and $S_1, \ldots, S_n$ are sets that satisfy $\langle S_1, \ldots, S_n \rangle = L$, we say $S_1, \ldots, S_n$ generate L (*linearly*).

Observe that

$$\langle S_1, \ldots, S_n \rangle = \langle S_1 \cup \cdots \cup S_n \rangle,$$

since a set is a linear extension of $S_1, \ldots, S_n$ if and only if it is a linear extension of $S_1 \cup \cdots \cup S_n$. (Compare Section 3.13, the paragraph next to the last.)

Finally observe that

$$\langle S_1, \ldots, S_m \rangle \supset T_1, \ldots, T_n$$

implies

$$\langle S_1, \ldots, S_m \rangle \supset \langle T_1, \ldots, T_n \rangle. \tag{1}$$

This is useful in proving relations of the form (1). It can be proved in the same way as its analogue for convex hulls (Section 3.13, last paragraph).

The notion of linear hull of a pair of sets is applied in the next two theorems.

Theorem 6.13. $\langle A, B \rangle = \langle AB \rangle = \langle A/B \rangle$ *provided* $A, B \neq \varnothing$.

PROOF. Note that $\langle A, B \rangle \supset AB$. Thus $\langle A, B \rangle \supset \langle AB \rangle$. The proof of the reverse inclusion employs $AB/A \supset B$ [Corollary 4.7(b)]. Using Corollaries 4.5 and 4.1.2 we have

$$\langle AB \rangle \supset AB/AB = (AB/A)/B \supset B/B \supset B.$$

By symmetry $\langle AB \rangle \supset A$. Thus $\langle AB \rangle \supset \langle A, B \rangle$, and we conclude $\langle A, B \rangle = \langle AB \rangle$.

A similar method is used to prove $\langle A, B \rangle = \langle A/B \rangle$. As above, $\langle A, B \rangle \supset \langle A/B \rangle$. By Corollary 4.7(a), $B(A/B) \supset A$. Then

$$\langle A/B \rangle \supset (A/B)/(A/B) = A/B(A/B) \supset A/A \supset A.$$

Similarly, using $A/(A/B) \supset B$ [Corollary 4.7(c)],

$$\langle A/B \rangle \supset (A/B)/(A/B) = (A/(A/B))/B \supset B/B \supset B.$$

Thus $\langle A/B \rangle \supset \langle A, B \rangle$, and the theorem holds. □

Remark. If $A = a$, $B = b$, $a \neq b$, the theorem says that segment ab has the line $\langle a, b \rangle$ as its linear hull. Similarly for ray a/b.

Theorem 6.14. *Let A and B be linear and $A \approx B$. Then $\langle A, B \rangle = A/B$.*

(Compare Theorem 6.5.)

PROOF. A/B is linear (Theorem 6.5). Hence by the last theorem,

$$\langle A, B \rangle = \langle A/B \rangle = A/B.$$ □

Remark. The theorem strengthens Theorem 6.5 (that A/B is linear on the same hypothesis) by converting it into an instrument for linear hull construction. Note that the condition $A \approx B$ is essential.

Two corollaries follow: a symmetry result and an alternative form for $\langle A, B \rangle$.

Corollary 6.14.1. *Let A and B be linear and $A \approx B$. Then $A/B = B/A$.*

Corollary 6.14.2. *Let A and B be linear and contain p. Then $\langle A, B \rangle = AB/p = p/AB$.*

PROOF. By Lemma 6.5, $B = p/B$. Then

$$\langle A, B \rangle = A/B = A/(p/B) \subset AB/p \subset \langle A, B \rangle.$$

Thus $\langle A, B \rangle = AB/p$. Similarly for the second conclusion. □

1. Prove: $\langle A, \langle B, C \rangle \rangle = \langle \langle A, B \rangle, C \rangle = \langle A, B, C \rangle$.

2. Prove: If $B \subset \langle A \rangle$, then $\langle A, B \rangle = \langle A \rangle$.

3. Prove: $\langle A_1, \ldots, A_n \rangle = \langle A_1 \ldots A_n \rangle = [A_1 \ldots A_n]/[A_1 \ldots A_n]$, provided $A_1, \ldots, A_n \neq \emptyset$.

4. Prove: If A and B are linear and $A \approx B$, then $A/(A/B) = A/B$.

5. Prove: If A and B are linear, $p \subset A \cup B$ and $A \approx B$, then $\langle A, B \rangle = AB/p = p/AB$.

6.13 Linear Hulls of Interiors and Closures

Having introduced the new operation of taking linear hulls, we naturally ask what happens when it is combined with the interior and closure operations. For convenience closure is considered first.

Theorem 6.15. *Let A be convex. Then $\langle \mathcal{C}(A) \rangle = \langle A \rangle$.*

PROOF. Using Theorem 2.19,

$$A \subset \mathcal{C}(A) \subset \langle \mathcal{C}(A) \rangle.$$

Thus $\langle A \rangle \subset \langle \mathcal{C}(A) \rangle$. We show

$$\mathcal{C}(A) \subset \langle A \rangle. \tag{1}$$

Let $a \subset \mathcal{C}(A)$. By Theorem 2.20,

$$ap \subset A, \quad \text{where } p \subset A. \tag{2}$$

The relation (2) implies $a \subset A/p \subset \langle A \rangle$ and (1) holds. But (1) implies $\langle \mathcal{C}(A) \rangle \subset \langle A \rangle$. Thus $\langle \mathcal{C}(A) \rangle = \langle A \rangle$. $\qquad\square$

Corollary 6.15.1. *Let A be convex. Then $\mathcal{C}(A) \subset \langle A \rangle$.*

Corollary 6.15.2. *Let A be convex and L linear. Then $A \subset L$ implies $\mathcal{C}(A) \subset L$.*

Corollary 6.15.3. *Any linear set is closed.*

The result for the interior now falls into place.

Theorem 6.16. *Let A be convex. Then $\langle \mathcal{I}(A) \rangle = \langle A \rangle$, provided $\mathcal{I}(A) \neq \emptyset$.*

PROOF. By Theorem 2.27, $\mathcal{C}(\mathcal{I}(A)) = \mathcal{C}(A)$. This and the last theorem imply

$$\langle \mathcal{I}(A) \rangle = \langle \mathcal{C}(\mathcal{I}(A)) \rangle = \langle \mathcal{C}(A) \rangle = \langle A \rangle. \qquad\square$$

6.14 Geometric Relations of Points—Linear Dependence and Independence

Now a new theme is introduced. We shift from considering operations on points or sets of points to relations of points. In high school geometry the strong emphasis placed on figures tends to obscure relations of points. But they are present at least implicitly. References, for example, to collinear or coplanar points indicate geometric relations of points.

As a simple example consider this statement in Euclidean geometry:

$$\text{Point } a \text{ is in line } bc. \tag{1}$$

Rewriting (1) as $a \subset \langle b, c \rangle$, $b \neq c$, and dropping the restriction, we have

$$a \subset \langle b, c \rangle. \tag{2}$$

This expresses a typical linear relation of the points a, b, c.

Similarly the statement "a is in plane bcd" suggests

$$a \subset \langle b, c, d \rangle, \tag{3}$$

a linear relation of the four points a, b, c, d. Generalizing (2) and (3) to n points, we have

$$a_1 \subset \langle a_2, \ldots, a_n \rangle,$$

which expresses a linear relation of the points $a_1, \ldots, a_n$. Relations of this type are given a name.

Definition. Suppose points $a_1, \ldots, a_n$ satisfy the condition

$$a_i \subset \langle a_1, \ldots, a_{i-1}, a_{i+1}, \ldots, a_n \rangle$$

for some i, $1 \leqslant i \leqslant n$. Then we say $a_1, \ldots, a_n$ are *linearly related* or *linearly dependent*. Points $a_1, \ldots, a_n$ are *linearly unrelated* or *linearly independent* if they are not linearly dependent, that is, for each i, $1 \leqslant i \leqslant n$,

$$a_i \not\subset \langle a_1, \ldots, a_{i-1}, a_{i+1}, \ldots, a_n \rangle.$$

Remarks on the Definition. For the cases $n = 1, 2, 3$ observe: A single point a is always linearly independent; two points a, b are linearly independent if and only if they are distinct; three points a, b, c are linearly independent if and only if they are distinct and no one is in the line generated by the other two.

Euclidean Illustration. Here are the basic properties of linear dependence in a Euclidean 3-space.

(1) a, b, c are linearly dependent if and only if they are collinear (Figure 6.6).

(2) a, b, c, d are linearly dependent if and only if they are coplanar (Figure 6.7).

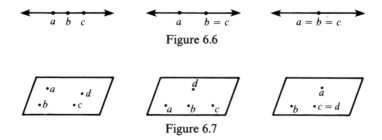

Figure 6.6

Figure 6.7

(3) $a_1, \ldots, a_n$ are linearly dependent if $n > 4$, that is, five or more points are always linearly dependent.

Corresponding properties of linear independence are not hard to get.

Alternatives to the Definition of Linear Dependence. There are other definitions of linear dependence in use which are rooted in elementary geometric relations of points. Consider the statement that a, b, c are collinear. This suggests: There exist two points x, y such that

$$a, b, c \subset \langle x, y \rangle.$$

Generalized to n points, this yields the following condition for linear dependence of $a_1, \ldots, a_n$:

(A) There exist points $x_1, \ldots, x_{n-1}$ such that

$$a_1, \ldots, a_n \subset \langle x_1, \ldots, x_{n-1} \rangle.$$

Similarly the statement "line ab intersects plane cde" leads to another condition for linear dependence of $a_1, \ldots, a_n$:

(B) There exists a permutation $i_1, \ldots, i_n$ of the integers $1, \ldots, n$ and an integer r, $1 \leqslant r < n$, such that

$$\langle a_{i_1}, \ldots, a_{i_r} \rangle \approx \langle a_{i_{r+1}}, \ldots, a_{i_n} \rangle.$$

What are the relations between the adopted definition of linear dependence and the two alternatives (A) and (B)? All three are equivalent in a Euclidean geometry and in $\mathbb{R}^n$. This follows from Theorem 11.13, that the three definitions are equivalent in exchange join geometries, which include Euclidean geometries and $\mathbb{R}^n$. In an arbitrary join geometry the adopted definition implies both alternatives. (See Exercise 4 at the end of Section 6.17 below.) But no two of the three definitions are equivalent. (See Exercises 1, 2 at the end of Section 11.12.)

6.15 Properties of Linear Independent Points

Two elementary properties of linearly independent points are now established.

Theorem 6.17. *If $a_1, \ldots, a_n$ are linearly independent, they are distinct.*

PROOF. Suppose the theorem is false. Then $a_i = a_j$ for some pair of distinct indices i, j. Hence

$$a_i = a_j \subset \langle a_1, \ldots, a_{i-1}, a_{i+1}, \ldots, a_n \rangle,$$

contrary to hypothesis. Thus the theorem holds. ☐

The definition of linear independence of points $a_1, \ldots, a_n$ involves implicitly the idea that $a_1, \ldots, a_n$ form a sequence of elements. Since, however, the points $a_1, \ldots, a_n$ are now known to be distinct and their order is immaterial to the defining condition, it is natural (and sometimes quite convenient) to refer to a linearly independent *set* of points. Thus we introduce the following definition.

Definition. Let S be a finite set of points. Suppose $x \subset S$ implies $x \not\subset \langle S - x \rangle$. Then S is said to be *linearly independent*. If S is not linearly independent, S is said to be *linearly dependent*.

Note by the definition that $\emptyset$ is linearly independent. This is very useful in certain contexts.

The notion of linear independence of a finite set is essentially equivalent to that of linear independence of a finite sequence. For if $a_1, \ldots, a_n$ are linearly independent, the *set* $\{a_1, \ldots, a_n\}$ is linearly independent. The converse holds with a slight but essential restriction. If the set $\{a_1, \ldots, a_n\}$ is linearly independent and the a's are *distinct*, then $a_1, \ldots, a_n$ are linearly independent.

For the sake of convenience we adopt the convention that whenever the phrase "$\{a_1, \ldots, a_n\}$ is linearly independent" is employed, it is understood that the a's are distinct.

Theorem 6.18. *Let $\{a_1, \ldots, a_n\}$ be linearly independent. Then any subset of $\{a_1, \ldots, a_n\}$ is linearly independent.*

PROOF. Suppose $\{a_1, \ldots, a_n\}$ is linearly independent. The result is trivial if $n = 1$. Assume $n > 1$. Any subset of $\{a_1, \ldots, a_n\}$ can be gotten by deleting points from it one at a time. It suffices to show then that if a single point is deleted from $\{a_1, \ldots, a_n\}$, a linearly independent set is obtained. Since the order of the a's is immaterial, let us delete a_1, obtaining the subset $\{a_2, \ldots, a_n\}$. Suppose $\{a_2, \ldots, a_n\}$ is not linearly independent. Then for some i, $2 \leqslant i \leqslant n$,

$$a_i \subset \langle a_2, \ldots, a_{i-1}, a_{i+1}, \ldots, a_n \rangle.$$

Hence

$$a_i \subset \langle a_1, \ldots, a_{i-1}, a_{i+1}, \ldots, a_n \rangle.$$

This contradicts the hypothesis, and the theorem is established. $\square$

6.16 Simplexes

Now the idea of linear independence is used to define an important family of polytopes, which generalizes the closed segments, closed triangular regions and closed tetrahedral regions of Euclidean geometry.

Definition. A polytope is called a *simplex* if its vertices (extreme points) form a linearly independent set.

Recall that any polytope is generated (convexly) by its set of vertices (Theorem 4.21). Thus the vertices of a simplex T form a linearly independent set of generators of T. This property characterizes the vertices of T.

Theorem 6.19. *The vertices of a simplex T form the only linearly independent set of generators of T.*

PROOF. Let E be the set of vertices of T. Let $\{a_1, \ldots, a_n\}$ be a linearly independent set of generators of T. We show

$$\{a_1, \ldots, a_n\} = E. \tag{1}$$

Suppose $a_i \not\subset E$, that is, a_i is not an extreme point of T. By Theorem 4.20, point a_i is redundant in the set of generators $\{a_1, \ldots, a_n\}$, that is, T is generated by

$$\{a_1, \ldots, a_n\} - a_i = \{a_1, \ldots, a_{i-1}, a_{i+1}, \ldots, a_n\}.$$

Thus $T = [a_1, \ldots, a_{i-1}, a_{i+1}, \ldots, a_n]$. Hence

$$a_i \subset [a_1, \ldots, a_{i-1}, a_{i+1}, \ldots, a_n] \subset \langle a_1, \ldots, a_{i-1}, a_{i+1}, \ldots, a_n \rangle,$$

contrary to the linear independence of the a's. Hence $a_i \subset E$, and we infer $\{a_1, \ldots, a_n\} \subset E$. But by Theorem 4.19, $E \subset \{a_1, \ldots, a_n\}$. Thus (1) is verified and the theorem is proven. $\square$

How can we construct or generate simplexes? A natural procedure is to take linearly independent points $a_1, \ldots, a_n$ and form the polytope $[a_1, \ldots, a_n]$. Will this be a simplex; will $a_1, \ldots, a_n$ be its vertices?

Corollary 6.19. *Let* $a_1, \ldots, a_n$ *be linearly independent. Then* $[a_1, \ldots, a_n]$ *is a simplex and* $a_1, \ldots, a_n$ *are its vertices.*

PROOF. By Theorem 4.19 the vertices of $[a_1, \ldots, a_n]$ form a subset of $\{a_1, \ldots, a_n\}$ and so are linearly independent (Theorem 6.18). By definition $[a_1, \ldots, a_n]$ is a simplex. Now apply the theorem. ☐

6.17 Linear Dependence and Intersection of Joins

The linear dependence of $a_1, \ldots, a_n$ is intimately related to the condition that two joins of the a's intersect. For example, if $ab \approx bcd$, then a, b, c, d can be shown to be linearly dependent. Conversely, if a, b, c, d are linearly dependent, then it can be shown that some pair of joins of points of a, b, c, d must intersect, for example, $ab \approx cd$. The situation requires clarification, since there are "trivial" relations like $ab \approx ba$ which hold for any two points. Such a relation imposes no restriction on a and b and certainly does not imply the linear dependence of a, b, c, d.

Definition. Let points $a_1, \ldots, a_n$ be given. Then two joins of the a's, $a_{i_1} \ldots a_{i_r}$ and $a_{j_1} \ldots a_{j_s}$, are said to be *formally identical* if the sets of indices $\{i_1, \ldots, i_r\}$ and $\{j_1, \ldots, j_s\}$ are identical. If the sets of indices are distinct, $a_{i_1} \ldots a_{i_r}$ and $a_{j_1} \ldots a_{j_s}$ are *formally distinct.*

Suppose for example a_1, a_2, a_3 are given. Then $a_1a_2a_3$ and $a_2a_1a_3a_1$ are formally identical joins of the a's—but $a_1a_2a_3$ and a_2a_3 are formally distinct joins of the a's. Observe that the relation $a_1a_2a_3 \approx a_2a_1a_3a_1$ is trivial in the sense that it puts no limitation at all on the points a_1, a_2, a_3. On the other hand $a_1a_2a_3 \approx a_2a_3$ imposes a restriction on a_1, a_2, a_3. A similar situation arises in algebra: compare the equations $a + b = b + a$ and $a + b = c + d$.

Note. If two joins are formally distinct, this does not mean that when interpreted concretely the resulting sets must be distinct. For example (Figure 6.8), a_1a_2 and $a_1a_2a_3$ are formally distinct, but in a Euclidean line $a_1a_2 = a_1a_2a_3$ if $a_3 \subset a_1a_2$.

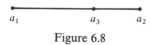

Figure 6.8

Theorem 6.20. $a_1, \ldots, a_n$ *are linearly dependent if and only if there exists a pair of formally distinct joins of the* a's *which intersect.*

PROOF. Suppose $a_1, \ldots, a_n$ linearly dependent. Then for some i, $1 \leqslant i \leqslant n$,

$$a_i \subset \langle a_1, \ldots, a_{i-1}, a_{i+1}, \ldots, a_n \rangle.$$

By Theorem 6.9,

$$a_i \subset a_1 \ldots a_{i-1}a_{i+1} \ldots a_n / a_1 \ldots a_{i-1}a_{i+1} \ldots a_n,$$

so that

$$a_1 \ldots a_n \approx a_1 \ldots a_{i-1}a_{i+1} \ldots a_n. \tag{1}$$

Certainly the joins in (1) are formally distinct.

Conversely suppose $a_{i_1} \ldots a_{i_r}$ and $a_{j_1} \ldots a_{j_s}$ satisfy

$$a_{i_1} \ldots a_{i_r} \approx a_{j_1} \ldots a_{j_s} \tag{2}$$

and are formally distinct. It is not restrictive to assume in (2) that the i's and the j's are distinct. Observe that

$$\{i_1, \ldots, i_r\} \neq \{j_1, \ldots, j_s\}. \tag{3}$$

Then in (3) there must be a number in one of the sets that is not in the other. It is not restrictive to assume i_1 is such a number. Note that i_1 is distinct from the remaining i's and from all the j's. For convenience let $i_1 = k$, so that $a_{i_1} = a_k$.

In (2) suppose $r = 1$. Then

$$a_k \approx a_{j_1} \ldots a_{j_s} \subset \langle a_1, \ldots, a_{k-1}, a_{k+1}, \ldots, a_n \rangle,$$

and $a_1, \ldots, a_n$ are linearly dependent. Finally suppose $r > 1$. Then (2) implies

$$a_k \approx a_{j_1} \ldots a_{j_s} / a_{i_2} \ldots a_{i_r} \subset \langle a_1, \ldots, a_{k-1}, a_{k+1}, \ldots, a_n \rangle.$$

Again the a's are linearly dependent and the proof is complete. □

The result is rather striking, since one does not expect "linearity" relations such as $a \subset \langle b, c, d \rangle$ to be equivalent to join intersection relations such as $ab \approx bcd$, which seem much simpler.

Corollary 6.20.1. $a_1, \ldots, a_n$ *are linearly independent if and only if every two joins of the a's that are formally distinct also are disjoint.*

The theorem yields a characterization of simplexes.

Corollary 6.20.2. *Let P be a polytope and $a_1, \ldots, a_n$ its vertices. Then P is a simplex if and only if every two joins of the a's that are formally distinct also are disjoint.*

PROOF. If P is a simplex, $a_1, \ldots, a_n$ are linearly independent by definition. The disjointness condition follows by the preceding corollary. Conversely the disjointness condition implies $a_1, \ldots, a_n$ linearly independent (Corollary 6.20.1), and P is a simplex by definition. □

To get a somewhat deeper view of Corollary 6.20.2, note that $P = [a_1, \ldots, a_n]$. Then the polytope join formula (Theorem 3.2) implies

$$P = a_1 \cup \cdots \cup a_n \cup a_1 a_2 \cup \cdots \cup a_{n-1}a_n \cup \cdots \cup a_1 \ldots a_n. \tag{1}$$

Note that any join of the a's is formally identical to some term in the right member of (1) and that each two of these terms are formally distinct.

Restatement of Corollary 6.20.2. Let $a_1, \ldots, a_n$ be the vertices of polytope P. Then P is a simplex if and only if each pair of terms in the right member of (1) are disjoint.

This corollary indicates the extent to which a simplex deserves its name: Not only are its generating vertices linearly unrelated, but the joins of its vertices, of which it is composed, are "unrelated" as sets. Roughly put: Simplexes are to polytopes as triangles to polygons.

It is hard to see how to get an explicit formula for the frontier of a polytope; for a simplex it is easy.

Corollary 6.20.3. *Let P be a simplex and $a_1, \ldots, a_n$ its vertices. Then*

$$\mathcal{F}(P) = a_1 \cup \cdots \cup a_n \cup a_1 a_2 \cup \cdots \cup a_{n-1} a_n \cup \cdots$$

$$\cup a_1 \ldots a_{n-1} \cup \cdots \cup a_2 \ldots a_n.$$

PROOF. Using Theorem 2.30(b) and Corollary 4.30, we have

$$\mathcal{F}(P) = P - \mathcal{G}(P) = [a_1, \ldots, a_n] - a_1 \ldots a_n.$$

The conclusion follows by applying the last corollary. $\square$

EXERCISES (LINEAR INDEPENDENCE)

1. Prove: If $a_1, \ldots, a_n$ are linearly independent and $i_1, \ldots, i_n$ is any permutation of the integers $1, \ldots, n$, then

$$[a_{i_1}, \ldots, a_{i_r}] \not\approx [a_{i_{r+1}}, \ldots, a_{i_n}], \qquad 1 \leqslant r < n. \tag{1}$$

2. Prove that the following converse of the proposition in Exercise 1 is not valid: If $a_1, \ldots, a_n$ have the property that (1) holds whenever $i_1, \ldots, i_n$ is a permutation of $1, \ldots, n$, then $a_1, \ldots, a_n$ are linearly independent.

3. Let P be a polytope and $a_1, \ldots, a_n$ its vertices. Prove that P is a simplex if and only if

$$\mathcal{F}(P) = a_1 \cup \cdots \cup a_n \cup a_1 a_2 \cup \cdots \cup a_{n-1} a_n \cup \cdots$$

$$\cup a_1 \ldots a_{n-1} \cup \cdots \cup a_2 \ldots a_n.$$

4. Prove: If $a_1, \ldots, a_n$ are linearly dependent, then
   (a) there exist points $x_1, \ldots, x_{n-1}$ such that
   $$a_1, \ldots, a_n \subset \langle x_1, \ldots, x_{n-1} \rangle;$$
   (b) there exists a permutation $i_1, \ldots, i_n$ of the integers $1, \ldots, n$ and an integer r, $1 \leqslant r < n$, such that
   $$\langle a_{i_1}, \ldots, a_{i_r} \rangle \approx \langle a_{i_{r+1}}, \ldots, a_{i_n} \rangle.$$

5. Prove: If a linear set is finitely generated, it has a linearly independent set of generators.

6.18 Covering in the Family of Linear Sets—Hyperplanes

Now we introduce a very useful idea which is suggested by the relation in Euclidean geometry of a line to one of its points, or a 3-space to a plane contained in it.

Definition. Let A and B be linear sets. Suppose $A \supset B$, $A \neq B$ and no linear set "lies between" A and B—that is, $A \supset X \supset B$, where X is linear, implies $X = A$ or $X = B$. Then we say A *covers* B or B *is covered by* A (*in the family of linear sets*). Also we call B a *hyperplane* of A.

In $\mathbb{R}^n$ a linear subspace of dimension $n - 1$—which necessarily is covered by $\mathbb{R}^n$—is called a hyperplane. The term was chosen to indicate that such a subspace plays a role in $\mathbb{R}^n$ analogous to that of a plane in a Euclidean 3-space.

If A covers B, then A is spanned linearly by B and a point.

Theorem 6.21 (Point Spanning Principle). *Let linear set A cover linear set B and $p \subset A - B$. Then $A = \langle B, p \rangle$.*

PROOF. $A \supset \langle B, p \rangle \supset B$ and $\langle B, p \rangle \neq B$. By definition of covering, $\langle B, p \rangle = A$. $\qquad\square$

EXERCISES (ON THE JOIN GEOMETRY $\mathbb{R}^n$)

(Compare exercises at the end of Section 5.9.)

1. Prove: S is linear if and only if $S \supset a, b$ implies
$$S \supset \lambda a + \mu b$$
for all λ, μ satisfying $\lambda + \mu = 1$.

2. Prove: $\langle a, b \rangle$ is the set of all x expressible in the form
$$x = \lambda a + \mu b, \qquad \lambda + \mu = 1.$$
[Compare Exercise 5(b) at the end of Section 5.9.]

3. Prove: $\langle a, b \rangle = ab \cup a/b \cup b/a \cup a \cup b$.

4. Prove: $\langle a_1, a_2, a_3 \rangle$ is the set of all x that are expressible in the form
$$x = \lambda_1 a_1 + \lambda_2 a_2 + \lambda_3 a_3, \qquad \lambda_1 + \lambda_2 + \lambda_3 = 1.$$

5. Generalize Exercises 2 and 4 to m points.

6. (a) Prove: The set of solutions $x = (x_1, \ldots, x_n)$ of a linear equation
$$a_1 x_1 + \cdots + a_n x_n = k$$
is a linear set of $\mathbb{R}^n$.
(b) Is the result in (a) valid for a family of linear equations?

7. Prove: $a_1, \ldots, a_m$ are linearly dependent if and only if there exist $\lambda_1, \ldots, \lambda_m$ not all 0 such that

$$\lambda_1 a_1 + \cdots + \lambda_m a_m = 0$$

and

$$\lambda_1 + \cdots + \lambda_m = 0.$$

8. In $\mathbb{R}^2$ let $a = (0, 0)$, $b = (1, 0)$, $c = (0, 1)$. Determine and represent graphically: ab, ac, bc, a/b, b/a, a/c, c/a, b/c, c/b, abc, a/bc, b/ac, c/ab, ab/c, bc/a, ac/b, and finally $\langle a, b, c \rangle$.

9. Prove: $\langle a, b, c \rangle = a \cup b \cup c \cup ab \cup bc \cup ac \cup abc \cup a/b \cup b/a \cup b/c \cup c/b \cup a/c \cup c/a \cup a/bc \cup b/ac \cup c/ab \cup ab/c \cup bc/a \cup ac/b$.

10. Prove that $\mathbb{R}^n$ contains $n + 1$ linearly independent points which generate it (linearly).

In the following exercises use the definition of hyperplane and side of a hyperplane in Section 5.10 (the paragraph next to the last).

11. Prove:
   (a) Any hyperplane of $\mathbb{R}^n$ is linear.
   (b) Any open or closed halfspace determined by a hyperplane of $\mathbb{R}^n$ is convex.

12. Let H be a hyperplane of $\mathbb{R}^n$ and H_1, H_2 its sides. Prove:
   (a) $\mathbb{R}^n = H \cup H_1 \cup H_2$.
   (b) If $a \subset H_1$, $b \subset H_2$ then $ab \approx H$.
   (c) If $a \subset H_1$, $b \subset H_2$ then $H_1 = H/b$, $H_2 = H/a$.

13. Prove: If H is a hyperplane of $\mathbb{R}^n$, then $\mathbb{R}^n$ covers H, that is, "H is a hyperplane of $\mathbb{R}^n$" in the sense of the definition in the present section.

EXERCISES (COVERING IN THE FAMILY OF CONVEX SETS)

Definition. Let A and B be convex. Suppose $A \supset B$, $A \neq B$ and no convex set "lies between" A and B, that is, $A \supset X \supset B$, where X is convex implies $X = A$ or $X = B$. Then we say A *covers* B or B *is covered by* A (*in the family of convex sets*).

1. In a Euclidean geometry find examples of a convex set that is covered by (a) no convex set; (b) exactly one convex set; (c) exactly two convex sets; (d) an infinitude of convex sets.

2. In a Euclidean geometry find a sequence of convex sets $A_1, \ldots, A_n$ such that A_{i+1} covers A_i for $i = 1, \ldots, n - 1$. Can you find an infinite sequence of this type?

3. Prove: If A and B are convex and A covers B, then $A = B \cup a$ for some a that is absorbed by B.

4. Prove: If B is convex, $a \not\subset B$ and B absorbs a, then $B \cup a$ is convex and covers B.

6.19 The Height of a Linear Set—Approach to a Theory of Dimension

In this section a rudimentary theory of dimension for linear sets or linear spaces is developed.[1]

We shall look on the dimension of a linear space simply as an integer assigned to the linear space which indicates its relative complexity. As a minimal restriction this dimension number should satisfy the following condition:

If A and B are linear sets, $A \supset B$ and $A \neq B$, then the dimension of A should be greater than the dimension of B.

In Euclidean geometry the principle can be indicated symbolically as follows:

Sequence of linear sets: 3-space $\supset$ plane $\supset$ line $\supset$ point;
Sequence of dimension numbers: 3 $>$ 2 $>$ 1 $>$ 0.

The principle suggests that the dimension number of a linear set is related to the number of terms in a sequence of linear sets of the type indicated above.

Now we make more precise the kind of sequence we use.

Definition. Let $A_1, \ldots, A_r$ be linear sets such that $A_i \supset A_{i+1}$, $A_i \neq A_{i+1}$ for $1 \leq i \leq r - 1$. Then we call $A_1, \ldots, A_r$ a *decreasing sequence of linear sets* and $r - 1$ the *length* of the sequence.

Note that the length of the sequence counts, not the number of terms, but the number of individual downward steps from A_1 to A_r.

Now the notion of height of a linear set—which may be considered a measure of its complexity and is a sort of dimension or rank number—can be defined.

Definition. Let A be a linear set. Suppose the lengths of all decreasing sequences of linear sets whose first term is A have a maximum m. Then m is called the *height* of A and is denoted functionally by $h(A)$. If $h(A)$ exists, we say linear set A is *of finite height*; otherwise A is *of infinite height*. A join geometry $(J, \cdot)$ is of *finite (infinite) height* if J as a linear set is of finite (infinite) height.

Note that if A has height m, any decreasing sequence of linear sets with first term A and length m must terminate in $\varnothing$.

[1] See Chapter 11 for a definitive treatment of dimension of linear sets in join geometries that satisfy an additional postulate.

The height of $\varnothing$ is 0, and of a point a is 1. The height of a line, if it exists, is at least 2 and can exceed 2. (See Exercise 7 below.) In a Euclidean join geometry every line has height 2, every plane height 3—in general, a linear set's height is its dimension plus 1. In an arbitrary join geometry the dimension of a linear set A of finite height could be defined to be $h(A) - 1$.

If a linear set has a height, a "lower" linear set has a lesser height.

Theorem 6.22. *Suppose A and B are linear, $A \supset B$, $A \neq B$, and A is of finite height. Then B is of finite height and $h(A) > h(B)$.*

PROOF. Every decreasing sequence of linear sets with first term B can be enlarged to form a "longer" decreasing sequence of linear sets with first term A. The result follows without difficulty.   □

The following theorem will be found quite useful in the sequel.

Theorem 6.23. *Any linear set of finite height is finitely generated (linearly).*

PROOF. Suppose linear set A has height m. By definition there exists a decreasing sequence of linear sets

$$A = A_1, \ldots, A_{m+1}.$$

If $m = 0$, the sequence consists of A, and A must be $\varnothing$, which is finitely generated. Suppose $m \geqslant 1$. Then A_1 covers A_2, for otherwise there would exist a decreasing sequence of linear sets with first term A and length exceeding m. In general—by the same argument—A_i covers A_{i+1}, $1 \leqslant i \leqslant m$. Now let $p_i \subset A_i - A_{i+1}$, $1 \leqslant i \leqslant m$. By the point spanning principle (Theorem 6.21),

$$A_i = \langle p_i, A_{i+1} \rangle, \qquad 1 \leqslant i \leqslant m.$$

Applying this repeatedly and noting that $A_{m+1} = \varnothing$, we have

$$A_1 = \langle p_1, A_2 \rangle = \langle p_1, \langle p_2, A_3 \rangle \rangle = \langle p_1, p_2, A_3 \rangle = \cdots$$
$$= \langle p_1, p_2, \ldots, p_m, A_{m+1} \rangle = \langle p_1, p_2, \ldots, p_m \rangle.$$

Thus $A = A_1$ is finitely generated (linearly) and the theorem is proved.   □

Corollary 6.23. *Let A be a linear set of finite height. Then any linear subset of A is finitely generated (linearly).*

PROOF. Use the theorem in conjunction with the previous theorem.   □

EXERCISES (LINEAR SETS AND HEIGHTS)

1. Prove: If A and B are linear sets of finite height, $A \supset B$ and $h(A) = h(B) + 1$, then A covers B.

2. Prove: If $h(A) = n$ for linear set A, then there exist $a_1, \ldots, a_n$ such that $A = \langle a_1, \ldots, a_n \rangle$ and

$$\langle a_1, \ldots, a_{i+1} \rangle \text{ covers } \langle a_1, \ldots, a_i \rangle, \qquad 1 \leqslant i \leqslant n - 1.$$

3. Prove: If linear set A is generated (linearly) by a set of n independent points and $h(A)$ exists, then $h(A) \geqslant n$.

4. (a) Prove: If $h(A) = n$ for linear set A, then A is generated (linearly) by n distinct points.
   (b) Can A be generated by fewer than n points?

5. Prove: In $(\mathbb{R}^n, \cdot)$, if the linear set $\mathbb{R}^n$ has a height, its height is at least $n + 1$.

6. Prove: In $(\mathbb{R}^*, \cdot)$ the linear set $\mathbb{R}^*$ does not have finite height. (See Section 5.12.)

7. For the join geometry JG16 show that J is a line and that $h(J) = 3$.

6.20 Linear Sets and the Interior Operation

There are interrelations between linear sets and the interior operation on convex sets which yield new results for both.

First a set containment criterion for a point of a convex set to be an interior point (see Figure 6.9).

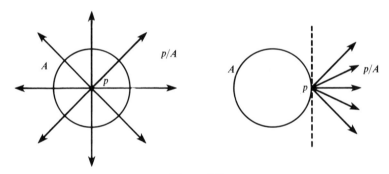

Figure 6.9

Theorem 6.24. *Let A be convex and $p \subset A$. Then $p \subset \mathcal{I}(A)$ if and only if $A \subset p/A$.*

PROOF. Suppose $p \subset \mathcal{I}(A)$. Then $p \subset xA$ for each $x \subset A$ (Theorem 4.24). Thus $x \subset p/A$ for each $x \subset A$. Hence $A \subset p/A$. Retrace steps for the converse.   □

Now the theorem is used to establish a new formula for the linear hull of a convex set (see Figure 6.10).

Theorem 6.25. *Let A be convex. Then $p \subset \mathcal{I}(A)$ implies $\langle A \rangle = p/A = A/p$.*

(Compare Corollary 6.11.1, Lemma 6.5.)

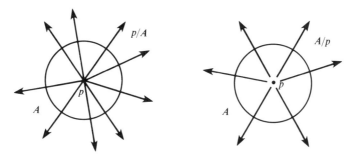

Figure 6.10

PROOF. Note that $p \subset A$. By the last theorem, $A \subset p/A$. Then using Corollary 6.11.1,

$$\langle A \rangle = A/A \subset (p/A)/A = p/AA = p/A \subset \langle A \rangle.$$

Thus $\langle A \rangle = p/A$. Similarly

$$\langle A \rangle = A/A \subset A/(p/A) \subset AA/p = A/p \subset \langle A \rangle,$$

and $\langle A \rangle = A/p$.   □

The theorem is a significant improvement on the result that $\langle A \rangle = A/A$ for convex A (Corollary 6.11.1), since it permits the replacement of either A in the right member by a single point. It affords a simple construction for linear sets: "divide" a convex set "by" or "into" one of its interior points. Finally note its intuitive geometric significance: The rays p/x, $x \subset A$ form a linear set. Similarly for the rays x/p, $x \subset A$.

Corollary 6.25. *Let A be convex and $p \subset \mathcal{I}(A)$. Then p/A and A/p are linear.*

The last theorem and corollary give linearity consequences of $p \subset \mathcal{I}(A)$. We seek conversely linearity *conditions* for $p \subset \mathcal{I}(A)$.

Theorem 6.26. *Let A be convex and $p \subset A$. Then $p \subset \mathcal{I}(A)$ if (a) p/A is linear or (b) A/p is linear.*

PROOF. (a) $p/A \supset p/p = p$. Note that $p/(p/A) \supset A$ [Corollary 4.7(c)]. Since p/A is linear,

$$p/A \supset p/(p/A) \supset A,$$

and $p \subset \mathcal{I}(A)$ by Theorem 6.24.
    (b) Let $x \subset A$. We show

$$p \subset xA \qquad\qquad (1)$$

and conclude $p \subset \mathcal{I}(A)$ by Theorem 4.24. Note that $A/p \supset p$. Since A is convex $A \supset px$ and $A/p \supset x$. Since A/p is linear,

$$A/p \supset p/x.$$

J6 implies $Ax \approx p$, so that (1) holds and the proof is complete.   □

Next a converse of Theorem 6.25.

Corollary 6.26.1. *Let A be convex and $p \subset A$. Then $p \subset \mathcal{I}(A)$ if (a) $\langle A \rangle = p/A$ or (b) $\langle A \rangle = A/p$.*

Corollary 6.26.2. *Let A be convex and $p \subset A$. Then $p/A = A/p$ if either p/A or A/p is linear.*

PROOF. By Theorems 6.26 and 6.25.   □

6.21 Applications: Interiority Properties

The results derived have direct application to interiority questions. First an elementary property of joins to interior points.

Theorem 6.27. *Let A be convex and $p \subset \mathcal{I}(A)$. Then $q \subset \langle A \rangle$ if and only if (a) $pq \approx A$ or (b) $pq \approx \mathcal{I}(A)$.*

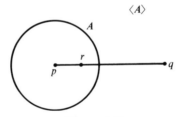

Figure 6.11

PROOF. (a) $p \subset \mathcal{I}(A)$ implies $\langle A \rangle = A/p$ (Theorem 6.25). Then $q \subset \langle A \rangle$ is equivalent to $q \approx A/p$ and so to $pq \approx A$.

(b) Suppose $q \subset \langle A \rangle$. By (a), $pq \approx A$; let r be a common point. Then $pq \supset r$ implies $pq \supset pr$. By Theorem 2.15, $pr \subset \mathcal{I}(A)$ and so $pq \approx \mathcal{I}(A)$. The converse is trivial by (a).   □

The result is related to Theorem 2.23, that if $p \subset \mathcal{I}(A)$ and $q \subset \mathcal{C}(A)$, then $pq \subset \mathcal{I}(A)$. It gives new information when $q \subset \langle A \rangle$ but $q \not\subset \mathcal{C}(A)$. This suggests the following

Definition. Let A be convex. Then $\langle A \rangle - \mathcal{C}(A)$ is called the *exterior* of A and is denoted by $\mathcal{E}(A)$. Any point of $\mathcal{E}(A)$ is called an *exterior point* of A.

Now we can assert the following corollary to Theorem 6.27.

Corollary 6.27. *The join of an interior point and an exterior point of a convex set always meets its interior.*

In Euclidean geometry or in $\mathbb{R}^n$, where the points of a line correspond to the real numbers, the conclusion can be strengthened: The join of an interior point and an exterior point of a convex set must meet its boundary. This is not valid in the theory of join geometries. JG14 provides a counterexample (the convex set may be taken to be the set of positive real numbers).

Note that if A is linear $\mathcal{E}(A) = \varnothing$. In Euclidean geometry it is intuitively evident that every nonlinear convex set has nonempty exterior. The following theorem is useful in proving such results.

Theorem 6.28. *Suppose A is convex, $\mathcal{I}(A) \neq \varnothing$ and A is not linear. Then $\mathcal{E}(A) \neq \varnothing$.*

PROOF. Suppose $\mathcal{E}(A) = \varnothing$. Then $\langle A \rangle - \mathcal{C}(A) = \varnothing$, so that $\langle A \rangle \subset \mathcal{C}(A)$. Thus $\langle A \rangle = \mathcal{C}(A)$ (Corollary 6.15.1). Note that $\langle A \rangle$ is open (Corollary 6.2.2) and $\mathcal{I}(\mathcal{C}(A)) = \mathcal{I}(A)$ (Theorem 2.28). Then

$$A \subset \langle A \rangle = \mathcal{I}(\langle A \rangle) = \mathcal{I}(\mathcal{C}(A)) = \mathcal{I}(A) \subset A.$$

Thus $A = \langle A \rangle$ and is linear, which contradicts the hypothesis and establishes the theorem. □

Corollary 6.28.1. *Suppose A is convex, $\mathcal{I}(A) \neq \varnothing$ and $\mathcal{B}(A) \neq \varnothing$. Then $\mathcal{E}(A) \neq \varnothing$.*

PROOF. $\mathcal{B}(A) \neq \varnothing$ implies $\mathcal{C}(A) \neq \mathcal{I}(A)$. Hence by Corollaries 6.2.2 and 6.15.3, A is not linear. □

Corollary 6.28.2. *Let A be a polytope and not a point. Then $\mathcal{E}(A) \neq \varnothing$.*

PROOF. $\mathcal{F}(A) \neq \varnothing$ (Corollary 4.22). Thus $\mathcal{B}(A) \neq \varnothing$. Certainly $\mathcal{I}(A) \neq \varnothing$. □

Remark on Theorem 6.28. The theorem fails if the proviso $\mathcal{I}(A) \neq \varnothing$ is omitted. The justification of this follows from Exercise 3 at the end of Section 5.14. The exercise involves a convex set K of the join geometry $(\mathbb{R}^*, \cdot)$ which satisfies $\mathcal{I}(K) = \varnothing$ and $\mathcal{C}(K)$ is closed under extension. $\mathcal{I}(K) = \varnothing$ implies K is not open and so not linear. Note that $\mathcal{C}(K)$ is linear and $\mathcal{C}(K) = \langle K \rangle$. Thus $\mathcal{E}(K) = \varnothing$ and the theorem fails.

In Chapter 2 the monotonic property for convex sets A and B, $A \subset B$ implies $\mathcal{I}(A) \subset \mathcal{I}(B)$, was proved with the proviso (a) $A \approx \mathcal{I}(B)$. (See Theorem 2.17). Now this property is justified under the proviso (b) $\langle A \rangle = \langle B \rangle$, which is an interesting and useful supplement to (a).

Theorem 6.29 (Monotonic Property of Interior). *Let the convex sets A and B satisfy $\langle A \rangle = \langle B \rangle$. Then $A \subset B$ implies $\mathcal{I}(A) \subset \mathcal{I}(B)$.*

PROOF. Let $p \subset \mathcal{I}(A)$ (Figure 6.12). Note that $p \subset B$. Then using Theorem 6.25,

$$\langle B \rangle = \langle A \rangle = p/A \subset p/B \subset \langle B \rangle.$$

Hence $\langle B \rangle = p/B$, so that Corollary 6.26.1 implies $p \subset \mathcal{I}(B)$. Hence $\mathcal{I}(A) \subset \mathcal{I}(B)$. $\qquad\square$

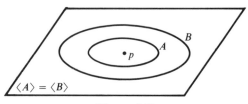

Figure 6.12

The theorem implies that the interior of a convex set K contains every open subset of K that has the same linear hull as K. This suggests the following corollary.

Corollary 6.29.1. *Let K be a convex set. Let U be the union of the open subsets of K that have the same linear hull as K. Then $\mathcal{I}(K) = U$.*

PROOF. By the theorem, $\mathcal{I}(K) \supset U$. We show $\mathcal{I}(K) \subset U$. This is trivial if $\mathcal{I}(K) = \varnothing$. If $\mathcal{I}(K) \neq \varnothing$, then $\langle \mathcal{I}(K) \rangle = \langle K \rangle$ by Theorem 6.16, and $\mathcal{I}(K) \subset U$. $\qquad\square$

The case where K is pathological (Section 5.11) and the case where K is not pathological are now considered.

Corollary 6.29.2. *Let K be a pathological convex set. Then there are no open sets in K that have the same linear hull as K.*

Corollary 6.29.3. *Let K be a nonpathological convex set. Then $\mathcal{I}(K)$ is the greatest member of the family of open sets in K that have the same linear hull as K.*

PROOF. This is trivial if $K = \varnothing$. If $K \neq \varnothing$, then $\mathcal{I}(K) \neq \varnothing$ and $\langle \mathcal{I}(K) \rangle = \langle K \rangle$ by Theorem 6.16. Thus $\mathcal{I}(K)$ is one of the open sets described, and by Corollary 6.29.1, the greatest. $\qquad\square$

The theorem yields information on boundaries of convex sets (Figure 6.13).

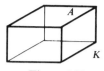

Figure 6.13

Theorem 6.30. *Suppose A and K are nonpathological convex sets, $A \subset \mathcal{B}(K)$, $K \neq \emptyset$. Then $\langle A \rangle$ is a proper subset of $\langle K \rangle$.*

PROOF. First we show $\langle A \rangle \neq \langle K \rangle$. Suppose on the contrary

$$\langle A \rangle = \langle K \rangle. \qquad (1)$$

$A \subset \mathcal{B}(K)$ implies

$$A \subset \mathcal{C}(K). \qquad (2)$$

By Theorem 6.15, $\langle \mathcal{C}(K) \rangle = \langle K \rangle$, and (1) implies

$$\langle A \rangle = \langle \mathcal{C}(K) \rangle. \qquad (3)$$

Applying Theorem 6.29 to (2) and (3) and using Theorem 2.28, we have

$$\mathcal{I}(A) \subset \mathcal{I}(\mathcal{C}(K)) = \mathcal{I}(K).$$

Since $K \neq \emptyset$, (1) implies $A \neq \emptyset$. Hence $\mathcal{I}(A) \neq \emptyset$. But $A \subset \mathcal{B}(K)$ yields $\mathcal{I}(A) \subset \mathcal{B}(K)$ and so $\mathcal{I}(A) \subset \mathcal{I}(K) \cap \mathcal{B}(K)$. Thus $\mathcal{I}(K) \cap \mathcal{B}(K) \neq \emptyset$, contrary to Theorem 2.30(c), so that (1) is false. It only remains to prove $\langle A \rangle \subset \langle K \rangle$. Using (2) and Theorem 6.15, we have $\langle A \rangle \subset \langle \mathcal{C}(K) \rangle = \langle K \rangle$ and the proof is complete. $\square$

Remarks on the Theorem. First note that $\mathcal{B}(K) \subset \langle K \rangle$. Then the theorem says, roughly speaking, that $\mathcal{B}(K)$ is not as densely spread in the linear space $\langle K \rangle$ as K is—for no convex portion A of $\mathcal{B}(K)$ can generate linearly the whole of $\langle K \rangle$. $\mathcal{B}(K)$ is not as "solid" as K. If $\langle K \rangle$ has finite height, the relations may be stated precisely in the form: $\langle A \rangle$ has lesser height than $\langle K \rangle$ (Theorem 6.22). An illustration of the relation is the intuition in Euclidean geometry that the boundary of a 3-dimensional polytope such as a solid cube consists of convex 2-dimensional pieces.

EXERCISES

1. Suppose $A \subset B$, where A and B are convex. Prove that $\mathcal{I}(A) \subset \mathcal{I}(B)$ if and only if $\langle A \rangle \approx \mathcal{I}(B)$ or $\mathcal{I}(A) = \emptyset$. (Compare Corollary 2.17.2.)

2. Prove: If A is an open subset of convex set B and $\langle A \rangle \approx \mathcal{I}(B)$, then $A \subset \mathcal{I}(B)$.

3. Prove: If A and B are convex and $A \subset \mathcal{B}(B)$, then $\langle A \rangle \not\approx \mathcal{I}(B)$.

4. Suppose A is convex, $A \neq \emptyset$ and p/A is linear.
   (a) Prove: $p \subset A$.
   (b) Infer $p \subset \mathcal{I}(A)$ and $p/A = A/p = \langle A \rangle$.

5. The same as Exercise 4 if in the hypothesis p/A is replaced by A/p.

6. Prove: If A is convex and $p/A = A/p$, then p/A is linear and $p/A = \langle A \rangle$.

7. (a) Prove: If A is open and $p \subset \langle A \rangle$, then $Ap \supset A$.
   (b) Prove: If A and B are open, $A \approx \langle B \rangle$ and $B \approx \langle A \rangle$, then $AB \supset A, B$.

8. Prove: If A is a convex set, then

$$\langle A \rangle = \mathcal{I}(A) \cup \mathcal{B}(A) \cup \mathcal{E}(A),$$

and no two of $\mathcal{I}(A)$, $\mathcal{B}(A)$, $\mathcal{E}(A)$ meet.

6.22 The Prevalence of Nonpathological Convex Sets

The existence of pathological convex sets in a certain join geometry $(\mathbb{R}^*, \cdot)$ was demonstrated in Chapter 5 (Sections 5.11–5.13). The join geometry $(\mathbb{R}^*, \cdot)$ was there described (Section 5.12) as *infinite dimensional*. We take this to mean $(\mathbb{R}^*, \cdot)$ has infinite height (Section 6.19), which is justified in the Remark following Corollary 6.32 below.

This illustration turns out to be essentially typical. In effect pathological convex sets do not exist in finite dimensional join geometries. We prove in the following theorems that the family of nonpathological convex sets of a join geometry includes:

(a) All convex sets whose linear hulls are finitely generated;
(b) All convex subsets of a linear set of finite height.

Theorem 6.31. *Let A be convex and $\langle A \rangle$ finitely generated (linearly). Then A is not pathological.*

Proof. The result is trivial if $A = \varnothing$. Suppose $A \neq \varnothing$. By hypothesis

$$\langle A \rangle = \langle p_1, \ldots, p_n \rangle. \tag{1}$$

First we show $\langle A \rangle$ is generated by a finite subset of A. By (1) and Corollary 6.11.1,

$$p_i \subset \langle A \rangle = A/A, \qquad 1 \leqslant i \leqslant n.$$

Hence $p_i \subset q_i/r_i$ where $q_i, r_i \subset A$. Thus

$$\langle A \rangle = \langle p_1, \ldots, p_n \rangle \subset \langle q_1, r_1, \ldots, q_n, r_n \rangle \subset \langle A \rangle$$

so that

$$\langle A \rangle = \langle q_1, r_1, \ldots, q_n, r_n \rangle. \tag{2}$$

We can rewrite (2) in the form

$$\langle A \rangle = \langle a_1, \ldots, a_m \rangle, \tag{3}$$

where the a's are in A. Thus

$$A \supset a_1 \ldots a_m. \tag{4}$$

By (3) and Corollary 6.11.2,

$$\langle A \rangle = \langle a_1 \ldots a_m \rangle. \tag{5}$$

Note that $a_1 \ldots a_m$ is nonempty (Corollary 2.6) and open (Theorem 4.28). Applying Theorem 6.29 to (4) and (5), we have

$$\mathcal{I}(A) \supset \mathcal{I}(a_1 \ldots a_m) = a_1 \ldots a_m \neq \varnothing,$$

and the theorem holds.   □

Theorem 6.32. *A linear set of finite height has no pathological subsets.*[2]

PROOF. Let L be a linear set of finite height and A a convex subset of L. Consider $\langle A \rangle$. By Corollary 6.23, $\langle A \rangle$ is finitely generated, and A is not pathological by the last theorem.   □

Corollary 6.32. *A join geometry of finite height has no pathological subsets.*

Significance for Euclidean Geometry and $\mathbb{R}^n$. A Euclidean 3-space has height 4, since it contains only five types of linear sets: $\varnothing$, point, line, plane, 3-space. Thus the corollary implies it has no pathological subsets. That $\mathbb{R}^n$ has no pathological subsets follows from a result of Chapter 11: $\mathbb{R}^n$ is finitely generated (Exercise 10 in the set on the Join Geometry $\mathbb{R}^n$ at the end of Section 6.18) and so has finite height (Corollary 11.16 and Exercise 6 at the end of Section 11.2).

Remark. The corollary implies that the join geometry $(\mathbb{R}^*, \cdot)$ of Section 5.12 has infinite height, since it contains a pathological convex set (Section 5.13, Example I).

6.23 The Linear Hull of a Pair of Convex Sets

In this section some results in Section 6.12 on the linear hull of a pair of linear sets are generalized.

Theorem 6.33. *Let A and B be convex and $\mathcal{I}(A) \approx \mathcal{I}(B)$. Then $\langle A, B \rangle = A/B$.*

(Compare Theorem 6.14.)

PROOF. Let $p \subset \mathcal{I}(A), \mathcal{I}(B)$ (Figure 6.14). Then using Theorems 6.14 and 6.25,

$$= \langle \langle A \rangle, \langle B \rangle \rangle$$
$$= \langle A \rangle / \langle B \rangle = (A/p)/(B/p) \subset Ap/Bp \subset A/B.$$

Thus $\langle A, B \rangle = A/B$.   □

---

[2] Compare Rockafellar [1], p. 45, Theorem 6.2.

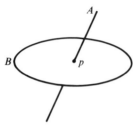

Figure 6.14

This is a strong generalization of Theorem 6.14, upon which the proof depends. It has several corollaries. First, analogues of the corollaries of Theorem 6.14.

Corollary 6.33.1. *Let A and B be convex and $\mathcal{I}(A) \approx \mathcal{I}(B)$. Then A/B is linear and $A/B = B/A$.*

(Compare Theorem 6.5, Corollary 6.14.1.)

Corollary 6.33.2. *Let A and B be convex and $\mathcal{I}(A)$ and $\mathcal{I}(B)$ contain p. Then $\langle A, B \rangle = AB/p = p/AB$.*

(Compare Corollary 6.14.2.)

PROOF. Adapt the method of Corollary 6.14.2—use Theorem 6.25 instead of Lemma 6.5.   □

Let A and B in the theorem and the two corollaries be open sets. Then we have

Corollary 6.33.3. *Let the open sets A and B contain p. Then*
$$\langle A, B \rangle = A/B = B/A = AB/p = p/AB.$$

6.24 The Interior of the Extension of Two Convex Sets

An analogue, for extension, of Theorem 4.31—the formula for the interior of the join of two convex sets—is now established.

Theorem 6.34. *Let A and B be convex. Then $\mathcal{I}(A/B) = \mathcal{I}(A)/\mathcal{I}(B)$, provided $\mathcal{I}(A)$ and $\mathcal{I}(B)$ are nonempty.*

(Compare Theorem 4.31.)

PROOF. Note $\mathcal{I}(A/B)$ is defined, since A/B is convex (Theorem 4.16). First we prove
$$\mathcal{I}(A/B) \subset \mathcal{I}(A)/\mathcal{I}(B). \tag{1}$$

Let $p \subset \mathcal{I}(A/B)$. Choose x to satisfy $x \subset \mathcal{I}(A)/\mathcal{I}(B)$. Then $x \subset A/B$. Hence $p \subset x(A/B)$ by Theorem 4.24. Thus using Corollary 2.16,

$$p \subset (\mathcal{I}(A)/\mathcal{I}(B))(A/B) \subset \mathcal{I}(A)A/\mathcal{I}(B)B = \mathcal{I}(A)/\mathcal{I}(B),$$

and (1) is established.

Conversely we show

$$\mathcal{I}(A)/\mathcal{I}(B) \subset \mathcal{I}(A/B). \tag{2}$$

Suppose

$$p \subset \mathcal{I}(A)/\mathcal{I}(B). \tag{3}$$

Note that $p \subset A/B$. Theorem 6.26 is applied to prove $p \subset \mathcal{I}(A/B)$, by showing $(A/B)/p$ is linear. By (3) and the formula for the interior of a join of convex sets (Theorem 4.31),

$$\mathcal{I}(A) \approx \mathcal{I}(B)p = \mathcal{I}(Bp).$$

Thus $\mathcal{I}(A) \approx \mathcal{I}(Bp)$, and A/Bp is linear (Corollary 6.33.1). But $A/Bp = (A/B)/p$. Thus $(A/B)/p$ is linear, so that $p \subset \mathcal{I}(A/B)$ (Theorem 6.26), and (2) holds. Then (1) and (2) establish the theorem. □

As a bonus we have the analogue for extension of the Open Set Join Theorem (Theorem 4.27).

Theorem 6.35 (The Open Set Extension Theorem). *Let A and B be open. Then A/B is open.*

PROOF. The result is immediate if A or B is empty. In the contrary case

$$\mathcal{I}(A/B) = \mathcal{I}(A)/\mathcal{I}(B) = A/B,$$

and A/B is open. □

Corollary 6.35.1. $a_1 \ldots a_m / b_1 \ldots b_n$ *is open.*

Corollary 6.35.2. *Let L be linear. Then L/p is open; in particular any ray is open.*

PROOF. Note that L is open (Corollary 6.2.2). □

EXERCISES

1. Prove: If A_1 and A_2 are convex and nonempty, and A_1/A_2 is linear, then $A_1 \approx A_2$.

2. Suppose A_1 and A_2 are convex and have nonempty interiors, and A_1/A_2 is linear.
   (a) Prove: $\mathcal{I}(A_1) \approx \mathcal{I}(A_2)$.
   (b) Infer $A_1/A_2 = A_2/A_1 = \langle A_1, A_2 \rangle$.

3. Prove: If A_1 and A_2 are convex and nonempty, and $A_1/A_2 = A_2/A_1$, then A_1/A_2 is linear and $A_1/A_2 = \langle A_1, A_2 \rangle$.

4. Prove: Any fraction in $a_1, \ldots, a_n$ is open. For the definition of fraction see Exercise 16 at the end of Section 4.13.

5. Prove: If A and B are open and $p \subset AB$, then $\langle A, B \rangle = AB/p = p/AB$.

6. Prove: If A and B are convex, $A \napprox B$ and $A \cup B$ is linear, then A is open if and only if B is closed.

7. Prove: If A and B are convex and B is nonempty, then $\mathcal{C}(A/B) \supset \mathcal{C}(A)$. (Compare Exercise 4 at the end of Section 2.17.)

8. Prove: If A and B are convex then $\mathcal{C}(A/B) \supset \mathcal{C}(A)/\mathcal{C}(B)$. (Compare Exercise 5 at the end of Section 2.17.)

Extremal Structure of Convex Sets: Components and Faces[1]

7

In this chapter we introduce the concept of extreme subset of a convex set. The idea is suggested by and is a generalization of the classical notion of vertices, edges and faces of a polyhedron. Properties of extremeness are studied, and two types of extreme subset of a convex set, called components and faces, are singled out for special attention. The components and faces of a convex set K are intimately related, since the interior operation converts the family of faces of K into the family of its components. The components of K are essentially its maximal open subsets and form a partition of K. The faces of K play a major role in the study of its structure and can be characterized as sections of K by linear spaces which do not "cut" K and are said to be *extremal* to it. An application of the theory is made to the characterization of the components and faces of a polytope.

7.1 The Notion of an Extreme Set of a Convex Set

The idea of extreme set has its origin essentially in the familiar notion of face of a geometric solid such as a tetrahedron or cube in Euclidean geometry. A face of such a solid may be described as a closed planar region which is a "complete" portion of the frontier of the solid. To illustrate the idea, let T be a tetrahedron $[a, b, c, d]$ (Figure 7.1). Then the closed triangular region $[a, b, c]$ is a face of tetrahedron T. Similarly if $a, b,$

---

[1] Although Chapter 6 is a prerequisite for this chapter, Sections 7.1 through 7.26 (except for certain portions of Sections 7.8, 7.10, 7.17 and 7.25) do not depend on Chapter 6 and may be read before it.

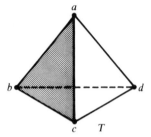
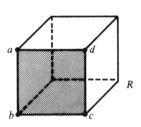

Figure 7.1

c and d are distinct vertices of a rectangular solid R and $[a, b, c, d]$ is a subset of its frontier, then the closed rectangular region $[a, b, c, d]$ is a face of R. Such faces are important examples of extreme sets of the solids T and R. But they are not the only ones. Open planar regions can also be examples of extreme sets of T and R. Thus the join abc is an extreme set of T, and $abcd$ is an extreme set of R.

These planar examples should not divert us into rejecting the possibility of nonplanar extreme sets. Points and segments are potential members of the family of extreme sets of convex figures. For example, the vertices a, b, c and d are extreme sets of T and of R, as are the segment ab and the closed segment $[c, d]$. Every point of the frontier of a solid sphere (Figure 7.2) is an extreme set of it.

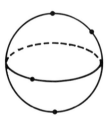

Figure 7.2

How can we formalize the idea of an extreme set for an arbitrary convex set in a join geometry? This is not as hard as it may seem. Consider the rectangular solid R described above and diagrammed again in Figure 7.3, and take $A = abcd$ as an extreme set of R. How can we characterize the relation of A to R? It will be helpful to consider some nonextreme subsets of R, for example, take $B = cdef$ or $C = cgh$. To describe nonextreme sets it is convenient to introduce a new term: A segment is said to *cut* a set S if it contains a point in S and a point not in S. B has the property that some segments of R cut it. C also has this property. But A does not: no segment of R cuts A. Thus we may assert: *Any segment of R that meets A must be contained in A.* We take this condition to be an essential characteristic of an extreme subset of a convex set.

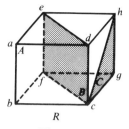

Figure 7.3

7.2 The Definition of Extreme Set of a Convex Set

Now the definition of extreme set is easily formulated.

Definition. Let K be a convex set. Let A be a convex subset of K that satisfies

Property (E). If A meets a segment of K, A contains the segment.

Then A is said to be *extreme in* K, and A is called an *extreme subset* of K or simply an *extreme set* of K.

Property (E) will usually be employed in the following form:

$$A \approx xy \text{ implies } A \supset xy \quad \text{for } xy \subset K.$$

Observe if K is convex, $\varnothing$ and K are extreme sets of K. They are called *trivial* extreme sets of K. Furthermore any extreme point of K (Section 4.14) is an extreme set of K, and conversely, if an extreme set of K consists of a point p, then p is an extreme point of K (Exercise 6 at the end of Section 7.3).

7.3 Remarks on the Definition

First observe that Property (E) is an elementary geometrical condition and that the extreme set concept might have been introduced earlier. It would be a mistake however to underestimate the importance of the concept. The very simplicity of the definition gives the notion broad coverage.

Note next that in Property (E) the points x, y are not required to belong to K; only the open segment xy is required to be contained in K. You may wonder if a different theory would result if x, y were required to belong to K, so that an extreme set would be defined in a fashion more analogous to an extreme point. The answer, given by Theorem 7.7, is that the same theory would result—the family of extreme sets of K would not be altered.

In algebraic form Property (E) has high deductive power, although it does not correspond closely to any standard algebraic principle. In a certain sense it is a closure property and a very powerful one—since it demands that A contain every point of xy if it contains a single one. As a formal principle it has a neat and attractive quality that is matched by its intuitive geometric salience: In K no segment cuts an extreme set.

Although the idea had its origin in the notion of a face of a polyhedron, the definition is completely free of dimensional restriction: An extreme set of K can be of any dimensional type from the empty set up to and including K itself. However, an extreme set can be a closed convex set, an open convex set or a convex set which is in a sense "between" the two and is neither open nor closed (Figure 7.4).

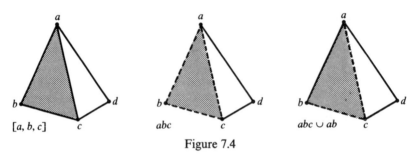

Figure 7.4

The term *extreme set*, like its special case, extreme point, suggests a subset of K which is peripheral to it, one that is contained in its *frontier*. However, it will appear in the exercises below that the interior of a convex set can be an extreme set of it. Indeed, Theorem 7.5 asserts this is true for every convex set. We employ the term *extreme set* since it is difficult to find a simple term that is not misleading and since most important examples of extreme sets of K *are* peripheral subsets of K.

EXERCISES

1. Find in Euclidean geometry all 7 examples of extreme sets of the closed segment $[a, b]$.

2. Find in Euclidean geometry several different examples of extreme sets of the closed triangular region $[a, b, c]$. There are 58 examples in all.

3. Describe in Euclidean geometry all extreme sets of a closed circular region; a closed spherical region.

4. Find in Euclidean geometry several different examples of extreme sets of the closed tetrahedral region $[a, b, c, d]$.

5. Prove: If K is a convex set and S and T are sets of points which satisfy Property (E), then so do $S \cap T$, $S \cup T$ and $S - T$.

6. Prove: A point is an extreme point of a convex set K if and only if it is an extreme set of K.

7.4 Elementary Properties of Extreme Sets

Our first result states that the relation extremeness is transitive.

Theorem 7.1. *If A is extreme in B and B is extreme in C, then A is extreme in C.*

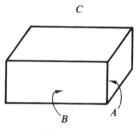

Figure 7.5

PROOF. Suppose (Figure 7.5)

$$A \approx xy, \qquad xy \subset C.$$

We show $A \supset xy$. Since $A \subset B$, we have

$$B \approx xy, \qquad xy \subset C.$$

Thus $B \supset xy$ by definition, since B is extreme in C. Then

$$A \approx xy, \qquad xy \subset B$$

implies $A \supset xy$, since A is extreme in B. Thus since A is a convex subset of C, A is extreme in C by definition. □

Now we examine the intersection of two extreme sets of a convex set.

Theorem 7.2. *The intersection of two extreme sets of convex set K is also an extreme set of K.*

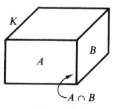

Figure 7.6

PROOF. Suppose A and B are extreme in K (Figure 7.6). Let

$$A \cap B \approx xy, \qquad xy \subset K.$$

Then $A \approx xy$, and by definition of extreme set, $A \supset xy$. Similarly $B \supset xy$. Hence $A \cap B \supset xy$, and since $A \cap B$ is convex, it is an extreme set of K by definition.   □

Similarly we can prove

Theorem 7.3. *The intersection of any nonempty family of extreme sets of convex set K is an extreme set of K.*

Next, an intersection property of extremeness.

Theorem 7.4 (Preservation of Extremeness under Intersection). *If A is extreme in B and X is convex, then $A \cap X$ is extreme in $B \cap X$.*

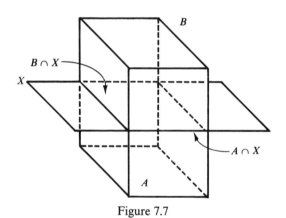

Figure 7.7

PROOF. Suppose (Figure 7.7)

$$A \cap X \approx xy, \qquad xy \subset B \cap X.$$

Then

$$A \approx xy, \qquad xy \subset B.$$

Hence $A \supset xy$, since A is extreme in B. Certainly $X \supset xy$. Hence

$$A \cap X \supset xy,$$

and $A \cap X$ is extreme in $B \cap X$ by definition.   □

The theorem says that the relation of extremeness is preserved under intersection by a convex set. In elementary geometrical terms it says: If a point of A lies on a segment of B only when that segment is in A, then the same relation holds for $A \cap X$ and $B \cap X$.

Corollary 7.4.1. *If A is extreme in B and C is extreme in D, then A ∩ C is extreme in B ∩ D.*

PROOF. By the theorem, "*A* is extreme in *B*" implies

$$A \cap C \text{ is extreme in } B \cap C. \tag{1}$$

Similarly, "*C* is extreme in *D*" implies

$$B \cap C \text{ is extreme in } B \cap D. \tag{2}$$

The conclusion follows from (1) and (2) by Theorem 7.1.   □

Corollary 7.4.2. *Suppose A is extreme in B, A ⊂ X ⊂ B and X is convex. Then A is extreme in X.*

PROOF. Direct application of the theorem.   □

Restatement. If *A* is extreme in *B*, then *A* is extreme in any convex set that is "between" *A* and *B*.

Corollary 7.4.3. *If A and B are extreme in C and A ⊂ B, then A is extreme in B.*

Note that all the results in this section depend only on the definition of extreme set—none of the postulates for a join geometry have been used in the proofs.

7.5 The Interior Operation Is Applied to Extreme Sets

Theorem 7.5. *Let K be a convex set. Then $\mathcal{I}(K)$ is extreme in K.*

PROOF. Suppose (Figure 7.8)

$$\mathcal{I}(K) \approx xy, \qquad xy \subset K.$$

Let $p \subset \mathcal{I}(K)$, xy. Since xy is open, Theorem 4.25 gives $xy = pxy$. Then by Theorem 2.16

$$\mathcal{I}(K) = pK \supset pxy = xy,$$

and the result follows by definition.   □

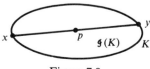

Figure 7.8

Corollary 7.5.1. *If A is extreme in K, then $\mathcal{I}(A)$ is extreme in K.*

Corollary 7.5.2. *If K is convex, then $\mathcal{I}(K)$ is extreme in $\mathcal{C}(K)$.*

PROOF. Trivial if $\mathcal{I}(K) = \varnothing$. Suppose $\mathcal{I}(K) \neq \varnothing$. By Theorem 2.28, $\mathcal{I}(K) = \mathcal{I}(\mathcal{C}(K))$, and $\mathcal{I}(\mathcal{C}(K))$ is extreme in $\mathcal{C}(K)$ by the theorem.   □

Remark. A convex set K is not necessarily extreme in $\mathcal{C}(K)$. In Euclidean join geometry JG4 let p, q and r be noncollinear: $K = pqr \cup a$, $a \subset qr$ (Figure 7.9). Then $\mathcal{C}(K) = [p, q, r]$ and $K \approx qr$, $\mathcal{C}(K) \supset qr$, but $K \not\supset qr$.

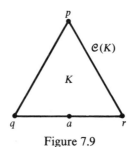

Figure 7.9

7.6 The Closure Operation Is Applied to Extreme Sets

The interior of an extreme set of K is itself an extreme set of K by Corollary 7.5.1. We naturally ask: Is the closure of an extreme set of K also an extreme set of K? The answer must be no, since the closure need not even be a subset of K. However, the portion of the closure which lies in K will be extreme in K.

Theorem 7.6. *If A is extreme in K, then $\mathcal{C}(A) \cap K$ is extreme in K.*

PROOF. Suppose (Figure 7.10)

$$\mathcal{C}(A) \cap K \approx xy, \qquad xy \subset K. \tag{1}$$

Let $p \subset \mathcal{C}(A) \cap K$, xy. Since $p \subset \mathcal{C}(A)$, there exists $q \subset A$ such that $pq \subset A$ (Theorem 2.20). Since $p \subset xy$, we have $pq \subset xyq$. Hence

$$A \approx xyq; \qquad \text{say } r \text{ is in both.} \tag{2}$$

But $r \subset xyq$ implies $r \subset xs$, where $s \subset yq$. Then (2) implies

$$A \approx xs,$$

but

$$xs \subset xyq \subset K.$$

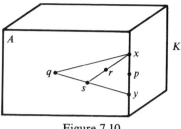

Figure 7.10

By definition of extreme set, $A \supset xs$. Thus $x \subset \mathcal{C}(A)$. Similarly $y \subset \mathcal{C}(A)$. Then $xy \subset \mathcal{C}(A)$, so that $\mathcal{C}(A) \cap K \supset xy$, and $\mathcal{C}(A) \cap K$ is extreme in K by definition. □

Corollary 7.6. *If A is extreme in K, then $\mathcal{C}(A) \cap K$ has the following property:*

$$\mathcal{C}(A) \cap K \approx xy \quad implies \quad \mathcal{C}(A) \cap K \supset x, y \quad for \ x, y \subset K.$$

PROOF. Essentially the proof of the theorem covers the corollary. Suppose $\mathcal{C}(A) \cap K \approx xy$ for $x, y \subset K$. Then in the proof above (1) holds and $\mathcal{C}(A) \supset x, y$ follows. Thus $\mathcal{C}(A) \cap K \supset x, y$, and the corollary is proved. □

The property in the corollary will take on importance in the later development (Theorem 7.21 below).

7.7 Other Characterizations of Extremeness

Extreme sets are now characterized in two additional ways.

Theorem 7.7. *Let K be a convex set and A a convex subset of K. Then the following statements are equivalent:*

(a) *A is extreme in K.*
(b) *$A \approx xy$ implies $A \supset xy$ for $x, y \subset K$.*
(c) *$A \approx U$ implies $A \supset U$ for U an open subset of K.*

PROOF. Let (a) hold. Suppose $A \approx xy$ for $x, y \subset K$. Since K is convex, $xy \subset K$. So $A \approx xy$ and $xy \subset K$. By (a), $A \supset xy$ and (b) holds.

Let (b) hold. Suppose $A \approx U$ where U is an open subset of K (Figure 7.11). Let $p \subset A \cap U$ and suppose $u \subset U$. Since $p, u \subset U$ and U is an open subset of K, Theorem 4.47 implies $p, u \subset xy$ where $x, y \subset K$. Then $A \approx xy$, and (b) implies $A \supset xy \supset u$. Thus $A \supset U$ and (c) holds.

Finally let (c) hold. Suppose $A \approx xy$ for $xy \subset K$. Since xy is open, (c) implies $A \supset xy$. Thus (a) holds. □

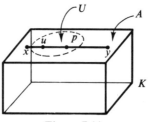

Figure 7.11

Discussion of Theorem. Let us compare the theorem with the definition of "*A* is extreme in *K*" (Section 7.2). Recall Property (E) of the definition:

(E) $A \approx xy$ implies $A \supset xy$ for $xy \subset K$.

Condition (b) is weaker than (E)—it demands less of *A*. Its conclusion requires $A \supset xy$, not for all segments *xy* of *K*, but only for those that join two points of *K*. Condition (b) then characterizes an extreme set by an intrinsic relation to *K*—the relation would be unchanged if the surrounding space were blotted out and *K* were the only geometric figure in existence. This condition is useful in proving extremeness since it is sometimes much easier to establish than (E). Condition (c) requires that *A* contain *any* open subset of *K* which it meets, not merely the segments of *K*, and is a strong generalization of (E). Condition (c) is not infrequently a useful consequence of extremeness.

The theorem has several nontrivial consequences. First a restatement of the result that (a) implies (c).

Corollary 7.7.1. *An extreme set of a convex set K contains every open set of K which it meets.*

Next a special case of this which generalizes Property (E) of the definition of extreme set (Section 7.2).

Corollary 7.7.2. *An extreme set of a convex set K contains every join of points $a_1 \ldots a_n \subset K$ which it meets.*

PROOF. Every join of points is open (Theorem 4.28). ☐

Another form of the first corollary:

Corollary 7.7.3. *Let K be a convex set, A an extreme set of K, and B a convex subset of K. Then $A \approx \mathcal{I}(B)$ implies $A \supset \mathcal{I}(B)$.*

PROOF. $\mathcal{I}(B)$ is open (Corollary 4.23). ☐

Now a pair of corollaries involving two extreme sets.

Corollary 7.7.4. *Let K be a convex set, and A and B extreme sets of K. Then $\mathcal{G}(A) \approx \mathcal{G}(B)$ implies $\mathcal{G}(A) = \mathcal{G}(B)$.*

PROOF. A is extreme in K implies $\mathcal{G}(A)$ is extreme in K (Corollary 7.5.1). By the preceding corollary, $\mathcal{G}(A) \approx \mathcal{G}(B)$ implies $\mathcal{G}(A) \supset \mathcal{G}(B)$. Similarly $\mathcal{G}(B) \supset \mathcal{G}(A)$. Thus $\mathcal{G}(A) = \mathcal{G}(B)$. $\qquad\square$

Corollary 7.7.5. *If A and B are open extreme sets of a convex set K, then $A \approx B$ implies $A = B$.*

PROOF. Since A and B are open, $A = \mathcal{G}(A)$ and $B = \mathcal{G}(B)$ (Theorem 4.23). $\qquad\square$

The corollary can be restated in two ways:

(1) *As a uniqueness principle*: A point of a convex set K is in at most one open extreme set of K.
(2) *As a disjointness principle*: Two distinct open extreme sets of a convex set have no common point.

Finally a result on extreme sets of an open set.

Corollary 7.7.6. *If K is an open convex set, its only extreme sets are K and $\varnothing$.*

PROOF. Let A be extreme in K. If $A \neq \varnothing$, then $A \approx K$, which is an open subset of K. By Corollary 7.7.1, $A \supset K$, so that $A = K$. $\qquad\square$

EXERCISES

1. Prove: If A and B are convex, then
   (a) AB is extreme in $[A, B]$;
   (b) $A \cap (AB)$ is extreme in A.

2. Prove: If L is linear and a is any point, then
   (a) L and La are extreme in $La \cup L$;
   (b) L and L/a are extreme in $L/a \cup L$.

3. Prove: If L and M are nonintersecting linear sets, then L and M are extreme in $[L, M]$.

4. Using Theorem 7.7(c), supply a shorter proof for Theorem 7.6. Can you do this in such a way as to also yield a proof of Corollary 7.6?

5. Prove: If $a \neq b$ then the extreme sets of $[a, b]$ are precisely the sets $\varnothing$, a, b, ab, $a \cup ab$, $b \cup ab$ and $[a, b]$. (Compare Exercise 1 at the end of Section 7.3.)

6. Prove: If A and $K - A$ are extreme in convex set K, then one is contained in $\mathcal{G}(K)$.

7.8 Is Extremeness Preserved under Join and Extension?

We seek properties for the operations of join and extension which are analogues of Theorem 7.4. The exact analogues are not valid. For example, the statement: A is extreme in B and X is convex imply AX is extreme in BX, does not hold as is indicated in Figure 7.12 where X, A and B are coplanar. Yet analogues do hold if a suitable condition involving the notion of linear independence of two sets is assumed. We proceed to define the concept.

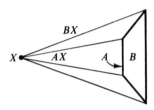

Figure 7.12

Definition. Let A and B be sets of points. Suppose that whenever $a_1, \ldots, a_m$ are linearly independent points of A and $b_1, \ldots, b_n$ are linearly independent points of B, it follows that $a_1, \ldots, a_m, b_1, \ldots, b_n$ are linearly independent. Then we say A and B are *linearly independent* or A is *linearly independent* of B.

For example, in Euclidean 3-space (JG5) two segments are linearly independent if and only if they lie on skew lines.

Theorem 7.8 (Preservation of Extremeness under Join and Extension). *If A is extreme in B and X is convex and linearly independent of B, then*

(a) AX *is extreme in* BX;
(b) A/X *is extreme in* B/X;
(c) X/A *is extreme in* X/B.

PROOF. Consider (a) (Figure 7.13). Suppose

$$AX \approx pq, \qquad p, q \subset BX. \qquad (1)$$

We show (1) implies

$$AX \supset pq \qquad (2)$$

and infer the result by Theorem 7.7. By (1)

$$p \subset b_1 x_1, \qquad q \subset b_2 x_2,$$

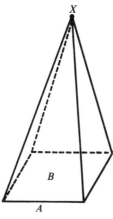

Figure 7.13

where b_1, $b_2 \subset B$ and x_1, $x_2 \subset X$. Then $pq \subset b_1 x_1 b_2 x_2$, and (1) implies

$$AX \approx b_1 b_2 x_1 x_2.$$

Hence

$$ax' \approx bx, \qquad (3)$$

where $a \subset A$, $x' \subset X$, $b \subset b_1 b_2$, $x \subset x_1 x_2$. But (3) and Corollary 6.20.1 imply the linear dependence of a, b, x, x'. Since B and X are linearly independent, we must have that a, b or x, x' are linearly dependent, that is, $a = b$ or $x = x'$. We show $a = b$. Suppose $x = x'$; then (3) becomes $ax \approx bx$, so that Corollary 6.20.1 yields that a, b, x are linearly dependent. Thus $a = b$ by the linear independence of B and X. But $a = b$ gives

$$A \supset a \approx b_1 b_2, \qquad b_1, b_2 \subset B,$$

and so the assumption that A is extreme in B implies $A \supset b_1 b_2$ (Theorem 7.7). Thus

$$AX \supset b_1 b_2 x_1 x_2 \supset pq,$$

so that (2) holds, and (a) follows by Theorem 7.7.

Similar proofs can be given for (b) and (c). $\qquad\qquad\qquad\square$

Corollary 7.8. *Suppose A is extreme in B, C is extreme in D, and B and D are linearly independent. Then* (a) *AC is extreme in BD, and* (b) *A/C is extreme in B/D.*

PROOF. (a) The linear independence of B and D implies that of B and C, since $C \subset D$. Then the theorem implies

$$AC \text{ is extreme in } BC. \qquad (1)$$

Similarly we obtain

$$BC \text{ is extreme in } BD. \qquad (2)$$

The result is immediate from (1) and (2).

(b) This can be proved in the same way. $\qquad\qquad\qquad\square$

EXERCISES

1. (a) Prove: If a and B are linearly independent, then $a \not\subset \langle B \rangle$.
   (b) Show that the converse of (a) is false.

2. Prove: If A and B are linearly independent and $x \subset AB$, then $x \subset ab$ for unique $a \subset A$ and $b \subset B$. What can be said if $x \subset A/B$? Justify this.

3. Prove Theorem 7.8(b) and (c).

4. Prove: If A is extreme in B, and X is convex and linearly independent of B, then $[A, X]$ is extreme in $[B, X]$.

5. Prove: If A and B are linearly independent convex sets, then A and B are extreme in $[A, B]$.

7.9 Extreme Sets with a Preassigned Interior Point

A substantial introduction to the theory of extreme sets of a convex set K has been given—but no evidence for the existence of extreme subsets of K other than K, $\varnothing$ and $\mathcal{I}(K)$. The theory would seem trivial if these were the only extreme sets for every convex set K. The following theorem establishes the existence of an extreme set of K which contains a preassigned point p of K in its interior and—contrary to what one might have expected —presents a simple formula for the extreme set in terms of p and K.

Theorem 7.9. *Let K be a convex set and p a point of K. Then $(p/K) \cap K$ is an extreme set of K, and p is in its interior.*

Note. The diagram in Figure 7.14 is intended to help you grasp the intuitive basis for the theorem, but not the proof.

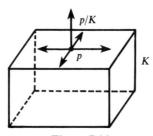

Figure 7.14

PROOF. We show $(p/K) \cap K$ is extreme in K by Theorem 7.7(b). Certainly $(p/K) \cap K \subset K$ and is convex. Suppose

$$(p/K) \cap K \approx xy \quad \text{and} \quad x, y \subset K.$$

Then

$$p/K \approx xy$$

and
$$p/Kx \approx y. \tag{1}$$
But $p/K \supset p/Kx$, and (1) implies
$$p/K \supset y.$$
By symmetry $p/K \supset x$, so that $p/K \supset xy$. Thus $(p/K) \cap K \supset xy$ and is extreme in K [Theorem 7.7(b)].

Finally suppose
$$u \subset (p/K) \cap K.$$
Then $p \subset uK$ and $p \subset uv$ for some $v \subset K$. Thus
$$v \subset (p/u) \cap K \subset (p/K) \cap K.$$
Note $p \subset (p/K) \cap K$. Then $p \subset \mathcal{I}((p/K) \cap K)$ by definition. □

Afterword. The theorem is unmotivated—it seems pulled out of a hat. How might it have been discovered? Let A be an extreme set of K and $p \subset \mathcal{I}(A)$ (Figure 7.15). Suppose $x \subset A$. Then $p \subset xy$ where $y \subset A$. Thus
$$x \subset p/y \subset p/K.$$
Then $x \subset (p/K) \cap K$, and we conclude $A \subset (p/K) \cap K$. It may be remarked that the derivation of this relation merely used the fact that A is a convex subset of K. Nevertheless, if one were sufficiently curious to examine $(p/K) \cap K$, the theorem would be revealed.

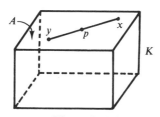

Figure 7.15

Corollary 7.9. *Let K be a convex set and p a point of K. Then $(p/K) \cap K$ is an extreme set of K and satisfies the following condition:*

$(p/K) \cap K \approx xy$ *implies* $(p/K) \cap K \supset x, y$ *for $x, y \subset K$.*

(Compare Corollary 7.6.)

PROOF. Examine closely the proof of the theorem. □

An important result on unique existence follows from Theorem 7.9.

Theorem 7.10. *Let K be a convex set and p a point of K. Then there is a unique open extreme set of K that contains p, namely, $\mathcal{I}((p/K) \cap K)$.*

PROOF. By the last theorem and Corollary 7.5.1, $\mathcal{I}((p/K) \cap K)$ is an open extreme set of K that contains p. Uniqueness is immediate (Corollary 7.7.5). □

The unique existence property can be generalized from a point to any nonempty open set.

Corollary 7.10. *Let K be a convex set and A a nonempty open set of K. Then there is a unique open extreme set of K that contains A.*

PROOF. Let $p \subset A$. By the theorem, p is contained in a unique open extreme set U of K. Then $U \approx A$ and $U \supset A$, since A is open (Theorem 7.7). Uniqueness easily follows.   □

EXERCISES

1. Prove the converse of Corollary 7.7.6: If K is convex and its only extreme sets are $\varnothing$ and itself, then K is open.

2. Suppose convex set K is not open, that is, $K \neq \mathcal{I}(K)$. Prove that K has an extreme set distinct from $\varnothing$, K and $\mathcal{I}(K)$.

3. (a) Prove: If K is convex and $p \subset K$, then pK is an extreme set of K.
   (b) Must $p \subset pK$; $p \subset \mathcal{I}(pK)$? Explain.
   (c) Can pK be open? Must it be open? Explain.

4. (a) Prove: If K is convex and $p \subset K$, then $(pK) \cap (p/K)$ is an extreme set of K.
   (b) Does p belong to $(pK) \cap (p/K)$; to its interior?
   (c) Is $(pK) \cap (p/K)$ open? Justify your answer.
   (d) How do you conjecture $(pK) \cap (p/K)$ is related to $(p/K) \cap K$? Try to prove your conjecture.
   (e) Show (a) holds even if $p \subset K$ is not assumed. [*Hint.* What does $(pK) \cap (p/K)$ equal if $p \not\subset K$?]

5. Suppose K is convex and p is any point. Is $(K/p) \cap K$ an extreme set of K? Justify your answer. What does $(K/p) \cap K$ equal when $p \subset K$?

6. Let K be convex and $p \subset K$. Let A be the union of all segments that contain p and join two points of K. How do you conjecture A is related to p and K? Try to prove your conjecture.

7. Let K be convex and $p \subset K$. Let A be the intersection of all extreme sets of K that contain p. How do you conjecture A is related to p and K? Try to prove your conjecture.

7.10 Classifying Extreme Sets

Having introduced the concept of extreme set and developed its basic properties, it becomes desirable to sort out the extreme subsets of a given convex set. A simple example will help to show the need for this. In the Euclidean join geometry JG5, let K be the (solid) tetrahedron $[a, b, c, d]$ (Figure 7.16). Here are listed some of the extreme sets of K:

$$[a, b, c], \, d, \, ab, \, abc, \, bcd \cup c, \, abc \cup ab.$$

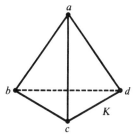

Figure 7.16

The correctness of the list can be seen intuitively: for example, if a segment of the tetrahedron meets the triangular region $[a, b, c]$, it must be contained in $[a, b, c]$. A formal justification can be based on Corollary 7.20.5 below.

Consider the three sets in Figure 7.17. They are intimately related. They have the same interior, namely, abc; the same closure, $[a, b, c]$; and the same linear hull, $\langle a, b, c \rangle$. These three extreme sets of K, together with others such as $abc \cup ab \cup ac$, $abc \cup ab \cup b$ and so on, form a class of extreme sets of K which are in a sense *equivalent* (see Section 7.28 for a formal treatment of this concept). So there is no need to become very involved with them individually—it would seem that the closed triangular region $[a, b, c]$ and the open triangular region abc could be taken as essential representatives of the whole class.

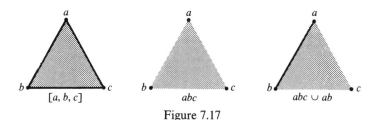

Figure 7.17

Finally we point out that some light is shed on the illustrative example by the theory, since we have shown that if A is extreme in K, both $\mathcal{C}(A) \cap K$ and $\mathcal{I}(A)$ are extreme in K (Theorem 7.6, Corollary 7.5.1). For example, let $A = abc \cup ab$. Then

$$\mathcal{C}(A) \cap K = \mathcal{C}(abc \cup ab) \cap [a, b, c, d]$$

$$= [a, b, c] \cap [a, b, c, d]$$

$$= [a, b, c].$$

And that $\mathcal{I}(A) = abc$ has been observed above. [Exercise 4(b) below asks for a proof in the theory.]

The discussion suggests that the two types of extreme set be singled out for special examination. They would correspond to the open and closed triangular regions in the example above or to the extreme sets $\mathcal{I}(A)$ and $\mathcal{C}(A) \cap K$ when A is extreme in some convex set K.

EXERCISES

1. Prove: If a, b, c are linearly independent, then $ab, ab \cup a$ and $[a, b]$ are extreme sets of $[a, b, c]$.

2. Prove: If a, b, c, d are linearly independent, then $ab, ab \cup a$ and $[a, b]$ are extreme sets of $[a, b, c, d]$.

3. Prove: If a, b, c, d are linearly independent, then $abc, abc \cup ab$ and $[a, b, c]$ are extreme sets of $[a, b, c, d]$.

4. Prove:
   (a) $\mathcal{I}(ab \cup a) = ab$;
   (b) $\mathcal{I}(abc \cup ab) = abc$.

7.11 Open Extreme Sets: Components

Now we take up the first of the two dominating families of extreme sets: open extreme sets, to be called components.

Definition. An open extreme set of a convex set K is called an *open component* of K or simply a *component* of K.

Note that $\varnothing$ is a component of each convex set K. K is a component of K if and only if it is open. A point is a component of K if and only if it is an extreme point of K.

Our first theorem on components is essentially a summary of previously obtained results expressed as properties of components.

Theorem 7.11. *Let K be a convex set. Then the following statements hold:*

(a) *A component of K is an open set of K (definition).*
(b) *A component of K contains every open set of K which it meets (Corollary 7.7.1).*
(c) *If two components of K intersect, they are identical (Corollary 7.7.5).*
(d) *Each point of K is contained in a unique component of K (Theorem 7.10).*
(e) *Each nonempty open set of K is contained in a unique component of K (Corollary 7.10).*
(f) *If A is a component of B and B is extreme in C, then A is a component of C (Theorem 7.1).*

(g) *If A is extreme in K then $\mathcal{G}(A)$ is a component of K (Corollary 7.5.1).*
(h) *If K is open its only components are K and $\varnothing$ (Corollary 7.7.6).*

Property (c) is covered by Property (d) but is of sufficient importance to merit separate inclusion.

Property (d), which is not valid for extreme sets of arbitrary type, is a crucial property in the study of the structure of convex sets, as will soon become evident.

Property (f) is merely the special case of the transitivity of extremeness (Theorem 7.1) in which the first set is open.

7.12 The Partition Theorem for Convex Sets

The components of convex set K are important because they form in a natural sense a family of building blocks for K—precisely, they constitute a *partition* of K in the sense of the following

Definition. Let S be an arbitrary set, not necessarily a set of points. A family F of subsets of S is a *partition* of S or it *partitions* S if it satisfies the following conditions:

(a) S is the union of the sets of F;
(b) the sets of F are disjoint in pairs, that is, if A and B are distinct sets of F, then $A \cap B = \varnothing$.

Equivalently, a family F of subsets of S is a partition of S if each point of S is in just one set of F.

Note that set S has two trivial partitions: The first consists just of S, the second of all one-element subsets of S.

Note also that we have permitted the empty set to be a member of a partition of set S. It is customary not to permit this, but we have done so anyway in order to facilitate the statement of some results. Obviously if the partition F contains $\varnothing$, one may remove $\varnothing$ from F and the resulting family will still be a partition of S.

Theorem 7.12 (The Partition Theorem for Convex Sets). *The family of components of a convex set K is a partition of K.*[2]

Proof. Each component of K is a subset of K, and by Theorem 7.11(d) each point of K is in a unique component of K.   □

---

[2] Compare Rockafellar [1], p. 164, Theorem 18.2.

7.13 The Component Structure of a Convex Set

As a consequence of the Partition Theorem, components play a deep and pervasive role in the structure of a convex set.

First, they shed light on the interior–frontier decomposition [Theorem 2.30(d)].

Theorem 7.13. $\mathcal{I}(K)$ *is a component of the convex set* K. $\mathcal{F}(K)$ *is the union of all other components of* K.

PROOF. $\mathcal{I}(K)$ is a component of K by definition, since it is open and an extreme set of K (Theorem 7.5).

Since $K = \mathcal{I}(K) \cup \mathcal{F}(K)$ and $\mathcal{I}(K) \not\approx \mathcal{F}(K)$, the Partition Theorem implies that $\mathcal{F}(K)$ is the union of all components of K other than $\mathcal{I}(K)$.   □

Remark. By Theorem 2.30 the pair of sets $\mathcal{I}(K)$ and $\mathcal{F}(K)$ form a partition of K. The theorem is an improvement on this result. It shows in effect that the components of K distinct from $\mathcal{I}(K)$ form a partition of $\mathcal{F}(K)$, so that all the components of K form a "refinement" of the original two set partition of K. It may be noted that the existence of the partition of $\mathcal{F}(K)$ into components of K is one of the few nontrivial results we have on $\mathcal{F}(K)$.

The next theorem uses the component concept to characterize extreme sets of K and includes a sort of generalization of the Partition Theorem, namely, that every extreme set of K has the property of being expressible as a union of K's components.

Theorem 7.14. *Let* K *be a convex set. Then* A *is an extreme set of* K *if and only if* A *is convex and a union of components of* K.

PROOF. Suppose A is extreme in K. Then A is convex and by the Partition Theorem the union of its own components. But any component of A is a component of K [Theorem 7.11(f)]. Hence A is a union of components of K.

Conversely, let A be convex and a union of components of K. Suppose $A \approx xy$ for $xy \subset K$. Then for some component U of K we have

$$A \supset U \quad \text{and} \quad U \approx xy.$$

Thus $U \supset xy$, so that $A \supset xy$ and A is extreme in K.   □

Note that an extreme set A of K is the union of the family F of those components of K which are contained in A—and this family is, of course, a partition of A.

The components of K were introduced to play the role of intrinsic building blocks for K. The theorem indicates that they play the same role for the extreme sets of K.

7.14 Components as Maximal Open Subsets

How do the components of convex set K fit into the family of open subsets of K? The family of open subsets of K is a fairly rich aggregate: It contains all points of K, all finite joins of points of K and all interiors of convex subsets of K. Given an open subset U of K, one might naturally ask whether U is contained in a larger open subset of K. This suggests the idea of a *maximal* open subset of K. Since a component of K contains any open subset of K which it intersects, we expect components of K to be intimately related to maximal open subsets of K.

Definition. Let F be a family or collection of sets (not necessarily sets of points). If M is a member of F which is not contained in (does not contain) any other member of F, we call M a *maximal* (*minimal*) member of F.

Observe that if a greatest (least) member of family F exists, then it is a maximal (minimal), and indeed the only maximal (minimal), member of F.

To illustrate the definition, consider the example where F is the family of all nonempty proper subsets of a given set S. Then any member of F consisting of one point of S is a minimal member of F, whereas any member of F consisting of all points of S but one is a maximal member of F. Thus if S has at least two points, F has minimal and maximal members, but no least and no greatest member.

Definition. Let K be convex and F the family of open subsets of K. Then a maximal member of F is called a *maximal open subset*, or simply a *maximal open set*, of K.

Components of K, excluding $\varnothing$, are characterized as a type of open subset of K independently of the notion extreme set:

Theorem 7.15 (Maximality Characterization of Components). *U is a component of convex set K if and only if $U = \varnothing$ or U is a maximal open subset of K.*

PROOF. Suppose U is a component of K and $U \neq \varnothing$. Suppose

$$U \subset V \subset K,$$

where V is open. Then $U \approx V$ and $U \supset V$ [Theorem 7.11(b)]. Thus $U = V$, and U is a maximal open subset of K.

Conversely suppose U is a maximal open subset of K. Suppose

$$U \approx xy, \qquad xy \subset K.$$

Since U and xy are open

$$Uxy \supset U, \qquad Uxy \supset xy \qquad \text{(Corollary 4.25.2).} \qquad (1)$$

But Uxy is open (Theorem 4.27). Thus (1) implies successively, $Uxy = U$ and $U \supset xy$. Hence U is an extreme set of K and so a component of K. Finally, if $U = \varnothing$ it certainly is a component of K.   □

7.15 Intersection Properties of Components

Now we prove the analogue for components of Corollary 7.4.1.

Theorem 7.16. *If A is a component of B, and C is a component of D, then $A \cap C$ is a component of $B \cap D$.*

(Compare Corollary 7.4.1.)

PROOF. $A \cap C$ is extreme in $B \cap D$ (Corollary 7.4.1). $A \cap C$ is open (Theorem 4.26). By definition $A \cap C$ is a component of $B \cap D$.   □

The theorem may be restated so: Let K_1 and K_2 be convex sets. Then the intersection of a component of K_1 and a component of K_2 is always a component of $K_1 \cap K_2$.

A characteristic property of components of the intersection of two convex sets is now given.

Theorem 7.17. *Let K_1 and K_2 be convex sets. Then U is a component of $K_1 \cap K_2$ if and only if U is the intersection of a component of K_1 and a component of K_2.*

PROOF. Suppose U is a component of $K_1 \cap K_2$. If $U = \varnothing$, then U is the intersection of a component of K_1 and a component of K_2, since both these components may be taken to be $\varnothing$. Assume $U \neq \varnothing$. By Theorem 7.11(e), $U \subset U_1$, a component of K_1. Similarly $U \subset U_2$, a component of K_2. Thus

$$U \subset U_1 \cap U_2 \subset K_1 \cap K_2. \qquad (1)$$

By Theorem 7.15, U is a maximal open subset of $K_1 \cap K_2$. Since $U_1 \cap U_2$ is open (Theorem 4.26), (1) implies $U = U_1 \cap U_2$, and the forward implication is established. The reverse is covered by Theorem 7.16, and our proof is complete.   □

Corollary 7.17.1. *Let K_1 and K_2 be convex sets and U a nonempty component of $K_1 \cap K_2$. Then U is uniquely represented in the form*

$$U = U_1 \cap U_2,$$

where U_1 and U_2 are components of K_1 and K_2 respectively.

PROOF. Suppose $U = U_1 \cap U_2 = V_1 \cap V_2$, where U_1, V_1 are components of K_1, and U_2, V_2 are components of K_2. Since $U \neq \varnothing$, we have $U_1 \approx V_1$ and $U_2 \approx V_2$. Then $U_1 = V_1$ and $U_2 = V_2$ [Theorem 7.11(c)].   □

Corollary 7.17.2. *If two convex sets have finitely many components, so has their intersection.*

Several exercises in the following set involve additional analogues for components of properties of extreme sets derived in Sections 7.4 and 7.8.

EXERCISES

1. Describe all the components of each of the following convex sets in Euclidean geometry:
   (a) a closed triangular region;
   (b) a closed tetrahedral region;
   (c) a closed rectangular region;
   (d) a closed cubical region;
   (e) a closed circular region;
   (f) a closed spherical region;
   (g) a closed cylindrical region.

2. Prove: If A is a component of B, and X is convex, then $A \cap \mathcal{G}(X)$ is a component of $B \cap X$. (Compare Theorem 7.4.)

3. Prove: The intersection of two components of convex set K is a component of K. (Compare Theorem 7.2.) Interpret this result in some examples. Is the result of much value?

4. If A is a component of B, and B is a component of C, is A a component of C? (Compare Theorem 7.1.) Justify your conclusion and interpret it in some examples. Is the conclusion of much value?

5. Prove: If A is a component of B, and X is open and linearly independent of B, then AX is a component of BX, A/X is a component of B/X, and X/A is a component of X/B. (Compare Theorem 7.8.)

6. Prove: If A is a component of B, and C is a component of D, then AC is a component of BD, and A/C is a component of B/D, provided B and D are linearly independent. (Compare Corollary 7.8.)

7. Can each component of a convex set K be expressed as $\mathcal{G}(X)$ for some extreme set X of K? Justify your answer.

8. Prove: If A and B are components of some convex set and $A \cup B$ is convex, then A absorbs B or B absorbs A (Section 2.14). Find all cases when "A absorbs B" and "B absorbs A" both hold.

9. Give an example of a family of sets which has
   (a) maximal members but no minimal members;
   (b) minimal members but no maximal members;
   (c) neither maximal nor minimal members.

10. Prove: If F is a partition of set S, and $\varnothing$ is not a member of F, then every member of F is both a maximal and a minimal member of F.

11. Prove: Each nonempty component of a convex set K is a minimal member of the family of all nonempty extreme sets of K.

*12. Prove:
  (a) A pathological convex set has infinitely many components.
  (b) If a convex set K has only finitely many components, then no extreme set of K is pathological.

13. Show in Euclidean plane geometry (JG4) that Theorem 7.17 fails if "component" is replaced by "extreme set".

14. Suppose A is a convex subset of convex set K, and U is the union of all components of K which meet A. Prove that U is an extreme set of K, and moreover, U is the least extreme set of K that contains A.

*15. Prove: If K_1 and K_2 are convex and contain no pathological extreme sets, and U is a component of $K_1 K_2$, then U is the join of a component of K_1 and a component of K_2.

7.16 Components and Perspectivity of Points

Rather remarkably, the notion of a component of a convex set is related closely to the notion of perspectivity of points introduced earlier (Section 4.26).

Theorem 7.18. *Let K be a convex set. Then a and b are in the same component of K if and only if $a \equiv b(K)$, that is, a and b are mutually perspective in K.*

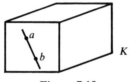

Figure 7.18

PROOF. (See Figure 7.18.) By Theorem 4.47, (i) $a \equiv b(K)$ is equivalent to (ii) a and b are in an open set of K. Using Theorem 7.11(e), we see that (ii) is equivalent to (iii) a and b are in a component of K. Thus (i) is equivalent to (iii), and the theorem holds.[3]

The theorem yields a formula for components of a convex set.

Corollary 7.18.1. *Let p be a point of convex set K. Then the component of K that contains p is given by $(pK) \cap (p/K)$.*

PROOF. (See Figure 7.19.) Let U be the component of K that contains p. Then the following form a chain of equivalent statements: (1) $x \subset U$; (2)

---
[3] See Rockafellar [1], p. 164, Theorem 18.2 and the last paragraph.

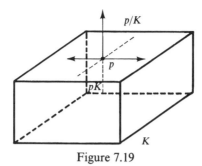

Figure 7.19

$x \equiv p(K)$; (3) $x \subset pK$ and $p \subset xK$; (4) $x \subset (pK) \cap (p/K)$. Thus $U = (pK) \cap (p/K)$. $\qquad\square$

The formula is essentially exhaustive:

Corollary 7.18.2. *Let K be a convex set and p any point. Then $(pK) \cap (p/K)$ is always a component of K.*

PROOF. If $p \subset K$, apply the preceding corollary. If $p \not\subset K$, $(pK) \cap (p/K) = \varnothing$. $\qquad\square$

Observe that if $K = J$, the set of all points, and $J \neq \varnothing$, then $(pK) \cap (p/K)$ fails to represent the component $\varnothing$ of K. For $K \neq J$ the formula represents all components of K.

By Theorem 4.46 we know that the relation perspectivity in K is an equivalence relation. It is now possible (and very easy) to characterize the component concept in terms of this equivalence relation. To facilitate the statement of the characterization we introduce the notion of equivalence class of an equivalence relation.

Definition. Let R be an equivalence relation in a set S (not necessarily a set of points). If a is an element of S, let $S(a)$ be the set of all elements x of S such that $x \, R \, a$, that is, x has the relation R to a. Then $S(a)$ is called an *equivalence class* of R and is said *to be determined* by a.

If R is an equivalence relation in set S and a and b are elements of S it is not hard to show that

(1) a is a member of $S(a)$;
(2) $a \, R \, b$ holds if and only if $S(a) = S(b)$;
(3) $S(a) \neq S(b)$ implies $S(a) \cap S(b) = \varnothing$.

These properties imply that the family of equivalence classes of the relation R is a partition of set S.[4]

----

[4] Equivalence relations and equivalence classes are discussed in Paley and Weichsel [1], Section 1.8.

Now the characterization of the component concept.

Theorem 7.19. *Let K be a convex set. Then A is a component of K if and only if $A = \varnothing$ or A is an equivalence class of the relation perspectivity in K.*[5]

PROOF. Suppose A is a component of K, $A \neq \varnothing$. Let $p \subset A$. Let $K(p)$ be the equivalence class of the relation perspectivity in K determined by p. Using the last theorem, we see that $x \subset A$ if and only if $x \equiv p(K)$. By definition $x \subset K(p)$ if and only if $x \equiv p(K)$. Thus $x \subset A$ if and only if $x \subset K(p)$ and $A = K(p)$.

Conversely, suppose A is an equivalence class of the relation perspectivity in K, say $A = K(p)$. Let B be the component of K that contains p. By the argument above with A replaced by B we have $B = K(p) = A$. Thus A is a component of K. The latter is certainly true if $A = \varnothing$, and the theorem holds. □

EXERCISES

1. Let A be an open set in convex set K. Derive a formula for the component of K that contains A.

2. Prove statements (1), (2) and (3) above.

3. Prove: If U and V are components of convex set K, and $p < q(K)$ for some $p \subset U$ and some $q \subset V$, then $x < y(K)$ for all $x \subset U$ and all $y \subset V$.

*4. Prove: For p and q points of convex set K, the following statements are equivalent:
   (a) $p \equiv q(K)$;
   (b) $pK = qK$;
   (c) $p/K = q/K$;
   (d) $K/p = K/q$.

7.17 Components of Polytopes

In this section the components and extreme sets of a polytope are characterized in terms of a finite set of generators, which of course can be taken to be its set of vertices (Section 4.14). The components of a simplex are explicitly determined and its extreme sets are characterized.

Definition. Let S be a set of points. Let F be the family of joins of points of S. Then a maximal set of F is called a *maximal join of points of S*.

Theorem 7.20. *Let P be a polytope and S a finite set of generators of P. Then A is a nonempty component of P if and only if A is a maximal join of points of S.*

[5] See Footnote 3.

PROOF. Suppose A is a maximal join of points of S. We show A is a component of P by using Theorem 7.7. Certainly A is an open set of P. Suppose

$$A \approx xy, \quad \text{where } x, y \subset P. \tag{1}$$

Then $x \subset B$, $y \subset C$, where B and C are joins of points of S (Corollary 3.7.1). Hence (1) implies

$$A \approx BC. \tag{2}$$

Note that BC is open since B and C are (Theorem 4.27). Then since A is open, (2) implies

$$ABC \supset A, BC \quad \text{(Corollary 4.25.2)}. \tag{3}$$

Since ABC is a join of points of S and A is a maximal such join, (3) implies

$$A = ABC \supset BC \supset xy.$$

Thus A is extreme in P (Theorem 7.7), and so a component of P. Clearly $A \neq \varnothing$.

Conversely, suppose A is a component of P; $A \neq \varnothing$. Let $p \subset A$. Then p is in some join of points of S. But the family of joins of points of S is finite. Hence p is in B, a maximal join of points of S. By the preceding paragraph, B is a component of P. Since $A \approx B$, we infer $A = B$ [Theorem 7.11(c)], and the theorem holds. $\square$

Corollary 7.20.1. *The number of components of a polytope is finite. Indeed, if it has n vertices, it has at most 2^n components.*

PROOF. Apply the principle that the number of subsets of a set of n elements is 2^n. $\square$

Corollary 7.20.2. *Let P be a polytope and S a finite set of generators of P. Then A is an extreme set of P if and only if A is convex and a union of maximal joins of points of S.*

PROOF. Apply Theorem 7.14. $\square$

The theorem is applicable to simplexes (Section 6.16).

Corollary 7.20.3. *Let P be a simplex and $a_1, \ldots, a_n$ its vertices. Then A is a nonempty component of P if and only if A is a join of the a's.*

PROOF. P is generated by $a_1, \ldots, a_n$. Hence if A is a component of P, it is a join of the a's, by the theorem.

Conversely, suppose A is a join of the a's. Let B be a join of the a's. Then $A = B$ or $A \cap B = \varnothing$ by Corollary 6.20.1. Thus A is a maximal join of the a's and so, by the theorem, a component of P. $\square$

Next a characterization of simplexes in terms of the notion of components.

Corollary 7.20.4. *Let P be a polytope and $a_1, \ldots, a_n$ its vertices. Then P is a simplex if and only if*

(i) *every join of the a's is a component of P, and*
(ii) *every nonempty component of P is expressible as the join of a unique set of the a's.*

PROOF. Suppose P is a simplex. Then (i) holds by the preceding corollary.

To prove (ii) let A be a nonempty component of P. By the preceding corollary, A is expressible in the form

$$A = a_{i_1} \ldots a_{i_r}, \qquad 1 \leqslant i_1 < i_2 < \cdots < i_r \leqslant n. \qquad (1)$$

By Corollary 6.20.2, A cannot be expressed as the join of a set of a's distinct from $\{a_{i_1}, \ldots, a_{i_r}\}$. Thus (ii) holds.

Conversely, suppose (i) and (ii) hold. Let A and B be formally distinct joins of the a's. By (i), A and B are components of P. But A and B are joins of distinct sets of the a's. Thus by (ii), $A \neq B$. Hence $A \not\approx B$ [Theorem 7.11(c)] and P is a simplex (Corollary 6.20.2). □

Finally we characterize extreme sets of a simplex.

Corollary 7.20.5. *Let P be a simplex and $a_1, \ldots, a_n$ its vertices. Then A is an extreme set of P if and only if A is convex and a union of joins of the a's.*

PROOF. Apply Theorem 7.14 and Corollary 7.20.3. □

EXERCISES

1. Find a counterexample to show that Corollary 7.20.4 fails if condition (ii) is removed. (*Hint.* Try JG16.)

2. Suppose a, b, c are linearly independent, so that $[a, b, c]$ is a simplex with vertices a, b, c (Corollary 6.19). Use Corollary 7.20.5 to verify your solution to Exercise 2 at the end of Section 7.3. Compare your proofs in Exercise 1 at the end of Section 7.10 with the development that results in Corollary 7.20.5.

3. Suppose a, b, c, d are linearly independent, so that $[a, b, c, d]$ is a simplex with vertices a, b, c, d. As in Exercise 2, reconsider Exercises 4 at the end of Section 7.3 and 2 and 3 at the end of Section 7.10 in the light of Corollary 7.20.5.

4. Suppose a, b, c, d are distinct and $(ab) \cap (cd)$ is a point.
   (a) Prove: The vertices of $[a, b, c, d]$ are a, b, c, d, but condition (i) of Corollary 7.20.4 fails to hold.
   (b) Use Theorem 7.20 to find all the components of $[a, b, c, d]$.

5. Prove: A polytope has no pathological extreme sets.

6. Suppose P_1 and P_2 are polytopes.
   (a) Prove: $P_1 \cap P_2$ has finitely many components.
   (b) Show $P_1 \cap P_2$, even if nonempty, need not be a polytope. (*Hint.* Consider JG16.)

7.18 The Concept of a Face of a Convex Set

Recall that in Section 7.10 we suggested that two types of extreme set deserved special examination. The first type, open extreme sets or components, has received a great deal of attention—and indeed will merit more. The second type, which is a sort of dual to the first, is now considered.

A component of a convex set K is an extreme set which is open. It is natural to inquire about an opposite type of extreme set—an extreme set A of K that is closed, one that satisfies the condition $A \supset \mathcal{C}(A)$ (Section 4.22). This however may be asking a bit too much. If A is a convex subset of K, it hardly can contain those of its contact points that are not in K. Thus it seems natural to impose the condition that A contains all its contact points that are in K. This property is given formal status in the following definition and then used to define the new type of extreme set.

Definition. Let K be convex and A a convex subset of K. Then A is said to be *relatively closed in K* or *closed relative to K* or just *closed in K* if A contains all its contact points that are in K, that is, if $A \supset \mathcal{C}(A) \cap K$.

Note that $\varnothing$ and K are closed in K. Moreover, if A is closed in K, then $A = \mathcal{C}(A) \cap K$, since $A \subset \mathcal{C}(A)$. Finally observe that A is closed in J if and only if A is closed (Section 4.22).

Definition. If an extreme set of convex set K is closed in K, it is called a *face* or *closed component* of K. $\varnothing$ and K satisfy the defining property and are called *trivial* faces of K. A face of K properly contained in K is called a *proper* face of K.

Observe that A is a face of K if and only if A is an extreme set of K and $A \supset \mathcal{C}(A) \cap K$ or $A = \mathcal{C}(A) \cap K$.

A point is a face of K if and only if it is an extreme point of K (Section 4.14).

Observe that the property "K is a face of K" specifically depends on the requirement in the definition that a face of K is closed *in* K. For example, if K is the interior of a circle in Euclidean geometry (Figure 7.20), K is not closed, but is of course closed in itself.

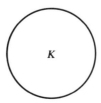

Figure 7.20

At this point a few words may be in order to compare the theory of extreme sets presented here with current treatments. Usually the only type of extreme subset of a convex set studied is face.[6] In our treatment the concept of extreme set has been introduced as a generalization of the standard notion of a face of a convex set in order to give a balanced theory of the extremal properties of convex sets involving faces *and* components.

7.19 The Nonseparation Property of a Face

A face of a convex set K is an extreme set of K which is closed in K. We have the following characterization of such a set:

Theorem 7.21 (The Nonseparation Property of Faces). *Let K be a convex set. Then A is a face of K if and only if A is a convex subset of K that satisfies*

$$A \approx xy \quad \text{implies} \quad A \supset x, y \quad \text{for } x, y \subset K. \tag{1}$$

PROOF. Suppose A is a face of K. Then A is extreme in K, thus convex, and $A = \mathcal{C}(A) \cap K$. By Corollary 7.6, A satisfies (1).

Conversely, suppose A is a convex subset of K that satisfies (1). Then (1) and the convexity of A yield: $A \approx xy$ for $x, y \subset K$ implies $A \supset xy$. By Theorem 7.7(b), A is an extreme set of K. It remains to show A is closed in K. Suppose $p \subset \mathcal{C}(A) \cap K$. Then $pq \subset A$ for some $q \subset A$ (Theorem 2.20). Hence $A \approx pq$, where $p, q \subset K$. By (1), $A \supset p$. Thus $A \supset \mathcal{C}(A) \cap K$, and the proof is complete. $\square$

The theorem is interesting and important. It recalls the definition of extreme set (Section 7.2) and the characterization of extreme sets in Theorem 7.7(b). The characteristic property (1) in the theorem can be employed with the ease and facility of Property (E) in the definition of extreme set, but has greater deductive power. In application the theorem has the advantage over the definition of face that the closure operation is dispensed with.

To interpret the theorem geometrically, recall the notion of separation of points by a point (Section 4.25). Then property (1) can be stated as a nonseparation property: If a point of A separates two points of K, these points must be in A. Thus a point of A can never separate two points of $K - A$ or a point of $K - A$ and a point of A.

Finally observe that property (1) implies a stronger property: $A \approx xy$ implies $A \supset [x, y]$ for $x, y \subset K$. Thus property (1) is a sort of closure

---

[6] It should be noted that the notion of component appears in Rockafellar [1], p. 164, Theorem 18.2.

condition—a powerful one—since it implies that if A contains a single point of xy ($x, y \subset K$), it contains all points of xy, and indeed of $[x, y]$.

This completes the discussion of the theorem—it's time to apply it.

Recall that if A is extreme in K, then $\mathcal{I}(A)$ and $\mathcal{C}(A) \cap K$ also are extreme in K (Corollary 7.5.1, Theorem 7.6). $\mathcal{I}(A)$ is a component of K [Theorem 7.11(g)]. Does $\mathcal{C}(A) \cap K$ have any special status as an extreme set of K?

Theorem 7.22. *If A is extreme in the convex set K, then $\mathcal{C}(A) \cap K$ is a face of K.*

PROOF. Certainly $\mathcal{C}(A) \cap K$ is a convex subset of K. But by Corollary 7.6, $\mathcal{C}(A) \cap K$ satisfies the Nonseparation Property: $\mathcal{C}(A) \cap K \approx xy$ implies $\mathcal{C}(A) \cap K \supset x, y$ for $x, y \subset K$. By Theorem 7.21, $\mathcal{C}(A) \cap K$ is a face of K. $\qquad\square$

Corollary 7.22. *If A is a component of convex set K, then $\mathcal{C}(A) \cap K$ is a face of K.*

EXERCISES

1. Describe all the faces of each of the following convex sets in Euclidean geometry:
    (a) a closed triangular region;
    (b) a closed tetrahedral region;
    (c) a closed rectangular region;
    (d) a closed cubical region;
    (e) a closed circular region;
    (f) a closed spherical region;
    (g) a closed cylindrical region.
    (Compare Exercise 1 at the end of Section 7.15.)

2. Prove: If A is a closed convex set and B is convex, then $A \cap B$ is closed in B.

3. Prove: If A is closed in B, and B is closed, then A is closed.

4. Prove: If A is closed in B, and B is closed in C, then A is closed in C.

5. Prove: If A is closed in B, and X is convex, then $A \cap X$ is closed in $B \cap X$.

6. Prove: If A and B are closed in K, then $A \cap B$ is closed in K.

7. If A and B are closed in K, is AB closed in K; A/B; $(A/B) \cap K$? Justify your answers.

*8. Prove: If A is closed in B, and X is convex and linearly independent of B, then AX, A/X and X/A are closed in BX, B/X and X/B respectively.

9. Can each face of convex set K be expressed as $\mathcal{C}(X) \cap K$ for some extreme set X of K? Justify your answer. (Compare Exercise 7 at the end of Section 7.15.)

10. Prove: If A is a face of K and $A \approx a_1 \ldots a_n$, where $a_1, \ldots, a_n \subset K$, then $a_1, \ldots, a_n \subset A$.

11. Prove: If A and B are faces of some convex set and $A \cup B$ is convex, then $A \supset B$ or $B \supset A$. (Compare Exercise 8 at the end of Section 7.15.)

12. Prove Theorem 7.22 without employing Theorem 7.21.

7.20 Elementary Properties of Faces of Convex Sets

The results of Section 7.4 on extreme sets have valid analogues for faces. For example, Theorem 7.1 on the transitivity of extremeness has the following analogue:

Theorem 7.23. *If A is a face of B, and B is a face of C, then A is a face of C.*

(Compare Theorem 7.1.)

PROOF. An extremely simple proof of this can be based on Theorem 7.21. Suppose

$$A \approx xy, \quad x, y \subset C. \tag{1}$$

We show $A \supset x, y$ and infer A is a face of C by Theorem 7.21. Since $A \subset B$, we have

$$B \approx xy, \quad x, y \subset C. \tag{2}$$

Thus $B \supset x, y$ by Theorem 7.21, since B is a face of C. Then

$$A \approx xy, \quad x, y \subset B$$

implies $A \supset x, y$ again by Theorem 7.21, since A is a face of B. Thus A is a face of C by Theorem 7.21. □

Remark on the Proof. The proof is exactly analogous to that of Theorem 7.1 with Theorem 7.21, the characteristic property of faces, playing the role of Property (E) in the definition of extreme set.

You may wonder whether a proof can be based directly on the definition of face. It can, and the problem is left as an exercise.

We continue by presenting additional analogues, for faces, of results of Section 7.4 on extreme sets. First come analogues of Theorems 7.2 and 7.3, for which proofs exactly analogous to those of Theorems 7.2 and 7.3 can be constructed by using Theorem 7.21.

Theorem 7.24. *The intersection of two faces of convex set K is also a face of K.*

(Compare Theorem 7.2, Exercise 3 at the end of Section 7.15.)

Theorem 7.25. *The intersection of any nonempty family of faces of convex set K is a face of K.*

(Compare Theorem 7.3.)

Next an analogue of Corollary 7.4.2.

Theorem 7.26. *Suppose A is a face of convex set K, $A \subset X \subset K$ and X is convex. Then A is a face of X.*

(Compare Corollary 7.4.2.)

PROOF. A is extreme in K, so that A is extreme in X (Corollary 7.4.2). Since A is a face of K, by definition $A \supset \mathcal{C}(A) \cap K$. Certainly $A \supset \mathcal{C}(A) \cap X$. By definition A is a face of X.   □

Several additional analogues, for faces, of extreme set properties are covered in the exercises below. They may be compared with corresponding analogues for components in the exercises at the end of Section 7.15.

Finally a property of faces which fails for extreme sets in general.

Theorem 7.27. *If A is a face of convex set K, then $K - A$ is convex and extreme in K.*

PROOF. First we show $K - A$ is convex. Let $u, v \subset K - A$. Then $uv \subset K$. Suppose $A \approx uv$. By Theorem 7.21, $A \supset u, v$, which is impossible. Thus $A \not\approx uv$ and $uv \subset K - A$.

Next, to complete the proof, we show $K - A$ satisfies Property (E). Let $K - A \approx xy$ for $xy \subset K$. Suppose $A \approx xy$. Then since A satisfies Property (E), we have $A \supset xy$, which is impossible. Thus $A \not\approx xy$ and $xy \subset K - A$. □

EXERCISES

1. Prove: If A is a face of B, and X is convex, then $A \cap X$ is a face of $B \cap X$. (Compare Theorem 7.4.)

2. Prove: If A is a face of B, and C is a face of D, then $A \cap C$ is a face of $B \cap D$. (Compare Corollary 7.4.1.)

3. Prove: If A is a face of B, and X is convex and linearly independent of B, then AX is a face of BX, A/X is a face of B/X, and X/A is a face of X/B. (Compare Theorem 7.8.)

4. Prove: If A is a face of B, and C is a face of D, then AC is a face of BD and A/C is a face of B/D, provided B and D are linearly independent. (Compare Corollary 7.8.)

5. Prove Theorem 7.23 without employing Theorem 7.21.

6. Prove Theorem 7.24 by employing Theorem 7.21; without employing Theorem 7.21.

7. Prove: If A is a face of convex set K, and $p \subset K$, then $(A/p) \cap K$ equals $\varnothing$ or A.

8. Prove: If A is a convex subset of convex set K which has the property

$$p \subset K \quad \text{implies} \quad (A/p) \cap K \subset A,$$

then A is a face of K.

9. Prove: If A and B are faces of convex set K, then $A - B$ is convex and extreme in K.

7.21 Additional Properties of Faces

The forward implication in Theorem 7.21 can be restated so: a face of convex set K contains any closed segment $[a, b]$ of K whose interior it meets. This is now generalized.

Theorem 7.28. *A face of a convex set K contains any convex subset of K whose interior it meets.*[7]

[Compare Theorems 7.7(c) and 7.21.]

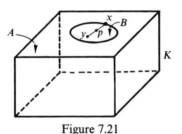

Figure 7.21

PROOF. Suppose A is a face of K, B a convex set of K and $A \approx \mathcal{I}(B)$ (Figure 7.21). We show $A \supset B$. Let $p \subset A$, $\mathcal{I}(B)$. Suppose $x \subset B$. Then

$$p \subset xy, \quad \text{where} \quad y \subset B.$$

Thus

$$A \approx xy, \quad x, y \subset K. \tag{1}$$

By the nonseparation property (Theorem 7.21), (1) implies $A \supset x$. Thus $A \supset B$ and the result is proved. □

Remark. We know by Theorem 7.7(c) that an extreme set of K is a sort of predatory subset that ingests any open set of K which it meets. The theorem shows that a face of K is even more predatory.

---

[7] The theorem is well known for $\mathbb{R}^n$, see Rockafellar [1]; p. 163, Theorem 18.1.

Corollary 7.28. *If two faces of a convex set K meet in a common interior point, they are identical.*

[Compare Theorem 7.11(c).]

7.22 Facial Structure of Convex Sets

The faces of a convex set play an important role in its structure—though one which is not as obvious as that of its components. A convex set K always is the union of its faces—but in a trivial fashion, since K itself is a face of K. Furthermore, faces of K need not be disjoint—so a partition theorem for faces can't hold. Nevertheless the faces of K are building blocks of a sort for K: they are strongly involved in its structure. This will be seen in the results of the section and become increasingly evident as the theory is developed.

To begin consider a most simple question: If p is a point of convex set K, is there a face of K that contains p? There is, of course, a unique component of K that contains p. There are, in general, many faces of K that contain p. So we sharpen the question to inquire: Is there a least face of K that contains p?

The question can be generalized to an arbitrary subset of K and still easily answered:

Theorem 7.29. *Let A be a subset of convex set K. Then there is a least face of K that contains A.*

PROOF. Let F be the family of faces of K that contain A. F is not empty, since K is a member of it. By Theorem 7.25 the intersection M of the faces of F is a face of K. Clearly $M \supset A$. Certainly M is the least member of F. $\square$

If A is a point, a sharper result and a formula for the least face can be obtained.

Theorem 7.30. *Let p be a point of a convex set K. Then p is an interior point of the least face of K that contains p; and this face is $(p/K) \cap K$.*

(Compare Theorem 7.9.)

PROOF. By Theorem 7.9, $p \subset \mathcal{I}((p/K) \cap K)$. We show $(p/K) \cap K$ is a face of K. By Corollary 7.9, $(p/K) \cap K$ satisfies the Nonseparation Property:

$$(p/K) \cap K \approx xy \quad \text{implies} \quad (p/K) \cap K \supset x, y \quad \text{for } x, y \subset K.$$

By Theorem 7.21, $(p/K) \cap K$ is a face of K.

Finally suppose X is a face of K and $X \supset p$. We show

$$X \supset (p/K) \cap K. \tag{1}$$

Let $z \subset (p/K) \cap K$. Then $z \subset p/k$ where $k \subset K$. Thus

$$X \supset p \approx zk.$$

Then $z, k \subset K$ implies $X \supset z$. Thus (1) follows, and $(p/K) \cap K$ is the least face of K that contains p. □

A strong generalization of the first conclusion of Theorem 7.30 is now obtained.

Theorem 7.31. *Let A be a convex subset of a convex set K. Then $\mathcal{F}(A)$ is contained in the interior of the least face of K that contains A.*

PROOF. Let B be the least face of K that contains A. We show

$$\mathcal{F}(A) \subset \mathcal{F}(B). \tag{1}$$

Suppose $p \subset \mathcal{F}(A)$. Let C be the least face of K that contains p. Then

$$C \approx \mathcal{F}(A),$$

so that $C \supset A$ (Theorem 7.28). This implies, since C is a face of K,

$$C \supset B.$$

Similarly we argue that $B \supset p$ implies

$$B \supset C,$$

and we have

$$C = B. \tag{2}$$

But $p \subset \mathcal{F}(C)$ (Theorem 7.30). By (2), $p \subset \mathcal{F}(B)$ and (1) holds. □

Corollary 7.31. *Any nonpathological convex subset of convex set K is contained in a nonpathological face of K.*

PROOF. In the theorem if A is nonpathological, then B is nonpathological. □

An interesting relation between components and faces follows from the theorem.

Theorem 7.32 (Component–Face Pairing Theorem). *Let U be a component of convex set K, and F the least face of K that contains U. Then*

$$F = \mathcal{C}(U) \cap K \quad and \quad U = \mathcal{F}(F).$$

PROOF. $\mathcal{C}(U) \cap K \supset U$ and is a face of K (Corollary 7.22). Hence

$$\mathcal{C}(U) \cap K \supset F. \tag{1}$$

Moreover, F is closed in K (by definition), and $\mathcal{C}(F) \supset \mathcal{C}(U)$ (Theorem 2.21). Thus

$$F \supset \mathcal{C}(F) \cap K \supset \mathcal{C}(U) \cap K,$$

and (1) implies $F = \mathcal{C}(U) \cap K$.

If $U = \varnothing$, then $F = \varnothing$ and $U = \mathcal{G}(F)$. Suppose $U \neq \varnothing$. By Theorem 7.31, $U \subset \mathcal{G}(F)$. But $\mathcal{G}(F)$ is a component of K [Theorem 7.11(g)]. Hence $U = \mathcal{G}(F)$. $\square$

Corollary 7.32.1. *Let U be a component of convex set K. Then* (i) $\mathcal{C}(U) \cap K$ *is the least face of K that contains U; and* (ii) $\mathcal{G}(\mathcal{C}(U) \cap K) = U$.

Corollary 7.32.2. *Let F be a nonpathological face of convex set K. Then*

(i) $\mathcal{G}(F)$ *is a component of K;*
(ii)   *F is the least face of K that contains $\mathcal{G}(F)$; and*
(iii) $F = \mathcal{C}(\mathcal{G}(F)) \cap K$.

PROOF. If $F = \varnothing$, the result holds. Suppose $F \neq \varnothing$. Theorem 7.11(g) implies (i). To prove (ii) let G be a face of K, $G \supset \mathcal{G}(F)$; then $G \approx \mathcal{G}(F)$ and $G \supset F$ (Theorem 7.28). Conclusion (iii) follows from (ii) by the preceding corollary. $\square$

EXERCISES

1. What relation can you discover between U and F if U is the component of K that contains a given point p of K and F is the least face of K that contains p?

2. Prove: If p is a point of convex set K, then the least face of K that contains p consists of precisely those points of K which p precedes with respect to K (Section 4.26).

3. Find a formula for the least face of a convex set K that contains subset A of K.

4. Prove: If F is a face of convex set K, A a convex subset of K and $F \approx \mathcal{G}(A)$, then $F \supset \mathcal{C}(A) \cap K$.

5. Prove: If K_1 and K_2 are convex and F is a nonpathological face of $K_1 \cap K_2$, then there exist nonpathological faces F_1 of K_1 and F_2 of K_2 such that $F = F_1 \cap F_2$.

6. Prove:
   (a) If A is an open subset of convex set K, then $\mathcal{G}(\mathcal{C}(A) \cap K) = A$.
   (b) If A is a relatively closed nonpathological convex subset of convex set K, then $A = \mathcal{C}(\mathcal{G}(A)) \cap K$.

7. If A is extreme in convex set K, is $\mathcal{C}(A) \cap K$ the least face of K that contains A? Justify your answer. (Compare Exercise 3.)

7.23 Facial Structure Continued

We improve on the triviality that a convex set is the union of its faces.

Theorem 7.33. *Any convex set K is a union of nonpathological faces of K.*

(Compare Theorem 7.12, the Partition Theorem.)

PROOF. Let K be convex. We only need show that each point p of K is contained in a nonpathological face of K. Since p itself is nonpathological, Corollary 7.31 applies, and the proof is complete.   □

Remark on the Theorem. There is a slight informality in the statement of the theorem, since there is no reference to a family of faces whose union is formed. It means that K is the union of some family F of nonpathological faces of K. The theorem can be made specific by taking F to be the family of all such faces of K. The theorem holds for certain subfamilies of this family. If K is not pathological, it is the union of the single term K. The substance of the theorem lies in the case where K is pathological, and it gives nontrivial information in this case.

Corollary 7.33. *Any face of a convex set K is a union of nonpathological faces of K.*

(Compare Theorem 7.14.)

PROOF. Let F be a face of K. Then by the theorem F is a union of nonpathological faces of F. But the transitivity property (Theorem 7.23) ensures that each face of F is a face of K.   □

Theorem 7.34. *Let K be a convex set. Then $\mathcal{F}(K)$ is the union of all proper faces of K; and $\mathcal{F}(K)$ is a union of proper nonpathological faces of K.*

PROOF. It suffices to show

(1) $\mathcal{F}(K)$ contains every proper face of K, and
(2) every point of $\mathcal{F}(K)$ is in a proper nonpathological face of K.

To prove (1) let A be a proper face of K. If $A \approx \mathcal{I}(K)$, then by Theorem 7.28 $A \supset K$, so that $A = K$. Hence $A \not\approx \mathcal{I}(K)$ and $A \subset \mathcal{F}(K)$.

To prove (2) let $p \subset \mathcal{F}(K)$. By Theorem 7.30, $p \subset \mathcal{I}(A)$, where A is a face of K. Clearly A is not pathological. Finally, A is a proper face of K, for $A = K$ implies $p \subset \mathcal{I}(K)$.   □

7.24 A Correspondence Between Components and Faces

If A is a component of K, we know $\mathcal{C}(A) \cap K$ is a face of K (Corollary 7.22). Thus there is a correspondence

$$A \rightarrow \mathcal{C}(A) \cap K$$

which assigns to each component A of K a unique face of K, $\mathcal{C}(A) \cap K$. The correspondence merits attention, since it relates the two basic types of extreme set.

Theorem 7.35. *Let K be a convex set. Let A be an arbitrary component of K. Then*

$$A \rightarrow \mathcal{C}(A) \cap K \qquad (1)$$

is a one to one correspondence between the family F of components of K and the family G of nonpathological faces of K.

PROOF. First we show that for each member A of F the correspondent, $\mathcal{C}(A) \cap K$, is a member of G. This is certainly so if $A = \varnothing$. Suppose $A \neq \varnothing$. Then

$$\mathcal{G}(\mathcal{C}(A) \cap K) = A \qquad \text{(Corollary 7.32.1)}.$$

Thus $\mathcal{C}(A) \cap K$ is nonpathological, is a face of K (Theorem 7.22), and thus is a member of G.

Next we show that each member of G is the correspondent of some member of F. Let B be a member of G. We must find a component X in F such that B is the correspondent of X, that is,

$$\mathcal{C}(X) \cap K = B. \qquad (2)$$

Suppose (2) holds. Then by Corollary 7.32.1,

$$X = \mathcal{G}(\mathcal{C}(X) \cap K) = \mathcal{G}(B), \qquad (3)$$

in other words, if (2) has a solution X, it must be $\mathcal{G}(B)$. Since B is a nonpathological face of K, Corollary 7.32.2 yields $\mathcal{G}(B)$ is a component of K, that is, $\mathcal{G}(B)$ is in F, and $B = \mathcal{C}(\mathcal{G}(B)) \cap K$. This proves that B is the correspondent of the member $\mathcal{G}(B)$ of family F.

Finally to show that the correspondence (1) is one to one, we must prove B is the correspondent of only one member of F. Suppose P and Q are members of F whose correspondent as determined by (1) is B. We show $P = Q$. Now (2) holds with P or with Q in place of X. Thus (3) holds for X replaced by P or by Q. We conclude then that $P = Q$. Thus B is the correspondent of a unique member of F, and the theorem is established. $\square$

7.25 Faces of Polytopes

Having discussed components of polytopes (Section 7.17), now we consider their faces. The treatment is based on the characterization (Theorem 7.20) of a nonempty component of a polytope as a maximal join $a_1 \ldots a_n$ of points of a finite set of generators. One might then expect $[a_1, \ldots, a_n]$ to be a face of the polytope. However, care must be exercised. The problem encountered is that it is possible to have

$$p_1 \cdots p_m = q_1 \cdots q_n$$

where

$$\{p_1, \ldots, p_m\} \neq \{q_1, \ldots, q_n\},$$

but moreover

$$[p_1, \ldots, p_m] \neq [q_1, \ldots, q_n].$$

(JG16 furnishes an example, see Exercises 6 and 7 at the end of Section 2.24.) Therefore, in order to use Theorem 7.20, as indicated above, to characterize faces, we must impose a condition on $a_1 \ldots a_n$ which ensures that $[a_1, \ldots, a_n]$ will be the convex hull we desire.

Theorem 7.36. *Let P be a polytope and S a finite set of generators of P. Then A is a nonempty face of P if and only if A is expressible in the form*

$$A = [a_1, \ldots, a_n]$$

where the a's are distinct points of S and $a_1 \ldots a_n$ is a maximal join of points of S expressed with a maximum number of factors n.

(Compare Theorem 7.20.)

(The number of factors n is said to be a *maximum* if $a_1 \ldots a_n = b_1 \ldots b_p$ for distinct $b_1, \ldots, b_p$ also in S implies $n \geqslant p$.)

PROOF. Suppose A is a face of P, $A \neq \varnothing$. Let $T = A \cap S$. We show A is generated by T. Certainly

$$[T] \subset A. \tag{1}$$

Let $x \subset A$. Then $x \subset P = [S]$ and

$$x \subset s_1 \ldots s_m, \quad \text{where } s_1, \ldots, s_m \subset S \quad \text{(Corollary 3.7.1).} \tag{2}$$

Thus

$$A \approx s_1 \ldots s_m = \mathcal{G}[s_1, \ldots, s_m] \quad \text{(Theorem 4.30).}$$

Hence since A is a face of P and $[s_1, \ldots, s_m] \subset P$,

$$A \supset [s_1, \ldots, s_m] \supset s_1, \ldots, s_m \quad \text{(Theorem 7.28).}$$

Thus $s_1, \ldots, s_m \subset T$, so that (2) implies $x \subset [T]$ and we have

$$A \subset [T]. \tag{3}$$

The relations (1) and (3) give $A = [T]$. Suppose $T = \{a_1, \ldots, a_n\}$, where the a's are distinct. Then

$$A = [a_1, \ldots, a_n]. \tag{4}$$

But $\mathfrak{I}(A) = a_1 \ldots a_n$ is a nonempty component of P [Theorem 7.11(g)] and so by Theorem 7.20 a maximal join of points of S.

Next we prove the maximum property of n. Suppose

$$a_1 \ldots a_n = b_1 \ldots b_p, \tag{5}$$

where the b's are distinct and in S. Equations (4) and (5) imply

$$A \supset b_1 \ldots b_p. \tag{6}$$

Since A is a face of P and $b_1 \ldots b_p = \mathfrak{I}[b_1, \ldots, b_p]$, (6) implies

$$A \supset [b_1, \ldots, b_p] \qquad \text{(Theorem 7.28)}.$$

Thus $A \supset b_i$, $1 \leqslant i \leqslant p$, and

$$b_i \subset A \cap S = T = \{a_1, \ldots, a_n\}.$$

Therefore in (5) each of the b's equals one of the a's, and the distinctness of the b's implies $n \geqslant p$. This completes the verification of the forward implication.

Conversely suppose

$$A = [a_1, \ldots, a_n],$$

where the a's are distinct points of S, and $a_1 \ldots a_n$ is a maximal join of points of S expressed with a maximum number of factors n. We show A is a face of P by applying Theorem 7.21. Suppose

$$A \approx xy, \qquad x, y \subset P.$$

We have

$$xy \approx a_{i_1} \ldots a_{i_r}, \qquad 1 \leqslant i_j \leqslant n, \quad 1 \leqslant j \leqslant r.$$

Note that

$$x \subset b_1 \ldots b_s, \qquad y \subset c_1 \ldots c_t \tag{7}$$

where the b's and c's are in S. Then

$$b_1 \ldots b_s c_1 \ldots c_t \approx a_{i_1} \ldots a_{i_r},$$

$$b_1 \ldots b_s c_1 \ldots c_t a_1 \ldots a_n \approx a_1 \ldots a_n,$$

and

$$b_1 \ldots b_s c_1 \ldots c_t a_1 \ldots a_n \supset a_1 \ldots a_n \qquad \text{(Corollary 4.25.2)}.$$

But $a_1 \ldots a_n$ is a component of P (Theorem 7.20), so that

$$a_1 \ldots a_n \supset b_1 \ldots b_s c_1 \ldots c_t a_1 \ldots a_n$$

and

$$a_1 \ldots a_n = b_1 \ldots b_s c_1 \ldots c_t a_1 \ldots a_n. \tag{8}$$

Thus $a_1 \ldots a_n$ is expressed as a join of points of S containing more than n factors. Hence in (8) each of the b's must equal one of the a's, and (7) implies that x is in some join of $a_1, \ldots, a_n$. Thus $x \subset [a_1, \ldots, a_n]$ and $x \subset A$. Similarly $y \subset A$, and we infer that A is a face of P (Theorem 7.21).   □

Corollary 7.36.1. *A nonempty face of a polytope P is a polytope whose vertices are vertices of P.*

Corollary 7.36.2. *The number of faces of a polytope is finite. Indeed, if it has n vertices, it has at most 2^n faces.*

(Compare Corollary 7.20.1.)

Corollary 7.36.3. *A polytope has no pathological faces.*

Corollary 7.36.4. *Let P be a simplex. Then A is a face of P if and only if $A = [T]$, where T is a subset of the set of vertices of P.*

PROOF. For the reverse implication apply Theorem 7.20 and Corollary 7.20.3; the maximum condition for n in the theorem follows readily from the linear independence of the vertices of P.   □

7.26 Covering in the Family of Faces of a Convex Set

The faces of a convex set are naturally "ordered" by the set containment relation. For example, in a closed solid cube, a vertex is contained in an edge, an edge in a "face" and a "face" in the cube itself (Figure 7.22). In this example each face has been included in a *covering* face. The notion of covering for faces of a convex set is analogous to the notion of covering for linear spaces (Section 6.18) and is now formalized.

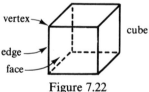

Figure 7.22

Definition. Let A and B be faces of convex set K. Suppose $A \supset B$, $A \neq B$ and no face of K "lies between" A and B, that is, $A \supset X \supset B$, where X is a face of K implies $X = A$ or $X = B$. Then we say A *covers* B or B *is covered by* A (*in the family of faces of* K).

Now an important formal criterion for covering of faces.

Theorem 7.37 (The Facial Covering Theorem). *Let A and B be faces of convex set K. Suppose $A \supset B$ and $\mathcal{G}(B) \neq \emptyset$. Then A covers B if and only if $\mathcal{G}(A) = (A - B)\mathcal{G}(B)$.*

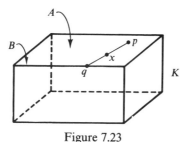

Figure 7.23

PROOF. Suppose A covers B (Figure 7.23). First we show

$$\mathcal{G}(A) \supset (A - B)\mathcal{G}(B). \tag{1}$$

Let $x \subset (A - B)\mathcal{G}(B)$; then

$$x \subset pq, \quad \text{where } p \subset A - B, q \subset \mathcal{G}(B).$$

Note that $[p, q] \subset A$. By Theorem 7.29 there is a least face F of K such that

$$F \supset [p, q]. \tag{2}$$

This relation and $q \subset \mathcal{G}(B)$ give $F \approx \mathcal{G}(B)$. Applying Theorem 7.28, we have $F \supset B$. Certainly $A \supset F$, so that $A \supset F \supset B$. Hence the fact that A covers B implies

$$F = B \tag{3}$$

or

$$F = A. \tag{4}$$

But (3) cannot hold in view of (2), since $p \not\subset B$. Thus (4) holds, and A is the least face of K containing $[p, q]$. By Theorems 7.31 and 4.30,

$$\mathcal{G}(A) \supset \mathcal{G}[p, q] = pq \supset x.$$

Thus (1) holds.

Next we prove the reverse inclusion

$$\mathcal{G}(A) \subset (A - B)\mathcal{G}(B). \tag{5}$$

Let $x \subset \mathcal{G}(A)$. Since $\mathcal{G}(B) \neq \emptyset$, there exists $y \subset \mathcal{G}(B)$. Then $y \subset A$, and

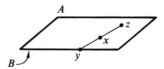

Figure 7.24

there exists $z \subset A$ such that (Figure 7.24)

$$x \subset yz. \tag{6}$$

We show

$$z \subset A - B. \tag{7}$$

Suppose $z \subset B$; then by Theorem 2.15,

$$yz \subset \mathcal{G}(B). \tag{8}$$

The relations (6) and (8) give

$$\mathcal{G}(A) \approx \mathcal{G}(B).$$

Thus by Corollary 7.28, $A = B$, a contradiction to the covering hypothesis. Therefore (7) holds and relation (6) now yields (5). By (1) and (5) we conclude

$$\mathcal{G}(A) = (A - B)\mathcal{G}(B). \tag{9}$$

Conversely, suppose (9) holds. We show A covers B. First $A \neq B$, since $A = B$ implies $\mathcal{G}(A) = \mathcal{G}(B) \neq \varnothing$, but with (9) implies $\mathcal{G}(A) = \varnothing$. Next suppose X is a face of K and

$$A \supset X \supset B. \tag{10}$$

Assume $X \neq B$ (Figure 7.25). Then there exists $x \subset X - B$. Again since $\mathcal{G}(B) \neq \varnothing$, there exists $y \subset \mathcal{G}(B) \subset X$, so that

$$xy \subset X.$$

But (10) implies $x \subset A - B$, so that by (9)

$$xy \subset \mathcal{G}(A).$$

Hence

$$X \approx \mathcal{G}(A).$$

By Theorem 7.28, $X \supset A$, and by (10), $X = A$. Therefore A covers B. □

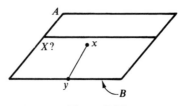

Figure 7.25

Corollary 7.37. *If A covers B (B ≠ ∅) in the family of faces of a convex set and B is not pathological, then A is not pathological.*

1. Prove: A convex set K is open if and only if its only faces are K and $\varnothing$.

2. Are your solutions to Exercise 1 at the end of Section 7.15 and Exercise 1 at the end of Section 7.19 in agreement with Theorem 7.35?

3. Prove: If K is a convex set and A is an arbitrary nonpathological face of K, then
$$A \rightarrow \mathcal{G}(A)$$
is a one to one correspondence between the family of nonpathological faces of K and the family of components of K.

4. Is the correspondence $X \rightarrow \mathcal{C}(X) \cap K$ a one to one correspondence between the family of open subsets of convex set K and the family of relatively closed nonpathological convex subsets of K? Justify your answer.

5. Prove: If K is a pathological convex set, then K has infinitely many faces; infinitely many nonpathological faces.

6. Prove: If K is convex and $\langle K \rangle$ has finite height (Section 6.19), then the family of components of K and the family of faces of K are in one to one correspondence via $X \rightarrow \mathcal{C}(X) \cap K$ for arbitrary component X of K.

7. Find a counterexample to show that Theorem 7.36 fails if we drop the assumption that n is a maximum.

8. Prove: If F_1 and F_2 are faces of a polytope and $F_1 \cap F_2 \neq \varnothing$, then the set of vertices of $F_1 \cap F_2$ is the intersection of the sets of vertices of F_1 and F_2.

9. Suppose P is a polytope, S is a finite set of generators of P and $a_1, \ldots, a_n$ are distinct points of S.
   (a) Prove without using Theorem 7.36 that if $a_1 \ldots a_n$ is a maximal join of points of S expressed with a maximum number of factors n, then
   $$\mathcal{C}(a_1 \ldots a_n) \cap P = [a_1, \ldots, a_n].$$

   (b) Prove: If S is the set of vertices of P and $\mathcal{C}(a_1 \ldots a_n) \cap P = [a_1, \ldots, a_n]$, then as a join of points of S, $a_1 \ldots a_n$ is expressed with a maximum number of factors n.

10. Prove: The number of faces of a polytope equals the number of its components.

11. Prove that the proviso $B \neq \varnothing$ may be dropped from Corollary 7.37. That is, prove that if A covers $\varnothing$ in the family of faces of a convex set, then A is not pathological.

12. Prove: If A and B are faces of a convex set and $\varnothing \neq \mathcal{G}(A) = (A - B)\mathcal{G}(B)$, then A covers B.

7.27 Extreme Sets and Extremal Linear Spaces

Extreme sets of convex set K have been studied essentially as internally related to K—as if no points existed outside K. But K *is* immersed in the universe J. What light does this shed on the extreme sets of K? More specifically, how are these extreme sets related to the linear subspaces of J?

A simple start is made by examining the linear space generated by an extreme set.

Theorem 7.38. *Let A be extreme in convex set K. Then $\langle A \rangle \cap K$ is a face of K, namely, $\mathcal{C}(A) \cap K$.*

(Compare Theorem 7.22.)

PROOF. $\mathcal{C}(A) \subset \langle A \rangle$ by Corollary 6.15.1 (Figure 7.26). Hence

$$\mathcal{C}(A) \cap K \subset \langle A \rangle \cap K. \tag{1}$$

Conversely, let $x \subset \langle A \rangle \cap K$. Then

$$x \subset \langle A \rangle = A/A \qquad \text{(Corollary 6.11.1)}$$

and $x \subset A/a$, where $a \subset A$. Thus $a, x \subset K$ and $A \approx ax$. By Theorem 7.7(b), $A \supset ax$ and so $x \subset \mathcal{C}(A)$. Thus $x \subset \mathcal{C}(A) \cap K$, and we have

$$\langle A \rangle \cap K \subset \mathcal{C}(A) \cap K. \tag{2}$$

The relations (1) and (2) yield $\langle A \rangle \cap K = \mathcal{C}(A) \cap K$, which is a face of K (Theorem 7.22). $\qquad \square$

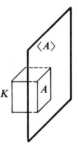

Figure 7.26

Corollary 7.38. *If A is a face of a convex set K, then $\langle A \rangle \cap K = A$.*

PROOF. $\mathcal{C}(A) \cap K = A$. $\qquad \square$

Let a convex set K be given. The last theorem says that if we choose an extreme set of K, its linear hull always intersects K in a face. Suppose we reverse the situation and choose a linear space L. When will L intersect K

in an extreme set of K? When in a face of K? The next two theorems provide answers to these questions.

Theorem 7.39. *Suppose K is convex and L is linear. Then $L \cap K$ is extreme in K if and only if L satisfies*

$$L \approx xy \quad implies \quad L \supset xy \quad for \ xy \subset K. \tag{1}$$

PROOF. A straightforward set theoretic argument shows that L satisfies (1) if and only if $L \cap K$ satisfies (1). Thus by definition $L \cap K$ is extreme in K if and only if L satisfies (1).   □

That property (1) of L in the theorem is precisely Property (E) in the definition of an extreme set of K (Section 7.2) indicates a close relation between the convex subsets of K and the linear spaces that have property (1) in common. To signalize this relation we introduce the following

Definition. Let K be a convex set and L a linear space. Suppose $L \approx xy$ implies $L \supset xy$ for $xy \subset K$. Then we say that L is an *extremal linear space to K* or is *extremal to K* or is an *extremal of K*.

Note that any linear set which contains K or (at the other extreme) fails to meet K is extremal to K, and that any extremal to K which is contained in K is an extreme set of K.

Corollary 7.39. *If A is extreme in convex set K, then $\langle A \rangle$ is extremal to K.*

PROOF. By Theorem 7.38, $\langle A \rangle \cap K$ is a face of K, so it is certainly extreme in K.   □

Now the precise nature of the connection between the extreme sets of a convex set K and its extremal linear spaces is established. First a characterization of extremals to K which is a sort of generalization of Theorem 7.38.

Theorem 7.40. *Let K be a convex set and L a linear space. Then L is extremal to K if and only if $L \cap K$ is a face of K.*

PROOF. Suppose L is extremal to K. Suppose

$$L \cap K \approx xy, \quad x, y \subset K.$$

Then $L \approx xy$, $xy \subset K$ and by definition $L \supset xy$. By Corollary 6.15.2,

$$L \supset \mathcal{C}(xy) \supset x, y.$$

Then $L \cap K \supset x, y$, and $L \cap K$ is a face of K by Theorem 7.21.

The converse is covered by Theorem 7.39, and the theorem is proved. □

Corollary 7.40.1. *Let K be a convex set and L a linear space. Then L is extremal to K if and only if L satisfies*

$$L \approx xy \quad implies \quad L \supset x, y \quad for \ x, y \subset K. \tag{1}$$

(Compare Theorem 7.21.)

PROOF. Show that L satisfies (1) if and only if $L \cap K$ satisfies (1). Apply the theorem and Theorem 7.21.   □

Now a new characterization of faces.

Corollary 7.40.2. *A is a face of convex set K if and only if $A = L \cap K$ for some linear space L which is extremal to K.*

PROOF. Let A be a face of K. Then $A = \langle A \rangle \cap K$ (Corollary 7.38), and $\langle A \rangle$ is extremal to K by Corollary 7.39. The reverse implication holds by the theorem.   □

Note how the corollary supplements and contrasts with Theorem 7.21. The latter characterizes a face of K internally in terms of its relation to the points of K—the former externally as a section of K by linear spaces which in general are not subsets of K. Each states a nontrivial geometric property of faces of K.

This characterization of faces singles them out from the family of extreme sets by a new and important property. You may wonder why other extreme sets do not enjoy this property of being a section by an extremal linear space. The question may seem silly: The theorem says they cannot. But sometimes after a result is proved and there is no question of its validity, it still may seem alien—not assimilated to our other knowledge. An answer to the question can be given in this way: Let L be linear and K convex. By Corollary 6.15.3, L is *closed*, that is, $L \supset \mathcal{C}(L)$. Certainly then $L \cap K$ should be closed in K (Exercise 2 at the end of Section 7.19). So if $L \cap K$ is an extreme set of K, it cannot be other than a face.

EXERCISES

1. Prove: If K is convex and L is linear, then $K - L$ is convex if and only if L is extremal to K.

2. Prove: If K is convex, then A is a face of K if and only if $K - \langle A \rangle$ is convex and $K \cap \langle A \rangle = A$.

3. Prove: The intersection of any family of extremals to a convex set K is an extremal to K.

4. Prove: An extremal of a convex set K contains the closure of any convex subset of K whose interior it meets.

5. Prove: If A is a face of B and X is convex and contained in $\langle A \rangle$, then AX is a face of BX, A/X is a face of B/X, and X/A is a face of X/B.

6. Prove: If A is a component of B, and X is open and contained in $\langle A \rangle$, then AX is a component of BX, A/X is a component of B/X, and X/A is a component of X/B.

7. Suppose A is extreme in B, and X is convex and contained in $\langle A \rangle$. Is AX extreme in BX; A/X extreme in B/X; X/A extreme in X/B? Justify your answers.

7.28 Associated Extreme Sets

Let A be an arbitrary extreme set of convex set K. Recall that A gives rise to a component of K, $\mathcal{G}(A)$, and a face of K, $\mathcal{C}(A) \cap K$. Moreover, by Corollary 7.39, A gives rise to an extremal linear space to K, namely, $\langle A \rangle$. If two extreme sets of K give rise to the same extremal, they give rise to the same component and the same face:

Theorem 7.41. *Let A and B be extreme in convex set K. Then $\langle A \rangle = \langle B \rangle$ implies* (a) $\mathcal{C}(A) \cap K = \mathcal{C}(B) \cap K$, *and* (b) $\mathcal{G}(A) = \mathcal{G}(B)$.

PROOF. (a) Using Theorem 7.38 we have

$$\mathcal{C}(A) \cap K = \langle A \rangle \cap K = \langle B \rangle \cap K = \mathcal{C}(B) \cap K.$$

(b) The result certainly holds if both $\mathcal{G}(A)$ and $\mathcal{G}(B)$ are empty. Assume that one, say $\mathcal{G}(A)$, is not empty. Then

$$\mathcal{G}(A) \approx A \subset \langle A \rangle = \langle B \rangle = B/B,$$

so that

$$\mathcal{G}(A)B \approx B.$$

Thus

$$\mathcal{G}(A)b \approx B, \quad \text{where } b \subset B. \tag{1}$$

Since $\mathcal{G}(A)b$ is an open subset of K (Theorem 4.27), (1) implies

$$\mathcal{G}(A)b \subset B \quad [\text{Theorem 7.7(c)}]. \tag{2}$$

Using Theorem 6.16, we have

$$b \subset \langle A \rangle = \langle \mathcal{G}(A) \rangle = \mathcal{G}(A)/\mathcal{G}(A),$$

so that

$$\mathcal{G}(A)b \approx \mathcal{G}(A). \tag{3}$$

But $\mathcal{G}(A)$ is extreme in K (Corollary 7.5.1). Hence (3) implies

$$\mathcal{G}(A)b \subset \mathcal{G}(A). \tag{4}$$

By (2) and (4), $\mathcal{G}(A) \approx B$, so that

$$\mathcal{G}(A) \subset B. \tag{5}$$

Since $\langle \mathcal{G}(A) \rangle = \langle B \rangle$, (5) yields

$$\mathcal{G}(A) = \mathcal{G}(\mathcal{G}(A)) \subset \mathcal{G}(B) \qquad \text{(Theorem 6.29)}.$$

Thus $\mathcal{G}(A) \approx \mathcal{G}(B)$, and the result follows by Corollary 7.7.4. $\qquad\square$

Corollary 7.41. *Suppose A and B are extreme in convex set K, and $\langle A \rangle = \langle B \rangle$. Then A is pathological if and only if B is pathological.*

The theorem calls to mind Section 7.10 on classifying or sorting out extreme sets of a convex set. It is the basis for an effort to group or to associate extreme sets of a convex set.

Definition. If A and B are extreme sets of convex set K and $\langle A \rangle = \langle B \rangle$, we say A and B are *equivalent* or *associated* in K and write $A \equiv_K B$.

The properties of $\equiv_K$ are investigated in the exercises at the end of Section 7.29.

7.29 Covering of Extremals Arising from Faces

Suppose A and B are faces of a convex set K so that $\langle A \rangle$ and $\langle B \rangle$ are extremals of K. Is there any connection between the relation A covers B (as faces of K) and the relation $\langle A \rangle$ covers $\langle B \rangle$ (as linear spaces)?

Theorem 7.42. *Let A and B be faces of convex set K. Then A properly contains B if and only if $\langle A \rangle$ properly contains $\langle B \rangle$.*

PROOF. Suppose A properly contains B. Then $\langle A \rangle \supset \langle B \rangle$. Suppose $\langle A \rangle = \langle B \rangle$. By Corollary 7.38,

$$A = \langle A \rangle \cap K = \langle B \rangle \cap K = B,$$

a contradiction. Thus $\langle A \rangle$ properly contains $\langle B \rangle$.

Conversely, suppose $\langle A \rangle$ properly contains $\langle B \rangle$. Then certainly $A \neq B$. Let $x \subset B$. Then

$$x \subset \langle B \rangle \subset \langle A \rangle = A/A.$$

Thus $xA \approx A$, so that $xy \approx A$ for some $y \subset A$. But A is a face of K and $x, y \subset K$. Hence $x \subset A$, and we conclude that $A \supset B$, completing the proof. $\qquad\square$

Now we obtain a connection between the two types of covering relation.

Theorem 7.43. *Let A and B be faces of convex set K. Then if $\langle A \rangle$ covers $\langle B \rangle$ in the family of linear spaces, A covers B in the family of faces of K.*

PROOF. By the preceding theorem, A properly contains B. Suppose $A \supset X \supset B$, where X is a face of K. Then $\langle A \rangle \supset \langle X \rangle \supset \langle B \rangle$, so that $\langle A \rangle = \langle X \rangle$ or $\langle B \rangle = \langle X \rangle$. Then $\langle A \rangle \cap K = \langle X \rangle \cap K$ or $\langle B \rangle \cap K = \langle X \rangle \cap K$ and Corollary 7.38 implies $A = X$ or $B = X$. Thus A covers B.   □

EXERCISES

1. Prove $\equiv_K$ is an equivalence relation in the family of extreme sets of a convex set K.

2. Let A and B be extreme in convex set K. Prove: $A \equiv_K B$ implies $\mathcal{C}(A) = \mathcal{C}(B)$. Does the converse hold? Justify your answer.

3. Prove: If A is extreme in convex set K, then $A \equiv_K \mathcal{C}(A) \cap K$. What then can we conclude about $\mathcal{G}(\mathcal{C}(A) \cap K)$?

4. Let A be extreme in convex set K. Let F be the family of extreme sets X of K which satisfy $X \equiv_K A$. Prove:
   (a) $\mathcal{G}(A) \subset X \subset \mathcal{C}(A) \cap K$.
   (b) $\mathcal{C}(A) \cap K$ is a member of F, and so is $\mathcal{G}(A)$ if A is not pathological.
   (c) $\mathcal{C}(A) \cap K$ is the greatest member of F, and $\mathcal{G}(A)$ is the least member if A is not pathological.
   (d) Every member of F is pathological or every member of F is nonpathological.

5. Prove: If K_1 and K_2 are convex with $\langle K_1 \rangle = \langle K_2 \rangle$ and every component of K_1 is a component of K_2, then

$$K_1 \subset K_2 \subset \mathcal{C}(K_1).$$

6. Show that the converse of Theorem 7.43 is false in the Euclidean join geometries JG4 and JG5.

7.30 Extremal Hyperplanes and Exposed Faces

Now we return to the discussion of the relation between faces and extremals of a convex set begun in Section 7.27.

Let A be a given face of convex set K. Recall that $A = \langle A \rangle \cap K$ (Corollary 7.38). Are there linear spaces other than $\langle A \rangle$ whose intersection with K is A? In formal terms, what can be said about the number of linear spaces L which satisfy

$$A = L \cap K? \tag{1}$$

EXAMPLE I. In JG4, the planar Euclidean join geometry, let K be the closed triangular region $[a, b, c]$ indicated in Figure 7.27, and $A = a$. It is evident that there are infinitely many lines L which satisfy (1).

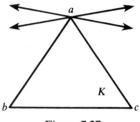

Figure 7.27

EXAMPLE II. Take K and A as above in JG5, the 3-dimensional Euclidean join geometry.

Query. What new solutions L of (1) arise?

Answer: Planes.

The examples show that for a given face A of a convex set K there may be infinitely many linear spaces, even linear spaces of different dimension, whose intersection with K is A. All of these linear spaces though are extremal to K (Theorem 7.40).

Suppose then a convex set K is given and we wish to obtain faces of K by sectioning K with linear spaces, which will be, of course, extremals to K. If we take all extremals, we get all faces (Corollary 7.40.2). But as the examples show, some faces may be obtained many times over. What happens if we choose a subfamily of the extremals? How should a subfamily be chosen?

Consider a concrete case. Let K be a convex set in a Euclidean space. Suppose K is "planar" or 2-dimensional, that is, $\langle K \rangle$ is a plane (Figure 7.28). Then it seems natural to choose lines as the sectioning extremals. Similarly suppose K is "solid" or 3-dimensional, that is, $\langle K \rangle$ is a 3-dimensional space. Then it is natural to take the extremals to be planes.

These considerations suggest, for the case where K is in an arbitrary join geometry, that extremals to K that are hyperplanes (Section 6.18) of $\langle K \rangle$ be taken as a natural subfamily to be used to obtain or generate faces of K. It turns out in general that not all faces of K can be generated by hyperplanes extremal to K. The ones that can be so generated are given a name.

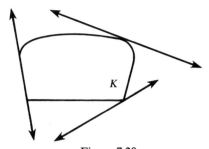

Figure 7.28

Definition. Suppose A is a face of convex set K such that $A = H \cap K$, where H is a hyperplane of $\langle K \rangle$. Then we say A is an *exposed face* of K. If the face A is a point, we say A is an *exposed point* of K.

Observe that an exposed point of K is always an extreme point of K, since any single-pointed face is an extreme point (Section 7.18). Note that A is an exposed face of convex set K if and only if A is the intersection of K with a hyperplane of $\langle K \rangle$ which is extremal to K. You probably will find that your understanding of exposed points and faces will be heightened as you encounter and examine counterexamples: nonexposed extreme points and nonexposed faces (see the exercises at the end of Section 7.31).

7.31 Supporting and Tangent Hyperplanes

The notion of a supporting hyperplane to a convex set is related to the idea of a tangent line to a plane curve or a tangent plane to a surface. A supporting hyperplane to a convex K "bounds" K in the ambient space: it is in contact with K, but does not cut K; it is extremal to K. Figure 7.29 illustrates in Euclidean 3-space a supporting plane L to each of three convex solids. If you think of L as the plane of a table top on which the solids rest, you will have the origin of this usage of the term "supporting". The concept is easily formalized using the notion of extremal linear space to a convex set.

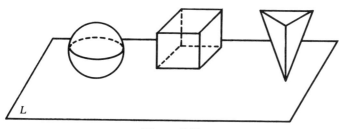

Figure 7.29

Definition. Let K be a convex subset of linear space M, and L a hyperplane of M which is extremal to K and meets $\mathcal{C}(K)$. Then L is called a *supporting hyperplane in M* to K, and we say L *supports K in M*. If $M = \langle K \rangle$, the qualifying phrase "in M" is dropped, and we refer to L simply as a *supporting hyperplane* to K and say L *supports K*.

Remark. If L is a supporting hyperplane to K, then the face $L \cap K$ of K is a proper face; otherwise $L \cap K = K$, and so $K \subset L$, contradicting the fact that L is a hyperplane of $\langle K \rangle$. Then by Theorem 7.34, $L \cap K \subset \mathcal{F}(K)$, so that $L \napprox \mathcal{F}(K)$. Since $L \approx \mathcal{C}(K)$ is required of L, it follows immediately that $L \cap \mathcal{C}(K)$ is a nonempty subset of $\mathfrak{B}(K)$. Note that $L \cap K$ is an exposed face of K. Conversely, note if A is a nonempty exposed face of K, then there exists a supporting hyperplane to K that contains A.

The definition of supporting hyperplane in M to K covers the case where $\langle K \rangle$ is a proper subset of M, as in Section 7.30, Example II, where M is taken to be Euclidean 3-space. In such a case there may exist a *trivial* supporting hyperplane in M to K, that is, one which contains K. This situation arises in the usual treatment of convex set theory in $\mathbb{R}^n$, since the only hyperplanes considered are those of $\mathbb{R}^n$.

Finally we consider the notion of a tangent hyperplane to a convex set.

Definition. Let K be convex and $p \subset \mathcal{C}(K)$ (Figure 7.30). Suppose there exists a unique supporting hyperplane L to K that contains p. Then L is said to be the *tangent hyperplane to K at p* or to be *tangent to K at p*; p is a *point of tangency* of L.

Note, by the remark above, that a point of tangency of L must be in $\mathfrak{B}(K)$.

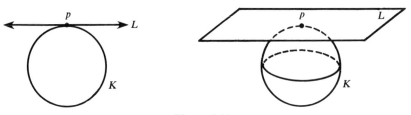

Figure 7.30

EXERCISES

1. Find the nonexposed extreme points (if any) for each of the closed convex planar regions diagrammed in Figure 7.31.

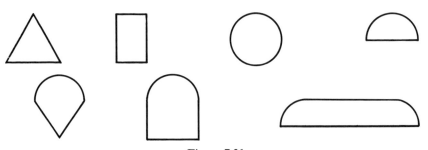

Figure 7.31

2. Find the nonexposed faces (if any) for each of the closed convex solid regions diagrammed in Figure 7.32.

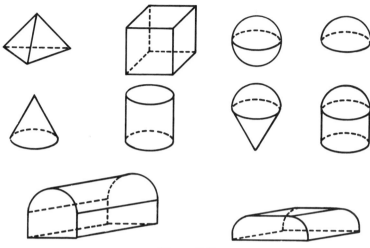

Figure 7.32

3. Must a tangent hyperplane to a convex set have a unique point of tangency? Justify your answer.

4. Find, among the diagrams of Exercise 2, examples of convex sets containing
   (a) an exposed point at which there exists a tangent hyperplane;
   (b) a nonexposed extreme point at which there exists a tangent hyperplane;
   (c) an exposed point at which there does not exist a tangent hyperplane;
   (d) a nonexposed extreme point at which there does not exist a tangent hyperplane.

5. Prove: If A is a nonempty face of a convex set K such that $\langle A \rangle$ is a hyperplane of $\langle K \rangle$, then
   (a) $\langle A \rangle$ is a supporting hyperplane to K;
   (b) A is an exposed face of K;
   (c) $\mathcal{I}(A) \supset p$ implies $\langle A \rangle$ is the tangent hyperplane to K at p.

6. Suppose K is a nonpathological convex set and L is a linear subspace of $\langle K \rangle$ such that $L \not\approx \mathcal{I}(K)$. Prove that if L is a hyperplane of $\langle K \rangle$, then L is extremal to K. Does the conclusion hold if the hypothesis is not assumed? Justify your answer.

7. Prove: If K is a nonpathological convex set and L is a linear space, then L is a supporting hyperplane to K if and only if L is a supporting hyperplane to $\mathcal{C}(K)$.

8

Rays and Halfspaces [1]

This chapter falls into two well-defined parts. First a study is made of rays (halflines). The emphasis is on rays which have a common endpoint. Among the concepts taken up is that of opposite rays which is suggested by the idea of opposite directions from a point. Secondly the notion of ray is generalized to that of halfspace. Then halfspaces which have a common bounding linear space or edge are studied in a treatment that parallels that given for rays.

8.1 Elementary Properties of Rays

The term ray was introduced in Section 4.1 as a synonym for an extension a/b. But no systematic study of rays has yet been given. We begin by introducing a convenient terminology.

Definition. A ray o/a is called an *o-ray*.

The term o-ray is a contraction for "ray with endpoint o" or "ray with origin o" and is particularly useful in dealing with rays that have a common endpoint—our major concern here.

By definition (Section 4.1) there is only one degenerate or improper o-ray, namely, $o/o = o$. All other o-rays o/a, $a \neq o$, are proper o-rays. Each o-ray is a convex set (Corollary 4.16), indeed, an open convex set (Corollary 6.35.2). Moreover, a proper o-ray is also "open" in the sense that it does not contain its endpoint o (Corollary 4.9).

---

[1] Sections 8.1–8.6 may be read independently of Chapter 6.

The results of this section correspond to familiar properties of rays in Euclidean geometry but are slightly more general since they cover improper rays. The first theorem asserts that if two rays with a common endpoint intersect, they are identical.

Theorem 8.1. *If $o/a \approx o/b$, then $o/a = o/b$.*

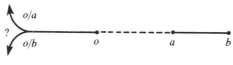

Figure 8.1

PROOF. Suppose $o/a \approx o/b$ (Figure 8.1). Then

$$ob \approx oa$$

and

$$b \subset oa/o.$$

Hence

$$o/b \subset o/(oa/o) \subset oo/oa = o/oa = (o/o)/a = o/a,$$

so that $o/b \subset o/a$. By symmetry $o/a \subset o/b$, and we conclude $o/a = o/b$. $\qquad\square$

Restatement of Theorem 8.1. If A and B are o-rays and $A \approx B$, then $A = B$.

Given a point p, we now ask, is there a ray with endpoint o that contains p (Figure 8.2)?

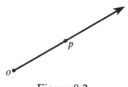

Figure 8.2

Theorem 8.2. *Let points o and p be given. Then there exists a unique ray with endpoint o that contains p.*

PROOF. Existence: We want to find a ray with endpoint o that contains p. Suppose o/a is to be such a ray (Figure 8.3). How can we find a point a with this property? We have $p \subset o/a$. Then $pa \approx o$ and $a \approx o/p$. Thus to

Figure 8.3

find a ray with endpoint o which is to contain p, we begin by choosing a to satisfy $a \subset o/p$. Does such a point exist? Yes. By J5, $o/p \neq \varnothing$. Then $a \approx o/p$, so that $ap \approx o$ and $p \approx o/a$. Thus o/a is a ray with endpoint o that contains p.

Uniqueness: Suppose $p \approx o/b$. Then

$$o/a \approx o/b,$$

and Theorem 8.1 implies

$$o/a = o/b.$$

Thus o/a is the only ray with endpoint o that contains p.   □

Definition. Let o and p be any points. Then the unique o-ray which contains p is denoted by $\overrightarrow{op}$ (read "op arrow") and is called the o-ray *determined* by p, the *ray op*, the *halfline op* or the *side of o which contains p*.

Corollary 8.2.1. *Let A be an o-ray. Then $A \supset a$ implies $A = \overrightarrow{oa}$.*

Corollary 8.2.2. *Any o-ray can be represented in the form $\overrightarrow{oa}$.*

PROOF. Let A be an o-ray. Then $A \neq \varnothing$ and contains a point a. By the preceding corollary $A = \overrightarrow{oa}$.   □

Corollary 8.2.3. *If $o \approx ab$, then $o/a = \overrightarrow{ob}$ and $o/b = \overrightarrow{oa}$.*

PROOF. $o \approx ab$ implies both $o/a \supset b$ and $o/b \supset a$ (Figure 8.4). Apply the first corollary.   □

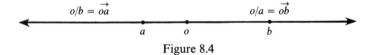

Figure 8.4

Remark on Terminology. We have employed both terms ray and halfline since they are in common usage. The term halfline was suggested by the principle in Euclidean geometry that a point on a line separates it into two rays. However, the term may be misleading since this separation principle may not hold in a join geometry—JG16 furnishes a counterexample (see Exercise 5 below).

EXERCISES

1. Prove: If $\overrightarrow{oa} = o/b$, then $\overrightarrow{ob} = o/a$ and $o \approx ab$ (Figure 8.5).

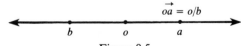

Figure 8.5

2. Prove: $\overrightarrow{oa} = \overrightarrow{ob}$ if and only if $o/a = o/b$ (Figure 8.6).

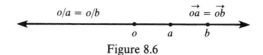

Figure 8.6

3. Prove: $\overrightarrow{oa} = \overrightarrow{ob}$ if and only if $oa \approx ob$.

4. Verify the results of Exercises 1, 2 and 3 in JG16.

5. In JG16 show that the basic set J is a line. Show further that for any point o, J contains exactly 9 o-rays (8 proper o-rays).

6. Prove: In the join geometries $\mathbb{R}^n$ (Sections 5.6–5.9), $\overrightarrow{ab}$ is the set of all points x satisfying

$$x = a + \lambda(b - a),$$

where λ is positive.

8.2 Elementary Operations on Rays

The operations we have in mind are forming products and quotients of a ray and its endpoint. If A is an o-ray, then we have three basic combinations to consider: oA, o/A and A/o. For the purpose of organizing results we find it convenient to consider o/A first.

Theorem 8.3. *If A is an o-ray, then o/A is an o-ray. In fact $o/A = o/a$ for $a \subset A$.*

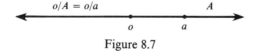

Figure 8.7

PROOF. Let a be a point of A (Figure 8.7). It suffices to prove

$$o/A = o/a, \tag{1}$$

for this implies o/A is an o-ray. Since $A \supset a$, certainly

$$o/A \supset o/a. \tag{2}$$

To show the reverse inclusion let $x \subset o/A$. Then $Ax \approx o$ and

$$A \approx o/x.$$

By Theorem 8.1,

$$A = o/x,$$

and $a \approx o/x$ follows. Then $xa \approx o$ and $x \approx o/a$. Thus

$$o/A \subset o/a,$$

which with (2) implies (1), and the theorem holds.   □

Several corollaries follow which are easily interpreted geometrically, including a formula for $\overrightarrow{oa}$.

Corollary 8.3.1. $o/\overrightarrow{oa} = o/a$.

Corollary 8.3.2. $\overrightarrow{oa} = o/(o/a)$.

PROOF. By the theorem, $o/(o/a)$ is an o-ray; by Corollary 4.7(c), $o/(o/a)$ $\supset a$.   □

Corollary 8.3.3. $o/(o/(o/a)) = o/a$.

PROOF. Apply the two preceding corollaries.   □

Theorem 8.4. *Let A be an o-ray. Then*

$$oA = A = A/o.$$

PROOF. First we show $A/o = A$. Let $A = o/p$. Then

$$A/o = (o/p)/o = o/po = (o/o)/p = o/p = A.$$

Next we have

$$oA = o(o/p) \subset oo/p = o/p = A,$$

so that $oA \subset A$. To prove the reverse inclusion we have

$$oA = o(A/o) \supset A \qquad [\text{Corollary 4.7(a)}].$$

Now conclude $oA = A$.   □

Remark. The theorem asserts that an o-ray is invariant under the operations of joining to o and extending from o. We may think of o as an "identity element" for these operations when applied to o-rays. Note the intuitive geometrical significance of the result.

Corollary 8.4. $\overrightarrow{oa} \supset oa \cup a \cup a/o$.

PROOF. $\overrightarrow{oa} = o(\overrightarrow{oa}) \supset oa$. Similarly $\overrightarrow{oa} = \overrightarrow{oa}/o \supset a/o$.   □

Remark. In a Euclidean join geometry the result becomes the equality (Figure 8.8)

$$\overrightarrow{oa} = oa \cup a \cup a/o.$$

This can be made the basis for a definition of ray in Euclidean or more

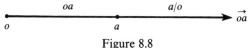

Figure 8.8

generally in classical ordered linear geometry. However, the equality does
not hold in general (see Exercise 3 below). It does hold (Theorem 12.3) in
an ordered join geometry (Section 12.1), a join theoretic analogue of an
ordered linear geometry.

Now a new formula for $\overrightarrow{oa}$ can be derived.

Theorem 8.5. $\overrightarrow{oa} = o/(o/a) = oa/o$.

PROOF. By Corollary 8.3.2, $\overrightarrow{oa} = o/(o/a)$. It suffices to show

$$\overrightarrow{oa} = oa/o. \tag{1}$$

We have

$$\overrightarrow{oa} = o/(o/a) \subset oa/o \qquad \text{(Theorem 4.13).} \tag{2}$$

To prove the reverse inclusion we have

$$oa \subset \overrightarrow{oa} \qquad \text{(Corollary 8.4),}$$

so that

$$oa/o \subset \overrightarrow{oa}/o = \overrightarrow{oa} \qquad \text{(Theorem 8.4).}$$

This with (2) yields (1) as desired. $\qquad\qquad\square$

EXERCISES

1. Prove Corollary 8.3.3 directly (without using the theory of rays).

2. Verify Corollaries 8.3.1, 8.3.2 and 8.3.3 in JG4 and in JG16.

3. Show JG16 is a counterexample to the contention that equality holds in
   Corollary 8.4.

4. Prove: In $\mathbb{R}^n$ (Sections 5.6–5.9),
   $$\overrightarrow{oa} = oa \cup a \cup a/o.$$

5. Prove: If o/A is an o-ray, then A is contained in some o-ray.

6. Prove: $\overrightarrow{ab} \supset \overrightarrow{bc}$ if $b \approx ac$.

7. If $\overrightarrow{ab} \supset \overrightarrow{bc}$, must $b \subset ac$ or $b = c$ hold? Justify your assertion.

8. Prove: $b \approx ac$ and $b \approx ad$ imply $\overrightarrow{ac} = \overrightarrow{ad}$ and $\overrightarrow{bc} = \overrightarrow{bd}$.

9. Prove: $b \approx ad$ and $c \approx ad$ imply $\overrightarrow{ab} = \overrightarrow{ac}$ and $a/b = a/c$.

8.3 Opposite Rays

The notion of ray may be considered a formalization of the intuitive idea of a direction in space at a given point. One naturally wonders if the idea of opposite directions at a point can be formalized.

Definition. Let A be an o-ray. Suppose A' is an o-ray such that

$$AA' \approx o.$$

Then we say o-ray A' is *opposite to* o-ray A, or that A and A' are *opposite* o-rays.

Note that if o-ray A' is opposite to o-ray A, then A is opposite to A'.

Observe that the o-rays A and A' are opposite if and only if there exist points a and a' in A and A' respectively such that $aa' \approx o$; or equivalently if and only if o separates some point of A and some point of A' (Section 4.25).

Note that o, the degenerate o-ray, is opposite to itself. Can a proper o-ray be opposite to itself, or to o?

Theorem 8.6. *Let A and A' be opposite o-rays. Then both are proper and $A \not\approx A'$, or both are improper and $A = A' = o$.*

PROOF. We must show that it is impossible for one of the rays, say A, to be proper and the other to be improper. Suppose then that A is proper and $A' = o$. Then by definition

$$o \approx AA' = Ao$$

and

$$o = o/o \approx A.$$

Since A and o are o-rays, Theorem 8.1 implies $A = o$, a contradiction. Thus A and A' are both proper or both improper. In the latter case $A = A' = o$, while in the former case $A \not\approx A'$, for otherwise $A = A'$, giving $o \approx AA' = AA = A$, that is, $A = o$. □

Corollary 8.6. *The only o-ray that is opposite to itself is o, the degenerate o-ray.*

The following theorem gives a class of examples of opposite o-rays.

Theorem 8.7. $\overrightarrow{oa}$ *and o/a are opposite o-rays.*

PROOF. Let $x \subset o/a$ (Figure 8.9). Note that $a \subset \overrightarrow{oa}$. We have $xa \approx o$, and the conclusion is immediate. □

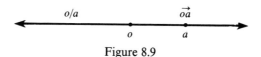

Figure 8.9

Next a property of unique existence.

Theorem 8.8. *If A is an o-ray, it has a unique opposite o-ray. In fact the o-ray opposite to A is o/A.*

(Compare Theorem 8.3.)

PROOF. Existence: Certainly o/A is an o-ray (Theorem 8.3). Let $x \subset o/A$. Then $xA \approx o$, so that o/A is opposite to A.

Uniqueness: Let o-ray A' be opposite to A. Then $A'A \approx o$. Thus

$$A' \approx o/A.$$

Hence $A' = o/A$ by Theorem 8.1, and the proof is complete. □

The theorem throws light on Theorem 8.7. Recall that any o-ray can be put in the form $\overrightarrow{oa}$ (Corollary 8.2.2). Hence the theorem implies that $\overrightarrow{oa}$ and o/a are not just an example of a pair of opposite o-rays, but a "universal" example: Any two opposite o-rays can be expressed in the form $\overrightarrow{oa}$ and o/a.

Corollary 8.8. *Suppose* $ab \approx o$. *Then*

(a) $\overrightarrow{oa}$ *and* $\overrightarrow{ob}$ *are opposite o-rays;*
(b) o/a *and* o/b *are opposite o-rays.*

(Compare Corollary 8.2.3.)

Figure 8.10

EXERCISES

1. Prove: If $\overrightarrow{oa}$ and $\overrightarrow{ob}$ are opposite o-rays, then $o \approx ab$.

2. Prove: If o/a and o/b are opposite o-rays, then $o \approx ab$.

3. Prove: If $\overrightarrow{oa}$ and $\overrightarrow{ob}$ are opposite o-rays, then $\overrightarrow{ao} = \overrightarrow{ab}$ and $\overrightarrow{bo} = \overrightarrow{ba}$. Show in JG10 that the converse of the result fails.

4. Prove: If A and A' are opposite o-rays then $A/A' = A$.

5. Prove: If A and A' are opposite o-rays and B and B' are opposite o-rays, then
   (a) $A/B = AB'/o$;
   (b) $o/AB = A'B'/o$.
   Interpret these results geometrically.

8.4 Separation of Two Rays by a Common Endpoint

This section is concerned with separation properties of opposite rays. To start with we generalize the notion of separation of two points by a point (Section 4.25) to the case of separation of two sets of points by a point.

Definition. Let A and B be nonempty sets of points. Suppose p separates each point of A from each point of B. Then we say p *separates A and B* (or *A from B*) or *A and B are separated by p*. If in addition $p \not\approx A, B$, we say p *strictly* separates A and B (or *A* from *B*).

Observe that p strictly separates A and B if and only if p strictly separates each point of A and each point of B.

Obviously the notion of strict separation is the important case—it is easy to deduce that if p separates A and B but not strictly, then $A = B = p$ (Exercise 3 at the end of Section 8.5). We permit nonstrict separation only to enable us to cover degenerate cases with ease. Compare Theorem 8.9 and Corollary 8.9.2 below.

Theorem 8.9. *Let A and A' be o-rays. Then A and A' are opposite o-rays if and only if o separates A and A'.*

PROOF. Suppose o separates A and A'. Then $AA' \approx o$, and by definition A and A' are opposite o-rays.

Conversely, suppose A and A' are opposite o-rays. Let $a \subset A$, $b \subset A'$. We show $o \approx ab$. Note $A = \overrightarrow{oa}$ (Corollary 8.2.1). But $\overrightarrow{oa}$ has a unique opposite o-ray (Theorem 8.8), which must be o/a (Theorem 8.7). Thus $A' = o/a$. Hence $b \subset o/a$ and $o \approx ab$. By definition o separates A and A'. ∎

Corollary 8.9.1. *Let A and A' be opposite o-rays. Then o separates each point of A from each point of A'.*

The corollary asserts that if A and B are o-rays and o separates a particular point of A from a particular point of B, then o separates every point of A from every point of B.

Corollary 8.9.2. *Let A and A' be o-rays. Then A and A' are proper opposite o-rays if and only if o strictly separates A and A'.*

PROOF. Use the theorem in conjunction with Theorem 8.6.   □

8.5 The Partition of Space into Rays

The rays with a given endpoint *o* fill out *J*, the set of all points, without overlapping. To state this result precisely we employ the definition of partition of a set (Section 7.12). Since this chapter is independent of Chapter 7, we restate the definition.

Definition. Let *S* be an arbitrary set, not necessarily a set of points. A family *F* of subsets of *S* is a *partition* of *S* or *partitions S* if it satisfies the following conditions:

(a) *S* is the union of the sets of *F*;
(b) The sets of *F* are disjoint in pairs, that is, if *A* and *B* are distinct sets of *F*, then $A \cap B = \varnothing$.

Equivalently a family *F* of subsets of *S* is a partition of *S* if each point of *S* is in just one set of *F*.

Theorem 8.10 (Partition Theorem for Rays). *Let o be a point. Then the family of o-rays is a partition of J, the set of all points.*

PROOF. *J* is the union of the family of *o*-rays (Figure 8.11), since each point is in some *o*-ray (Theorem 8.2). Any two distinct *o*-rays *A* and *B* are disjoint, since the contrary ($A \approx B$) implies $A = B$ (Theorem 8.1).   □

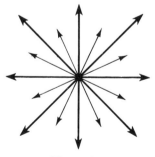

Figure 8.11

Remark. This simple and attractive result would not hold if a ray were defined in a more conventional way to include its endpoint or the concept did not cover the degenerate case of a "point" ray.

EXERCISES

1. Prove without applying the theory of rays: If o separates a and b, then o separates
   (a) oa and ob;
   (b) a/o and b/o;
   (c) o/a and o/b.

2. Let A and B be nonempty. Prove that o separates A and B if and only if there exist opposite o-rays R and S such that $A \subset R$ and $B \subset S$.

3. Prove: If p separates A and B, and $p \subset A$ or $p \subset B$, then $A = B = p$.

4. Verify the Partition Theorem (Theorem 8.10) in JG5 and in JG16.

5. Prove: A set of points S is linear (Section 6.2) if and only if $o \subset S$ implies that the family of o-rays in S is a partition of S.

8.6 Closed Rays

In Euclidean geometry and in the theory of convex sets one often considers "closed" rays, that is, rays which include their endpoints, rather than the "open" rays which we have been studying.

Definition. Let A be an o-ray. Then $A \cup o$ is called an *augmented* or *closed* $(o$-$)$ray or halfline; and o is its *endpoint*. $A \cup o$ is called *proper* if A is proper; otherwise it reduces to o and is *improper* or *degenerate*.

Theorem 8.11. *Any closed ray is convex.*

PROOF. Let A be an o-ray. Then using Theorem 8.4, we have
$$(A \cup o)(A \cup o) = A \cup Ao \cup oA \cup o = A \cup o. \qquad \square$$

Remarks on the Definition. The term closed ray has been retained because it is the familiar term of Euclidean geometry. However, like our usage of some other basic Euclidean terms, it may be misleading. It may suggest that a closed ray is not merely "closed" in the sense of containing its endpoint—but is a *closed* convex set (Section 4.22), as it would be in Euclidean geometry. This is not so, as evidenced by the following

Counterexample. In JG16 let $o = (0, 0)$, $a = (-1, -1)$ and $A = o/a$ (Figure 8.12). Then A is the set of points (x_1, x_2) such that $x_1, x_2 > 0$. $A \cup o$ is not closed, since, for example, any point $(x, 0)$ where $x > 0$ is in $\mathcal{C}(A \cup o)$ but not in $A \cup o$.

Recall that the problem encountered here has already been observed for closed segments—a closed segment need not be a closed convex set (Section 4.22).

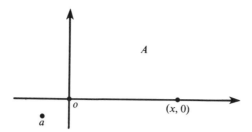

Figure 8.12

A Side Remark. The closure of an o-ray always contains the augmented o-ray. That is, if A is an o-ray, then

$$\mathcal{C}(A) \supset A \cup o. \tag{1}$$

(See Exercise 9(a) at the end of Section 4.11.) Whenever *equality* holds in (1), we have by Corollary 4.34.2 that a closed ray is a closed convex set. Certainly in Euclidean geometry (1) becomes an equality, and more generally this is the case (Exercise 9 at the end of Section 12.6) in ordered join geometries (Section 12.1).

8.7 The Linear Hull of a Ray

It is immediate from Theorem 6.13 that the linear hull of a proper ray is a line (Figure 8.13), that is, for $a \neq b$,

$$\langle a/b \rangle = \langle a, b \rangle = \text{line } ab.$$

Besides the ray, what else does this line contain?

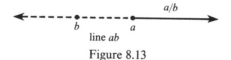

Figure 8.13

Theorem 8.12. *Let A and A' be opposite o-rays. Then*

$$\langle A \rangle = \langle A' \rangle \supset AA' \supset A \cup A' \cup o. \tag{1}$$

PROOF. First we show

$$AA' \supset A \cup A' \cup o. \tag{2}$$

By definition $AA' \supset o$. Thus since A is convex,

$$AA' = AAA' \supset Ao = A \qquad \text{(Theorem 8.4)}.$$

By symmetry $AA' \supset A'$, and (2) is established.
    Next we show

$$\langle A \rangle = \langle A' \rangle. \tag{3}$$

By (2), Theorem 8.8, Theorem 8.4 and Corollary 6.11.1 we have

$$A' \subset AA' = A(o/A) \subset Ao/A = A/A = \langle A \rangle.$$

Hence $\langle A' \rangle \subset \langle A \rangle$ follows. By symmetry $\langle A \rangle \subset \langle A' \rangle$, so that (3) holds.
    Finally observe that (3) yields

$$\langle A \rangle = \langle A' \rangle \supset AA'. \tag{4}$$

Then (2) and (4) imply (1), and the theorem is proven.   □

    Remark. In Euclidean geometry (1) becomes a complete equality (Figure 8.14):

$$\langle A \rangle = \langle A' \rangle = AA' = A \cup A' \cup o. \tag{5}$$

However, (5) does not hold in general, as neither (2) nor (4) above can
be strengthened (see Exercises 6 and 7 below). Yet (5) does hold in an
ordered join geometry (see Exercise 15 at the end of Section 12.6, and also
note the related result, Corollary 12.4.2). Compare the remark following
Corollary 8.4.

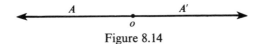

A A'

o

Figure 8.14

EXERCISES

  1. Can a closed ray be open? What is the interior of a closed ray? Prove your
     assertions.

  2. Find in JG16 the closure of a closed ray. Be sure to consider all cases.

  3. Prove: In $\mathbb{R}^n$ (Sections 5.6–5.9) a closed ray is a closed convex set.

  4. Prove: In $\mathbb{R}^n$ equality holds throughout (1) of Theorem 8.12.

  5. Prove: If equality holds throughout (1) of Theorem 8.12, then $A \cup o$ is a closed
     convex set.

  6. Find a counterexample to the contention that equality holds in (2) of the proof
     of Theorem 8.12.

 *7. In JG16 let C (Figure 8.15) be the interior of the unit circle, that is, C is the set
     of all ordered pairs of real numbers (x_1, x_2) for which $x_1^2 + x_2^2 < 1$. A join

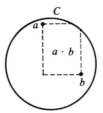

C

a

$a \cdot b$

b

Figure 8.15

operation $\cdot$ is defined in C by relativizing the join operation in JG16 to C. That is, for $a, b \subset C$,

$$a \cdot b = (ab) \cap C,$$

where the join on the right is that in JG16. Show that $(C, \cdot)$ is a join geometry, and that it furnishes a counterexample to the contention that equality holds throughout (4) of the proof of Theorem 8.12.

8. Prove:
   (a) If A and A' are opposite o-rays, then $[A, A'] = AA'$.
   (b) If $A_1, \ldots, A_n$ are o-rays and $o \approx A_1 \ldots A_n$, then $[A_1, \ldots, A_n] = A_1 \ldots A_n$.

*9. Let L and M be Euclidean lines with $L \cap M = o$ (Figure 8.16). The usual join operation on a Euclidean line is denoted by $\cdot$. A join operation $\circ$ is defined in $L \cup M$ as follows:
   (1) If $a, b \subset L$, then $a \circ b = a \cdot b$.
   (2) If $a, b \subset M - o$, then
       (i) if $a \cdot b \not\supset o$, $a \circ b = a \cdot b$,
       (ii) if $a \cdot b \supset o$, $a \circ b = (a \cdot b) \cup L$.
   (3) If $a \subset L$, $b \subset M - o$, then $a \circ b = o \cdot b$.
   (4) If $a \subset M - o$, $b \subset L$, then $a \circ b = a \cdot o$.
   Show that $(L \cup M, \circ)$ is a join geometry and that all points of L are endpoints of the same ray. Note then—a ray need not have a unique endpoint.

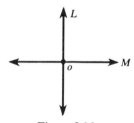

Figure 8.16

8.8 The Halfspaces of a Linear Set

Now the theory of rays is generalized to a theory of halfspaces; these include the rays and form a family of convex sets that is quite important.

In the historical development of Euclidean geometry halfspaces, as the name suggests, were defined in terms of the separation of a linear set. Thus a halfline or a ray was defined as either of the two sets into which a line is separated by one of its points; and a halfplane similarly in terms of the separation of a plane by one of its lines (Figure 8.17).

This type of characterization is not appropriate here. (Recall the remark on terminology in Section 8.1.) Our postulates J1–J7 are far too weak to imply the needed separation properties. (In Chapter 12 the postulate set is

Figure 8.17

strengthened and separation results of the kind indicated here are obtained in Section 12.23.) Even in Euclidean geometry the proof of a rather complicated separation theorem would be required in order to treat the special case of a halfplane.[2]

What shall we do? The answer is easily given. Our theory of rays is readily generalized. To illustrate this, let L be a line in Euclidean 3-space (JG5) (Figure 8.18). Let a be a point not in L. Then L/a consists of all points that are on the opposite side of L from a, and is a halfplane with edge L. By the proper choice of a we can obtain any given halfplane whose edge is L. Observe that the characterization requires no knowledge of the concept "plane"—it could be understood by someone who was innocent of the idea of a plane.

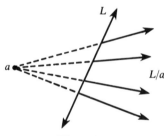

Figure 8.18

Here is the formalization of the concept of a halfspace.

Definition. Let L be a nonempty linear set. Then L/a is called a *halfspace* of L or simply a *halfspace*, and L is its *edge*.

Remark i. If L is a point o, then a halfspace of L is an o-ray. Thus the definition of halfspace is a direct generalization of that of ray.

Remark ii. Although we use the term halfspace, we do not mean to imply that such a set is "half" of anything.

Remark iii. The definition imposes no restriction on the point a. What if a is in L? Then L/a should correspond to the case of an improper ray. By

---

[2] See Prenowitz and Jordan [1], p. 225, Theorem 4.

Lemma 6.5, $a \subset L$ implies $L/a = L$. But on the other hand, since L is nonempty, $L/a = L$ yields $L/a \approx L$, from which $a \approx L/L = L$ follows. Hence $a \subset L$ if and only if $L/a = L$.

Definition. Let L be a nonempty linear set and $a \subset L$. Then the halfspace $L = L/a$ is said to be an *improper* or *degenerate* halfspace of L. All other halfspaces of L are said to be *proper*.

A few more observations: Any halfspace, be it proper or improper, is convex (Theorem 4.16)—in fact, open (Corollary 6.35.2). The family of all halfspaces includes all rays and all nonempty linear spaces.

8.9 A Point of Terminology

The idea of a halfspace is a strong generalization of the idea of a ray. Nevertheless the basic theory of rays that has been developed in this chapter applies to halfspaces. In order to highlight the analogy between halfspaces and rays and present it succinctly, it is desirable to introduce a contraction for "halfspace of L" analogous to the use of o-ray for "ray with endpoint o". Since a halfspace of L will behave like an o-ray in the theory we are developing, it is very convenient to use the term "L-side" or "L-ray" for halfspace of L.

This terminology may seem strange. But its usefulness is indicated by a certain use of the term "side" in Euclidean geometry. A ray is sometimes called a "side of a point", as in the phrase "the right side of point p on line L". Similarly a halfplane, that is, a halfspace of a line, is called a "side of a line", and the term "side of a plane" is used for a halfspace of a plane in Euclidean 3-space. This usage of the term side for halfspaces of different dimension (or halfspaces of linear spaces of different dimensions) has its origin in everyday language, as when we say two points are on the same side (or on opposite sides) of a point, or a line, or a plane (Figure 8.19).

With no more ado we adopt the

Definition. A halfspace of a nonempty linear set L is called an *L-ray*, or a *side of L*, or an *L-side*.

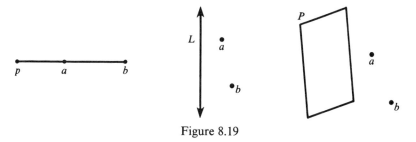

Figure 8.19

The term L-ray may sound somewhat barbarous, as ray has a strong 1-dimensional connotation. But even though this may be so, and even though we have made a case for the use of the term L-side, we will prefer the use of L-ray. Its use may serve as a reminder that halfspaces encompass and generalize o-rays. We feel that as you use it and perceive its advantages it will seem less unnatural, and predict, or at least hope, it will achieve a kind of legitimacy in your lexicon of technical terms.

8.10 Elementary Properties of L-Rays

We proceed to develop a theory for halfspaces of an arbitrary nonempty linear space L that almost exactly parallels the theory of halfspaces of a point, that is, the theory of o-rays.

First an analogue of Theorem 8.1.

Theorem 8.13. *Let L be a nonempty linear space. If $L/a \approx L/b$, then $L/a = L/b$.*

(Compare Theorem 8.1.)

PROOF. Suppose $L/a \approx L/b$. Then

$$Lb \approx La$$

and

$$b \subset La/L.$$

Hence

$$L/b \subset L/(La/L) \subset LL/La = L/La = (L/L)/a = L/a,$$

so that $L/b \subset L/a$. By symmetry $L/a \subset L/b$, and we conclude $L/a = L/b$.   □

Restatement of Theorem 8.13. If A and B are L-rays and $A \approx B$, then $A = B$.

Observe that the steps in the proof are the same as those for Theorem 8.1 with point o replaced by the nonempty linear set L, and that the justification for a step is the same as, or a direct generalization of, the justification of a corresponding step in the proof of Theorem 8.1.

It may be remarked that throughout the development of the theory of o-rays (Sections 8.1–8.7), the only properties of o employed which do not hold for an arbitrary set of points are $o = oo = o/o$ and $o \neq \varnothing$—and these do hold for any nonempty linear set (Theorems 2.9, 6.1).

Corollary 8.13. *A proper L-ray does not meet its edge L.*

The corollary asserts that a proper L-ray is "open" in the sense that it contains no portion of its edge L.

Now an analogue of Theorem 8.2.

Theorem 8.14. *Let a nonempty linear space L and a point p be given. Then there exists a unique L-ray that contains p.*

(Compare Theorem 8.2.)

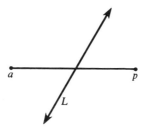

Figure 8.20

PROOF. Existence: $L/p \neq \varnothing$. Let $a \subset L/p$. Then $a \approx L/p$ so that $ap \approx L$ and $p \approx L/a$. Thus L/a is an L-ray that contains p.

Uniqueness: Suppose $p \approx L/b$. Then

$$L/a \approx L/b,$$

and Theorem 8.13 implies

$$L/a = L/b.$$

Thus L/a is the only L-ray that contains p.   □

Observe, as for the last theorem, the exact parallelism between the steps in the proof and the essential steps in the proof of Theorem 8.2, the o-ray analogue.

Continuing to parallel the treatment for o-rays, we introduce the following

Definition. Let L be a nonempty linear space and p a point. Then the unique L-ray which contains p is denoted by $\overrightarrow{Lp}$ (read "Lp arrow") and called the L-ray *determined* by p, the *ray Lp*, the *halfspace Lp* or the *side of L which contains p*.

Note that if L is a point o, the definition of $\overrightarrow{Lp}$ is exactly the same as the definition (Section 8.1) of $\overrightarrow{op}$.

The following corollaries are analogues of, and can be proved by the methods of, the corollaries to Theorem 8.2.

Corollary 8.14.1. *Let A be an L-ray. Then $A \supset a$ implies $A = \overrightarrow{La}$.*

(Compare Corollary 8.2.1.)

Corollary 8.14.2. *Any L-ray can be represented in the form $\overrightarrow{La}$.*

(Compare Corollary 8.2.2.)

Corollary 8.14.3. *If $L \approx ab$, then $L/a = \overrightarrow{Lb}$ and $L/b = \overrightarrow{La}$.*

(Compare Corollary 8.2.3.)

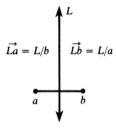

Figure 8.21

8.11 Elementary Operations on *L*-Rays

The operations considered are forming products and quotients of a half-space and its edge.

Theorem 8.15. *If A is an L-ray, then L/A is an L-ray. In fact $L/A = L/a$ for $a \subset A$.*

(Compare Theorem 8.3.)

PROOF. Apply the method of Theorem 8.3.   □

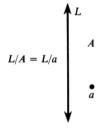

Figure 8.22

Corollary 8.15.1. *If L is linear and not empty, $L/\overrightarrow{La} = L/a$.*

(Compare Corollary 8.3.1.)

Corollary 8.15.2. *If L is linear and not empty, $\overrightarrow{La} = L/(L/a)$.*

(Compare Corollary 8.3.2.)

Corollary 8.15.3. *If L is linear and not empty, $L/(L/(L/a)) = L/a$.*

(Compare Corollary 8.3.3.)

Theorem 8.16. *Let A be an L-ray. Then*
$$LA = A = A/L.$$
(Compare Theorem 8.4.)

PROOF. Apply the method of Theorem 8.4.   □

Corollary 8.16. *If L is linear and not empty,*
$$\overrightarrow{La} \supset La \cup a \cup a/L.$$
(Compare Corollary 8.4.)

Theorem 8.17. *If L is linear and not empty,*
$$\overrightarrow{La} = L/(L/a) = La/L.$$
(Compare Theorem 8.5.)

PROOF. Apply the method of Theorem 8.5.   □

8.12 Opposite Halfspaces

Now the definition of opposite o-rays is generalized.

Definition. Let A be an L-ray. Suppose A' is an L-ray such that
$$AA' \approx L.$$
Then we say L-ray A' is *opposite to* L-ray A, or that A and A' are *opposite L-rays*.

Note that if L-ray A' is opposite to L-ray A, then A is opposite to A'.
A few observations. The L-rays A and A' are opposite if and only if there exist points a and a' in A and A' respectively such that $aa' \approx L$. The

degenerate L-ray L (Section 8.8) is opposite to itself. If L is a point o, the definition reduces to that for opposite o-rays adopted earlier (Section 8.3).

Theorem 8.18. *Let A and A' be opposite L-rays. Then both are proper and $A \not\approx A'$, or both are improper and $A = A' = L$.*

(Compare Theorem 8.6.)

PROOF. Apply the method of Theorem 8.6.   □

Corollary 8.18. *The only L-ray that is opposite to itself is L, the degenerate L-ray.*

(Compare Corollary 8.6.)

Theorem 8.19. $\overrightarrow{La}$ *and* L/a *are opposite L-rays.*

(Compare Theorem 8.7.)

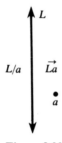

Figure 8.23

PROOF. Apply the method of Theorem 8.7.   □

Theorem 8.20. *If A is an L-ray, it has a unique opposite L-ray. In fact the L-ray opposite to A is L/A.*

(Compare Theorems 8.8, 8.15.)

PROOF. Apply the method of Theorem 8.8.   □

Corollary 8.20. *Suppose L is linear and $ab \approx L$. Then*

(a) $\overrightarrow{La}$ *and* $\overrightarrow{Lb}$ *are opposite L-rays;*
(b) L/a *and* L/b *are opposite L-rays.*

(Compare Corollaries 8.8, 8.14.3.) (See Figure 8.24.)

For opposite o-rays A and A' the condition $AA' \approx o$ is trivially equivalent to $AA' \supset o$. This equivalence is still true for opposite L-rays but is not

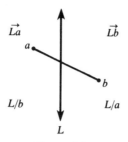

Figure 8.24

a triviality—it represents a real strengthening of the definition and deserves to be a theorem interrupting the parallel development.

Theorem 8.21. *If A and A' are opposite L-rays, then AA' ⊃ L.*

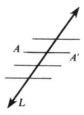

Figure 8.25

PROOF. We have (Figure 8.25)

$$AA' = A(L/A) \qquad \text{(Theorem 8.20)}$$

and

$$A(L/A) \supset L \qquad [\text{Corollary 4.7(a)}].$$

Thus $AA' \supset L$. ☐

EXERCISES

1. Prove: If L is a nonempty linear space and $\overrightarrow{La} = L/b$, then $\overrightarrow{Lb} = L/a$ and $L \approx ab$. (Compare Exercise 1 at the end of Section 8.1.)

2. Prove: If L is a nonempty linear space, then the following are equivalent:
   (a) $\overrightarrow{La} = \overrightarrow{Lb}$;   (b) $L/a = L/b$;
   (c) $L/a \approx L/b$;   (d) $La \approx Lb$.
   (Compare Exercises 2 and 3 at the end of Section 8.1.)

3. If L is a nonempty linear space, is $\overrightarrow{La} = \overrightarrow{Lb}$ equivalent to $a/L \approx b/L$? Justify your assertion. Does equivalence hold if L is a point?

4. Does equality hold in Corollary 8.16 if L is a line in JG4; a plane in JG5?

5. Prove: If L/A is an L-ray, then A is contained in some L-ray. (Compare Exercise 5 at the end of Section 8.2.)

6. Prove: If L and M are nonempty linear spaces, then $L \subset M$ or $L \subset Ma$ implies $\overrightarrow{La} \subset \overrightarrow{Ma}$. Interpret the result geometrically.

7. Prove: If $\overrightarrow{La}$ and $\overrightarrow{Lb}$, or L/a and L/b, are opposite L-rays, then $L \approx ab$. (Compare Exercises 1 and 2 at the end of Section 8.3.)

8. Prove: If A and A' are opposite L-rays, then $A/A' = A$. (Compare Exercise 4 at the end of Section 8.3.)

9. Prove: If A and A' are opposite L-rays and B and B' are opposite L-rays, then
   (a) $A/B = AB'/L$;
   (b) $L/AB = A'B'/L$.
   Interpret these results when L is a line in JG5. (Compare Exercise 5 at the end of Section 8.3.)

10. Prove: If $A_1, \dots, A_n$ are L-rays and $L \approx A_1 \dots A_n$, then $L \subset A_1 \dots A_n$.

8.13 Separation of Two Halfspaces by a Common Edge

The notion of separation of two sets of points by a set of points is now introduced and applied to the study of opposite L-rays.

Definition. Let A and B be nonempty sets of points. Suppose a set of points S has the property that $a \subset A$, $b \subset B$ implies $ab \approx S$. Then we say S *separates A and B* (or *A from B*), or A and B *are separated by S*. If in addition $S \not\approx A, B$, we say S *strictly* separates A and B (or A from B).

Note that if S is a point p, the definition of "S (strictly) separates A and B" is equivalent to that of "p (strictly) separates A and B". Moreover, S (strictly) separates A and B if and only if S (strictly) separates each point of A and each point of B.

Observation on Strict Separation. If S strictly separates A and B, then $A \not\approx B$; for $p \subset A$, B implies $p = pp \approx S$, contrary to $A \not\approx S$. Thus for strict separation the sets S, A and B are disjoint in pairs—no two meet.

Opposite L-rays can be characterized in terms of the relation of separation by L.

Theorem 8.22. *Let A and A' be L-rays. Then A and A' are opposite L-rays if and only if L separates A and A'.*

(Compare Theorem 8.9.)

PROOF. Use the method of Theorem 8.9.   □

Corollary 8.22.1. *Let A and A' be opposite L-rays. Then L separates each point of A from each point of A'.*

(Compare Corollary 8.9.1.)

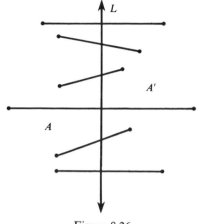

Figure 8.26

Corollary 8.22.2. *Let A and A' be L-rays. Then A and A' are proper opposite L-rays if and only if L strictly separates A and A'.*

(Compare Corollary 8.9.2.)

8.14 The Partition of Space into Halfspaces

Theorem 8.23 (Partition Theorem for Halfspaces). *Let L be a nonempty linear space. Then the family of L-rays is a partition of J, the set of all points.*

(Compare Theorem 8.10.)

PROOF. Apply the method of Theorem 8.10.   □

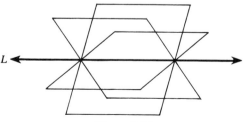

Figure 8.27

8.15 Closed Halfspaces

In Euclidean geometry and in the theory of convex sets, "closed" half-spaces—halfspaces which include their edges—play an important role. In Chapter 13 we will characterize polytopes as intersections of closed half-spaces.

Definition. Let H be a halfspace of nonempty linear set L. Then $H \cup L$ is called an *augmented* or *closed* halfspace (of L) of an *augmented* or *closed* L-ray; and L is its *edge*. $H \cup L$ is called *proper* if H is proper; otherwise it reduces to L and is *improper* or *degenerate*.

Theorem 8.24. *Any closed halfspace is convex.*

(Compare Theorem 8.11.)

PROOF. Apply the method of Theorem 8.11.   □

Remark. A closed halfspace is not necessarily a closed set. Recall the remarks on the definition in Section 8.6.

8.16 The Linear Hull of a Halfspace

We finish the chapter with the analogue of Theorem 8.12. Observe that the proof requires some care.

Theorem 8.25. *Let A and A' be opposite L-rays. Then*
$$\langle A \rangle = \langle A' \rangle \supset AA' \supset A \cup A' \cup L.$$

(Compare Theorem 8.12.)

PROOF. Apply the method of Theorem 8.12, but use Theorem 8.21 for $AA' \supset L$.   □

Remark. In Euclidean geometry the result becomes a complete equality:
$$\langle A \rangle = \langle A' \rangle = AA' = A \cup A' \cup L.$$
Although this equality does not hold in general, it does hold in ordered join geometries (see Exercise 4 at the end of Section 12.23).

EXERCISES

1. Prove without employing the theory of L-rays: If the nonempty linear space L separates a and b, then L separates
   (a) La and Lb;
   (b) a/L and b/L;
   (c) L/a and L/b.
   (Compare Exercise 1 at the end of Section 8.5.)

2. Let A and B be nonempty. Prove that the nonempty linear space L separates A and B if and only if there exist opposite L-rays R and S such that $A \subset R$ and $B \subset S$. (Compare Exercise 2 at the end of Section 8.5.)

3. Prove: If L is linear and separates A and B, then $A \approx L$ or $B \approx L$ implies $L \supset A, B$. (Compare Exercise 3 at the end of Section 8.5.)

4. Prove: If L is a nonempty linear space contained in the linear space M, then there exists a family of L-rays which is a partition of M.

5. Prove: If H is an L-ray, then

$$H \cup L \subset \mathcal{C}(H).$$

6. Prove: If H is an L-ray, then

$$\mathcal{S}(H \cup L) = H.$$

7. Prove: If H is an L-ray, A is an open set and $A \subset H \cup L$, then $A \subset H$ or $A \subset L$.

8. Prove: If equality holds throughout the conclusion of Theorem 8.25, then $A \cup L$ is a closed convex set. (Compare Exercise 5 at the end of Section 8.7.)

9. Prove: If $A_1, \ldots, A_n$ are L-rays and $L \approx A_1 \ldots A_n$, then $[A_1, \ldots, A_n] = A_1 \ldots A_n$. (Compare Exercise 8 at the end of Section 8.7.)

9 Cones and Hypercones

This chapter builds on the preceding chapter, and like its predecessor consists of two well-defined parts—a study of cones followed by a parallel study of their generalization, hypercones. A cone is defined to be a union of a family of rays that have a common endpoint. Cones which are convex, and of these, especially those which are "pointed", are of most importance. The problem of how convex cones are generated by their rays receives a major share of our attention. Naturally the case of the polyhedral cone—a convex cone generated by a finite family of rays—is considered. The notion of an extreme ray of a convex cone, as an analogue of an extreme point of an arbitrary convex set, is clarified and examined closely. Then the study of cones concludes with the determination of conditions for a polyhedral cone to be generated by its extreme rays.

9.1 Cones

The theory of cones is a natural outgrowth of the study of rays.

Definition. The union C of a family F of o-rays is called an o-*cone* or a *cone*; o is its *apex*. The family F is said to *determine* C.

The examples in Figure 9.1 indicate that a cone need not be convex or come to a point as one might normally expect. Our definition, having been chosen for its simplicity, is much too general to require this. The concepts of convex and pointed cones will be introduced and studied later in the chapter.

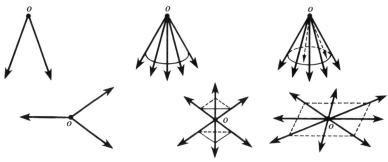

Figure 9.1

Note that any o-ray is an o-cone, and so are J (Theorem 8.10) and $\varnothing$.

In Euclidean geometry it is customary to define a cone so that it always contains its apex. The definition adopted here does not require this condition and does not preclude it—for o is an o-ray contained in no other o-ray. However, if C is an o-cone, $C \cup o$ and $C - o$ always are o-cones. Very often it is immaterial whether or not an o-cone contains its apex— but in some situations it is very convenient to employ the type without apex, since then each point of the cone is contained in some proper o-ray of the cone.

There are many useful criteria that a set of points be a cone.

Theorem 9.1. *A set of points A is an o-cone if and only if it satisfies one of the following conditions:*

(a) *$A = o/S$ for some set S;*
(b) *$A = oT/o$ for some set T.*

PROOF. Condition (a) is equivalent to: A is the union of the o-rays o/x where $x \subset S$. Similarly, condition (b) is equivalent to: A is the union of the o-rays ox/o where $x \subset T$ (Theorem 8.5). $\qquad\square$

Theorem 9.2. *A is an o-cone if and only if $A \supset x$ implies $A \supset \overrightarrow{ox}$.*

PROOF. If A is a union of o-rays and $A \supset x$, then x is in an o-ray R such that $A \supset R$. By Corollary 8.2.1, $R = \overrightarrow{ox}$ and so $A \supset \overrightarrow{ox}$.

For the converse observe that if $A \supset x$ implies $A \supset \overrightarrow{ox}$, then A is the union of the o-rays $\overrightarrow{ox}$ for $x \subset A$. $\qquad\square$

Corollary 9.2.1. *A is an o-cone if and only if $A \supset x$ implies (i) $A \supset ox/o$ or (ii) $A \supset o/(o/x)$.*

PROOF. $\overrightarrow{ox} = ox/o = o/(o/x)$ (Theorem 8.5). $\qquad\square$

Corollary 9.2.2. *Suppose A is an o-cone and R is an o-ray. Then $A \approx R$ implies $A \supset R$.*

PROOF. Let x be common to A and R. Then $R = \overrightarrow{ox}$, and the theorem applies.   □

Theorem 9.3. *A is an o-cone if and only if*

$$oA = A = A/o. \tag{1}$$

PROOF. Suppose A is a union of o-rays. Then (1) certainly holds if $oR = R = R/o$ holds for each o-ray R contained in A. But Theorem 8.4 verifies this.

Conversely suppose (1). Then

$$A = A/o = oA/o,$$

and Theorem 9.1(b) yields that A is an o-cone.   □

Corollary 9.3. *A is an o-cone if and only if* (i) $A = oA/o$, *or equivalently* (ii) $A \supset oA/o$.

PROOF. Conditions (i) and (ii) are equivalent, since $A \subset oA/o$ by Corollary 4.7. The rest now follows by the theorem and Theorem 9.1(b).   □

Given a set of points S, we can obtain a smallest o-cone that contains S simply by constructing o-rays through the points of S.

Theorem 9.4. *If S is any set of points, then there is a unique least o-cone that contains S, namely, oS/o.*

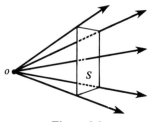

Figure 9.2

PROOF. By Theorem 9.1(b), oS/o is an o-cone (Figure 9.2). By Corollary 4.7, $oS/o \supset S$. If an o-cone $A \supset S$, then by the last corollary,

$$A = oA/o \supset oS/o.$$

Hence oS/o is a least o-cone containing S, and uniqueness follows as usual.   □

Definition. Let S be a set of points. Then oS/o, the least o-cone that contains S, is called the o-cone *generated by* S, and is denoted by S_o or $\overrightarrow{oS}$.

Note the consistency of notation—if $S = x$, a point, then the least o-cone containing x is denoted by $\overrightarrow{ox}$ and is in fact the o-ray determined by x. Note also the new notation x_o for the o-ray determined by x.

EXERCISES

1. Prove: That A is an o-cone is equivalent to each of the following:
   (a) $A = o/(o/S)$ for some set of points S.
   (b) $A \supset x$ implies $A \supset ox$ and $A \supset x/o$.
   (c) $A = o/(o/A)$, or equivalently, $A \supset o/(o/A)$.
   (d) $A \approx R$, where R is an o-ray, implies $A \supset R$.

2. Prove: If A is an o-cone, then o/A is an o-cone which contains precisely those o-rays which are opposite to o-rays contained in A.

3. Prove: If A and B are o-cones, then $A \cup B$, $A \cap B$ and $A - B$ are o-cones.

4. Let A and B be o-rays. Verify AB is an o-cone in the Euclidean join geometry JG4. Show that in general AB need not be an o-cone. (*Hint.* Consider JG8.)

5. Prove: In $\mathbb{R}^n$ (Sections 5.6–5.9), if A and B are o-rays, then AB is an o-cone.

6. Prove: If A is an o-cone, then A/S is an o-cone for any set of points S. Interpret the result geometrically when A is an o-ray and S a point; A and S are o-rays; etc. Examine Exercise 5 at the end of Section 8.3 in this regard.

7. (a) Find in Euclidean geometry examples of cones which have a unique apex and examples of cones which do not.
   (b) Prove: The set of apexes of a cone is a linear set.

8. Let F and G be families of o-rays which determine, respectively, the o-cones A and B. Prove
$$A = B \quad \text{if and only if} \quad F = G.$$
(The result implies that an o-cone is the union of a unique family of o-rays.)

9. Is the least o-cone which contains set S equal to (a) $o/(o/S)$; (b) the intersection of the family of all o-cones containing S; (c) the union of the family of all o-rays meeting S? Prove your assertions.

10. Prove: $S_o \supset o$ if and only if $S \supset o$.

9.2 Convex Cones

Hereafter we shall be concerned almost exclusively with convex cones, that is, cones which are convex sets.

The first theorem—that any halfspace is a convex cone—is neither glamorous nor trivial.

Theorem 9.5. *Let A be a halfspace of the nonempty linear space L (that is, A is an L-ray). Then $o \subset L$ implies A is a convex o-cone.*

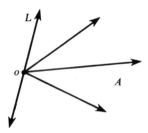

Figure 9.3

PROOF. We have, by using Theorem 8.16 twice, (Figure 9.3)

$$A = A/L = LA/L \supset oA/o,$$

so that A is an o-cone by Corollary 9.3. The result follows because any L-ray is convex.   □

The theorem yields three related types of cones.

Corollary 9.5.1. *Any linear space L is a convex o-cone for $o \subset L$.*

Corollary 9.5.2. *Any closed L-ray is a convex o-cone for $o \subset L$.*

Corollary 9.5.3. *Any union of L-rays is an o-cone for $o \subset L$.*

Is there a simple criterion for a set of points to be a convex o-cone? Suppose A is a convex o-cone. Then (1) $A \supset a, b$ implies $A \supset ab$; and (2) $A \supset S$ implies $A \supset oS/o$, the least o-cone containing S (Theorem 9.4). Putting (1) and (2) together, we have (3) $A \supset a, b$ implies $A \supset oab/o$. This suggests

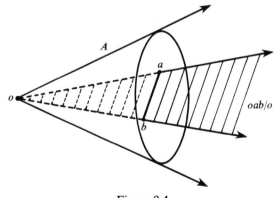

Figure 9.4

Theorem 9.6. *A set of points A is a convex o-cone if and only if $A \supset a, b$ implies $A \supset oab/o$.*

PROOF. The forward implication was established above. For the reverse implication suppose $A \supset a, b$ (Figure 9.4). Then $A \supset oab/o \supset ab$ (Corollary 4.7), and A is convex. Finally suppose $A \supset x$. Then $A \supset oxx/o = ox/o$, and A is an o-cone (Corollary 9.2.1). □

Remark. In Euclidean geometry, oab/o, the test set in the theorem, is in general the interior of an angle, the angle which is customarily denoted $\angle aob$.

The interior and closure operations come into play.

Theorem 9.7. *If A is a convex o-cone, then $\mathcal{I}(A)$ is a convex o-cone.*

PROOF. If $\mathcal{I}(A) = \varnothing$ the result holds. Suppose $\mathcal{I}(A) \neq \varnothing$. Then by Theorems 4.31, 9.3, 6.34 and 9.3 a second time,

$$o\mathcal{I}(A)/o = \mathcal{I}(oA)/o = \mathcal{I}(A)/o = \mathcal{I}(A/o) = \mathcal{I}(A).$$

Thus $\mathcal{I}(A)$ is an o-cone (Corollary 9.3), and the theorem follows. □

Corollary 9.7. *If A is a convex o-cone, then $\mathcal{F}(A)$ is an o-cone.*

PROOF. $\mathcal{F}(A) = A - \mathcal{I}(A)$. Apply the easily shown principle that the difference of two o-cones is an o-cone (Exercise 3 at the end of Section 9.1). □

Theorem 9.8. *If A is a convex o-cone, then $\mathcal{C}(A)$ is a convex o-cone.*

PROOF. Suppose $x \subset \mathcal{C}(A)$. We show $ox/o \subset \mathcal{C}(A)$. Suppose $z \subset ox/o$. We have $xy \subset A$ for some y. Then using Corollary 9.3 we have

$$zy \subset (ox/o)y \subset oxy/o \subset oA/o = A.$$

Thus $z \subset \mathcal{C}(A)$, so that $ox/o \subset \mathcal{C}(A)$. Then $\mathcal{C}(A)$ is an o-cone (Corollary 9.2.1), and the theorem follows. □

Corollary 9.8. *If A is a convex o-cone, then $\mathcal{B}(A)$ is an o-cone.*

PROOF. $\mathcal{B}(A) = \mathcal{C}(A) - \mathcal{I}(A)$. Use the method of proof of Corollary 9.7. □

9.3 Pointed Convex Cones

The cones in elementary geometry (and in everyday affairs, for that matter) are in a sense pointed or come to a point (Figure 9.5). In the present treatment it is desirable to distinguish between this type of cone and others such as linear spaces and closed halfspaces.

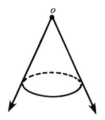
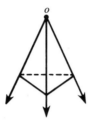
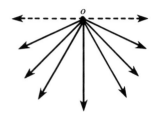

Figure 9.5

Definition. Let A be a convex o-cone. Suppose o is never contained in the join of two distinct points of A. Then we say A is *pointed* (*at* o) and call it a *pointed* (convex) o-cone or simply a *pointed cone*.

Observe that $\varnothing$, o and any proper o-ray are pointed o-cones.

Let A be a convex o-cone. Suppose $o \subset A$. Then A is a pointed o-cone if and only if o is an extreme point of A (Section 4.14). If $o \not\subset A$, certainly A is pointed. It follows that a convex o-cone A is pointed at o if and only if (1) o is an extreme point of A or (2) $o \not\subset A$.

Recall (Section 9.1, fifth paragraph) that if A is an o-cone, then $A \cup o$ and $A - o$ are o-cones. If further, A is a convex o-cone, then it is easily seen (Exercise 3 below) that $A \cup o$ is also a convex o-cone. But this principle fails for $A - o$, since, for example, if $ab \supset o$, $a \neq b$, then the line $\langle a, b \rangle$ is a convex o-cone (Corollary 9.5.1) but $\langle a, b \rangle - o$ is not convex. The correct result is

Theorem 9.9. *A is a pointed convex o-cone if and only if $A - o$ is a convex o-cone.*

PROOF. If $A \not\supset o$, all is trivial, as indicated above. If $A \supset o$, the matter is still essentially settled by the preceding discussion and Theorem 4.17. (Examine Exercise 9 below in this connection.) □

Another characterization of a pointed cone is given by

Theorem 9.10. *A convex o-cone A is a pointed o-cone if and only if it does not contain a pair of proper opposite o-rays.*

PROOF. The forward implication is essentially covered by Corollary 8.9.2. For the converse, suppose $ab \supset o$ where $a, b \subset A$. We show a, b are not distinct. By Corollary 8.8(a), $\overrightarrow{oa}$ and $\overrightarrow{ob}$ are opposite o-rays, which must be contained in A by Theorem 9.2. By hypothesis $\overrightarrow{oa}$ and $\overrightarrow{ob}$ are not a pair of *proper* opposite o-rays, so that Theorem 8.6 yields $\overrightarrow{oa} = \overrightarrow{ob} = o$. Then $a = o = b$ follows, and the theorem is proven. □

EXERCISES

1. Prove: S_o is a convex o-cone if and only if $S^2 \subset S_o$.

2. Prove: If A and B are convex o-cones, then so is $A \cap B$.

3. Prove: If A is a convex o-cone, then so is $A \cup o$.

4. Prove: If A is a convex o-cone and K is a convex set, then A/K is a convex o-cone.

5. Prove: If A is a nonempty convex o-cone, then $\mathcal{C}(A) \supset o$.

6. Prove: A nonempty convex o-cone A is linear if and only if $\mathcal{G}(A) \supset o$.

7. Prove: The only possible extreme point of a convex o-cone is o.

8. Prove: Any proper L-ray is a pointed o-cone for $o \subset L$.

9. Prove that for a convex o-cone A the following are equivalent:
   (a) A is a pointed o-cone.
   (b) $A \cup o$ is a pointed (convex) o-cone.
   (c) o is an extreme point of (convex) $A \cup o$.
   (d) $[A - o] \not\supset o$.

10. Prove: If K is convex, then oK/o is a pointed o-cone if and only if $o \not\subset K$ or o is an extreme point of K.

11. *Project.* Let o be a fixed point. Consider the *o-conar join* operation $*$ defined for arbitrary points a and b by

$$a * b = oab/o.$$

Compare the operation $*$ with join. Consider also the *o-conar extension* operation $/$ defined for arbitrary points a and b by a/b is the set of points x that satisfy $a \subset b * x$. Compare the operation $/$ with extension. Extend both $*$ and $/$ to operations on sets in the manner used for join and extension respectively. Do the operations $*$ and $/$ satisfy J1–J7? Investigate the family of o-rays under the operations $*$ and $/$. Can you find any interesting relationships?

9.4 Generation of Convex Cones

If a set of points is given, how can we form a convex o-cone from it? Is there an easiest or most natural way?

Theorem 9.11. *Let S be a set of points. Then there is a unique least convex o-cone that contains S, namely, $o[S]/o = [S]_o$.*

(Compare Theorem 9.4.)

PROOF. Certainly $o[S]/o$ is a convex o-cone. Moreover, $o[S]/o = [S]_o$, the o-cone generated by $[S]$, so surely $o[S]/o \supset S$. On the other hand suppose A is a convex o-cone and $A \supset S$. Then the convexity of A implies $A \supset [S]$, so that Theorem 9.4 yields $A \supset o[S]/o$, the least o-cone containing $[S]$. We infer $o[S]/o$ is *the* least convex o-cone that contains S, and the theorem holds.   □

It is desirable to generalize the theorem from a set S to a family F of sets. To do this in the natural way we generalize the convex hull operation from a set to an arbitrary family of sets.[1] Let F be a family of sets of points. We define a *convex extension* of F to be a convex set which contains each set of F. By a familiar method one shows the set of convex extensions of F has a least member. We call it the convex hull of (the family) F and denote it by $[F]$. Let U be the union of the sets of F. Then a set is a convex extension of F if and only if it is a convex extension of U. Thus $[F] = [U]$. Now the argument of the theorem yields the following result.

Corollary 9.11. *Let F be a family of sets of points. Then there is a unique least convex o-cone that contains each set of F, namely, $o[F]/o = [F]_o$. Moreover if U is the union of the sets of F, then $[F]_o = [U]_o$.*

Following a well-traveled path, we introduce a new type of hull concept.

Definition. Let S be a set of points. Then $o[S]/o = [S]_o$, the least convex o-cone that contains S, is called the convex o-cone *generated by* S or the *convex o-conar hull* of S. If S is finite and composed of $a_1, \ldots, a_n$, we say $[S]_o$ is *generated by* $a_1, \ldots, a_n$. Similarly if F is a family of sets of points $o[F]/o = [F]_o$, the least convex o-cone that contains each set of F, is called the convex o-cone *generated by* F or the *convex o-conar hull* of F. If F is finite and composed of $S_1, \ldots, S_n$, we say $[F]_o$ is *generated by* $S_1, \ldots, S_n$.

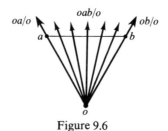

Figure 9.6

EXAMPLE. A very simple case of $[S]_o$ arises if we choose $S = \{a, b\}$, $a \neq b$ (Figure 9.6). Then the convex o-cone generated by $\{a, b\}$ is

$$[a, b]_o = o[a, b]/o = o(ab \cup a \cup b)/o = oab/o \cup oa/o \cup ob/o,$$

---

[1] In Section 3.13 we introduced the idea of the convex hull of a finite family of sets.

which would be described in Euclidean geometry as a closed angular region (*a*, *b*, *o* noncollinear) with vertex *o* omitted.

Note that if K is convex, then $[K]_o$, the convex *o*-cone generated by K, is merely K_o, the *o*-cone generated by K.

Remark. A set of points may be used as a set of generators in four different ways: to generate a convex set, a linear set, an *o*-cone or a convex *o*-cone. If the term generate is used without qualification, the context will indicate the type of generation intended: "convex", "linear", "*o*-conar" or "convex *o*-conar".

9.5 How Shall Convex *o*-Cones Be Generated?

The question may seem silly. A convex *o*-cone is generated by a set of points, as is a convex set. No question of this. There does remain a question: Since a convex *o*-cone A can possibly be generated by a family of sets of points, is there a conceivable advantage in choosing a family of geometric figures of some specific type to generate A? One does not have far to look. It may be more natural and possibly simpler to generate a convex *o*-cone by a family of *o*-rays rather than by a set of points.

For suppose $A = [S]_o$ for the set of points S (Figure 9.7). Then for any $x \subset S$, Theorem 9.2 yields $\overrightarrow{ox} \subset A$. Clearly then, if F is the family of *o*-rays determined by points of S, we have $A = [F]_o$. Consider $A = [F]_o$ as opposed to $A = [S]_o$. The rays of F are fundamental structural components of A. When one conceives of A as being generated by *o*-rays rather than points, the problem of redundant points in S, such as distinct p and q in S for which $\overrightarrow{op} = \overrightarrow{oq}$, is not present. The points of S are not the basic entities that must be reckoned with in the generation of A. The *o*-rays of F play roles in the generation of A comparable to the roles points play in the generation of convex sets.

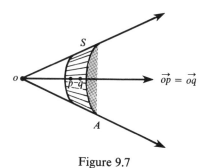

Figure 9.7

We also point out that given any convex o-cone A there is at least one family of o-rays which generates A (as a convex o-cone)—namely, the family F of o-rays that are contained in A. Any family of o-rays which generates A must be a subfamily of F.

We have then answered the question. Convex o-cones should be generated by families of o-rays.

9.6 Polyhedral Cones

We study the generation of convex o-cones by families of o-rays. Following a natural line of development, we consider first finite families.

Definition. Let C be a convex o-cone. If

$$C = [A_1, \ldots, A_n]_o,$$

where $A_1, \ldots, A_n$ are o-rays, we say C is a *polyhedral o-cone* or simply a *polyhedral* cone.

Remark. In the definition choose a point a_i in each o-ray A_i, $i = 1, \ldots, n$. Then clearly $C = [A_1, \ldots, A_n]_o$ if and only if $C = [a_1, \ldots, a_n]_o$. Thus C is "finitely generated" by o-rays if and only if it is "finitely generated" by points.

It is natural to seek analogues, for polyhedral cones, of basic properties of polytopes. Is there an analogue of Theorem 3.2, the formula for a polytope in terms of joins of its generating points? Specifically, can a polyhedral o-cone be expressed as a union of joins of its generating o-rays? No, not in general.

There is a simple counterexample. In JG8 let o be a point and A_1, A_2 two proper o-rays which are distinct and not opposite. J, the basic set of points, is the interior of a circle C (Figure 9.8). It is not hard to see that an o-ray in JG8 is a segment joining o to some point of C. Suppose A_1 is the segment oc_1 and A_2 is oc_2. Then o, c_1 and c_2 are three noncollinear points.

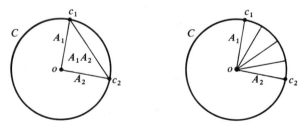

Figure 9.8

Question. Is $[A_1, A_2]_o$, the polyhedral o-cone generated by A_1 and A_2, representable in the form

$$[A_1, A_2]_o = A_1 \cup A_2 \cup A_1 A_2? \tag{1}$$

The answer is no. The right member of (1) is not an o-cone, as shown in the left hand part of Figure 9.8. The left member of (1) is represented by the shaded region in the right hand part. This diagram suggests a correct formula:

$$[A_1, A_2]_o = A_1 \cup A_2 \cup oA_1 A_2/o. \tag{2}$$

The difficulty in (1) is that $A_1 A_2$ in JG8 is not an o-cone (Exercise 4 at the end of Section 9.1), and so does not make a suitable contribution to the o-cone $[A_1, A_2]_o$. If we replace $A_1 A_2$ in (1) by the o-cone which it generates, namely $oA_1 A_2/o$, we get (2). Now invoking Theorem 8.4, we realize $oA_1 = A_1$. Thus $A_1 A_2/o$ is the o-cone generated by $A_1 A_2$, and (2) takes the slightly simpler form

$$[A_1, A_2]_o = A_1 \cup A_2 \cup A_1 A_2/o. \tag{3}$$

In a Euclidean join geometry, the join of the o-rays A_1 and A_2 would be an o-cone, and the formula (1) would be valid (see Exercise 4 at the end of Section 9.1). In this regard, a Euclidean join geometry is not a typical one.

Summary. The formula (2) and its simplification (3) above indicate a valid polyhedral cone analogue of Theorem 3.2, the join formula for a polytope, which is now stated and proved.

Theorem 9.12 (Polyhedral Cone Formula). *Let $A_1, \ldots, A_n$ be o-rays. Then*

$$[A_1, \ldots, A_n]_o = A_1 \cup \cdots \cup A_n \cup A_1 A_2/o$$

$$\cup \cdots \cup A_{n-1} A_n/o \cup \cdots \cup A_1 \ldots A_n/o.$$

(Compare Theorem 3.2 and Corollary 3.10.)

PROOF.

$$[A_1, \ldots, A_n]_o = o[A_1, \ldots, A_n]/o$$

$$= o(A_1 \cup \cdots \cup A_n \cup A_1 A_2 \cup \cdots \cup A_{n-1} A_n$$

$$\cup \cdots \cup A_1 \ldots A_n)/o \quad \text{(Corollary 3.10)}$$

$$= oA_1/o \cup \cdots \cup oA_n/o \cup oA_1 A_2/o$$

$$\cup \cdots \cup oA_{n-1} A_n/o \cup \cdots \cup oA_1 \ldots A_n/o$$

$$\text{(Distributive Laws)}$$

$$= A_1 \cup \cdots \cup A_n \cup A_1 A_2/o \cup \cdots \cup A_{n-1} A_n/o$$

$$\cup \cdots \cup A_1 \ldots A_n/o \quad \text{(Theorem 8.4).} \qquad \square$$

Note. The Polyhedral Cone Formula expresses a polyhedral o-cone as a union of convex o-cones.

Next an analogue of the Polytope Interior Theorem (Theorem 4.30).

Theorem 9.13 (Polyhedral Cone Interior Theorem). *Let* $A_1, \ldots, A_n$ *be* o-*rays. Then*

$$\mathcal{I}([A_1, \ldots, A_n]_o) = A_1 \ldots A_n / o.$$

(Compare Theorems 4.30 and 4.33.)

PROOF. First we record two preliminary results. Note that $A_1, \ldots, A_n$ are open and not empty. Then

$$\mathcal{I}[A_1, \ldots, A_n] = A_1 \ldots A_n \neq \varnothing \qquad \text{(Corollary 4.33.2)}. \qquad (1)$$

By (1) and Theorem 4.31

$$\mathcal{I}(o[A_1, \ldots, A_n]) = \mathcal{I}(o)\mathcal{I}[A_1, \ldots, A_n] = oA_1 \ldots A_n \neq \varnothing. \qquad (2)$$

Now we have

$$\begin{aligned}
\mathcal{I}([A_1, \ldots, A_n]_o) &= \mathcal{I}(o[A_1, \ldots, A_n]/o) \\
&= \mathcal{I}(o[A_1, \ldots, A_n])/\mathcal{I}(o) \qquad [\text{Theorem 6.34, (2)}] \\
&= oA_1 \ldots A_n / o \qquad\qquad\quad [(2)] \\
&= A_1 \ldots A_n / o. \qquad\qquad\qquad\qquad \square
\end{aligned}$$

Query. Are Theorems 9.7 and 9.13 consistent?

9.7 Finiteness of Generation of Convex Cones

An analogue of Corollary 3.7.3, the principle of finiteness of generation for convex sets:

Theorem 9.14 (Finiteness of Generation). *Let* F *be a family of* o-*rays. Then* $x \subset [F]_o$ *if and only if there are finitely many* o-*rays* $A_1, \ldots, A_n$ *of* F *such that*

$$x \subset [A_1, \ldots, A_n]_o.$$

(Compare Corollary 3.7.3.)

PROOF. Suppose $x \subset [F]_o$. Let U be the union of the rays of F. Then

$$x \subset [F]_o = [U]_o = o[U]/o \qquad \text{(Corollary 9.11)}.$$

Hence

$$x \subset oy/o, \qquad\qquad\qquad\qquad\qquad\qquad (1)$$

where $y \subset [U]$. By Corollary 3.7.3,

$$y \subset [a_1, \ldots, a_n], \quad \text{where } a_1, \ldots, a_n \subset U. \tag{2}$$

Thus for each i, $1 \leqslant i \leqslant n$, there is an o-ray A_i of F such that $a_i \subset A_i$. Then (2) implies

$$y \subset [A_1, \ldots, A_n]. \tag{3}$$

The conditions (1) and (3) imply

$$x \subset o[A_1, \ldots, A_n]/o = [A_1, \ldots, A_n]_o,$$

and the forward implication is proved. The reverse implication is immediate. $\qquad\square$

Corollary 9.14.1. *Let F be a family of o-rays. Then $[F]_o$ is the union of all polyhedral o-cones generated by finite subfamilies of F.*

(Compare Corollary 3.7.2.)

Corollary 9.14.2. *Let F be a family of o-rays. Then $x \subset [F]_o$ if and only if*

$$x \subset A_1 \ldots A_r/o, \tag{1}$$

where $A_1, \ldots, A_r$ are o-rays of F.

(Compare Corollary 3.7.1.)

PROOF. Apply the theorem and Theorem 9.12. $\qquad\square$

Note. The condition (1) of the corollary can be restated: x is in the o-cone generated by a finite join of rays of F.

EXERCISES

1. Does $[S]_o = [S_o]$? Explain.

2. Prove: If P is a polytope, then P_o is a polyhedral o-cone.

3. Prove: If C is a polyhedral o-cone, then $C = P_o$ for some polytope P.

4. Prove: If $A_1, \ldots, A_n$ are o-rays, then
$$[A_1, \ldots, A_n]_o = [A_1, \ldots, A_n]/o.$$

5. Prove: If the join of any two o-rays is an o-cone, then the join of any finite number of o-rays is an o-cone.

6. Prove: If $A_1, \ldots, A_n$ are o-rays, then $[A_1, \ldots, A_n]_o$ is linear if and only if $o \subset A_1 \ldots A_n$.

7. Show in JG16 that every nonempty convex o-cone is a polyhedral o-cone.

8. Prove: If L is a finitely generated linear set and $o \subset L$, then L is a polyhedral o-cone.

9. Prove: If L is a finitely generated linear set and $o \subset L$, then any augmented L-ray is a polyhedral o-cone.

10. Suppose L is a nonempty linear set which is not a point. Can a proper L-ray be a polyhedral cone? Can it not be? Explain.

9.8 Extreme Rays

Our object is to study rays in a convex cone which play a role somewhat analogous to that of extreme points (Section 4.14) in a convex set. The defining property for such an "extreme" ray is based on the notion that it cannot separate two points of the cone that are not in the ray itself. Put informally, the extreme ray does not "cut" the cone, it is "peripheral" to it. The extreme ray notion need not be restricted to cones—it can be defined in any convex set.

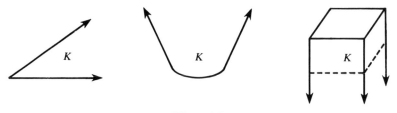

Figure 9.9

Definition. Let R be an o-ray in a convex set K. Then R is an *extreme o-ray* of K if

$$R \approx xy \quad \text{implies} \quad R \cup o \supset x, y \quad \text{for } x, y \subset K. \tag{1}$$

Notes. The definition does not require o to be a point in K (Figure 9.9). If K is the closed o-ray $R \cup o$, or if K is the o-ray R, then trivially R is an extreme o-ray of K. If R is the degenerate o-ray o, then the condition $R \cup o \supset x, y$ becomes $o = x = y$, so that o is an extreme o-ray of K if and only if it is an extreme point of K.

Remark. The condition (1) above, rather than the simpler condition

$$R \approx xy \quad \text{implies} \quad R \supset x, y \quad \text{for } x, y \subset K, \tag{2}$$

was chosen to permit a convex set K which contains o to contain proper extreme o-rays. For if $o \subset K$, it is easy to show that o is the only o-ray which can possibly satisfy (2). The conditions (1) and (2) are obviously equivalent if $o \not\subset K$.

Readers familiar with Chapter 7 may observe that if $o \not\subset K$ then an extreme o-ray is an example of a *face* of K. Relations between extreme rays in the present sense and extreme sets in Chapter 7 are considered in a second set of exercises following Section 9.12.

The defining property of an extreme o-ray can be generalized to any number of points.

Theorem 9.15. *Let R be an o-ray contained in convex set K. Then R is an extreme o-ray of K if and only if*

$$R \approx a_1 \ldots a_n \quad \text{for } a_1, \ldots, a_n \subset K \tag{1}$$

implies

$$R \cup o \supset a_1, \ldots, a_n. \tag{2}$$

(Compare Corollary 4.18.)

PROOF. Let R be an extreme o-ray of K. Suppose (1). We show $R \cup o \supset a_1$. This is trivial if $n = 1$. Assume $n > 1$. Then

$$R \approx a_1 a, \quad \text{where} \quad a \subset a_2 \ldots a_n \subset K.$$

By definition, $R \cup o \supset a_1$. By symmetry $R \cup o \supset a_i$, $2 \leqslant i \leqslant n$. Thus (2) holds.

Conversely, given that (1) implies (2), it is immediate that R is an extreme o-ray of K. Hence the theorem holds. ☐

Corollary 9.15. *Let R be an extreme o-ray of convex set K. Let $A_1, \ldots, A_n$ be o-rays of K. Then*

$$R \approx A_1 \ldots A_n \tag{1}$$

implies

$$A_i = R \quad \text{or} \quad A_i = o, \quad i = 1, \ldots, n. \tag{2}$$

PROOF. The relation (1) implies $R \approx a_1 \ldots a_n$, where $a_i \subset A_i$, $i = 1, \ldots, n$. By the theorem, $R \cup o \supset a_i$, $i = 1, \ldots, n$. Hence $R \cup o \approx A_i$, $i = 1, \ldots, n$, and (2) follows by Theorem 8.1. ☐

Now we give some results that throw light on the existence of extreme o-rays in a convex set.

Theorem 9.16. *If R is an extreme o-ray of convex set K and $K \supset o$, then o is an extreme point of K.*

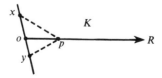

Figure 9.10

PROOF. Suppose (Figure 9.10)

$$o \subset xy, \tag{1}$$

where $x, y \subset K$. We show

$$o = x = y. \tag{2}$$

Let $p \subset R$. Then by (1), $op \subset xyp$. But $op \subset R$ (Corollary 8.4), so that

$$R \approx xyp.$$

Note $p \subset K$, so that Theorem 9.15 yields

$$R \cup o \supset x, y. \tag{3}$$

If one of x, y is o, it immediately follows from (1) that (2) holds. Hence in (3) we need only consider the case $R \supset x, y$. In this case

$$R \supset xy \supset o,$$

so that Theorem 8.1 yields $R = o$. Then (3) gives (2). Thus (2) holds, and since $o \subset K$, o is an extreme point of K by definition. ∎

Theorem 9.17. *If R is an extreme o-ray of convex set K, then $K - R$ is convex.*

(Compare Theorem 4.17.)

PROOF. Let

$$x, y \subset K - R. \tag{1}$$

Then

$$xy \subset K - R \tag{2}$$

unless

$$xy \approx R. \tag{3}$$

Suppose (3). Then since R is an extreme o-ray of K, we have

$$R \cup o \supset x, y.$$

Hence (1) yields $x = o$ and $y = o$. Then $x = y$, and (3) becomes $x = y \subset R$, a contradiction to (1). Thus (3) fails, (2) holds and $K - R$ is convex. ∎

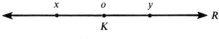

Figure 9.11

Remark. The converse of the theorem does not hold, that is, we can find an o-ray R in a convex set K for which $K - R$ is convex, but R is not an extreme o-ray of K. For example, let K be a Euclidean line which contains point o, and let R be a proper o-ray of K. As Figure 9.11 discloses, $K - R$ is convex, but R does not satisfy the defining condition for an extreme o-ray.

9.9 Extreme Rays of Convex Cones

We now restrict our attention to the case of extreme o-rays of convex o-cones. Our goal is to relate the concepts of extreme o-ray and convex o-conar hull in a fashion analogous to that done for extreme point and convex hull in Section 4.14. Not all convex o-cones have extreme o-rays.

Theorem 9.18. *Only pointed convex o-cones can have extreme o-rays.*

PROOF. Suppose C is a convex o-cone and R is an extreme o-ray of C. We must prove C is pointed at o. If $C \not\supset o$, then certainly C is pointed at o. Assume then that $C \supset o$. By Theorem 9.16, o is an extreme point of C. Again we conclude that C is pointed at o. $\qquad\square$

Corollary 9.18. *A convex o-cone C is pointed at o if and only if o is an extreme o-ray of C or $o \not\subset C$.*

We now begin the investigation of the relationship between extreme o-rays and convex o-conar hulls with an analogue of the property that an extreme point of a convex set K is a member of every set of generators of K.

Theorem 9.19. *Let R be an extreme o-ray of the convex o-cone C, and F a family of o-rays that generates C. Then R is a member of F.*

(Compare Theorem 4.19.)

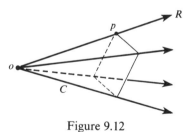

Figure 9.12

PROOF. Choose p in R (Figure 9.12). Then

$$p \subset C = [F]_o.$$

By Corollary 9.14.2 there exist o-rays $A_1, \ldots, A_r$ in F such that

$$p \subset A_1 \ldots A_r / o.$$

Thus

$$op \approx A_1 \ldots A_r.$$

But $op \subset R$ (Corollary 8.4), so that

$$R \approx A_1 \ldots A_r. \tag{1}$$

By Corollary 9.15,

$$A_i = R \quad \text{or} \quad A_i = o, \qquad i = 1, \ldots, r. \tag{2}$$

If in (2), $A_i = R$ for at least one i, then of course, R is a member of F. Suppose then

$$A_1 = \cdots = A_r = o.$$

Then (1) becomes $R \approx o$, so that

$$R = o = A_1 = \cdots = A_r$$

results, and again R is a member of F.   □

Corollary 9.19. *Let R be an extreme o-ray of the convex o-cone C, and F a family of o-rays that generates C. Then $F - \{R\}$, the family of o-rays of F other than R, does not generate C.*

(Compare Corollary 4.19.)

9.10 The Analogous Development Breaks Down

At this point in the development we naturally seek an analogue for Theorem 4.20. This theorem states that if a nonextreme point of convex set K is deleted from a set of generators of K, the resulting set still generates K. Theorem 4.20 played a crucial role in the study of generators of a polytope P—for it enabled us to delete from a generating set of P any nonextreme point of P, and led to Theorem 4.21: a polytope is generated by its extreme points. It would be hoped then that an analogue of Theorem 4.20 would play a similar role and lead to a theorem dealing with the generation of a polyhedral o-cone by its extreme o-rays.

The analogue of Theorem 4.20 for convex cones would assert:

Let F be a family of o-rays which generates the convex o-cone C. Let R be a nonextreme o-ray of C. Then $F - \{R\}$, the family of o-rays of F other than R, generates C.

This statement is not valid. For consider the following examples.

EXAMPLE I. In JG4 let R_1 and R_2 be proper opposite o-rays. Let $F = \{R_1, R_2\}$. Then $C = [R_1, R_2]_o$ is the line containing R_1, R_2. Neither R_1 nor R_2 is an extreme o-ray of C, but C is not generated by the o-ray remaining if either R_1 or R_2 is removed from F.

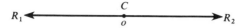

EXAMPLE II. In JG4 let R_1, R_2, R_3 be o-rays each two of which make an angle of 120° (Figure 9.13). Let $F = \{R_1, R_2, R_3\}$. Then $C = [R_1, R_2, R_3]_o$ is the entire plane. None of R_1, R_2, R_3 is an extreme o-ray of C, but if any one of them is deleted from F, the resulting family of two o-rays does not generate C.

EXAMPLE III. In JG4 let R_1, R_2 be proper opposite o-rays, and let the o-ray R_3 be perpendicular to both (Figure 9.14). Let $F = \{R_1, R_2, R_3\}$. Then $C = [R_1, R_2, R_3]_o$ is a closed halfplane. None of R_1, R_2, R_3 is an extreme o-ray of C, but if any one of them is deleted from F, the resulting family of two o-rays does not generate C.

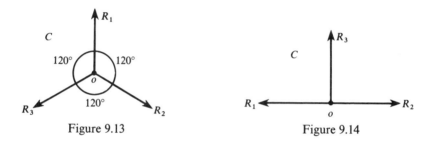

Figure 9.13   Figure 9.14

If we analyze the above examples, we see in all three of them that C is not pointed at o. Recalling Theorem 9.18, we realize that in a nonpointed convex o-cone, every o-ray is a nonextreme o-ray. Thus in a nonpointed convex o-cone the analogue of Theorem 4.20 asks far too much—for if it held, we would be able to remove *any* o-ray from a family of generators and still have a generating family. Examples I–III show that this is absurd.

Therefore we should restrict the analogue of Theorem 4.20 to *pointed* convex o-cones:

Let F be a family of o-rays which generates the pointed convex o-cone C. Let R be a nonextreme o-ray of C. Then $F - \{R\}$ generates C.

However, this restricted statement is also invalid, as is attested by the following example.

EXAMPLE IV. In JG16 let $o = (0, 0)$, $a = (1, 1)$, $b = (-1, 1)$ (Figure 9.15). Then $\overrightarrow{oa}$ is open quadrant I and $\overrightarrow{ob}$ is open quadrant II. Let $F = \{\overrightarrow{oa}, \overrightarrow{ob}\}$. Then $C = [\overrightarrow{oa}, \overrightarrow{ob}]_o$ is the open upper halfplane, and so $C \not\ni o$, so C is pointed at o. Neither $\overrightarrow{oa}$ nor $\overrightarrow{ob}$ is an extreme o-ray of C, for observe: both meet ab. But the deletion of either $\overrightarrow{oa}$ or $\overrightarrow{ob}$ from F leaves the other, which does not generate C.

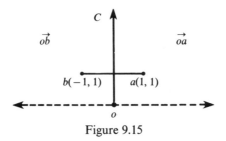

Figure 9.15

Example IV does more than establish the failure of the restricted analogue of Theorem 4.20. It shows that a pointed polyhedral o-cone need have no extreme o-rays, and so need not be generated by extreme o-rays. What shall we do? It is clear now that we cannot find for arbitrary polyhedral o-cones—nor even for pointed polyhedral o-cones—a theory of generation by extreme o-rays analogous to that for generation of polytopes by extreme points. Can we possibly set up some restricted theory with nontrivial content?

Examining the proof of Theorem 4.20, we find that the crucial point is this: If $p \approx pa_1 \dots a_m$, then $p \approx a_1 \dots a_m$. The corresponding property for o-rays $R, A_1, \dots, A_m$,

$$R \approx RA_1 \dots A_m \quad \text{implies} \quad R \approx A_1 \dots A_m, \tag{1}$$

cannot be proved, since in Example IV

$$\overrightarrow{oa} \approx \overrightarrow{oa} \cdot \overrightarrow{ob},$$

but $\overrightarrow{oa} \not\approx \overrightarrow{ob}$. Therefore the key to progress in our search for an analogue of Theorem 4.20 is to find a property whose development has interesting content but which allows us to bypass the invalid property (1). This is what we will do next.

EXERCISES

1. Suppose R is an o-ray in convex set K such that $R \approx xy$ implies $R \supset x, y$ for $x, y \subset K$. Prove: If $o \subset K$, then $R = o$.

2. Prove: If R is an extreme o-ray of convex set K, then R is an extreme o-ray of $[K, o]$.

3. Prove: If R is an extreme o-ray of convex set K, then R is an extreme o-ray of K_o.

4. Prove: The frontier of a convex set K contains its extreme o-rays provided K is neither an o-ray nor an augmented o-ray. (Compare Theorem 4.22.)

5. Suppose R is an o-ray in convex o-cone C. Prove: R is a member of each family of o-rays that generates C if and only if $C - R$ is convex.

6. Show that the converse of Theorem 9.17 fails for pointed convex o-cones. That is, find a pointed convex o-cone C and an o-ray $R \subset C$ such that $C - R$ is convex but R is not an extreme o-ray of C.

7. Suppose R is an o-ray in convex o-cone C. Prove: R is an extreme o-ray of C if and only if

$$R \approx A_1 A_2 \quad \text{for } o\text{-rays } A_1, A_2 \subset C$$

implies

$$A_i = R \quad \text{or} \quad A_i = o, \quad i = 1, 2.$$

8. Prove: If R is an extreme o-ray of convex o-cone C, and F is a family of o-rays which generates C, then

$$R \not\subset [F - \{R\}]_o,$$

or equivalently,

$$C \neq [F - \{R\}]_o.$$

9. Show that the converse of Theorem 9.19 fails. That is, find a convex o-cone C and an o-ray $R \subset C$ such that R is a member of each family of o-rays that generates C but R is not an extreme o-ray of C. (Compare Exercise 4 at the end of Section 4.14.) May such C be pointed at o? Must such C be pointed at o? Explain.

10. Find examples of polyhedral o-cones which contain some extreme o-rays but are not generated by their extreme o-rays.

11. Suppose $p \neq q$ but R is both a p-ray and a q-ray (see Exercise 9 at the end of Section 8.7). If R is contained in convex set K, can R be both an extreme p-ray and an extreme q-ray of K; an extreme p-ray but not an extreme q-ray of K? Explain.

12. For any polyhedral o-cone C, must there always exist a minimal family of o-rays which generates C? A least family? Explain. Do the answers change if C is also required to be pointed at o? [*Hint.* In the set T of all ordered triples (x_1, x_2, x_3) of real numbers, define a join operation $\cdot$ as follows: $(a_1, a_2, a_3) \cdot (b_1, b_2, b_3)$ is the set of all (x_1, x_2, x_3) such that $x_i \subset a_i \cdot b_i$, $i = 1, 2, 3$; and $\cdot$ here is as defined in JG13. Show that $(T, \cdot)$ is a join geometry, and compare it with JG16. If $o = (0, 0, 0)$, find and describe all o-rays in T. Study pointed polyhedral o-cones in T.]

9.11 Regularly Imbedded Rays

We consider in a convex o-cone those o-rays whose linear hulls intersect the cone essentially in the rays themselves.

Definition. Let R be an o-ray in a convex o-cone C (Figure 9.16). Then R is *regularly imbedded* in C, or briefly, is *regular* in C, if

$$\langle R \rangle \cap C \subset R \cup o.$$

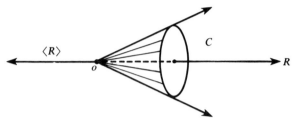

Figure 9.16

Notes. The definition does not require o to be in C, but is made to accommodate this case, as $\langle R \rangle \supset o$ always holds (Theorem 8.12). The degenerate o-ray o is regularly imbedded in any convex o-cone which contains it.

How are extreme o-rays imbedded in convex o-cones?

Theorem 9.20. *An extreme o-ray R of a convex o-cone C is regularly imbedded in C.*

PROOF. Let $x \subset \langle R \rangle \cap C$. By Corollary 6.11.1, $\langle R \rangle = R/R$, so that $x \subset y/z$ where $y, z \subset R$. Hence $y \approx xz$, so $R \approx xz$. Note that $x, z \subset C$. Since R is an extreme o-ray of C, it follows that $x \subset R \cup o$. Therefore

$$\langle R \rangle \cap C \subset R \cup o.$$

By definition, R is regularly imbedded in C.   □

An extreme o-ray of a convex o-cone is characterized by the property of being regularly imbedded and the fact that its removal from the cone leaves a convex set.

Theorem 9.21. *An o-ray R in a convex o-cone C is an extreme o-ray of C if and only if R is regular in C and $C - R$ is convex.*

PROOF. Suppose R is an extreme o-ray of C. Then by the previous theorem R is regular in C, and by Theorem 9.17, $C - R$ is convex.
    Conversely, assume R is regular in C and $C - R$ is convex. Suppose

$$R \approx xy \quad \text{for } x, y \subset C. \tag{1}$$

Since $C - R$ is convex, (1) makes it impossible for $x, y \subset C - R$ to hold. Hence at least one of x, y, say x, is in R. Then (1) yields

$$y \approx R/x \subset \langle R \rangle.$$

Hence, since R is regular in C,

$$y \subset \langle R \rangle \cap C \subset R \cup o.$$

Thus (1) implies $x, y \subset R \cup o$, that is, R is an extreme o-ray of C.   □

Remark. One may interpret the theorem so: for R regular in C, the properties "R is extreme in C" and "$C - R$ is convex" are equivalent. Examine again, in this light, the Remark that follows Theorem 9.17.

Are the o-rays of a pointed convex o-cone regularly imbedded in the cone? The answer to the question is no. In Example IV of Section 9.10 $\langle \overrightarrow{oa} \rangle$ and $\langle \overrightarrow{ob} \rangle$ equal the entire set of points J, so that neither $\overrightarrow{oa}$ nor $\overrightarrow{ob}$ is regularly imbedded in the pointed convex o-cone $C = \left[\overrightarrow{oa}, \overrightarrow{ob} \right]_o$.

The question suggests a sort of converse. Is a convex o-cone pointed if its o-rays are regularly imbedded? Yes, in fact a stronger result holds.

Theorem 9.22. *Suppose convex o-cone C is generated by the family of o-rays F, and each o-ray in F is regular in C. Then C is pointed at o.*

PROOF. Suppose

$$o \approx xy, \quad \text{where } x, y \subset C. \tag{1}$$

We show $x = y$. Since $C = [F]_o$, Corollary 9.14.2 implies

$$x \subset A_1 \ldots A_m/o, \quad y \subset A_{m+1} \ldots A_n/o, \tag{2}$$

where the A's are in F. By (1) and (2),

$$o \approx xy \subset (A_1 \ldots A_m/o)(A_{m+1} \ldots A_n/o) \subset A_1 \ldots A_n/o,$$

so that

$$o \approx A_1 \ldots A_n. \tag{3}$$

Note in (3) we can assume $n \geqslant 2$, so that (3) gives

$$o/A_1 \approx A_2 \ldots A_n \subset C. \tag{4}$$

But o/A_1 is the o-ray opposite to A_1 (Theorem 8.8). Thus $o/A_1 \subset \langle A_1 \rangle$ (Theorem 8.12). By (4) and the regularity of A_1 in C,

$$o/A_1 \approx \langle A_1 \rangle \cap C \subset A_1 \cup o,$$

from which $A_1 = o$ follows. By the symmetry in (3) we have

$$A_2 = \cdots = A_n = o.$$

Hence (2) gives $x = o$ and $y = o$. Therefore, $x = y$ and by definition C is pointed at o.   □

9.12 The Generation of Convex Cones by Extreme Rays

The concept of regularly imbedded o-rays yields the sought-for analogue of Theorem 4.20.

Theorem 9.23. *Let F be a family of o-rays that generates the pointed convex o-cone C. Let R be a nonextreme o-ray of C that is regular in C. Then $F - \{R\}$ generates C.*

(Compare Theorem 4.20.)

PROOF. We have to prove

$$C = [F - \{R\}]_o.$$

Since

$$C = [F]_o \supset [F - \{R\}]_o,$$

it remains to prove

$$C \subset [F - \{R\}]_o. \tag{1}$$

To establish (1) it suffices to show

$$R \subset [F - \{R\}]_o. \tag{2}$$

Since R is a nonextreme o-ray of C, there exist $p, q \subset C$ such that $R \approx pq$ but one of p, q, say p, satisfies

$$p \not\subset R \cup o. \tag{3}$$

Since $p \subset C = [F]_o$, Corollary 9.14.2 yields

$$p \subset A_1 \ldots A_m/o, \tag{4}$$

where the A's are in F. Similarly

$$q \subset A_{m+1} \ldots A_n/o, \tag{5}$$

where the A's are in F.

We claim that at least one A in (4) is distinct from both R and o. Suppose otherwise. Then

$$A_1, \ldots, A_m \subset R \cup o,$$

so that by Theorem 8.11,

$$A_1 \ldots A_m \subset R \cup o.$$

Hence (4) and Theorem 8.4 yield

$$p \subset (R \cup o)/o = R/o \cup o/o = R \cup o,$$

a contradiction to (3). Therefore the claim is established. Let us say

$$A_1 \neq R \quad \text{and} \quad A_1 \neq o. \tag{6}$$

Combining (4) and (5), we have

$$pq \subset (A_1 \ldots A_m/o)(A_{m+1} \ldots A_n/o) \subset A_1 \ldots A_n/o.$$

Since $R \approx pq$, we have

$$R \approx A_1 \ldots A_n/o.$$

Thus $Ro \approx A_1 \ldots A_n$, and Theorem 8.4 gives

$$R \approx A_1 \ldots A_n. \tag{7}$$

We want to show that R meets a join of o-rays of F all of which are distinct from R. This is trivial if in (7) no A is R. Suppose in (7) some A is R. By (6), R occurs among $A_2, \ldots, A_n$. Since the order of these A's is immaterial, let us suppose

$$R \neq A_1, \ldots, A_r \quad \text{but} \quad R = A_{r+1} = \cdots = A_n, \text{ where } 1 \leqslant r < n.$$

Then Theorem 2.9 and (7) imply

$$R \approx A_1 \ldots A_r R,$$

so that (Corollary 6.11.1)

$$\langle R \rangle = R/R \approx A_1 \ldots A_r. \tag{8}$$

But since $A_1 \ldots A_r \subset C$, and R is regular in C, (8) yields

$$A_1 \ldots A_r \approx \langle R \rangle \cap C \subset R \cup o.$$

Hence either $R \approx A_1 \ldots A_r$ (which we want to show) or

$$o \approx A_1 \ldots A_r. \tag{9}$$

Suppose the latter. Then $o \subset C$, and since C is pointed at o, Corollary 9.18 implies that o is an extreme o-ray of C. By Corollary 9.15 applied to (9),

$$o = A_1 = \cdots = A_r,$$

which is a contradiction to (6).

Therefore (9) fails, and R meets a join of o-rays of F all of which are distinct from R. Hence surely

$$R \approx [F - \{R\}]_o.$$

Then (2) follows by Corollary 9.2.2, and the theorem is proven. $\square$

Remark. Observe in the proof of the theorem how the use of the property of regularity allows us to bypass the invalid property (1) in the last paragraph of Section 9.10.

We proceed now to an analogue of Theorem 4.21. In Section 9.10, following Example IV, it was observed that a pointed polyhedral o-cone need not be generated by extreme o-rays. Yet a polyhedral o-cone must be pointed at o in order to possess extreme o-rays (Theorem 9.18). Thus we see that to get a class of polyhedral o-cones which are generated by their extreme o-rays, we must restrict our attention to some proper subclass of the pointed ones. Theorems 9.22 and 9.23 suggest such a subclass.

Theorem 9.24. *Suppose P is a polyhedral o-cone. Then the family E of extreme o-rays of P is finite. Furthermore, if P is generated by a family F of o-rays each of which is regular in P, then E generates P.*

(Compare Theorem 4.21.)

PROOF. By definition

$$P = [R_1, \ldots, R_m]_o \tag{1}$$

where $R_1, \ldots, R_m$ are o-rays. By Theorem 9.19,

$$E \subset \{R_1, \ldots, R_m\},$$

so that E is finite.

Assume next that

$$P = [F]_o \tag{2}$$

where F is a family of o-rays each of which is regular in P. First we show that P is generated by a finite subfamily F' of F. Let $p_i \subset R_i$, $i = 1, \ldots, m$. Then (1) and (2) yield $p_i \subset [F]_o$. By the Finiteness of Generation Principle (Theorem 9.14), $p_i \subset [F_i]_o$, where F_i is a finite subfamily of F. Let $F' = F_1 \cup \cdots \cup F_m$. Then $p_i \subset [F']_o$, and so $R_i \approx [F']_o$, for $i = 1, \ldots, m$. By Corollary 9.2.2,

$$R_1, \ldots, R_m \subset [F']_o,$$

so that (1) gives

$$P \subset [F']_o \subset [F]_o.$$

Hence in view of (2), $P = [F']_o$ for the finite subfamily F' of F.

By Theorem 9.19 again, $F' \supset E$. If $F' = E$, the theorem holds. Suppose then that F' contains nonextreme o-rays of P, and let $\{A_1, \ldots, A_n\}$ be the family of these o-rays. Theorem 9.22 yields that P is pointed at o. Hence Theorem 9.23 may be applied to the family F' and the o-ray A_1, which is regular in P. We obtain

$$F' - \{A_1\} \text{ generates } P. \tag{3}$$

But A_2 is regular in P, so that (3) and Theorem 9.23 imply

$$(F' - \{A_1\}) - \{A_2\} = F' - \{A_1, A_2\} \text{ generates } P.$$

Continuing in this way, we obtain finally

$$E = F' - \{A_1, \ldots, A_n\} \text{ generates } P,$$

so that the theorem is proven.   □

The regularity condition in the theorem which ensures that a polyhedral o-cone is generated by its extreme o-rays actually characterizes o-cones that are generated in this way.

Theorem 9.25. *A polyhedral o-cone P is generated by its extreme o-rays if and only if P is generated by a family of o-rays each of which is regular in P.*

PROOF. The forward implication is covered by Theorem 9.20, the reverse implication by Theorem 9.24.   □

Final Remarks. We conclude our investigation of the connection between the concepts of extreme o-ray and convex o-conar hull by returning to a question in Section 9.11: Are the o-rays of a pointed convex o-cone regularly imbedded in the cone? The answer given was no. But the example which confirmed the answer is not from Euclidean geometry. Our Euclidean intuition would have us say that the answer to the question is yes. What is it about Euclidean geometry that ensures that the o-rays of a pointed o-cone are regular in the cone? The answer is not hard to find.

In Euclidean geometry if R is an o-ray, then

$$\langle R \rangle = R \cup R' \cup o, \tag{1}$$

where R' is the o-ray opposite to R. If R is contained in the pointed convex o-cone C, it follows from (1) (Exercise 8 below) that R is regular in C.

Thus in Euclidean geometry, or more generally in any join geometry where (1) holds for o-rays, each o-ray of a pointed convex o-cone is regular in the cone. In such a join geometry the assumption of regularity in Theorem 9.23 is then superfluous, so that the restricted analogue of Theorem 4.20 holds as stated preceding Example IV in Section 9.10. Furthermore, in such a join geometry a pointed polyhedral o-cone is obviously generated by a family of regularly imbedded o-rays, so that by Theorem 9.24, a pointed polyhedral o-cone is generated by its extreme o-rays. In Chapter 12 ordered join geometries are studied, and it can be shown that (1) holds in an ordered join geometry (Exercise 15 at the end of Section 12.6). Thus the above pertains to ordered join geometries.

EXERCISES

1. (a) Find in the Euclidean join geometry JG5 examples of convex o-cones
    (i) all of whose o-rays are regularly imbedded;
    (ii) none of whose proper o-rays are regularly imbedded;
    (iii) some of whose proper o-rays are regularly imbedded, while some are not.
   (b) In part (a), for each of the three types of cones, can you find examples which are pointed at o and also examples which are not? Explain.
   (c) In part (a), for each of the three types of cones, can you find examples which are polyhedral o-cones and also examples which are not? Explain.

2. Do Exercise 1 with JG16 in place of JG5.

3. Suppose R is an o-ray in the convex o-cone C. Prove: R is regular in C if and only if

$$\langle R \rangle \cap C = (R \cup o) \cap C.$$

4. Prove: If $R_1, \ldots, R_n$ are regularly imbedded o-rays in convex o-cone C, and $o \approx R_1 \ldots R_n$, then

$$o = R_1 = \cdots = R_n.$$

5. Suppose convex o-cone C is generated by a family of o-rays each of which is regular in C. Must every o-ray in C be regular in C? Explain.

6. Suppose C is a convex o-cone and F is the family of o-rays of C that are regular in C. Must $[F]_o$ be pointed at o? Explain.

7. Suppose P is a polyhedral o-cone that is generated by a family of o-rays each of which is regular in P. Prove: P is open if and only if P is an o-ray. (Compare Exercise 1 at the end of Section 4.15.)

8. Suppose R is an o-ray for which

$$\langle R \rangle = R \cup R' \cup o,$$

where R' is the o-ray opposite to R. Prove: If R is contained in a pointed convex o-cone C, then R is regular in C.

9. Suppose R is an o-ray for which

$$\langle R \rangle = R \cup R' \cup o,$$

where R' is the o-ray opposite to R. Let R be contained in the pointed convex o-cone C. Is R an extreme o-ray of C if
(a) $C - R$ is convex;
(b) R is contained in each family of o-rays that generates C?
Explain.

10. *Project.* Consider a family F of o-rays which satisfies

$$\langle R \rangle \approx R_1 \dots R_n \quad \text{for } R, R_1, \dots, R_n \text{ in } F$$

(A)

implies

$$R \cup o \approx R_1 \dots R_n.$$

Investigate families of o-rays which satisfy (A). In particular, prove:
(a) If a family of o-rays satisfies (A), then its convex o-conar hull is pointed at o.
(b) The family of extreme o-rays of a convex set satisfies (A).
(c) The family of regularly imbedded o-rays in a convex o-cone satisfies (A).
(d) If a polyhedral o-cone is generated by a family of o-rays satisfying (A), then it is generated by its family of extreme o-rays.

EXERCISES (Chapter 7 required)

1. Suppose R is an o-ray in convex set K. Prove:
(a) If $o \not\subset K$, then R is an extreme o-ray of K if and only if R is a face of K.
(b) If $o \subset K$, then R is an extreme o-ray of K if and only if the closed o-ray $R \cup o$ is a face of K.

2. Prove: An extreme o-ray of a convex set K is an extreme set of K, in fact, a component of K. Show that an o-ray which is an extreme set of K need not be an extreme o-ray of K.

3. Prove: If A is an extreme set of a convex o-cone, then A is a convex o-cone.

4. (a) Prove: If A is a face of B and $o \subset \langle A \rangle$, then A_o is a face of B_o.
(b) Prove that (a) holds with the term face replaced by component.
(c) Does (a) hold if the term face is replaced by extreme set?

5. (a) Prove: If K is convex and X is an extreme set of K_o, then $X = A_o$ for some extreme set A of K.
(b) Prove that (a) holds with the term extreme set replaced by face.
(c) Prove that (a) holds with the term extreme set replaced by component if in addition it is assumed K contains no pathological extreme set.

6. The same as Exercise 4, with the assumption $o \subset \langle A \rangle$ replaced by "o is linearly independent of B".

7. State and prove for polyhedral o-cones the analogue of
   (a) Theorem 7.20;
   (b) Theorem 7.36.
   What can be inferred as immediate corollaries?

9.13 Hypercones

Now the theory of cones is generalized to a theory of hypercones. For a given nonempty linear set L, objects called L-cones will be studied. They bear the same analogy to o-cones as L-rays do to o-rays.

Definition. Let L be a nonempty linear set (Figure 9.17). The union of a family F of L-rays is called an *L-cone* or generically a *hypercone*. The family F is said to *determine* the L-cone.

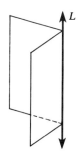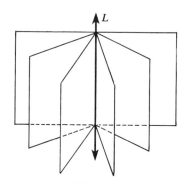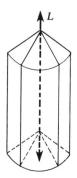

Figure 9.17

Note that any L-ray is an L-cone, and so are J (Theorem 8.23) and $\varnothing$.

Remark. If L is a point o, then an L-cone is an o-cone. Thus the definition of hypercone is a direct generalization of that of cone. Moreover, we have the chain of relations: rays are halfspaces are cones are hypercones.

We develop a theory of L-cones that parallels the theory of o-cones in the previous sections. The treatment is given in outline form and consists almost exclusively of definitions and of theorems and corollaries with proofs omitted. Each such theorem (corollary) will be an exact or restricted analogue of a theorem (corollary) in the theory of o-cones, and in most cases the reader can supply a proof by generalizing each step in the proof of the analogue.

9.14 Elementary Properties of Hypercones

Theorem 9.26 (Analogue, Theorem 9.1). *A set of points A is an L-cone if and only if it satisfies one of the following conditions:*

(a) $A = L/S$ *for some set S;*
(b) $A = LT/L$ *for some set T.*

Theorem 9.27 (Analogue, Theorem 9.2). *A is an L-cone if and only if $A \supset x$ implies $A \supset \overrightarrow{Lx}$.*

Corollary 9.27.1 (Analogue, Corollary 9.2.1). *A is an L-cone if and only if $A \supset x$ implies* (a) $A \supset Lx/L$ *or* (b) $A \supset L/(L/x)$.

Corollary 9.27.2 (Analogue, Corollary 9.2.2). *Suppose A is an L-cone and H is an L-ray. Then $A \approx H$ implies $A \supset H$.*

Theorem 9.28 (Analogue, Theorem 9.3). *A is an L-cone if and only if $LA = A = A/L$.*

Corollary 9.28 (Analogue, Corollary 9.3). *A is an L-cone if and only if* (a) $A = LA/L$, *or equivalently* (b) $A \supset LA/L$.

Theorem 9.29 (Analogue, Theorem 9.4). *If S is any set of points, then there is a unique least L-cone that contains S, namely, LS/L.*

Definition. Let L be a nonempty linear set and S a set of points. Then LS/L, the least L-cone that contains S, is called the L-cone *generated by* S, and is denoted by S_L or $\overrightarrow{LS}$.

Note the double consistency of notation—for either S or L may reduce to a point. Note also the new notation for an L-ray when S reduces to a point.

9.15 Convex Hypercones

A convex L-cone is, of course, an L-cone which is a convex set.

Theorem 9.30 (Analogue, Theorem 9.5). *Let A be an M-ray. Let L be a nonempty linear set such that $L \subset M$. Then A is a convex L-cone.*

Corollary 9.30.1 (Analogue, Corollary 9.5.1). *Any linear space M is a convex L-cone for a nonempty linear set $L \subset M$.*

Corollary 9.30.2 (Analogue, Corollary 9.5.2). *Any closed M-ray is a convex L-cone for a nonempty linear set $L \subset M$.*

Before stating the analogue of Corollary 9.5.3, we observe that Corollary 9.5.3 can be restated.

Restatement of Corollary 9.5.3. Any *L-cone is an o-cone for $o \subset L$.*

Corollary 9.30.3 (Analogue, Corollary 9.5.3). *Any M-cone is an L-cone for a nonempty linear set $L \subset M$.*

The restatement of Corollary 9.5.3 is interesting, since it indicates that in making the generalization from cones to hypercones we have not enlarged the original domain. However, it does not indicate that the theory of cones encompasses that of hypercones. For example, for o in linear set L, the L-cone S_L generated by the set S is in general not the o-cone generated by S. Moreover, if S is finite, S_L considered as an o-cone need not be generated by any finite set. The structure of an L-cone is highly dependent on the linear set L, and so the theory of o-cones can hardly yield much information on an L-cone's structure.

Theorem 9.31 (Analogue, Theorem 9.6). *A set of points A is a convex L-cone if and only if $A \supset a, b$ implies $A \supset Lab/L$.*

Theorem 9.32 (Analogue, Theorems 9.7, 9.8 and Corollaries 9.7, 9.8). *If A is a convex L-cone, then $\mathcal{G}(A)$ and $\mathcal{C}(A)$ are convex L-cones and $\mathcal{F}(A)$ and $\mathcal{B}(A)$ are L-cones.*

9.16 Tapered Convex Hypercones

Definition. Let L be a nonempty linear set and A a convex L-cone (Figure 9.18). Suppose

$$L \approx xy \text{ implies } L \supset x, y \quad \text{for } x, y \subset A. \tag{1}$$

Then we say A is *tapered* (*at L*) and call it a *tapered* (convex) *L-cone* or simply a *tapered hypercone*.

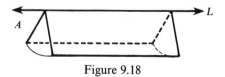

Figure 9.18

Observe that $\varnothing$, L and any proper L-ray are tapered L-cones.

If L reduces to a point o, then certainly the condition (1) is equivalent to A's being pointed at o. Thus the term tapered o-cone is an alternative for pointed o-cone. Also note that the condition (1) is the "correct" generalization of the defining condition for a pointed o-cone. The linear set L should not separate two points of the L-cone that are not both in L.

Readers familiar with Chapter 7 should realize that the condition (1) is equivalent to saying L is an extremal linear space to A (Section 7.27), so that in the case $A \supset L$, L is a face of A.

Theorem 9.33 (Analogue, Theorem 9.9). *A is a tapered convex L-cone if and only if $A - L$ is a convex L-cone.*

PROOF. For the reverse implication, use the fact that if $L \approx xy$ and $L \supset x$, then $y \approx L/x \subset L$, so that $L \supset y$ also.   □

Theorem 9.34 (Analogue, Theorem 9.10). *A convex L-cone A is a tapered L-cone if and only if it does not contain a pair of proper opposite L-rays.*

9.17 Generation of Convex Hypercones

Theorem 9.35 (Analogue, Theorem 9.11). *Let S be a set of points. Then there is a unique least convex L-cone that contains S, namely, $L[S]/L = [S]_L$.*

Corollary 9.35 (Analogue, Corollary 9.11). *Let F be a family of sets of points. Then there is a unique least convex L-cone that contains each set of F, namely, $L[F]/L = [F]_L$. Moreover if U is the union of the sets of F, then $[F]_L = [U]_L$.*

Definition. Let L be a nonempty linear set and S a set of points. Then $L[S]/L = [S]_L$, the least convex L-cone that contains S, is called the convex L-cone *generated by* S or the *convex L-conar hull* of S. If S is finite and composed of $a_1, \ldots, a_n$, we say $[S]_L$ is *generated by* $a_1, \ldots, a_n$. Similarly if F is a family of sets of points $L[F]/L = [F]_L$, the least convex L-cone that contains each set of F, is called the convex L-cone *generated by* F or the *convex L-conar hull* of F. If F is finite and composed of $S_1, \ldots, S_n$, we say $[F]_L$ is *generated by* $S_1, \ldots, S_n$.

9.18 Polyhedral Hypercones

Definition. Let L be a nonempty linear set and C a convex L-cone. If

$$C = [A_1, \ldots, A_n]_L,$$

where $A_1, \ldots, A_n$ are L-rays, we say C is a *polyhedral L-cone* or simply a *polyhedral* hypercone.

Theorem 9.36 (Polyhedral Hypercone Formula: Analogue, Theorem 9.12). *Let $A_1, \ldots, A_n$ be L-rays. Then*

$$[A_1, \ldots, A_n]_L = A_1 \cup \cdots \cup A_n \cup A_1 A_2 / L \cup \cdots \cup A_{n-1} A_n / L$$

$$\cup \cdots \cup A_1 \ldots A_n / L.$$

Theorem 9.37 (Polyhedral Hypercone Interior Theorem: Analogue, Theorem 9.13). *Let $A_1, \ldots, A_n$ be L-rays. Then*

$$\mathcal{G}\big([A_1, \ldots, A_n]_L\big) = A_1 \ldots A_n / L.$$

9.19 Finiteness of Generation of Convex Hypercones

Theorem 9.38 (Finiteness of Generation: Analogue, Theorem 9.14). *Let F be a family of L-rays. Then $x \subset [F]_L$ if and only if there are finitely many L-rays $A_1, \ldots, A_n$ of F such that*

$$x \subset [A_1, \ldots, A_n]_L.$$

Corollary 9.38.1 (Analogue, Corollary 9.14.1). *Let F be a family of L-rays. Then $[F]_L$ is the union of all polyhedral L-cones generated by finite subfamilies of F.*

Corollary 9.38.2 (Analogue, Corollary 9.14.2). *Let F be a family of L-rays. Then $x \subset [F]_L$ if and only if*

$$x \subset A_1 \ldots A_r / L,$$

where $A_1, \ldots, A_r$ are L-rays of F.

9.20 Extreme Halfspaces of Convex Hypercones

In order to make more rapid progress in the theory of hypercones, we define extreme L-rays only for convex L-cones.

Definition. Let L be a nonempty linear set. Suppose H is an L-ray in the convex L-cone C. Then H is an *extreme L-ray* of C if

$$H \approx xy \text{ implies } H \cup L \supset x, y \quad \text{for } x, y \subset C. \tag{1}$$

Note. If H is the degenerate L-ray L, then (1) becomes the defining condition for a tapered L-cone. Thus if $L \subset C$, then L is an extreme L-ray of C if and only if C is tapered at L.

Theorem 9.39 (Analogue, Theorem 9.15). *Let H be an L-ray in convex L-cone C. Then H is an extreme L-ray of C if and only if $H \approx a_1 \ldots a_n$ for $a_1, \ldots, a_n \subset C$ implies*

$$H \cup L \supset a_1, \ldots, a_n.$$

Corollary 9.39 (Analogue, Corollary 9.15). *Let H be an extreme L-ray of convex L-cone C. Let $A_1, \ldots, A_n$ be L-rays of C. Then $H \approx A_1 \ldots A_n$ implies*

$$A_i = H \quad or \quad A_i = L, \quad i = 1, \ldots, n.$$

Theorem 9.40 (Analogue, Theorem 9.17). *If H is an extreme L-ray of convex L-cone C, then $C - H$ is convex.*

Theorem 9.41 (Analogue, Theorem 9.18). *Only tapered convex L-cones can have extreme L-rays.*

PROOF. Suppose C is a convex L-cone and H is an extreme L-ray of C. We must show C is tapered at L. Suppose not. Then by Theorem 9.34, C contains a pair of proper opposite L-rays A_1, A_2. Therefore $L \approx A_1 A_2$, and so

$$H = HL \approx HA_1A_2. \tag{1}$$

Since $A_1, A_2 \neq L$, Corollary 9.39 and (1) yield $A_1 = H$ and $A_2 = H$, that is, $A_1 = A_2$. But this contradicts Theorem 8.18. Hence we have shown C is tapered at L. $\qquad\square$

Corollary 9.41 (Analogue, Corollary 9.18). *A convex L-cone C is tapered at L if and only if L is an extreme L-ray of C or $L \not\subset C$.*

A Side Remark. The result in Corollary 9.41 could have been introduced earlier. A proof can be given which rests on two earlier results: the note following the definition of extreme L-ray, in the case $L \subset C$; and Theorem 9.33, in the case $L \not\subset C$, since by Corollary 9.27.2, $L \not\subset C$ is equivalent to $L \not\approx C$.

Theorem 9.42 (Analogue, Theorem 9.19). *Let H be an extreme L-ray of the convex L-cone C, and F a family of L-rays that generates C. Then H is a member of F.*

Corollary 9.42 (Analogue, Corollary 9.19). *Let H be an extreme L-ray of the convex L-cone C, and F a family of L-rays that generates C. Then $F - \{H\}$, the family of L-rays of F other than H, does not generate C.*

9.21 Regularly Imbedded Halfspaces

Definition. Let L be a nonempty linear set. Suppose H is an L-ray in a convex L-cone C. Then H is *regularly imbedded* in C, or briefly, is *regular* in C, if

$$\langle H \rangle \cap C \subset H \cup L.$$

Theorem 9.43 (Analogue, Theorem 9.20). *An extreme L-ray H of a convex L-cone C is regular in C.*

Theorem 9.44 (Analogue, Theorem 9.21). *An L-ray H in a convex L-cone C is an extreme L-ray of C if and only if H is regular in C and $C - H$ is convex.*

Theorem 9.45 (Analogue, Theorem 9.22). *Suppose convex L-cone C is generated by the family of L-rays F, and each L-ray in F is regular in C. Then C is tapered at L.*

9.22 The Generation of Convex Hypercones by Extreme Halfspaces

Theorem 9.46 (Analogue, Theorem 9.23). *Let F be a family of L-rays that generates the tapered convex L-cone C. Let H be a nonextreme L-ray of C that is regular in C. Then $F - \{H\}$ generates C.*

Theorem 9.47 (Analogue, Theorem 9.24). *Suppose P is a polyhedral L-cone. Then the family E of extreme L-rays of P is finite. Furthermore, if P is generated by a family F of L-rays each of which is regular in P, then E generates P.*

Theorem 9.48 (Analogue, Theorem 9.25). *A polyhedral L-cone P is generated by its extreme L-rays if and only if P is generated by a family of L-rays each of which is regular in P.*

EXERCISES

Many of the exercises below are L-cone analogues of o-cone exercises found in the exercises at the end of Sections 9.1, 9.3, 9.7 and 9.10 and in the first group at the end of Section 9.12. However, there are other exercises contained in these groups that have generalizations to L-cones and that have not been selected. The reader is encouraged to find these, to generalize them and to solve them. The reader is also encouraged to supply proofs to results in Sections 9.14–9.22 that were stated without proof. Readers familiar with Chapter 7 are encouraged to generalize the exercises in the second group at the end of Section 9.12.

1. Prove: If A and B are L-cones, then $A \cup B$, $A \cap B$ and $A - B$ are L-cones. (Compare Exercise 3 at the end of Section 9.1.)

2. Prove: If A is an L-cone, then A/S is an L-cone for any set of points S. (Compare Exercise 6 at the end of Section 9.1.)

3. Prove: $S_L \supset L$ if and only if $S \approx L$. (Compare Exercise 10 at the end of Section 9.1.)

4. Prove: If A is a convex L-cone, then so is $A \cup L$. (Compare Exercise 3 at the end of Section 9.3.)

5. Prove: If A is a nonempty convex L-cone, then
   (a) $\mathcal{C}(A) \supset L$;
   (b) A is linear if and only if $\mathcal{S}(A) \supset L$.
   (Compare Exercises 5 and 6 at the end of Section 9.3.)

6. Prove: If A is a convex o-cone, L is linear and $o \subset L \subset A$, then A is a convex L-cone. Does the conclusion hold if A is a convex M-cone and $M \subset L \subset A$?

7. Prove: If A is a tapered L-cone and $o \subset L$, then A is a pointed o-cone if and only if $L \napprox A$ or $L = o$.

*8. Suppose A is an o-cone. Prove that there exists a greatest linear set L such that A is an L-cone. If in addition A is assumed convex, prove that A is a tapered L-cone.

9. Prove:
   (a) If P is a polytope, then P_L is a polyhedral L-cone.
   (b) If C is a polyhedral L-cone, then $C = P_L$ for some polytope P.
   (Compare Exercises 2 and 3 at the end of Section 9.7.)

10. Prove: If $A_1, \ldots, A_n$ are L-rays, then $[A_1, \ldots, A_n]_L$ is linear if and only if $L \approx A_1 \ldots A_n$. (Compare Exercise 6 at the end of Section 9.7.)

11. Prove: If L is a finitely generated linear set and $o \subset L$, then any polyhedral L-cone is a polyhedral o-cone. (Compare Exercises 8 and 9 at the end of Section 9.7.)

12. Prove: The frontier of a convex L-cone C contains its extreme L-rays provided C is neither an L-ray nor an augmented L-ray. (Compare Exercise 4 at the end of Section 9.10.)

13. Suppose H is an L-ray contained in convex L-cone C. Prove: H is a member of each family of L-rays that generates C if and only if $C - H$ is convex. (Compare Exercise 5 at the end of Section 9.10.)

14. Suppose P is a polyhedral L-cone that is generated by a family of L-rays each

of which is regular in P. Prove: P is open if and only if P is an L-ray. (Compare Exercise 7 in the first group at the end of Section 9.12.)

15. Suppose H is an L-ray such that

$$\langle H \rangle = H \cup H' \cup L,$$

where H' is the L-ray opposite to H. Let F be a finite family of such L-rays H; suppose $[F]_L$ is a tapered L-cone. Prove that $[F]_L$ is generated by its extreme L-rays. (See Exercise 4 at the end of Section 12.23.)

10 Factor Geometries and Congruence Relations[1]

In this chapter the family of halfspaces of a nonempty linear space is converted into a join system—called a factor geometry—by defining a join operation in it in a natural way. The theory of factor geometries has its roots in the problem of constructing a geometry out of the family of rays that emanate from a given point in a Euclidean space. Such a ray geometry is implicit in classical geometry and is closely related to spherical geometry. Factor geometries and join geometries share many common properties and can be studied by similar methods. Thus in a factor geometry convexity and linearity are treated in a familiar way. However a factor geometry, as an algebraic system, differs markedly from a join geometry since it contains an identity element and its elements have inverses. The development has strong—though unforced—analogies with algebraic theories of congruence relations and factor or quotient systems.

10.1 Congruence Relations Determined by Halfspaces

If M is a nonempty linear set in join geometry J, the family of halfspaces of M partitions J (Theorem 8.23). We find it pays to study the relationships among points that this partition induces.

[1] The prerequisites for this chapter are Chapters 6, 8, Sections 9.1–9.4 and Sections 9.13–9.17.

Definition. Let M be a nonempty linear set in join geometry J. Suppose a and b are in the same halfspace of M, or equivalently $\overrightarrow{Ma} = \overrightarrow{Mb}$. Then we say a is *congruent* to b *modulo* M and write $a \equiv b \pmod{M}$.

Geometric Interpretation. Let M be a line in JG5, the 3-dimensional Euclidean join geometry (Figure 10.1). Suppose $a \equiv b \pmod{M}$. Consider first the case where a, $b \not\subset M$. Then a and b are in a proper halfspace of M, and so are on the same "side" of M. Suppose one of a, b is in line M. Then both are, since M is a (degenerate) halfspace of M. Thus $a \equiv b \pmod{M}$ is seen to be equivalent to: a and b are on the same "side" of line M or are both in M.

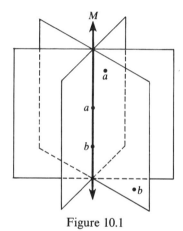

Figure 10.1

Theorem 10.1. *The relation congruence modulo M is an equivalence relation in set J:*

$a \equiv a \pmod{M}$,

$a \equiv b \pmod{M}$ *implies* $b \equiv a \pmod{M}$,

$a \equiv b \pmod{M}$, $b \equiv c \pmod{M}$ *imply* $a \equiv c \pmod{M}$.

PROOF. Apply the definition: $x \equiv y \pmod{M}$ means $\overrightarrow{Mx} = \overrightarrow{My}$. ☐

The following theorem gives a simple criterion for congruence which does not involve the idea of halfspace.

Theorem 10.2. $a \equiv b \pmod{M}$ *if and only if* $aM \approx bM$.

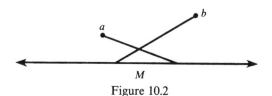

Figure 10.2

PROOF. Suppose $a \equiv b \pmod{M}$ (Figure 10.2). By definition

$$\overrightarrow{Ma} = \overrightarrow{Mb}.$$

Then

$$aM/M = bM/M \qquad \text{(Theorem 8.17)},$$

and the Transposition Law (Theorem 4.8) implies

$$aM \approx bM. \tag{1}$$

Conversely, suppose (1). Then

$$a \approx bM/M = \overrightarrow{Mb} \qquad \text{(Theorem 8.17)}.$$

Thus $\overrightarrow{Ma} = \overrightarrow{Mb}$ (Corollary 8.14.1), and $a \equiv b \pmod{M}$ by definition.   ☐

10.2 Pairs of Congruences

Suppose $a \equiv a' \pmod{M}$ and $b \equiv b' \pmod{M}$. What consequences follow?

Theorem 10.3. *Suppose $a \equiv a' \pmod{M}$ and $b \equiv b' \pmod{M}$. Then the segments ab and $a'b'$ are related in the following way: Each point of ab is congruent modulo M to some point of $a'b'$, and each point of $a'b'$ is congruent modulo M to some point of ab.*

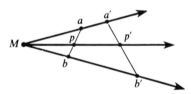

Figure 10.3

PROOF. Let $p \subset ab$ (Figure 10.3). We show $p \equiv p' \pmod{M}$ for some $p' \subset a'b'$. Since $a \equiv a' \pmod{M}$, the previous theorem yields $aM \approx a'M$, so that $a \subset a'M/M$. Similarly $b \subset b'M/M$. Then

$$p \subset ab \subset (a'M/M)(b'M/M) \subset a'b'M/M,$$

which gives $p \subset p'M/M$ for some $p' \subset a'b'$. Thus $pM \approx p'M$ and $p \equiv p'$ (mod M). The second part of the conclusion follows by symmetry.   ☐

10.3 Congruence Relations between Sets

The last theorem establishes a rather close relation between the sets ab and $a'b'$, which may be described by saying that they are *pointwise congruent modulo M*, and suggests the following

Definition. Let M be a nonempty linear set. Suppose sets A and B have the property that each point of A is congruent modulo M to some point of B and each point of B is congruent modulo M to some point of A. Then we say A is *congruent* to B *modulo* M and write $A \equiv B$ (mod M).

Note that if A and B reduce to single points, then the definition merely requires that A and B be congruent as points, that is, the definition is consistent with our agreement to identify $\{a\}$ with a. Also note that $A \equiv b$ (mod M) is equivalent to "every element of A is congruent to b modulo M."

In view of the definition, Theorem 10.3 may be stated in the following form:

Restatement of Theorem 10.3. Suppose $a \equiv a'$ (mod M) and $b \equiv b'$ (mod M). Then $ab \equiv a'b'$ (mod M).

Corollary 10.3.1. $a \equiv b$ (mod M) *implies* $ax \equiv bx$ (mod M).

Corollary 10.3.2. $a \equiv b$ (mod M) *and* $a \equiv c$ (mod M) *imply* $a \equiv bc$ (mod M).

In JG4, the join geometry of the Euclidean plane, let M be a point m and consider the restatement of Theorem 10.3. As a typical case suppose a, b and m are noncollinear as depicted in Figure 10.3. Then the rays $\overrightarrow{ma}$ and $\overrightarrow{mb}$ are distinct and form an angle, namely, $\angle amb$. Moreover a', b' are in the respective sides $\overrightarrow{ma}$, $\overrightarrow{mb}$ of the angle. What is the significance of $ab \equiv a'b'$ (mod m)? It asserts that each point p of segment ab lies in the same ray from m as some point p' of segment $a'b'$, and vice versa. It implies that any ray from the vertex m of $\angle amb$ which meets one segment joining the sides of the angle, meets all such segments. This is a subtle and important property of angles which is intimately related to the notion of angle interior. (See Prenowitz and Jordan [1], p. 246, Theorem 7.)

To sum up: $ab \equiv a'b'$ (mod m) if and only if the family of the m-rays determined by ab is the same as that determined by $a'b'$.

The result suggests the following theorem.

Theorem 10.4. *Let M be a nonempty linear set and A and B sets of points. Then $A \equiv B$ (mod M) if and only if the family of halfspaces of M determined by the points of A is identical to the family of halfspaces of M determined by the points of B.*

PROOF. Suppose $A \equiv B$ (mod M). Then each point of A is congruent modulo M to some point of B—and so determines the same halfspace as some point of B. Similarly each point of B determines the same halfspace as some point of A. Thus the indicated families of halfspaces are identical. The converse is proved as easily. ☐

The geometrical significance of congruence of sets is easily given. Let M be a point m in JG5, the 3-dimensional Euclidean join geometry (Figure 10.4). For simplicity suppose $m \not\subset A, B$. Then $A \equiv B \pmod{m}$ if and only if each ray from m which passes through a point of A also passes through a point of B, and vice versa. That is, $A \equiv B \pmod{m}$ if and only if the cones with apex m and "bases" A, B (that is, the m-cones generated by A, B) are identical. (See Exercise 8 below.)

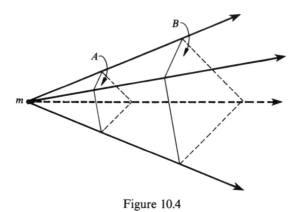

Figure 10.4

10.4 Properties of Modular Set Congruence

Theorem 10.5. *The relation of congruence modulo M for sets is an equivalence relation.*

PROOF. By the definition and Theorem 10.1, the corresponding result for points. ☐

Each point is congruent to the M-ray it determines.

Theorem 10.6. $a \equiv \overrightarrow{Ma} \pmod{M}$.

PROOF. a is congruent modulo M to every point of $\overrightarrow{Ma}$. ☐

Corollary 10.6.1. $aM \equiv a \pmod{M}$.

PROOF. $aM \subset \overrightarrow{Ma}$ (Corollary 8.16). ☐

Every point of the modulus M acts as an identity element of the operation join with respect to the relation of congruence.

Corollary 10.6.2. *If linear set $M \supset m$, then $am \equiv a \pmod{M}$.*

Corollary 10.6.3. $a/M \equiv a \pmod{M}$.

PROOF. $a/M \subset \overrightarrow{Ma}$ (Corollary 8.16).   □

Corollary 10.6.4. *If linear set $M \supset m$, then $a/m \equiv a \pmod{M}$.*

The following exercise set contains some deeper congruence properties which are not included as theorems because they are not used in the sequel.

EXERCISES (CONGRUENCE PROPERTIES)

1. If $a \equiv b \pmod{M}$ and $c \equiv d \pmod{M}$, does $a/c \equiv b/d \pmod{M}$ follow? Justify your answer.

2. If $a \equiv b \pmod{M}$, is $a/c \equiv b/c \pmod{M}$; $c/a \equiv c/b \pmod{M}$? Justify your assertions.

3. Let A and B be halfspaces of the nonempty linear set M and B' the halfspace of M opposite to B.
   (a) Prove: $A/B \equiv AB' \pmod{M}$.
   (b) Show that $A/B = AB'$ does not hold.

4. Let A and B be halfspaces of the nonempty linear set M.
   (a) Prove: $(M/A)(M/B) \equiv M/AB \pmod{M}$.
   (b) Show that $(M/A)(M/B) = M/AB$ does not hold.

*5. Prove:
   (a) $aM/b \equiv a(M/b) \pmod{M}$.
   (b) $ab/M \equiv a(b/M) \pmod{M}$.

6. Suppose $a \equiv b \pmod{M}$ and X is a halfspace of M.
   (a) Prove: $X/a = X/b$.
   (b) Is $a/X = b/X$?
   (c) Is $a/X \equiv b/X \pmod{M}$?

7. Prove: If $am \equiv a \pmod{M}$ for all a, then $m \subset M$.

8. Prove: $A \equiv B \pmod{M}$ is equivalent to each of the hypercone equalities:
   (a) $MA/M = MB/M$;
   (b) $M/A = M/B$.

9. Prove: If $A \equiv B \pmod{M}$ and $C \equiv D \pmod{M}$, then $AC \equiv BD \pmod{M}$.

10. Prove: If $A \equiv B \pmod{M}$, then $[A] \equiv [B] \pmod{M}$.

11. Prove: For any set A the family of all sets congruent to A modulo a nonempty linear set M contains a greatest member. Describe the greatest member.

12. Prove: For a nonempty convex set K, $aK \approx bK$ if and only if $a \equiv b \pmod{\langle K \rangle}$.

10.5 Families of Halfspaces

We shall find it necessary to study systematically families of halfspaces of a given linear set. The idea has appeared, of course, in Chapter 9 in formulating the definitions of o-cone and L-cone. Even if cones had not been introduced as a special object of study, there are situations in classical Euclidean geometry which involve the idea of a family of half-spaces. For example, the interior of $\angle aob$ is filled by the family of o-rays that are "between" $\overrightarrow{oa}$ and $\overrightarrow{ob}$, the sides of the angle (Figure 10.5). A halfplane similarly is filled by a family of o-rays.

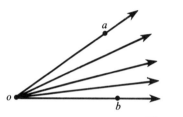 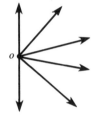

Figure 10.5

Moreover, in the study of spherical geometry a systematic procedure involving o-rays is employed to represent spherical figures by "linear" figures in 3-space. Suppose a sphere S has center o (Figure 10.6). Then each point x of S is represented by the ray $\overrightarrow{ox}$. To represent a figure T on the sphere S, take the family of all o-rays that represent the points of T. Thus, for example, an open minor arc $\overgroup{ab}$ of a great circle of S is represented by the family F of rays $\overrightarrow{ox}$ for all x in $\overgroup{ab}$. (Observe that F fills the interior of $\angle aob$.) Similarly the spherical triangle abc would be represented by the family of o-rays consisting of $\overrightarrow{oa}$, $\overrightarrow{ob}$, $\overrightarrow{oc}$ and all rays $\overrightarrow{ox}$ that are "between" any two of these rays.

Finally let us remark that this study of halfspaces not only will shed light on the theory of congruence relations but will yield a deeper and more abstract theory of cones.

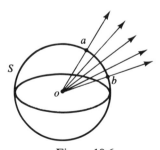

Figure 10.6

10.6 New Notations

First a notation for the family of all halfspaces of M or equivalently of all M-rays.

Definition. Let M be a nonempty linear set in join geometry J. Then the family of halfspaces of M, equivalently, the family of sets $\overrightarrow{Mx}$ for $x \subset J$ is denoted by $J : M$ (read J *modulo* M). Furthermore if A is any subset of J, $A : M$ denotes the family of halfspaces of M determined by the points of A, that is, the family of sets $\overrightarrow{Mx}$ for $x \subset A$.

The notation is applied in the following

Restatement of Theorem 10.4. Let M be a nonempty linear set and A and B sets of points. Then $A \equiv B \pmod{M}$ if and only if $A : M = B : M$.

Next an alternative to the arrow notation $\overrightarrow{Ma}$ which often will be more convenient in the sequel.

Definition. Let M be a nonempty linear set. Then $\overrightarrow{Ma}$, the halfspace of M determined by point a, is denoted by $|a|_M$, or if M is fixed throughout a discussion, simply by $|a|$. It is also convenient to employ a similar notation for families of halfspaces of M, so that if S is a subset of J we write $S : M$ as $|S|_M$ or simply as $|S|$.

Remark. It is desirable to compare the notation for a family of M-rays with that for an M-cone. Recall (Section 9.14) that the M-cone generated by a set of points S is denoted by S_M, whereas $|S|_M$ denotes the family of M-rays determined by points of S. Vertical bars serve to distinguish families of M-rays from M-cones. There is a close relation between S_M and $|S|_M$, as S_M is merely the union of the members of $|S|_M$. This relation is studied in Section 10.13 below. Observe though that when S reduces to a point a, then $|a|_M = a_M$ is the M-ray determined by a. In this case we retain the vertical bars to emphasize that $|a|_M$ is being considered as a member of $J : M$ and not merely as a subset of J.

10.7 A Join Operation in $J : M$

We introduce now a join operation in the family $J : M$ of halfspaces of M and so convert it into a join system, which turns out to have many similarities with join geometry J.

Definition. Suppose M is a nonempty linear set. Let $|a|$ and $|b|$ be members of the family of halfspaces $J : M$ (Figure 10.7). The *join* of $|a|$ and $|b|$, denoted $|a| \cdot |b|$, is defined to be the family of members $|x|$ of $J : M$ for which $x \subset ab$. Equivalently, $|a| \cdot |b|$ is the subfamily $|ab|$ of $J : M$.

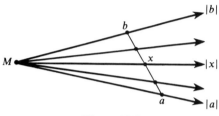

Figure 10.7

Critique of the Definition. The join of $|a|$ and $|b|$ is defined in terms of the specific points a and b which are chosen to determine the halfspaces $|a|$ and $|b|$. It is conceivable that if different determining points a' and b' were chosen, the resulting join might be different. We show this is not so. Suppose

$$|a| = |a'|, \qquad |b| = |b'|. \tag{1}$$

Then by definition $|a| \cdot |b| = |ab|$ and $|a'| \cdot |b'| = |a'b'|$. We prove $|ab| = |a'b'|$. By definition (1) implies

$$a \equiv a' \ (\text{mod } M), \qquad b \equiv b' \ (\text{mod } M),$$

and Theorem 10.3 yields

$$ab \equiv a'b' \ (\text{mod } M). \tag{2}$$

The congruence (2) implies, by Theorem 10.4, $ab : M = a'b' : M$ or, in the alternative notation introduced above, $|ab| = |a'b'|$.

Thus we may assert the following theorem.

Theorem 10.7. *The join of any two members of $J : M$ is independent of the choice of the points which determine them, and so is a uniquely determined subfamily of $J : M$.*

10.8 Factor Geometries

Definition. Let M be a nonempty linear set in a join geometry J. Then the family $J : M$ of halfspaces of M with join as defined above, that is, the pair $(J : M, \cdot)$, is called the *factor geometry* of J *with respect to* M and is said to arise by *reducing J modulo M.* In brief, $J : M$ will be called a *factor geometry.*

We proceed to study factor geometries $J : M$ in the same manner as join geometries J. In order to facilitate comparison we will refer to the halfspaces of M, that is, the members of the family $J : M$, as *elements* of $J : M$. $J : M$ will be considered the basic *set*, and subfamilies of $J : M$ will be called *subsets* of $J : M$ or simply *sets*. Observe that $\cdot$ is a join operation in $J : M$.

In studying $J : M$ we adopt the same notational conventions concerning the use of $\subset$, $\supset$, $\approx$ as in J (Sections 2.3 and 4.5). For example, in $J : M$, $|a| \subset |b| \cdot |c|$ or $|a| \approx |b| \cdot |c|$ means that $|a|$ is an element (M-ray) of the set (of M-rays) $|b| \cdot |c|$. Extension is defined in $J : M$ as in J.

Definition. If $|a|$ and $|b|$ are elements of the factor geometry $J : M$, then $|a|/|b|$, the *extension* of $|a|$ *from* $|b|$, is the set of elements $|x|$ of $J : M$ satisfying $|a| \subset |b| \cdot |x|$ (Figure 10.8).

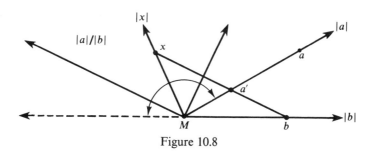

Figure 10.8

Since $|a| \cdot |b|$ is the set of elements $|x|$ of $J : M$ for which $x \subset ab$, one might conjecture that $|a|/|b|$ is the set of elements $|x|$ of $J : M$ for which $x \subset a/b$. This is not the case.

It turns out that

$$|a|/|b| \text{ is the set of elements } |x| \text{ of } J : M$$
$$\text{for which there exists } a' \text{ in } |a| \text{ with } x \subset a'/b. \tag{1}$$

Let us justify (1). By definition $|x| \subset |a|/|b|$ is equivalent to $|a| \subset |b| \cdot |x| = |bx|$. But $|a| \subset |bx|$ if and only if $a' \subset bx$ for some a' in $|a|$. Certainly the latter condition is equivalent to $x \subset a'/b$ for some a' in $|a|$. Thus (1) holds.

A final word. Note in the key step: $|a| \subset |bx|$ does not imply $a \subset bx$ (Figure 10.8).

The operations join $\cdot$ and extension $/$ are extended to subsets of $J : M$.

Definition. If $|A|$ and $|B|$ are subsets of the factor geometry $J : M$, then $|A| \cdot |B|$, the *join* of $|A|$ and $|B|$, is the set of all elements $|x|$ of $J : M$ satisfying $|x| \subset |a| \cdot |b|$ for some $|a|$ in $|A|$ and some $|b|$ in $|B|$; $|A|/|B|$, the *extension* of $|A|$ *from* $|B|$, is the set of all elements $|x|$ of $J : M$ satisfying $|x| \subset |a|/|b|$ for some $|a|$ in $|A|$ and some $|b|$ in $|B|$.

In view of these definitions, certain results on join geometries of Chapters 2, 3 and 4 that depend only on definitions automatically hold for $J : M$. These include corollaries to the monotonic laws for join and extension (Corollaries 2.1 and 4.1.1) and the Three Term Transposition Law (Theorem 4.4). Specifically, we have

$$|A| \subset |B| \text{ and } |C| \subset |D|$$
$$\text{imply } |A| \cdot |C| \subset |B| \cdot |D| \text{ and } |A|/|C| \subset |B|/|D|,$$

and

$$|A| \approx |B| \cdot |C| \quad \text{if and only if} \quad |A|/|B| \approx |C|.$$

We conclude with a theorem that relates join of sets in $J : M$ to join of sets in J.

Theorem 10.8. *Let* $|A|$ *and* $|B|$ *be sets in* $J : M$. *Then*

$$|A| \cdot |B| = |AB|. \tag{1}$$

PROOF. Note in (1) that $|A| \cdot |B|$ is a join in $J : M$, while AB is a join in J. Let $|x| \subset |A| \cdot |B|$. By definition of join of sets,

$$|x| \subset |a| \cdot |b| \tag{2}$$

where $|a|, |b|$ satisfy $|a| \subset |A|$, $|b| \subset |B|$. Consider the relation $|a| \subset |A|$. This asserts that $|a|$ is a member of $|A|$, the set of halfspaces (of M) determined by the points of A. Thus $|a|$ is determined by a point of A, that is, $|a| = |a'|$, where $a' \subset A$. Similarly $|b| = |b'|$, where $b' \subset B$. Then (2) implies

$$|x| \subset |a'| \cdot |b'| = |a'b'| \subset |AB|.$$

Conversely, suppose $|x| \subset |AB|$. Then $|x| = |x'|$, where $x' \subset AB$. Hence $x' \subset ab$, where $a \subset A$, $b \subset B$. Consequently

$$|x| = |x'| \subset |ab| = |a| \cdot |b| \subset |A| \cdot |B|,$$

and (1) is verified.   □

10.9 $J : M$ Satisfies J1–J6

Now we are prepared to substantiate the basic similarities between geometries $J : M$ and J.

Theorem 10.9. *A factor geometry* $J : M$ *satisfies* J1–J6.

PROOF. The postulates are verified in order.

   J1. By definition, $|a| \cdot |b| = |ab|$. By J1, (for geometry J) $ab \neq \varnothing$. Hence $|a| \cdot |b| \neq \varnothing$.

J2. $|a| \cdot |b| = |ab| = |ba| = |b| \cdot |a|$.

J3. Using Theorem 10.8 we have

$$(|a| \cdot |b|) \cdot |c| = |ab| \cdot |c| = |abc| = |a| \cdot |bc| = |a| \cdot (|b| \cdot |c|).$$

J4. $|a| \cdot |a| = |aa| = |a|$.

J5. Observe that if in J, $x \subset a/b$, then in $J : M$, $|x| \subset |a|/|b|$. Hence $|a|/|b| \neq \varnothing$.

J6. Suppose $|a|/|b| \approx |c|/|d|$. Then for some $|x|$

$$|a|/|b| \approx |x|, \qquad |c|/|d| \approx |x|.$$

Hence

$$|a| \approx |b| \cdot |x| = |bx|.$$

Thus $|a| = |a'|$, where a' satisfies

$$a' \subset bx. \tag{1}$$

Similarly $|c| = |c'|$, where c' satisfies

$$c' \subset dx. \tag{2}$$

The relations (1) and (2) in J imply $a'/b \approx c'/d$ and so $a'd \approx bc'$. This yields, in $J : M$, $|a'| \cdot |d| \approx |b| \cdot |c'|$ or $|a| \cdot |d| \approx |b| \cdot |c|$. Thus J6 is verified and the theorem proved. □

Remark. As you may have guessed, J7 cannot be proved for factor geometries (Corollary 10.16 below, see also Exercise 2 at the end of Section 10.10 below).

10.10 Geometric Interpretation of Factor Geometries

Let J be the Euclidean join geometry JG5 and M a point m in J. Then the family $J : M$ is the set of all rays from point m in Euclidean 3-space including m as an improper ray. We have converted the family $J : M$ into a factor geometry by introducing the operation of join in $J : M$. What is the specific geometrical significance of the join of two rays?

Let $|a|$ and $|b|$ be elements of $J : M$. Four cases may be considered (Figure 10.9):

Case I. $|a|$ and $|b|$ are in general position, that is, $|a|$ and $|b|$ are proper rays which are distinct and not opposite.

Case II. $|a|$ and $|b|$ are distinct and opposite.

Case III. One of $|a|$ or $|b|$ is m.

Case IV. $|a| = |b|$.

In Case I, $|a|$ and $|b|$ are the rays $\overrightarrow{ma}$ and $\overrightarrow{mb}$, which form $\angle\, amb$. The join $|a| \cdot |b|$ is the set of rays from m that meet ab—these are the rays that lie between $|a|$ and $|b|$ and cover the interior of $\angle\, amb$.

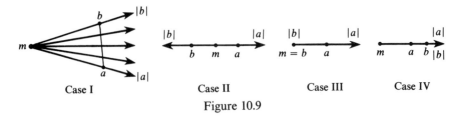

Case I   Case II   Case III   Case IV

Figure 10.9

In Case II,

$$|a| \cdot |b| = |ab| = |am \cup bm \cup m| = \{|a|, |b|, m\}.$$

Thus the join of two opposite rays consists of the two rays and m.

In Case III suppose, for example, $|b| = m = b$. Then

$$|a| \cdot |b| = |a| \cdot |m| = |am| = |a|.$$

Finally, in Case IV we have $|a| \cdot |a| = |a|$.

In summary: $J : M$, in the given instance, is the natural or intrinsic geometry of the family of rays from m in which the notion of join (or of betweenness of rays) is that naturally induced by the join operation in J. Briefly, $J : M$ is a Euclidean *ray geometry*.

The ray geometry $J : M$ can be represented graphically (Figure 10.10) by taking a sphere S centered at m, and projecting each ray from m into the point in which it pierces S. Then if a and b are points of S in general position (that is, a and b are distinct and not opposite to each other), the rays of the join $|a| \cdot |b|$ will pierce S in the points of $\widehat{ab}$, the (open) minor arc of a great circle joining a and b. Thus the ray geometry $J : M$ is representable essentially as the natural geometry of a sphere in which the join of two points (in general position) is the minor arc of a great circle joining them.

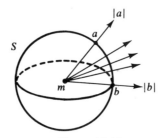

Figure 10.10

To characterize more precisely this "spherical" representation of $J : M$, observe first that there is no point of S that represents the improper ray m of $J : M$. To remedy this we adjoin point m, as a sort of ideal point, to the points of S forming a set S'. Then we define the join $a \cdot b$ in S' as follows:

(1) if a and b are points of S in general position $a \cdot b$ is the (open) minor arc of a great circle joining a and b;

(2) if a and b are opposite points of S, $a \cdot b = \{a, b, m\}$;

(3) $a \cdot a = a$;

(4) $a \cdot m = m \cdot a = a$.

This converts S' into what may be termed a "spherical join geometry" which is a representation of (or is isomorphic to) $J : M$. (For a join theoretic treatment of spherical geometries, see Prenowitz [2].)

EXERCISES

1. Find the significance of a/b in S', the spherical representation of factor geometry $J : M$ given above, when
   (a) a and b are points of S in general position;
   (b) a and b are opposite points of S;
   (c) $a = b$;
   (d) one of a or b is m.

2. Prove: If $|a|$ is an element of $J : M$, then $|a|/|a| \supset |a|$ and $|a|/|a| \supset M$. What can you conclude about the validity of J7 in $J : M$?

3. Prove: If M is a nonempty linear set, then $|A|_M = |B|_M$ if and only if $M/(M/A) = M/(M/B)$. (Compare Exercise 8 at the end of Section 10.4.)

4. Prove in $J : M$:
   (a) $|a|/|b| \supset |a/b|$;
   (b) $|a|/|b| = |aM/b|$;
   (c) $|A|/|B| \supset |A/B|$;
   (d) $|A|/|B| = |AM/B|$.

5. Prove in $J : M$:
   (a) $|A| \subset |B|$ and $|C| \subset |D|$ imply $|A| \cdot |C| \subset |B| \cdot |D|$ and $|A|/|C| \subset |B|/|D|$;
   (b) $|A| \cdot |B| = |B| \cdot |A|$;
   (c) $|A| \cdot (|B| \cdot |C|) = (|A| \cdot |B|) \cdot |C|$;
   (d) $|A| \approx |B| \cdot |C|$ if and only if $|A|/|B| \approx |C|$;
   (e) $|A|/|B| \approx |C|/|D|$ implies $|A| \cdot |D| \approx |B| \cdot |C|$.

6. Suppose J is JG3, the join geometry of the Euclidean line, and M is a point m.
   (a) Show that $J : M$ consists of precisely 3 elements.
   (b) Make a table for the join operation $\cdot$ applied to the elements of $J : M$.
   (c) The same as (b) for the extension operation $/$.
   (d) Is $J : M$ a join geometry? (Hint: Theorem 4.29.)

7. Suppose J is JG16 and M is a point m.
   (a) Show that $J : M$ consists of precisely 9 elements.
   (b) The same as 6(b).
   (c) The same as 6(c).
   (d) The same as 6(d).

8. Suppose J is JG5, Euclidean 3-space, and M is a line. Give a geometric interpretation of $J : M$ similar to that of $J : M$ in the text of Section 10.10.

10.11 Continuation of the Theory of Factor Geometries

Since J1–J4 hold in a factor geometry $J : M$ (Theorem 10.9), all the consequences of J1–J4 developed in Chapters 2 and 3 hold for $J : M$. (It is assumed, of course, that the definitions in Chapters 2 and 3 are adopted without change.) It follows that in $J : M$ the join operation can be extended to n terms and satisfies the associative and commutative laws of Section 2.7.

The principle $|A| \cdot |B| = |AB|$ (Theorem 10.8) can now be extended to n terms.

Theorem 10.10. *In* $J : M$, $|A_1| \cdot \ldots \cdot |A_n| = |A_1 \ldots A_n|$.

PROOF. By induction, using Theorem 10.8. $\qquad\square$

Corollary 10.10. *In* $J : M$, $|a_1| \cdot \ldots \cdot |a_n| = |a_1 \ldots a_n|$.

Up till now we have consistently used the notation $|a|$ and $|A|$ to represent elements and sets, respectively, of a factor geometry $J : M$. We will now use capital letters to represent both elements and sets of $J : M$. However, we will be careful to point out whether a specific capital letter represents an element of $J : M$ or a subset of $J : M$.

Let A be an element of $J : M$. Then A, being an M-ray, is a subset of J. Thus $|A|_M = |A|$ is significant and is, of course, the set of M-rays determined by points of A. But since each $a \subset A$ determines M-ray A itself, we have $|A|$ is the set consisting of the single M-ray A and we write $|A| = A$. We now have an immediate consequence of Theorem 10.8.

Theorem 10.11. *Let* A *and* B *be elements of* $J : M$. *Then* $A \cdot B = |AB|$.

PROOF. $A \cdot B = |A| \cdot |B| = |AB|$. $\qquad\square$

The theorem gives an alternative to the definition for finding the join of two M-rays—to join M-rays A and B in $J : M$, find their join in J and then form the set of all M-rays determined by points of this join.

Corollary 10.11. *If* $A_1, \ldots, A_n$ *are elements of* $J : M$, *then* $A_1 \cdot \ldots \cdot A_n = |A_1 \ldots A_n|$.

10.12 Differences between $J : M$ and J

We have been emphasizing properties a factor geometry $J : M$ shares with the join geometry J. There are some striking differences. First, $J : M$ has an identity element.

Definition. Let I be an element of a factor geometry $J : M$ such that $A \cdot I = A$ for each element A of $J : M$. Then I is an *identity element* of $J : M$.

Theorem 10.12. $J : M$ *has a unique identity element, namely,* M.

PROOF. Let A be an element of $J : M$. Certainly M is an element of $J : M$, and

$$
\begin{aligned}
A \cdot M &= |AM| \qquad \text{(Theorem 10.11)} \\
&= |A| \qquad \text{(Theorem 8.16)} \\
&= A.
\end{aligned}
$$

Thus M is an identity element of $J : M$.

Uniqueness is easily proved. For if I and I' are identities of $J : M$ then on the one hand $I' \cdot I = I'$. But on the other hand $I' \cdot I = I \cdot I' = I$. Thus $I = I'$. □

Since $J : M$ has an identity element, our experience with number systems leads us to wonder whether elements have inverses.

Definition. Let A and X be elements of a factor geometry $J : M$ such that $A \cdot X \supset M$, that is, M is an element of $A \cdot X$. Then X is an *inverse* of A, or A and X are *inverse* elements.

Remark. The requirement $A \cdot X \supset M$ in the definition is reasonable in the present context. Since the join $A \cdot X$ is a set, the condition $A \cdot X = M$ would be unduly restrictive.

Theorem 10.13. A *and* X *are inverse elements in* $J : M$ *if and only if in* J, A *and* X *are opposite halfspaces of* M.

PROOF. For elements A and X of $J : M$, Theorem 10.11 implies $A \cdot X \supset M$ is equivalent to $|AX| \supset M$. But the condition $|AX| \supset M$ holds in $J : M$ if and only if the condition $AX \approx M$ holds in J. By employing definitions, we obtain the result. □

Corollary 10.13.1. *In* $J : M$ *each element* A *has a unique inverse, which may be expressed as* M/A.

PROOF. A has a unique inverse, since it has a unique opposite (Theorem 8.20). Hence the relation $A \cdot X \supset M$ has a unique solution X, which by the definition of extension is M/A. □

Remark. If in the corollary one replaces the extension operation in $J : M$ by that in J, the corollary is still valid, since by Theorem 8.20, M/A is the halfspace opposite to A.

Definition. If A is any element of a factor geometry $J : M$, then A^{-1} denotes the inverse of A. Similarly, if S is any set of elements of $J : M$, S^{-1} denotes the set of inverses of the elements of S.

Corollary 10.13.2. $(A^{-1})^{-1} = A$ *for any element A of $J : M$.*

It seems rather remarkable that in $J : M$ division (extension) can be reduced to multiplication (join) as in elementary algebra:

Theorem 10.14. *Let A and B be elements of $J : M$. Then $A/B = A \cdot B^{-1}$.*

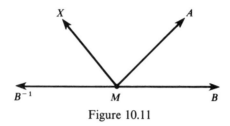

Figure 10.11

PROOF. There is now sufficient information available to formulate a proof by reasoning wholly in $J : M$. Let X denote an element of $J : M$ (Figure 10.11). Suppose

$$X \subset A/B. \tag{1}$$

By Corollary 10.13.1,

$$B^{-1} = M/B. \tag{2}$$

We eliminate B between (1) and (2). We have

$$A/X \approx B \approx M/B^{-1}.$$

Hence by J6 and the identity property of M (Theorem 10.12)

$$A \cdot B^{-1} \approx X \cdot M = X,$$

that is, $X \subset A \cdot B^{-1}$. Thus $A/B \subset A \cdot B^{-1}$.
  Conversely suppose

$$X \subset A \cdot B^{-1}. \tag{3}$$

We eliminate B^{-1} between (2) and (3), getting $M/B \approx X/A$, and therefore $M \cdot A \approx B \cdot X$. Hence $A \approx B \cdot X$ and $X \subset A/B$. Thus $A \cdot B^{-1} \subset A/B$, and the theorem holds.   □

Corollary 10.14. *Let S and T be subsets of $J : M$. Then $S/T = S \cdot T^{-1}$.*

We conclude this section by showing that the familiar algebraic principle for the inverse (or reciprocal) of a product is valid in $J : M$.

Theorem 10.15. *Let A and B be elements of $J : M$. Then*

$$(A \cdot B)^{-1} = A^{-1} \cdot B^{-1}.$$

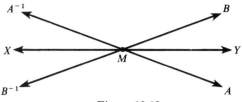

Figure 10.12

PROOF. Let X, Y denote inverse elements of $J : M$ (Figure 10.12). Suppose

$$X \subset (A \cdot B)^{-1}. \tag{1}$$

By definition $Y \subset A \cdot B$. Then $X \cdot Y \supset M$ implies

$$X \cdot A \cdot B \supset M. \tag{2}$$

We obtain from (2) by applying Corollary 10.13.1 and Theorem 10.14:

$$X \cdot A \approx M/B = B^{-1},$$
$$X \approx B^{-1}/A = B^{-1} \cdot A^{-1} = A^{-1} \cdot B^{-1}.$$

Thus (1) implies

$$X \subset A^{-1} \cdot B^{-1}. \tag{3}$$

Conversely, that (3) implies (1) is shown by essentially retracing steps, and so the theorem follows. ☐

Corollary 10.15. *Let S and T be subsets of $J : M$. Then*

$$(S \cdot T)^{-1} = S^{-1} \cdot T^{-1}.$$

Theorem 10.16. *Let A be an element of $J : M$. Then $A/A \supset \{A, A^{-1}, M\}$.*

PROOF. By Theorem 10.14, $A/A = A \cdot A^{-1}$. But

$$A \cdot A^{-1} \supset M.$$

Hence $A \cdot A \cdot A^{-1} \supset A \cdot M = A$, so that $A \cdot A^{-1} \supset A$. Similarly $A \cdot A^{-1} \supset A^{-1}$, and the theorem follows. ☐

Corollary 10.16. *J7 fails in $J : M$ if $M \neq J$.*

EXERCISES

Capital letters denote elements of a factor geometry $J : M$.

1. Verify Theorems 10.14 and 10.15 in S', the spherical representation of $J : M$ of Section 10.10. (See Exercise 1 at the end of Section 10.10.)

2. The same as Exercise 1 for $J : M$ of
   (a) Exercise 6,
   (b) Exercise 7,
   at the end of Section 10.10.

3. Prove: Join geometry J contains an identity if and only if J consists of a single point.

4. Prove in $J : M$: $A/A = A$ if and only if $A = M$.

5. Prove in $J : M$: $A/B = |A/B|$.

6. Prove in $J : M$:
   (a) $A/B \cdot C = (A/B)/C$;
   (b) $A \cdot (B/C) = A \cdot B/C$;
   (c) $A/(B/C) = A \cdot C/B$;
   (d) $(A/B) \cdot (C/D) = A \cdot C/B \cdot D$;
   (e) $(A/B)/(C/D) = A \cdot D/B \cdot C$.
   (The same conventions as regards omission of parentheses that are used in J are used in $J : M$.)

7. Prove in $J : M$: $A_1 \cdot \ldots \cdot A_n \supset M$ if and only if
$$A_1 \cdot \ldots \cdot A_{i-1} \cdot A_{i+1} \cdot \ldots \cdot A_n \supset A_i^{-1}$$
   for $1 \leqslant i \leqslant n$. Interpret geometrically.

8. Prove that the converse of J6 is valid in a factor geometry $J : M$.

9. (a) Let J be the Euclidean join geometry JG5. Verify in $J : M$ that $A/A = \{A, A^{-1}, M\}$ when M is a point; a line; a plane.
   (b) Find a factor geometry $J : M$ where $A/A \neq \{A, A^{-1}, M\}$.

10. Prove: If a join system satisfies J1–J6 and has an identity element, then it has a unique identity element, and moreover every element of the system has a unique inverse.

*11. Prove: If every line in join geometry J is ordered (Section 4.24) and M is a nonempty linear set in J, then in the factor geometry $J : M$ it follows that $A/A = \{A, A^{-1}, M\}$.

10.13 A Correspondence Relating $J : M$ and J

Any subset of a factor geometry $J : M$ is a family of M-rays or halfspaces of M, and its union by definition is an M-cone in join geometry J. This relation between $J : M$ and J is formalized in the following definition.

Definition. Let S be a subset of a factor geometry $J : M$. Then the union of the members of S is denoted by S^* (read S star).

The symbol S^* merely denotes the M-cone determined by S, and is introduced because the correspondence $S \to S^*$ yields interrelations between $J : M$ and J. As examples, observe first that if $S = \{A\}$, where A is an element of $J : M$, then $S^* = A$; and secondly that $(J : M)^* = J$.

Elementary properties of the * operation are now formulated and proved.

Theorem 10.17. *Let S_1 and S_2 be subsets of a factor geometry $J : M$. Then $S_1 \subset S_2$ if and only if $S_1^* \subset S_2^*$.*

PROOF. Suppose $S_1 \subset S_2$. Then $S_1^* \subset S_2^*$ is the set theoretic fact that the union of the members of S_1 is contained in the union of the members of S_2. Conversely, suppose $S_1^* \subset S_2^*$. Let A be a member of S_1. Then $A \subset S_1^*$, so that $A \subset S_2^*$. Let $p \subset A$. Then p is a point of the M-cone S_2^* and so belongs to some M-ray, say B, of S_2. Hence $A \approx B$, and $A = B$ follows. Thus A is a member of S_2, and we infer $S_1 \subset S_2$. $\qquad\square$

Theorem 10.18. *The mapping $(I) : S \to S^*$ effects a one to one correspondence between the family F of subsets of a factor geometry $J : M$ and the family G of M-cones of the join geometry J.*

PROOF. Let S be a member of F, so that S is a set of M-rays. Then S^* is clearly an M-cone of J, that is, a member of G. Thus (I) associates with each member of F a specific member of G.

Now let S' be a member of G. We show that S' is the correspondent [as determined by (I)] of some member of F. S' is an M-cone and by definition the union of some set S of M-rays. Thus S is a subset of $J : M$ and so a member of F. By definition $S^* = S'$, and S' is the correspondent of S [as determined by (I)]. Thus (I) effects a correspondence *between* the families F and G.

Finally we show the correspondence (I) is one to one. Suppose S_1 and S_2 are distinct members of F. Their correspondents in G are S_1^* and S_2^*. Assume $S_1^* = S_2^*$. By the previous theorem $S_1 = S_2$, contrary to supposition. Thus our assumption is false, and S_1^* and S_2^* are distinct. To summarize: distinct members S_1, S_2 of F have distinct correspondents S_1^*, S_2^* in G. Hence the correspondence effected by (I) is one to one, and the proof is finished. $\qquad\square$

A lemma is now introduced which relates the M-cone generated by a set of points S and the family of M-rays determined by the points of S.

Lemma 10.19. *Let S be a set of points of join geometry J and M a nonempty linear space. Then*

$$(|S|_M)^* = S_M.$$

PROOF. By definition, $S_M = MS/M$, and so is the union of all M-rays of the form Ms/M where $s \subset S$. But $(|S|_M)^*$ is precisely this union. Hence the lemma is proven. $\qquad\square$

The lemma asserts: Let S be a set of points, take the M-rays determined by the points of S and form their union. The result is the M-cone generated by S.

Theorem 10.19. *Let S be a subset of $J : M$. Then $|S^*|_M = S$.*

PROOF. Note that S^* is a subset of J; $|S^*|_M$ is significant and a subset of $J : M$. We apply the $*$ operation to $|S^*|_M$. By the lemma $(|S^*|_M)^* = (S^*)_M$. But since S^* is an M-cone, $(S^*)_M = S^*$, so that $(|S^*|_M)^* = S^*$. Since $|S^*|_M$ and S have the same correspondent under the star correspondence, the previous theorem implies $|S^*|_M = S$.   □

The theorem is not very deep. It simply says that the family of M-rays determined by points in the union of a given family of M-rays is precisely that given family.

Corollary 10.19. *Let S and T be subsets of $J : M$. Then $S \cdot T = |S^* T^*|_M$.*

PROOF. By the theorem, $S = |S^*|_M = |S^*|$ and $T = |T^*|$. By Theorem 10.8, $S \cdot T = |S^*| \cdot |T^*| = |S^* T^*|$.   □

The corollary gives a means for constructing the join of two sets in $J : M$—take the M-cones determined by the given sets, find their join in J, and form the set of all M-rays determined by points of this join.

Theorem 10.20. *Let A and B be elements of $J : M$. Then*

$$(A \cdot B)^* = (AB)_M. \tag{1}$$

PROOF. Note in (1) that in the left member $A \cdot B$ denotes a join of elements in $J : M$, while in the right member AB denotes a join of sets in J. By Theorem 10.11 and Lemma 10.19,

$$(A \cdot B)^* = (|AB|_M)^* = (AB)_M.   □$$

The theorem may be described so: the star operation converts the join of two elements of $J : M$ into the M-cone generated by their join as subsets of J.

Corollary 10.20. *Let $A_1, \ldots, A_n$ be elements of $J : M$. Then*

$$(A_1 \cdot \ldots \cdot A_n)^* = (A_1 \ldots A_n)_M.$$

PROOF. Apply the method of the theorem using Corollary 10.11 instead of Theorem 10.11.   □

10.14 Convexity in a Factor Geometry

As we have observed above (Section 10.11, first paragraph) the theory developed in Chapters 2 and 3 for join geometries J is valid for factor geometries $J : M$. Specifically, the results on convex sets, their interiors and closures and convex hulls in those chapters are valid in $J : M$. We derive now some relations between convexity in $J : M$ and convexity in J. But first we find it convenient to state the definition of convexity in $J : M$ and to set up notation for convex hulls.

Definition. Let S be a subset of a factor geometry $J : M$. Then S is *convex* (in $J : M$) if for A and B elements of S, $A \cdot B \subset S$. The *convex hull* of S (in $J : M$), denoted by $[S]$, is the least convex set in $J : M$ that contains S.

Note that, as in J, the convex hull of a set S in $J : M$ always exists and is the intersection of all convex sets containing S.

Theorem 10.21. *Let S be a subset of $J : M$. Then S is convex (in $J : M$) if and only if S^* is convex (in J).*

PROOF. Suppose S is convex. Let $a, b \subset S^*$. Then $|a|, |b| \subset S$. By the convexity of S, $|a| \cdot |b| \subset S$. Thus by Lemma 10.19 and Theorem 10.17,

$$ab \subset (ab)_M = (|ab|_M)^* = (|a| \cdot |b|)^* \subset S^*,$$

and S^* is convex by definition.

Conversely, suppose S^* convex. Let A and B be elements of S. Then $A, B \subset S^*$, and the convexity of S^* yields $AB \subset S^*$. Hence by Theorems 10.11 and 10.19,

$$A \cdot B = |AB|_M \subset |S^*|_M = S,$$

and S is convex by definition. $\qquad\square$

Corollary 10.21. $S \to S^*$ *effects a one to one correspondence between the family of convex subsets of a factor geometry $J : M$ and the family of convex M-cones of the join geometry J.*

PROOF. Apply Theorem 10.18. $\qquad\square$

Next a relation between convex hulls in $J : M$ and convex M-conar hulls in J (Section 9.17).

Theorem 10.22. *Let S be a subset of $J : M$. Then*

$$[S]^* = [S^*]_M .$$

PROOF. First we show

$$[S]^* \supset [S^*]_M . \tag{1}$$

$[S]$ is a convex set in $J : M$. Hence $[S]^*$ is a convex set in J (Theorem 10.21). Certainly $[S]^*$ is an M-cone. Since $[S] \supset S$, we have

$$[S]^* \supset S^* \qquad \text{(Theorem 10.17)}.$$

Thus by definition $[S]^*$ must contain the convex M-conar hull of S^*, that is, (1) holds.

To prove the reverse inclusion, observe by Theorem 10.18 that $[S^*]_M$ is the correspondent under the star correspondence of some subset T of $J : M$, that is,

$$T^* = [S^*]_M . \tag{2}$$

Note that $T^* \supset S^*$. Hence $T \supset S$ (Theorem 10.17). Since T^* is convex in J, T must be convex in $J : M$ (Theorem 10.21). Thus $T \supset [S]$ and $T^* \supset [S]^*$ (Theorem 10.17). By (2)

$$[S^*]_M \supset [S]^*,$$

and the theorem holds.   □

Thus the star operation converts the convex hull of a subset S of $J : M$ into the convex M-conar hull of S^*, the M-cone determined by S, or in the terminology of Section 9.17, into the convex M-cone generated by the family S of M-rays.

10.15 Linearity in a Factor Geometry

Linear set and linear hull are defined in a factor geometry as in a join geometry.

Definition. Let S be a subset of a factor geometry $J : M$. Then S is *linear* (in $J : M$) if for A and B elements of S, $A \cdot B \subset S$ and $A/B \subset S$. The *linear hull* of S (in $J : M$), denoted by $\langle S \rangle$, is the least linear set in $J : M$ that contains S.

Note that as in J, the linear hull of a set S in $J : M$ always exists and is the intersection of all linear sets containing S. Also the notion of linear hull extends, as in J, to n subsets of $J : M$. Thus if $S_1, \ldots, S_n$ are subsets of $J : M$, we use $\langle S_1, \ldots, S_n \rangle$ to denote the least linear set in $J : M$ containing $S_1, \ldots, S_n$, and call it the *linear hull* of $S_1, \ldots, S_n$.

It is interesting that the basic results on linear sets in J, Theorems 6.1–6.6 and 6.8–6.14 and corollaries, can be proved without appealing to Postulate J7 and so are valid in $J : M$ by Theorem 10.9. However, our treatment of subsequent material will not depend on the validity of these results for $J : M$. We now proceed to derive properties of linear sets in $J : M$.

Theorem 10.23. *In $J : M$, a set S is linear if and only if S is convex and $S = S^{-1}$.*

(Compare Theorem 6.1.)

PROOF. Suppose S is linear. By definition S is convex and closed under extension. If $S = \varnothing$ then $S = S^{-1}$. Suppose $S \neq \varnothing$; let A be an element of S. Then

$$S \supset A/A \supset M \qquad \text{(Theorem 10.16)}$$

and $S \supset M/S = S^{-1}$. But this implies $S^{-1} \supset (S^{-1})^{-1} = S$, so that $S = S^{-1}$.

Conversely, let S be convex and $S = S^{-1}$. By convexity of S and Corollary 10.14,

$$S \supset S \cdot S^{-1} = S/S.$$

Thus S is closed under extension and linear by definition. $\square$

Corollary 10.23.1. *In $J : M$ any nonempty linear set contains M.*

Corollary 10.23.2. *In $J : M$, let S and T be linear. Then*

(a) *$S \cdot T$ is linear;*
(b) *$\langle S, T \rangle = S \cdot T$ provided $S, T \neq \varnothing$;*
(c) *$S \cdot T = S/T$.*

(Compare Theorems 6.5, 6.14.)

PROOF. (a) $S \cdot T$ is convex, and $(S \cdot T)^{-1} = S^{-1} \cdot T^{-1} = S \cdot T$.
(b) $S \cdot T \supset S \cdot M = S$; similarly $S \cdot T \supset T$. The result follows using (a).
(c) $S \cdot T = S \cdot T^{-1} = S/T$. $\square$

Corollary 10.23.3. *Let S and T be nonempty subsets of $J : M$. Then $\langle S, T \rangle = \langle S \rangle \cdot \langle T \rangle$.*

PROOF. Using Corollary 10.23.2(b), we have

$$\langle S, T \rangle = \langle \langle S \rangle, \langle T \rangle \rangle = \langle S \rangle \cdot \langle T \rangle. \qquad \square$$

Theorem 10.24. *Let A be an element of $J : M$. Then A/A is linear; in fact, $\langle A \rangle = A/A$.*

PROOF. First we prove A/A is linear. We have

$$(A/A) \cdot (A/A) = A \cdot A^{-1} \cdot A \cdot A^{-1} = A \cdot A \cdot (A \cdot A)^{-1}$$
$$= A \cdot A^{-1} = A/A,$$

and so A/A is convex. Moreover

$$(A/A)^{-1} = (A \cdot A^{-1})^{-1} = A^{-1} \cdot A = A/A.$$

Thus A/A is linear (Theorem 10.23). By Theorem 10.16, $A/A \supset A$, and $\langle A \rangle = A/A$ follows. $\square$

Corollary 10.24. *Let $A_1, \ldots, A_n$ be elements of $J : M$. Then*

$$\langle A_1, \ldots, A_n \rangle = (A_1/A_1) \cdot \ldots \cdot (A_n/A_n).$$

(Compare Theorem 6.9.)

PROOF. By induction, using Corollary 10.23.3, we can prove

$$\langle A_1, \ldots, A_n \rangle = \langle A_1 \rangle \cdot \ldots \cdot \langle A_n \rangle.$$

Application of the theorem then yields the result.   □

EXERCISES

1. Prove: If S and T are subsets of $J : M$, then
   (a) $(S \cup T)^* = S^* \cup T^*$;
   (b) $(S \cap T)^* = S^* \cap T^*$;
   (c) $(S - T)^* = S^* - T^*$.

2. Prove: If A and B are elements of $J : M$, then
   (a) $(A \cdot B)^* = AB/M$;
   (b) $(A/B)^* = A/B$;
   (c) $(A/B)^{-1} = B/A$.

3. Prove: If S and T are subsets of $J : M$, then
   (a) $(S \cdot T)^* = S^* T^* / M$;
   (b) $(S/T)^* = S^*/T^*$;
   (c) $S/T = |S^*/T^*|_M$.

4. Prove: If $A_1, \ldots, A_n$ are elements of $J : M$ such that $A_1 \cdot \ldots \cdot A_n \supset M$, then $\langle A_1, \ldots, A_n \rangle = [A_1, \ldots, A_n] = A_1 \cdot \ldots \cdot A_n$. Interpret the result geometrically in JG5.

5. Let S be a subset of a factor geometry $J : M$. Prove S is linear (in $J : M$) if and only if S^* is linear (in J).

6. Prove: If S is a subset of $J : M$, then $\langle S \rangle^* = \langle S^* \rangle$.

7. Let S be a convex set in $J : M$.
   (a) Give in $J : M$ definitions for the *interior* of S, denoted by $\mathcal{I}(S)$, and the *closure* of S, denoted by $\mathcal{C}(S)$.
   (b) Prove: (i) $\mathcal{I}(S)^* = \mathcal{I}(S^*)$; (ii) $\mathcal{C}(S)^* = \mathcal{C}(S^*)$.
   (c) Give a proof of Theorem 9.32 based on the results in part (b).

8. Prove: If S is a subset of $J : M$, then
   (a) $[S]^{-1} = [S^{-1}]$;
   (b) $\langle S \rangle = \langle S^{-1} \rangle$.

10.16 The Cross Section Correspondence

We return now to work in a join geometry J. Our aim is to establish results in J that lead to interesting correspondences between factor geometries. This section is devoted to a type of correspondence between certain families of halfspaces, which has its origin in a simple sectioning operation on a halfspace. The correspondence is established in Theorem 10.27 and leads, in the next section, to an important isomorphism theorem. We begin with two theorems needed to prove Theorem 10.27.

The first theorem is a weak form of distributive law for intersection with respect to extension of sets. It is similar to the modular identity in lattice theory (see Mac Lane and Birkhoff [1], p. 491).

Theorem 10.25 (Modular Law). *In a join geometry let A and B be sets of points, L a linear space and $B \subset L$. Then*

$$(A/B) \cap L = (A \cap L)/B.$$

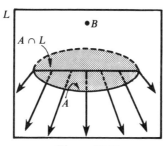

Figure 10.13

PROOF. First we show (Figure 10.13)

$$(A \cap L)/B \subset (A/B) \cap L. \tag{1}$$

Certainly $A \cap L \subset A$ implies

$$(A \cap L)/B \subset A/B. \tag{2}$$

Moreover the relations $A \cap L \subset L$ and $B \subset L$ yield

$$(A \cap L)/B \subset L. \tag{3}$$

The relations (2) and (3) imply (1).

To prove the reverse inclusion

$$(A/B) \cap L \subset (A \cap L)/B, \tag{4}$$

suppose

$$x \subset (A/B) \cap L. \tag{5}$$

Then

$$x \subset a/b, \tag{6}$$

where $a \subset A$, $b \subset B$ and also $x \subset L$. Solving (6) for a gives $a \approx bx \subset L$. Thus $a \subset A \cap L$, and (6) implies

$$x \subset (A \cap L)/B. \tag{7}$$

Thus (5) implies (7), so that (4) is verified and the theorem follows.   □

Remark. The conclusion of the theorem may be written, in view of $B \subset L$,

$$(A/B) \cap L = (A \cap L)/(B \cap L),$$

and so the theorem may be considered a weak form of distributive law for intersection with respect to extension. (See Exercise 1 at the end of Section 10.17 below for a second modular law for extension, and Section 12.9 for similar principles involving operations other than extension.)

Theorem 10.26 (The Cross Section Theorem). *In a join geometry let H be a halfspace of the nonempty linear space A. Let linear space X intersect H and A. Then $H \cap X$ is a halfspace of $A \cap X$.*

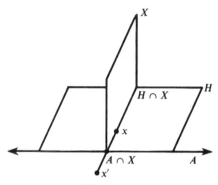

Figure 10.14

PROOF. Note that $A \cap X$ is a nonempty linear space (Figure 10.14). Let $x \subset H \cap X$. Choose x' such that $x' \subset (A \cap X)/x$. Then $xx' \approx A$ and $x \approx A/x'$. Thus $H \approx A/x'$, and $H = A/x'$ by Theorem 8.13. Note $x' \subset X$. By Theorem 10.25,

$$H \cap X = (A/x') \cap X = (A \cap X)/x'.$$

Thus $H \cap X$ is a halfspace of $A \cap X$ by definition.   □

Theorem 10.27. *In a join geometry let A and B be linear spaces such that $A \approx B$. Let H be in $B : A$. Then the cross section mapping*

$$H \to H \cap B \qquad\qquad (1)$$

effects a one to one correspondence between the family of halfspaces $B : A$ and the family of halfspaces $B : (A \cap B)$.

PROOF. Suppose H is in $B : A$, that is, H is a halfspace of A determined by a point of B (Figure 10.15). Then B intersects H as well as A, so that $H \cap B$ is a halfspace of $A \cap B$ (Theorem 10.26). Thus (1) associates with each member of $B : A$ a specific member of $B : (A \cap B)$.

Now consider any member H' of $B : (A \cap B)$. We show that H' is the correspondent [as determined by (1)] of some member of $B : A$. Suppose

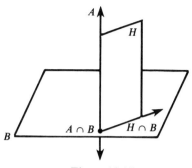

Figure 10.15

H' is determined by point b of B. Let H_1 be the halfspace of A determined by b. Then H_1 is in $B : A$, and its correspondent $H_1 \cap B$ is in $B : (A \cap B)$. But $b \subset H_1 \cap B$. Hence $H_1 \cap B = H'$ (Theorem 8.13), and H' is the correspondent of H_1 as determined by (1). Thus (1) effects a correspondence between the families $B : A$ and $B : (A \cap B)$.

Finally we show the correspondence is one to one. Suppose H_2 and H_3 are distinct elements of $B : A$. Their correspondents in $B : (A \cap B)$ are $H_2 \cap B$ and $H_3 \cap B$. Suppose

$$H_2 \cap B = H_3 \cap B. \tag{2}$$

Then $H_2 \approx H_3$, so that $H_2 = H_3$, contrary to supposition. Thus (2) is false, and $H_2 \cap B$ and $H_3 \cap B$ are distinct. Hence the correspondence effected by (1) is one to one, and the proof is finished. ▫

Next we show $\langle A, B \rangle : A = B : A$, so that the cross section mapping relates $\langle A, B \rangle : A$ and $B : (A \cap B)$.

Theorem 10.28. *In a join geometry let A and B be linear and $A \approx B$. Then*

$$\langle A, B \rangle : A = B : A.$$

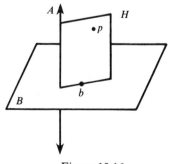

Figure 10.16

PROOF. First we show

$$\langle A, B \rangle : A \subset B : A. \qquad (1)$$

Let H be an element of $\langle A, B \rangle : A$ (Figure 10.16). By definition H is a halfspace of A that contains a point $p \subset \langle A, B \rangle$. Then by Theorem 6.14,

$$p \subset A/B \subset A/\langle A, B \rangle = A/(A/B).$$

Then for some $b \subset B$,

$$p \subset A/(A/b) = \overrightarrow{Ab} \qquad \text{(Theorem 8.17)}.$$

Hence $H \approx \overrightarrow{Ab}$ and $H = \overrightarrow{Ab}$. Thus H is a halfspace of A that contains a point of B, that is, H is an element of $B : A$ and (1) holds. The reverse inclusion $B : A \subset \langle A, B \rangle : A$ follows directly from $B \subset \langle A, B \rangle$. □

Theorem 10.29. *In a join geometry let A and B be linear and $A \approx B$. Then the cross section mapping $H \to H \cap B$ effects a one to one correspondence between $\langle A, B \rangle : A$ and $B : (A \cap B)$.*

PROOF. Apply the last two theorems. □

10.17 Isomorphism of Factor Geometries

First we observe that if L is a linear space in a join geometry J, and L contains the nonempty linear space M, then $L : M$ is indeed a factor geometry (as defined in Section 10.8) under the join operation $\cdot$ defined in Section 10.7. This result follows immediately from the observation that L is a join geometry with join and extension operations those of J, merely restricted in their application to points of L (Section 6.6). Hence if A and B are linear with $A \approx B$, then $\langle A, B \rangle : A$ and $B : (A \cap B)$ are factor geometries. The last theorem is now extended to show that the factor geometries $\langle A, B \rangle : A$ and $B : (A \cap B)$ have the same structure or are isomorphic. Recall that the notion of isomorphism was formulated in Section 5.5 for arbitrary join systems.

Theorem 10.30 (Isomorphism Theorem). *In a join geometry let A and B be linear spaces such that $A \approx B$. Then the factor geometries $\langle A, B \rangle : A$ and $B : (A \cap B)$ are isomorphic.*

PROOF. By the last theorem the crㅇss section mapping $H \to H \cap B$ effects a one to one correspondence between the sets $\langle A, B \rangle : A$ and $B : (A \cap B)$. We have to show only that the correspondence is an isomorphism between the factor geometries $\langle A, B \rangle : A$ and $B : (A \cap B)$.

Let H be an arbitrary element of $\langle A, B \rangle : A$. We have

$$\langle A, B \rangle : A = B : A \qquad \text{(Theorem 10.28)}.$$

Hence H is an element of $B : A$ and can be expressed in the form (Section 10.6)

$$H = |x|_A, \qquad x \subset B. \tag{1}$$

Now $H \cap B$, the correspondent of H, is an element of $B : (A \cap B)$, and so a halfspace of $A \cap B$ determined by some element of B. But (1) implies

$$H \cap B = |x|_A \cap B \supset x.$$

Thus

$$H \cap B = |x|_{A \cap B}.$$

The cross section mapping then takes on the simple form

$$|x|_A \rightarrow |x|_{A \cap B}, \tag{2}$$

where x is an arbitrary point of B. Suppose $b, b' \subset B$. Then

$$|b|_A \rightarrow |b|_{A \cap B}, \qquad |b'|_A \rightarrow |b'|_{A \cap B}.$$

Now consider the product $|b|_A \cdot |b'|_A$. By definition

$$|b|_A \cdot |b'|_A = |bb'|_A. \tag{3}$$

The right member of (3) is the set of $|x|_A$ for $x \subset bb'$. Similarly

$$|b|_{A \cap B} \cdot |b'|_{A \cap B} = |bb'|_{A \cap B}, \tag{4}$$

and the right member of (4) is the set of $|x|_{A \cap B}$ for $x \subset bb'$. Hence by (2) the cross section mapping effects a correspondence between the products $|b|_A \cdot |b'|_A$ and $|b|_{A \cap B} \cdot |b'|_{A \cap B}$, and is an isomorphism between $\langle A, B \rangle : A$ and $B : (A \cap B)$. This establishes the theorem. $\square$

Geometric Interpretation. Let J be JG5 the 3-dimensional Euclidean join geometry. Let A be a line and B a plane such that $A \cap B$ is a point (Figure 10.17). Then $\langle A, B \rangle = J$, and $\langle A, B \rangle : A$ is the natural geometry of the halfplanes in J with edge A. Similarly $B : (A \cap B)$ is the natural geometry of the rays in B with endpoint $A \cap B$. The cross section mapping is easily pictured, and the two factor geometries are seen to be isomorphic—essentially, each is the natural geometry of a circle. (Center a circle in plane B at

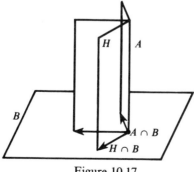

Figure 10.17

point $A \cap B$ and follow the procedure for the spherical representation of the ray geometry of Section 10.10. See also Exercise 8 at the end of Section 10.10.)

Note. The Isomorphism Theorem bears a close analogy to a well-known isomorphism theorem in the theory of groups. See Barnes [1], p. 55, Theorem 2.22, or Mac Lane and Birkhoff [1], p. 458, Theorem 2. For other analogies between the theory of join geometries and group theory see Prenowitz [3]. A join theoretic treatment covering several geometrical theories and the theory of abelian groups is found in Prenowitz and Jantosciak [1].

EXERCISES

1. Prove a second modular law: In a join geometry if A and B are sets of points, L a linear set and $A \subset L$, then

$$(A/B) \cap L = A/(B \cap L).$$

Interpret the result geometrically.

2. Convince yourself that the example of a factor geometry $J : M$ given in Section 10.10 is isomorphic to its spherical representation S'.

3. Prove: $J : M$ is isomorphic to itself under the correspondence $A \to A^{-1}$ for each element A of $J : M$.

4. Suppose every line in join geometry J is ordered (Section 4.24). Prove:
   (a) If $a \neq b$ then $\langle a, b \rangle : a$ contains exactly three elements.
   (b) If $a \neq b$ and $c \neq d$, then the factor geometries $\langle a, b \rangle : a$ and $\langle c, d \rangle : c$ are isomorphic.

5. Prove for any isomorphism between factor geometries $J : M$ and $J' : M'$ that
   (a) M' is the correspondent of M;
   (b) if the element A of $J : M$ corresponds to the element A' of $J' : M'$, then A^{-1} corresponds to $(A')^{-1}$.

Exchange Join Geometries—The Theory of Incidence and Dimension **11**

A new postulate is now introduced which permits "exchange of points" in the generation of linear hulls. The postulate is equivalent to a most basic property of Euclidean geometry: two points determine a line. Join geometries that satisfy the new postulate are called exchange join geometries or exchange geometries. It is in an exchange join geometry, rather than an arbitrary join geometry, that incidence relations among linear spaces can be satisfactorily studied. Under consideration will be the complex of relations suggested by familiar propositions of elementary geometry, such as, three noncollinear points belong to a unique plane or, in 3-space if two distinct planes meet their intersection is a line. The concept of dimension (or rank), which indicates the relative complexity of linear spaces and organizes them into a hierarchy, takes on a central position in the development. A key result involving it, The Dimension Principle (Theorem 11.12), generalizes familiar properties of lines and planes in Euclidean geometry. Finally various notions of linear dependence are shown to be equivalent and the concepts of rank and height of a linear space are identified.

11.1 Exchange Join Geometries

The familiar incidence properties of Euclidean geometry[1] cannot be derived from J1–J7 and do not hold in an arbitrary join geometry. Indeed, the property that two distinct points belong to a unique line is independent of J1–J7 (see Exercise 2 at the end of Section 11.2). It is necessary

---

[1] See Hilbert [1], pp. 3–4; Prenowitz and Jordan [1], Chapter 7.

therefore to postulate this property, or an equivalent, in order to obtain the familiar theory of incidence. It seems somewhat remarkable that the property is sufficient. Actually it is sufficient merely to assume the following property:

E. *Exchange Postulate*. If $c \subset \langle a, b \rangle$ and $c \neq a$, then $\langle a, b \rangle = \langle a, c \rangle$.

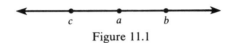

Figure 11.1

The Exchange Postulate says in effect that if c belongs to line ab, and is distinct from a, then line ab and line ac are the same (Figure 11.1). Postulate E may be described as an "exchange" principle, since it permits us to exchange b for c in $\langle a, b \rangle$ without affecting the result.

Join geometries that satisfy Postulate E deserve a name.

Definition. A join geometry that satisfies Postulate E is called an *exchange* join geometry or an *exchange* geometry.

Observe that Euclidean geometry satisfies Postulate E and that JG3, JG4 and JG5 are exchange geometries.

11.2 Two Points Determine a Line

Postulate E is equivalent to the familiar Euclidean property that two points "determine" a line. Specifically we have the following result.

Theorem 11.1. *A join geometry is an exchange geometry if and only if $a \neq b$ implies there exists a unique line containing a and b, namely, $\langle a, b \rangle$.*

PROOF. Suppose our join geometry is an exchange geometry, that is, Postulate E holds. Let $a \neq b$. Certainly $\langle a, b \rangle \supset a, b$ and is a line. Suppose line $L \supset a, b$. To show

$$L = \langle a, b \rangle, \tag{1}$$

let $L = \langle c, d \rangle$, $c \neq d$. Then $a, b \subset \langle c, d \rangle$. The relation $c \neq d$ implies $a \neq c$ or $a \neq d$, let us say $a \neq c$. By Postulate E,

$$L = \langle c, d \rangle = \langle c, a \rangle = \langle a, c \rangle. \tag{2}$$

Then $b \subset \langle a, c \rangle$ and $b \neq a$, so that Postulate E implies

$$\langle a, c \rangle = \langle a, b \rangle. \tag{3}$$

By (2) and (3), (1) holds and $\langle a, b \rangle$ is the unique line containing a and b.

Conversely, suppose $a \neq b$ implies $\langle a, b \rangle$ is the unique line containing a and b. Suppose $p \subset \langle q, r \rangle$ and $p \neq q$. Then $q \neq r$, since otherwise $p = q$

results. Hence $\langle q, r \rangle$ is a line and contains p, q. By uniqueness $\langle p, q \rangle = \langle q, r \rangle$, and Postulate E holds, so that our join geometry is an exchange geometry. $\qquad\qquad\qquad\qquad\qquad\qquad\qquad\qquad\qquad\qquad\qquad\qquad\qquad\quad$ □

Corollary 11.1. *In an exchange geometry, if a and b are distinct points, there exists a unique line containing a and b, namely, $\langle a, b \rangle$.*

EXERCISES

1. Convince yourself that all the join geometries JG1–JG15 are exchange geometries.

2. Show that JG16 is not an exchange geometry. Is the Exchange Postulate deducible from J1–J7? Do two distinct points determine a line in JG16?

3. Prove: In an exchange geometry, if two distinct lines meet, then their intersection is a point.

4. Prove: In an exchange geometry, points a, b, c are distinct and not contained in a line if and only if they are linearly independent.

5. Prove: A join geometry is an exchange geometry if and only if a, b, c are distinct points of a line implies

$$ab, ac, bc \subset abc.$$

6. Prove that $\mathbb{R}^n$ (Sections 5.6–5.9) is an exchange geometry. (You may want to use the exercises on the Join Geometry $\mathbb{R}^n$ at the end of Section 6.18.)

11.3 A Basis for a Linear Space

Suppose that L is a finitely generated linear space in an arbitrary join geometry and $a_1, \ldots, a_n$ generate L, that is,

$$L = \langle a_1, \ldots, a_n \rangle.$$

Then some of the a's may be redundant as generators of L, for L may be generated by fewer than all n of the a's. However, if $a_1, \ldots, a_n$ are linearly independent, then no fewer than all n of the a's generate L. This is easily seen. For suppose L is generated by r of the a's, where $r < n$. Since the order of the a's is immaterial, we may assume $L = \langle a_1, \ldots, a_r \rangle$. Then $a_n \subset L$ implies

$$a_n \subset \langle a_1, \ldots, a_r \rangle \subset \langle a_1, \ldots, a_{n-1} \rangle,$$

contrary to the linear independence assumption.

Conversely, if no fewer than all n of the a's generate L, then the a's are linearly independent. This is also easily seen. For suppose $a_1, \ldots, a_n$ are linearly dependent, that is, for some i,

$$a_i \subset \langle a_1, \ldots, a_{i-1}, a_{i+1}, \ldots, a_n \rangle.$$

Then

$$L = \langle a_1, \ldots, a_n \rangle = \langle a_1, \ldots, a_{i-1}, a_{i+1}, \ldots, a_n \rangle$$

results, contrary to the minimal nature of the a's.

Therefore a linearly independent set of generators of a linear space L can be characterized as a minimal set of generators of L.

Definition. In any join geometry let $\{a_1, \ldots, a_n\}$ be a linearly independent set of generators for linear space L. Then $\{a_1, \ldots, a_n\}$ is *basis* for L.

Remark. Recall from Section 6.15 the convention that if $\{a_1, \ldots, a_n\}$ is specified to be a linearly independent set, then the a's are assumed distinct. This convention is applied to a basis: if $\{a_1, \ldots, a_n\}$ is specified to be a basis for some linear space, the a's are assumed distinct.

Note that $\varnothing$ is a basis for the linear space $\varnothing$, and each point is a basis for itself as a linear space.

It is our objective to show that in an exchange geometry if L is linear and has a basis, then any two bases of L contain the same number of points. This is not valid in an arbitrary join geometry (see Exercise 2 at the end of Section 11.4 below.)

A linear space has a basis if it has a finite set of generators.

Theorem 11.2. *In any join geometry, if L is a finitely generated linear space, then L has a basis.*

PROOF. If $L = \varnothing$, the theorem holds. Thus suppose $L = \langle a_1, \ldots, a_n \rangle$. Consider those subsets of $\{a_1, \ldots, a_n\}$ which generate L, and let S be one with a minimum number of a's. Then no proper subset of S can generate L, that is, S is a minimal set of generators of L. Thus as we saw above (the paragraph preceding the definition), S is linearly independent, and so a basis for L.   □

11.4 Incidence Properties

We develop an incidence theory for exchange geometries. First Postulate E is generalized to apply to an arbitrary linear space.

Theorem 11.3. *In an exchange geometry let L be a linear space, $b \subset \langle L, a \rangle$ and $b \not\subset L$. Then $\langle L, a \rangle = \langle L, b \rangle$.*

PROOF. If $L = \varnothing$, then $b = a$ and the theorem holds. Suppose $L \neq \varnothing$. The relation $b \subset \langle L, a \rangle$ implies $\langle L, b \rangle \subset \langle L, a \rangle$. To prove the reverse inclusion we show

$$a \subset \langle L, b \rangle. \tag{1}$$

Let $a' \subset L$. Then using Theorem 6.14 we have

$$b \subset \langle L, a \rangle = \langle L, \langle a', a \rangle \rangle = L/\langle a', a \rangle.$$

Hence there exist p, q such that

$$b \subset p/q, \quad p \subset L, \quad q \subset \langle a', a \rangle.$$

Suppose $q = a'$. Then $b \subset p/q$ implies $b \subset L$, contrary to hypothesis. Hence $q \neq a'$, and we have, using Postulate E and $q \subset p/b$,

$$a \subset \langle a', a \rangle = \langle a', q \rangle \subset \langle a', p, b \rangle \subset \langle L, b \rangle.$$

Thus (1) holds, so that $\langle L, a \rangle \subset \langle L, b \rangle$, and the theorem follows. $\square$

Corollary 11.3. *In an exchange geometry, suppose* $b \subset \langle a_1, \ldots, a_m \rangle$ *but* $b \not\subset \langle a_1, \ldots, a_{m-1} \rangle$. *Then* $\langle a_1, \ldots, a_m \rangle = \langle a_1, \ldots, a_{m-1}, b \rangle$.

PROOF. Let $L = \langle a_1, \ldots, a_{m-1} \rangle$. Then

$$b \subset \langle a_1, \ldots, a_m \rangle = \langle \langle a_1, \ldots, a_{m-1} \rangle, a_m \rangle = \langle L, a_m \rangle,$$

but $b \not\subset L$. By the theorem,

$$\langle a_1, \ldots, a_m \rangle = \langle L, a_m \rangle = \langle L, b \rangle = \langle a_1, \ldots, a_{m-1}, b \rangle. \quad \square$$

The corollary is generalized in the following theorem.

Theorem 11.4 (Exchange Lemma). *In an exchange geometry, suppose* $b \subset \langle a_1, \ldots, a_m \rangle$ *but* $b \not\subset \langle a_1, \ldots, a_r \rangle$ *for some* r *satisfying* $1 \leqslant r < m$. *Then there is an* a_i, $r < i \leqslant m$, *which may be exchanged for* b *without altering* $\langle a_1, \ldots, a_m \rangle$, *that is,*

$$\langle a_1, \ldots, a_m \rangle = \langle a_1, \ldots, a_{i-1}, b, a_{i+1}, \ldots, a_m \rangle. \tag{1}$$

PROOF. The hypothesis implies the existence of integer i, $r < i \leqslant m$, such that

$$b \subset \langle a_1, \ldots, a_i \rangle, \quad b \not\subset \langle a_1, \ldots, a_{i-1} \rangle.$$

By the last corollary

$$\langle a_1, \ldots, a_i \rangle = \langle a_1, \ldots, a_{i-1}, b \rangle,$$

so that

$$\langle \langle a_1, \ldots, a_i \rangle, a_{i+1}, \ldots, a_m \rangle = \langle \langle a_1, \ldots, a_{i-1}, b \rangle, a_{i+1}, \ldots, a_m \rangle$$

from which (1) follows. $\square$

Corollary 11.4. *In an exchange geometry, if* $b \subset \langle a_1, \ldots, a_m \rangle$, *then for some* i, $1 \leqslant i \leqslant m$,

$$\langle a_1, \ldots, a_m \rangle = \langle a_1, \ldots, a_{i-1}, b, a_{i+1}, \ldots, a_m \rangle.$$

PROOF. If $b = a_1$, let $i = 1$. If $b \neq a_1$, then the theorem applies with $r = 1$. $\square$

EXERCISES

1. Interpret Corollary 11.3, Theorem 11.4 and Corollary 11.4 geometrically.

2. In the set T of all ordered triples (x_1, x_2, x_3) of real numbers, define a join operation · as follows:
   $(a_1, a_2, a_3) \cdot (b_1, b_2, b_3)$ is the set of all (x_1, x_2, x_3) such that $x_i \subset a_i \cdot b_i$, $i = 1, 2, 3$; and · here is as defined in JG13.
   Show that $(T, \cdot)$ is a join geometry, and compare it with JG16. Find two bases for T with different cardinal numbers.

3. Prove: In any join geometry, if $L = \langle a_1, \ldots, a_n \rangle$, then some subset of $\{a_1, \ldots, a_n\}$ is a basis for L.

4. Prove: In an exchange geometry three noncollinear points belong to a unique "plane". That is, if a_1, a_2, a_3 are distinct and not contained in a line and $a_1, a_2, a_3 \subset \langle b_1, b_2, b_3 \rangle$, then $\langle a_1, a_2, a_3 \rangle = \langle b_1, b_2, b_3 \rangle$.

5. In an exchange geometry, let S be a set of points, $b \subset \langle S, a \rangle$ and $b \not\subset \langle S \rangle$. Prove $\langle S, a \rangle = \langle S, b \rangle$.

6. In any join geometry suppose linear spaces A and B have bases $\{a_1, \ldots, a_m\}$ and $\{b_1, \ldots, b_n\}$, respectively. Does $\langle A, B \rangle$ have a basis? If so, explain how to obtain one.

11.5 The Exchange Theorem

In an exchange geometry, if L is a finitely generated linear space, say $L = \langle a_1, \ldots, a_n \rangle$, the Exchange Lemma (Theorem 11.4) enables us to change one point in the set of generators $\{a_1, \ldots, a_n\}$ of L and obtain a different set of generators. Sometimes we wish to change several or even all points in a set of generators of L to form a new set of generators. In Euclidean geometry, for example, let L be a plane $\langle a_1, a_2, a_3 \rangle$ (Figure 11.2). It is often necessary to re-express L as $\langle b_1, b_2, b_3 \rangle$ where b_1, b_2, b_3 are points of L. Clearly b_1, b_2, b_3 cannot be chosen arbitrarily—they must be distinct and noncollinear, and so independent. This and similar situations suggest the following Exchange Theorem, which is a key result in our theory of incidence and is a form of the well-known exchange theorem of Steinitz in the algebraic theory of dependence (see van der Waerden [1], p. 101, Corollary 4).

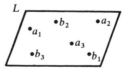

Figure 11.2

Theorem 11.5 (Exchange Theorem). *In an exchange geometry let*

$$\langle a_1, \ldots, a_m \rangle \supset b_1, \ldots, b_n,$$

where $b_1, \ldots, b_n$ *are linearly independent. Then there exist n of the a's,* $a_{i_1}, \ldots, a_{i_n}$, *which may be exchanged for* $b_1, \ldots, b_n$ *respectively without altering* $\langle a_1, \ldots, a_m \rangle$, *that is,*

$$\langle a_1, \ldots, a_m \rangle = \langle b_1, \ldots, b_n, a_{i_{n+1}}, \ldots, a_{i_m} \rangle,$$

where $a_{i_{n+1}}, \ldots, a_{i_m}$ *denote the a's other than* $a_{i_1}, \ldots, a_{i_n}$.

PROOF. Our procedure is to exchange a's for b's one at a time until n a's have been exchanged. Since $b_1 \subset \langle a_1, \ldots, a_m \rangle$, Corollary 11.4 implies there is an a_i which may be exchanged for b_1, so that

$$\langle a_1, \ldots, a_m \rangle = \langle a_1, \ldots, a_{i-1}, b_1, a_{i+1}, \ldots, a_m \rangle = \langle b_1, a_{j_2}, \ldots, a_{j_m} \rangle,$$

$$(1)$$

where $a_{j_2}, \ldots, a_{j_m}$ denote $a_1, \ldots, a_{i-1}, a_{i+1}, \ldots, a_m$.

Similarly we find an a to be exchanged for b_2. The relation (1) and the hypothesis imply

$$b_2 \subset \langle b_1, a_{j_2}, \ldots, a_{j_m} \rangle. \tag{2}$$

Since $b_1, \ldots, b_n$ are linearly independent, so are b_1, b_2 (Theorem 6.18), and thus $b_2 \not\subset \langle b_1 \rangle$. Hence the Exchange Lemma for $r = 1$ applies to (2), and we may exchange one of the a_j's for b_2, getting

$$\langle b_1, a_{j_2}, \ldots, a_{j_m} \rangle = \langle b_1, b_2, a_{k_3}, \ldots, a_{k_m} \rangle,$$

where $a_{k_3}, \ldots, a_{k_m}$ denote the "unexchanged" a's. By (1),

$$\langle a_1, \ldots, a_m \rangle = \langle b_1, b_2, a_{k_3}, \ldots, a_{k_m} \rangle.$$

Since b_1, b_2, b_3 are linearly independent, $b_3 \not\subset \langle b_1, b_2 \rangle$, so the Exchange Lemma can now be used with $r = 2$, and then next with $r = 3$, etc. So we keep repeating the procedure until we finally get

$$\langle a_1, \ldots, a_m \rangle = \langle b_1, \ldots, b_n, a_{i_{n+1}}, \ldots, a_{i_m} \rangle. \qquad \square$$

An important corollary is immediate.

Corollary 11.5.1. *In an exchange geometry let* $\langle a_1, \ldots, a_m \rangle \supset b_1, \ldots, b_n$, *where* $b_1, \ldots, b_n$ *are linearly independent. Then* $m \geqslant n$.

Thus in an exchange geometry the largest possible cardinal number of a linearly independent subset of the linear space $\langle a_1, \ldots, a_m \rangle$ is m.

The Exchange Theorem also yields a useful characterization of linear dependence of points.

Corollary 11.5.2. *In an exchange geometry, points* $a_1, \ldots, a_n$ *are linearly dependent if and only if one of them is contained in the linear hull of the* *preceding ones, that is, for some i, $1 < i \leqslant n$,*

$$a_i \subset \langle a_1, \ldots, a_{i-1} \rangle. \tag{1}$$

PROOF. Suppose (1). Then

$$a_i \subset \langle a_1, \ldots, a_{i-1}, a_{i+1}, \ldots, a_n \rangle,$$

and $a_1, \ldots, a_n$ are linearly dependent.

Conversely, suppose $a_1, \ldots, a_n$ are linearly dependent. Then since a_1 itself is linearly independent, there exists an i, $1 < i \leqslant n$, such that $a_1, \ldots, a_{i-1}$ are linearly independent but $a_1, \ldots, a_i$ are linearly dependent. Then for some j,

$$a_j \subset \langle a_1, \ldots, a_{j-1}, a_{j+1}, \ldots, a_i \rangle$$

and so

$$a_1, \ldots, a_{i-1} \subset \langle a_1, \ldots, a_{j-1}, a_{j+1}, \ldots, a_i \rangle.$$

Using the Exchange Theorem,

$$a_i \subset \langle a_1, \ldots, a_{j-1}, a_{j+1}, \ldots, a_i \rangle = \langle a_1, \ldots, a_{i-1} \rangle,$$

and (1) holds. $\qquad\qquad\square$

Corollary 11.5.3. *In an exchange geometry $a_1, \ldots, a_n$ are linearly independent if and only if $a_i \not\subset \langle a_1, \ldots, a_{i-1} \rangle$, $1 < i \leqslant n$.*

11.6 Rank and Dimension

In an exchange geometry the cardinal number of a basis for a linear set L is uniquely determined by L.

Theorem 11.6. *In an exchange geometry, if linear set L has a basis, then any two bases for L contain the same number of points.*

PROOF. The theorem holds if $L = \varnothing$. Suppose $L \neq \varnothing$ and $\{a_1, \ldots, a_m\}$ and $\{b_1, \ldots, b_n\}$ are bases for L. Then

$$\langle a_1, \ldots, a_m \rangle = L \supset b_1, \ldots, b_n.$$

But $b_1, \ldots, b_n$ are linearly independent. By Corollary 11.5.1, $m \geqslant n$. By symmetry $n \geqslant m$. Thus $m = n$ and the theorem holds. $\qquad\square$

The number of points in a basis is an important invariant of a linear space.

Definition. Let L be a linear space in an exchange geometry. Suppose L has a basis. Then the cardinal number of a basis for L is called the *rank* of L and is denoted by $r(L)$. The *dimension* of L, denoted by $d(L)$, is given by

$d(L) = r(L) - 1$. L is said to have a *rank* or a *dimension*, or to be of *finite rank* or *finite dimensional*. If $d(L) = n$, we call L an *n-space*; a 2-space is a *plane*. The *rank* (or *dimension*) of an exchange geometry $(J, \cdot)$ is merely the rank (or dimension) of J should J as a linear space have a basis.

Note that in an exchange geometry

(a) $r(L) = 0$ $[d(L) = -1]$ if and only if $L = \varnothing$;
(b) $r(L) = 1$ $[d(L) = 0]$ if and only if L is a point;
(c) $r(L) = 2$ $[d(L) = 1]$ if and only if L is a line.

Statements (b) and (c) assert that 0-space and 1-space are synonyms for point and line.

Note also that L is finite dimensional if and only if L is finitely generated (Theorem 11.2).

Remark. The notions of rank and dimension are interconvertible, and it is a matter of convenience which is employed. Each serves as a "measure" of the relative degree of complexity of a linear space.

The Exchange Theorem has further interesting and useful consequences.

Theorem 11.7. *In an exchange geometry let linear set L have rank n. Then*

(a) *any set $\{a_1, \ldots, a_n\}$ which generates L is a basis for L;*
(b) *any linearly independent subset of L with cardinal number n is a basis for L.*

PROOF. If $L = \varnothing$, (a) is meaningless and (b) is trivial. Assume $L \neq \varnothing$.

(a) Let $L = \langle a_1, \ldots, a_n \rangle$. We show $a_1, \ldots, a_n$ are linearly independent. Suppose not. Then for some a_i,

$$a_i \subset \langle a_1, \ldots, a_{i-1}, a_{i+1}, \ldots, a_n \rangle,$$

so that

$$L = \langle a_1, \ldots, a_{i-1}, a_{i+1}, \ldots, a_n \rangle. \tag{1}$$

Since L has rank n, there exist linearly independent points

$$b_1, \ldots, b_n \subset L. \tag{2}$$

But (1) and (2) contradict Corollary 11.5.1. Thus $\{a_1, \ldots, a_n\}$ is a linearly independent set and so a basis for L.

(b) Let $\{a_1, \ldots, a_n\}$ be a linearly independent subset of L. Since L has rank n,

$$L = \langle b_1, \ldots, b_n \rangle$$

for some points $b_1, \ldots, b_n$. Then

$$\langle b_1, \ldots, b_n \rangle \supset a_1, \ldots, a_n,$$

and Theorem 11.5 implies

$$L = \langle b_1, \ldots, b_n \rangle = \langle a_1, \ldots, a_n \rangle.$$

Thus $\{a_1, \ldots, a_n\}$ is a set of generators of L and so a basis for L. $\square$

The result (Corollary 11.1) that in an exchange geometry two distinct points determine a line is generalized—n points "in general position" determine an $(n - 1)$-space.

Theorem 11.8. *In an exchange geometry let $a_1, \ldots, a_n$ be linearly independent. Then there is a unique linear space of rank n that contains $a_1, \ldots, a_n$, namely, $\langle a_1, \ldots, a_n \rangle$.*

PROOF. Clearly $\langle a_1, \ldots, a_n \rangle \supset a_1, \ldots, a_n$ and has rank n. Suppose L is linear, $r(L) = n$ and $L \supset a_1, \ldots, a_n$. By (b) of the previous theorem $\{a_1, \ldots, a_n\}$ is a basis for L. Then $L = \langle a_1, \ldots, a_n \rangle$, and the uniqueness assertion holds.   □

EXERCISES

1. Convince yourself that the basic set J has a rank, and then find its rank for each of the exchange geometries JG1–JG15. (See Exercise 1 at the end of Section 11.2.)

2. Prove: In an exchange geometry if $a_1, \ldots, a_n$ are linearly independent and $a_{n+1} \not\subset \langle a_1, \ldots, a_n \rangle$, then $a_1, \ldots, a_{n+1}$ are linearly independent.

3. Prove: In an exchange geometry, $r(\langle a_1, \ldots, a_n \rangle) \leqslant n$, and equality holds if and only if $a_1, \ldots, a_n$ are linearly independent.

4. Prove: $d(\mathbb{R}^n) = n$. (See Exercise 6 at the end of Section 11.2.)

5. Find an exchange geometry that is not finite dimensional.

6. Prove: In an exchange geometry, if L is linear and $r(L) = n$, then any $n + 1$ points of L are linearly dependent and any $n - 1$ points of L do not generate L.

11.7 Rank and Linear Containment

We prove some basic results on the connection between set containment and ranks of linear spaces.

Theorem 11.9. *In an exchange geometry, suppose A and B are linear, $A \subset B$, and B has a rank. Then A has a rank and $r(A) \leqslant r(B)$. Furthermore, any basis for A is a subset of some basis for B.*

PROOF. The theorem holds if $A = \varnothing$. Assume $A \neq \varnothing$. Since any point of A is linearly independent, A contains linearly independent sets of points. But if points $a_1, \ldots, a_k$ of A are linearly independent, then since $a_1, \ldots, a_k \subset B$, Corollary 11.5.1 implies $r(B) \geqslant k$. Thus the number of points in any linearly independent subset of A can not exceed $r(B)$. It follows that there is a maximum m for the number of points in a linearly independent subset of A; that is, A contains linearly independent points

$a_1, \ldots, a_m$, and any finite subset of A consisting of more than m points is linearly dependent.

We claim $\{a_1, \ldots, a_m\}$ is a basis for A. Let $a \subset A$. Then the points $a_1, \ldots, a_m, a$ must be linearly dependent. Hence by Corollary 11.5.2,

$$a \subset \langle a_1, \ldots, a_m \rangle \tag{1}$$

or for some i, $1 < i \leqslant m$,

$$a_i \subset \langle a_1, \ldots, a_{i-1} \rangle. \tag{2}$$

But (2) and Corollary 11.5.2 imply $a_1, \ldots, a_m$ are linearly dependent. Thus (2) fails, (1) holds and we may conclude

$$A \subset \langle a_1, \ldots, a_m \rangle.$$

Thus $A = \langle a_1, \ldots, a_m \rangle$, and our claim is valid. We have established then that A has a rank and that $r(A) = m \leqslant r(B)$.

Suppose finally that $\{x_1, \ldots, x_m\}$ is a basis for A. Let $r(B) = n$ and $\{b_1, \ldots, b_n\}$ be a basis for B. Then $x_1, \ldots, x_m$ are linearly independent, and

$$x_1, \ldots, x_m \subset \langle b_1, \ldots, b_n \rangle = B.$$

By the Exchange Theorem

$$\langle x_1, \ldots, x_m, b_{i_{m+1}}, \ldots, b_{i_n} \rangle = B.$$

By Theorem 11.7(a) since $r(B) = n$, $\{x_1, \ldots, x_m, b_{i_{m+1}}, \ldots, b_{i_n}\}$ is a basis for B. $\qquad \square$

Corollary 11.9.1. *In an exchange geometry, suppose A and B are linear, $A \subset B$, and B has a rank. Then A has a rank, and $A = B$ if and only if $r(A) = r(B)$.*

PROOF. The theorem ensures that A has a rank. We need only show $r(A) = r(B)$ implies $A = B$. But this follows readily from the fact that a basis for A is a subset of some basis for B. $\qquad \square$

The proof of the theorem suggests a corollary that will be useful in Section 11.11.

Corollary 11.9.2. *In an exchange geometry, let A be a linear space. Then A is of finite rank if and only if there exists a maximum m for the cardinal numbers of the linearly independent subsets of A, and in this case $r(A) = m$.*

PROOF. The result is trivial if $A = \varnothing$. Assume $A \neq \varnothing$. The proof of the theorem shows that if such an m exists, then A has finite rank and $r(A) = m$. Conversely, suppose A has finite rank n. Then A is the linear hull of n points, and Corollary 11.5.1 implies that no linearly independent subset of A contains more than n points. Thus such an m exists: indeed, $m = n$. $\qquad \square$

11.8 Rank and Covering

The covering relation between linear spaces of finite rank (Section 6.18) is characterized in terms of their ranks.

Theorem 11.10 (Point Spanning Principle). *In an exchange geometry, linear space A covers linear space B if and only if $A = \langle B, a \rangle$ for some $a \subset A - B$.*

(Compare Theorem 6.21.)

PROOF. In view of Theorem 6.21, we need only assume $A = \langle B, a \rangle$ for some $a \subset A - B$, and prove A covers B. Certainly the assumption gives $A \supset B$ and $A \neq B$. Suppose for linear space X we have

$$A \supset X \supset B \quad \text{and} \quad X \neq B.$$

Let $x \subset X - B$. Then $x \subset A = \langle B, a \rangle$, and we can exchange a for x, getting $\langle B, a \rangle = \langle B, x \rangle$ (Theorem 11.3). Hence

$$A \supset X \supset \langle B, x \rangle = A,$$

so that $X = A$, and A covers B by definition. □

Remark. In view of Theorem 11.3, if $A = \langle B, a \rangle$ for some $a \subset A - B$, then $A = \langle B, a \rangle$ for all $a \subset A - B$.

Theorem 11.11 (Rank Covering Principle). *In an exchange geometry, let linear spaces A and B be of finite rank. Then A covers B if and only if $A \supset B$ and $r(A) = r(B) + 1$.*

PROOF. The theorem holds if $B = \varnothing$. Assume $B \neq \varnothing$. Suppose A covers B. Then $A = \langle B, a \rangle$, where $a \subset A - B$. Let $\{b_1, \ldots, b_n\}$ be a basis for B. Then

$$A = \langle \langle b_1, \ldots, b_n \rangle, a \rangle = \langle b_1, \ldots, b_n, a \rangle.$$

Hence A is finitely generated, and by Theorem 11.2 has a basis $\{a_1, \ldots, a_m\}$. Then $\langle b_1, \ldots, b_n, a \rangle \supset a_1, \ldots, a_m$. By Corollary 11.5.1, $r(A) = m \leqslant n + 1 = r(B) + 1$. Since we have $A \supset B$ and $A \neq B$, Theorem 11.9 and its first corollary imply $r(A) > r(B)$. Thus $r(A) = r(B) + 1$.

Conversely, suppose $A \supset B$ and $r(A) = r(B) + 1$. Let the set of points S be a basis for B. Then S is a subset of a basis T for A (Theorem 11.9). Since $r(A) = r(B) + 1$, we obtain $T = S \cup a$, where $a \not\subset S$. Note that $a \not\subset B = \langle S \rangle$, for otherwise T is a linearly dependent set of points. Hence

$$A = \langle T \rangle = \langle S, a \rangle = \langle \langle S \rangle, a \rangle = \langle B, a \rangle$$

and Theorem 11.10 yields that A covers B. □

EXERCISES

1. Prove: In an exchange geometry, if the linear spaces A and B both have a rank, then any linear subset of $\langle A, B \rangle$ has a rank.

*2. Prove: In an exchange geometry, if A and L are finite dimensional linear spaces and $A \subset L$, then there exists a finite dimensional linear space B such that $A \not\approx B$, $L = \langle A, B \rangle$ and $r(A) + r(B) = r(L)$. Interpret geometrically.

3. Prove: In an exchange geometry, if $a_1, \ldots, a_n$ are linearly independent points of a finite dimensional linear space L, then $\{a_1, \ldots, a_n\}$ is a subset of some basis for L. (Compare Exercise 3 at the end of Section 11.4.)

4. Prove: In an exchange geometry, if A and B are linear, A covers B, and B has a rank, then A has a rank.

*5. Prove: In any join geometry if A and B are linear and $A \approx B$ then A covers $A \cap B$ if and only if $\langle A, B \rangle$ covers B.

6. Prove: In an exchange geometry, if A and B are linear, then "A covers $A \cap B$" implies "$\langle A, B \rangle$ covers B". Does the converse implication hold? Justify your answer.

7. Prove: In an exchange geometry, if A, B and C are linear, $A \neq B$, and both A and B cover C, then $\langle A, B \rangle$ covers both A and B. (Do the given conditions determine C?)

11.9 The Dimension Principle

Familiar intersection properties of lines and planes in Euclidean geometry are covered and generalized in the following theorem.

Theorem 11.12 (The Dimension Principle). *In an exchange geometry let A and B be finite dimensional linear spaces. Then $\langle A, B \rangle$ and $A \cap B$ are finite dimensional and*

$$d(\langle A, B \rangle) + d(A \cap B) \leqslant d(A) + d(B). \qquad (1)$$

Furthermore, if $A \approx B$, then

$$d(\langle A, B \rangle) + d(A \cap B) = d(A) + d(B). \qquad (2)$$

PROOF. First we show (1). We have $A \cap B \subset A, B$. Hence by Theorem 11.9, $A \cap B$ is finite dimensional and $A \cap B, A, B$ have respective bases which take the forms

$$\{p_1, \ldots, p_s\}, \ \{p_1, \ldots, p_s, a_1, \ldots, a_t\}, \ \{p_1, \ldots, p_s, b_1, \ldots, b_u\}. \ (3)$$

Note that in (3) we will allow $s = 0$, that is, no p's need be present—the basis of $A \cap B$ may be $\varnothing$. Similarly (with obvious interpretation) $t = 0$ or

$u = 0$ may occur. We have

$$r(A \cap B) = s, \tag{4}$$

$$r(A) + r(B) = s + t + s + u = 2s + t + u. \tag{5}$$

Moreover,

$$\langle A, B \rangle = \langle \langle p_1, \ldots, p_s, a_1, \ldots, a_t \rangle, \langle p_1, \ldots, p_s, b_1, \ldots, b_u \rangle \rangle$$

$$= \langle p_1, \ldots, p_s, a_1, \ldots, a_t, b_1, \ldots, b_u \rangle.$$

Hence $\langle A, B \rangle$ is finitely generated. Thus by Theorem 11.2, $\langle A, B \rangle$ has a basis and is finite dimensional. By Corollary 11.5.1, $r(\langle A, B \rangle) \leqslant s + t + u$. This with (4) and (5) implies

$$r(\langle A, B \rangle) + r(A \cap B) \leqslant r(A) + r(B),$$

and (1) follows.

We now prove (2) under the given additional assumption $A \approx B$. Observe now that $s \geqslant 1$. We prove

$$r(\langle A, B \rangle) = s + t + u \tag{6}$$

by showing $p_1, \ldots, p_s, a_1, \ldots, a_t, b_1, \ldots, b_u$ are linearly independent. Suppose the latter false. Then by Corollary 11.5.2, one of the points in the sequence is in the linear hull of the preceding ones. Since $p_1, \ldots, p_s, a_1, \ldots, a_t$ are linearly independent, such a point must be one of the b's, say b_i. Thus

$$b_i \subset \langle p_1, \ldots, p_s, a_1, \ldots, a_t, b_1, \ldots, b_{i-1} \rangle$$

$$= \langle \langle p_1, \ldots, p_s, a_1, \ldots, a_t \rangle, \langle p_1, \ldots, p_s, b_1, \ldots, b_{i-1} \rangle \rangle$$

$$= \langle A, \langle p_1, \ldots, p_s, b_1, \ldots, b_{i-1} \rangle \rangle$$

$$= A / \langle p_1, \ldots, p_s, b_1, \ldots, b_{i-1} \rangle \quad \text{(Theorem 6.14).}$$

Hence

$$b_i \subset x/y, \tag{7}$$

where $x \subset A$ and

$$y \subset \langle p_1, \ldots, p_s, b_1, \ldots, b_{i-1} \rangle. \tag{8}$$

We have, by (7), $x \subset b_i\, y$. Since $b_i, y \subset B$, we have $x \subset B$. Thus

$$x \subset A \cap B \subset \langle p_1, \ldots, p_s, b_1, \ldots, b_{i-1} \rangle. \tag{9}$$

The relations (7), (8), (9) imply $b_i \subset \langle p_1, \ldots, p_s, b_1, \ldots, b_{i-1} \rangle$. This contradicts Corollary 11.5.3, since $p_1, \ldots, p_s, b_1, \ldots, b_u$ are linearly independent. Thus our supposition is false, and (6) holds. Equations (4), (5) and (6) imply

$$r(\langle A, B \rangle) + r(A \cap B) = r(A) + r(B),$$

and (2) follows. $\square$

EXERCISES

1. Prove: In an exchange geometry, if two distinct planes contained in a 3-space meet, their intersection is a line.

2. Prove: In an exchange geometry, if two distinct planes contained in a 4-space meet, then their intersection is a point or a line.

3. Prove: In an exchange geometry, if A and B are lines and $A \cap B$ is a point, then $\langle A, B \rangle$ is a plane. What can be said if A and B are planes and $A \cap B$ is a point?

Definition. In any join geometry, suppose A and B are linear spaces. Then A and B are *independent* or *skew* if whenever $a_1, \ldots, a_m$ are linearly independent points of A and $b_1, \ldots, b_n$ are linearly independent points of B, it follows that $a_1, \ldots, a_m, b_1, \ldots, b_n$ are linearly independent. If A and B are not skew and $A \not\approx B$, then A and B are said to be *parallel*.

4. Prove: In any join geometry, if linear spaces A and B are skew, then $A \not\approx B$.

5. Prove: In an exchange geometry, if A and B are finite dimensional linear spaces, then A and B are skew if and only if $r(\langle A, B \rangle) = r(A) + r(B)$.

6. Prove: In an exchange geometry, if A and B are finite dimensional linear spaces and $A \not\approx B$, then A and B are parallel if and only if $r(\langle A, B \rangle) < r(A) + r(B)$.

7. Prove: In an exchange geometry, if A and B are skew linear spaces and $a \subset A$, then $A \cap \langle B, a \rangle = a$.

8. Prove: In an exchange geometry, if A and B are linear, $A \not\approx B$, and $A \cap \langle B, a \rangle = a$ for some point a, then A and B are skew.

*9. Prove in $\mathbb{R}^n$ the following parallel property: If A is linear, $d(A) > 0$, and $p \not\subset A$, then there exists a linear space B containing p such that A and B are parallel and $d(A) = d(B)$. Is B unique? Explain.

11.10 Characterizations of Linear Dependence

In Section 6.14 alternatives to the chosen definition of linear dependence of points were considered. In an exchange geometry these alternatives are equivalent to the adopted definition.

Theorem 11.13. *In an exchange geometry, given the points* $a_1, \ldots, a_n$, *the following statements are equivalent:*

(1) $a_1, \ldots, a_n$ *are linearly dependent.*

(2) $\langle a_{i_1}, \ldots, a_{i_r} \rangle \approx \langle a_{i_{r+1}}, \ldots, a_{i_n} \rangle$ *for some permutation* $i_1, \ldots, i_n$ *of the integers* $1, \ldots, n$ *and some* r, $1 \leqslant r < n$.

(3) *There exist points* $x_1, \ldots, x_{n-1}$ *such that*

$$a_1, \ldots, a_n \subset \langle x_1, \ldots, x_{n-1} \rangle.$$

PROOF. Suppose (1). Then for some i,

$$a_i \subset \langle a_1, \ldots, a_{i-1}, a_{i+1}, \ldots, a_n \rangle,$$

so that

$$\langle a_i \rangle \approx \langle a_1, \ldots, a_{i-1}, a_{i+1}, \ldots, a_n \rangle,$$

and (2) holds.

Suppose (2). Since the order of the given points $a_1, \ldots, a_n$ is immaterial, we may assume

$$\langle a_1, \ldots, a_r \rangle \approx \langle a_{r+1}, \ldots, a_n \rangle.$$

Let x be a point common to both members of the last relation. Then by Corollary 11.4,

$$\langle a_1, \ldots, a_r \rangle = \langle a_1, \ldots, a_{i-1}, x, a_{i+1}, \ldots, a_r \rangle$$

and

$$\langle a_{r+1}, \ldots, a_n \rangle = \langle a_{r+1}, \ldots, a_{j-1}, x, a_{j+1}, \ldots, a_n \rangle.$$

Hence

$$a_1, \ldots, a_n \subset \langle a_1, \ldots, a_n \rangle$$

$$= \langle \langle a_1, \ldots, a_r \rangle, \langle a_{r+1}, \ldots, a_n \rangle \rangle$$

$$= \langle \langle a_1, \ldots, a_{i-1}, x, a_{i+1}, \ldots, a_r \rangle, \langle a_{r+1}, \ldots, a_{j-1}, x, a_{j+1}, \ldots, a_n \rangle \rangle$$

$$= \langle x, a_1, \ldots, a_{i-1}, a_{i+1}, \ldots, a_r, a_{r+1}, \ldots, a_{j-1}, a_{j+1}, \ldots, a_n \rangle,$$

and (3) holds.

Finally, suppose (3). Then Corollary 11.5.1 implies (1). Thus the three statements are equivalent.   □

11.11 Rank of a Set of Points

The notion of finite rank was defined for certain linear spaces in an exchange geometry. Could the idea be generalized to apply to certain sets in an arbitrary join geometry? Yes, Corollary 11.9.2 is a key. The characterization of finiteness of rank given there is applicable to sets in any join geometry.

Definition. Let S be a set of points in an arbitrary join geometry. Suppose there is a maximum m for the cardinal numbers of linearly independent subsets of S. That is, S has a linearly independent subset of m points, and no linearly independent subset of S contains more than m points. Then S

is said to be of *finite rank*, m is called the *rank* of S, and we write $r(S) = m$.

Note that $r(\varnothing) = 0$, $r(a) = 1$ for any point a, and $r(S)$ equals the number of points in S for any linearly independent set of points S.

Observe that in an exchange geometry this new notion of rank applied to a linear set coincides with the old notion (Corollary 11.9.2).

In an exchange geometry a set of points has the same rank as its linear hull.

Theorem 11.14. *In an exchange geometry, let S be a set of points. If either S or $\langle S \rangle$ is of finite rank, both are of finite rank and $r(S) = r(\langle S \rangle)$.*

PROOF. The theorem is trivial if $S = \varnothing$. Assume $S \neq \varnothing$. If $\langle S \rangle$ is of finite rank, then certainly S is of finite rank, since any linearly independent subset of S is a linearly independent subset of $\langle S \rangle$. Hence it suffices to show for nonempty S that S is of finite rank implies $\langle S \rangle$ is of finite rank and $r(S) = r(\langle S \rangle)$.

Suppose then that S is of finite rank and $r(S) = n$. Let $a_1, \ldots, a_n$ be linearly independent points of S. Suppose $x \subset S$. Then the points $a_1, \ldots, a_n, x$ are linearly dependent. By Corollary 11.5.2 some point p of this sequence is in the linear hull of the preceding ones. The linear independence of $a_1, \ldots, a_n$ implies $p = x$; hence

$$x \subset \langle a_1, \ldots, a_n \rangle.$$

It follows that

$$S \subset \langle a_1, \ldots, a_n \rangle \subset \langle S \rangle.$$

Thus

$$\langle S \rangle = \langle a_1, \ldots, a_n \rangle,$$

so that $\langle S \rangle$ is of finite rank and

$$r(\langle S \rangle) = n = r(S). \qquad \square$$

11.12 Rank and Height of Linear Spaces

In Section 6.19, a measure of the relative complexity of a linear space, known as height, was introduced. In this section the relationship between height and rank is settled.

Theorem 11.15. *In any join geometry, if linear space A is of finite height, then it is of finite rank and $r(A) \leqslant h(A)$.*

PROOF. Suppose $a_1, \ldots, a_n$ are linearly independent points of A. Consider the sequence of linear spaces

$$\langle a_1, \ldots, a_n \rangle, \langle a_1, \ldots, a_{n-1} \rangle, \ldots, \langle a_1, a_2 \rangle, \langle a_1 \rangle, \varnothing. \qquad (1)$$

We assert (1) is a decreasing sequence of linear spaces of A. Suppose

$$\langle a_1, \ldots, a_{i+1} \rangle = \langle a_1, \ldots, a_i \rangle, \qquad 1 \leqslant i \leqslant n - 1. \qquad (2)$$

Then

$$a_{i+1} \subset \langle a_1, \ldots, a_i \rangle \subset \langle a_1, \ldots, a_i, a_{i+2}, \ldots, a_n \rangle,$$

contrary to the linear independence of $a_1, \ldots, a_n$. Thus (2) is false, so that each term of (1) properly contains its successor, and our assertion holds. Note that (1) has length n. It follows from the definition of height of A that

$$n \leqslant h(A).$$

Thus there must be a maximum for the cardinal numbers of linearly independent subsets of A, and this maximum cannot exceed $h(A)$. By definition, A is of finite rank and $r(A) \leqslant h(A)$.   □

Theorem 11.16. *In an exchange geometry, let A be a linear space of finite height or finite rank. Then A is of both finite height and finite rank, and $r(A) = h(A)$.*

PROOF. In view of the last theorem we need only show that if A is of finite rank, it is of finite height and $h(A) \leqslant r(A)$. Suppose A is of finite rank. Let

$$A = A_0, A_1, \ldots, A_n$$

be a decreasing sequence of linear spaces. Let

$$a_1 \subset A_{n-1} - A_n, \qquad a_2 \subset A_{n-2} - A_{n-1}, \ldots, \quad a_n \subset A_0 - A_1.$$

Observe that

$$a_1, \ldots, a_i \subset A_{n-i}, \quad a_{i+1} \not\subset A_{n-i}, \qquad 1 \leqslant i < n.$$

Hence for $1 \leqslant i < n$,

$$\langle a_1, \ldots, a_i \rangle \subset A_{n-i}$$

and

$$a_{i+1} \not\subset \langle a_1, \ldots, a_i \rangle.$$

By Corollary 11.5.3, $a_1, \ldots, a_n$ are linearly independent (and so distinct). Thus $\{a_1, \ldots, a_n\}$ is a linearly independent subset of A with cardinal number n. By definition (and Corollary 11.9.2),

$$n \leqslant r(A).$$

By definition, A is of finite height and

$$h(A) \leqslant r(A).$$ □

Corollary 11.16. *In an exchange geometry, let A be a linear space. Then the following are equivalent: (a) A is finitely generated; (b) A is finite dimensional; (c) A is of finite height.*

EXERCISES

1. Prove: In the join geometry T of Exercise 2 at the end of Section 11.4, pairs of statements (1), (3) and (2), (3) of Theorem 11.13 are each nonequivalent pairs of statements.

2. In the set Q of all ordered quadruples of real numbers, define a join operation $\cdot$ analogous to that defined in Exercise 2 at the end of Section 11.4. Show that $(Q, \cdot)$ is a join geometry and that the pair of statements (1), (2) of Theorem 11.13 is a nonequivalent pair in Q.

3. Prove: In an exchange geometry, if linear set L is of finite rank and $a \not\subset L$, then La, L/a and a/L are all of finite rank and each has rank equal to $r(L) + 1$.

4. Prove: In an exchange geometry if sets S and T are of finite rank, then so are $S \cup T$ and $S \cap T$, and moreover

$$r(S \cup T) + r(S \cap T) \leqslant r(S) + r(T).$$

5. Prove: In an exchange geometry, if K is a convex set of finite rank, then $\mathcal{I}(K)$ and $\mathcal{C}(K)$ are of finite rank, and moreover

$$r(K) = r(\mathcal{I}(K)) = r(\mathcal{C}(K)).$$

Definition. In any join geometry, let $A_0, A_1, \ldots, A_n$ be a sequence of linear spaces such that A_{i-1} covers A_i, $i = 1, \ldots, n$. Then $A_0, A_1, \ldots, A_n$ is called a *chain* of *length* n that *connects* A_0 and A_n.

6. Prove (Jordan–Dedekind Chain Condition): In an exchange geometry, if A is a linear space of finite rank, then there exists a chain that connects A and $\varnothing$, and the length of any such chain equals $r(A)$.

7. Prove: In an exchange geometry, if A and B are linear spaces of finite rank and $A \approx B$, then the length of any chain that connects A and $A \cap B$ equals the length of any chain that connects $\langle A, B \rangle$ and B.

*8. In the join geometry $(\mathbb{R}^*, \cdot)$ (Section 5.12), let $o = (0, 0, \ldots)$ and $e_i = (0, \ldots, 0, 1, 0, 0, \ldots)$ (1 is the ith entry). Let $L_i = \langle o, e_i \rangle$, and let J be the union of all L_i, $i = 1, 2, \ldots$. A join operation $\circ$ is defined in J as follows:

(1) $a \circ o = o \circ a = a \cdot o$ for all $a \subset J$.
(2) If $a, b \subset J - o$, $a, b \subset L_i$ for some i, and $a \cdot b \not\approx o$, then $a \circ b = a \cdot b$.
(3) If $a, b \subset J - o$, $a, b \subset L_i$ for some i, and $a \cdot b \approx o$, then

$$a \circ b = a \cdot b \cup L_1 \cup \cdots \cup L_{i-1}.$$

(4) If $a, b \subset J - o$, $a \subset L_i$, $b \subset L_j$, and $i > j$, then $a \circ b = a \cdot o$.
(5) If $a, b \subset J - o$, $a \subset L_i$, $b \subset L_j$, and $i < j$, then $a \circ b = o \cdot b$.

(a) Show that $\circ$ is indeed a join operation in J and that $(J, \circ)$ is a join geometry (see Exercise 9 at the end of Section 8.7).
(b) Show that J is of finite rank by showing any three points of J are linearly dependent.
(c) Show that J is not of finite height.

12 Ordered Join Geometries[1]

In this chapter, as in the previous one, a new postulate is introduced. The postulate is equivalent to the familiar Euclidean property of linear order: If three distinct points are collinear, then one of the three is between the other two. Join geometries which satisfy the new postulate are called ordered join geometries or ordered geometries. Ordered geometries are exchange geometries studied in the last chapter, but the results there will not be used in the present investigation. A flood of results is produced. First come formulas for lines, rays and segments, expressing how they are divided into subrays and subsegments by their points. Next come many properties of polytopes familiar in Euclidean geometry and easily accessible to intuition. Then follow properties of convex sets, less familiar in classical geometry but no less important, the theorems of Radon, Helly and Caratheodory and related results. These flow from a sharpened expansion formula for the linear hull of a finite set mediated by a new type of dependence of points. This new type of dependence is defined in terms of the convex hull operation and is called convex dependence. It implies linear dependence in any join geometry and is equivalent to linear dependence in an ordered geometry. Finally, separation of linear spaces by linear subspaces is studied, and results familiar in Euclidean geometry are obtained for ordered geometries.

[1] Chapter 8 is a prerequisite for this chapter. However only Sections 8.1–8.7 are needed until Section 12.23.

456

12.1 Ordered Join Geometries

The following postulate states a familiar property of Euclidean lines in join theoretic language.

O. *Order Postulate.* Let a, b and c be distinct points of a line. Then

$$a \subset bc, \quad b \subset ac, \quad \text{or} \quad c \subset ab.$$

Postulate O may be described as a *comparability* property, since it asserts that one of a, b, c must bear an order or betweenness relation (Section 4.23) to the other two:

$$(bac), \quad (abc) \text{ or } (acb).$$

Note in view of Corollary 4.39 that exactly one of these betweenness relations holds. The postulate also is related to the notion of an ordered set of points (Section 4.24) and may be restated:

Any line is an ordered set of points.

For this reason Postulate O will be called the Postulate of Linear Order or simply the Order Postulate.

Join geometries that satisfy Postulate O deserve a name.

Definition. A join geometry that satisfies Postulate O is called an *ordered* join geometry or an *ordered* geometry.

Observe that a Euclidean geometry satisfies Postulate O and that JG3, JG4 and JG5 are ordered geometries.

12.2 Two Points Determine a Line

Theorem 12.1. *In an ordered geometry, if $a \neq b$, there is a unique line that contains a and b, namely, $\langle a, b \rangle$.*

PROOF. Certainly $\langle a, b \rangle \supset a, b$ and is a line. Suppose line $L \supset a, b$. Then $L \supset \langle a, b \rangle$. To show

$$L \subset \langle a, b \rangle, \tag{1}$$

let $x \subset L$. If $x = a$ or $x = b$, certainly $x \subset \langle a, b \rangle$. Suppose $x \neq a, b$. Then $x, a, b \subset L$, and Postulate O implies

$$x \subset ab, \quad a \subset xb \text{ or } b \subset xa.$$

In any case

$$x \subset ab \cup a/b \cup b/a \subset \langle a, b \rangle,$$

and (1) holds. Thus $L = \langle a, b \rangle$, and the theorem is proved.   □

Remark. In Chapter 11, exchange geometries were studied, and it was shown that the familiar theory of incidence and dimension of linear spaces in Euclidean geometry is valid for exchange geometries. The above theorem together with Theorem 11.1, which characterizes an exchange geometry as one where two points "determine" a line, imply that an ordered geometry is an exchange geometry. Consequently, all of the results of Chapter 11 are valid in an ordered geometry. This important fact, however, will not be used in the present chapter, which is intended to be independent of Chapter 11.

An interesting point is that exchange geometries need not be ordered geometries. The triode join geometry JG10 is an example. In JG10, it is easily seen that J is the only line—it is the linear hull of any pair of distinct points. Thus any pair of distinct points is contained in a unique line, the line J, and Theorem 11.1 yields that JG10 is an exchange geometry. But if in JG10 points a, b, c are taken (Figure 12.1), one from each of the three rays used to define J, then a, b, c are in line J, but $a \not\subset bc$, $b \not\subset ac$ and $c \not\subset ab$.[2]

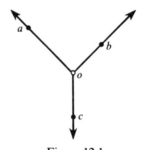

Figure 12.1

EXERCISES

1. Determine which of the join geometries JG1–JG16 are ordered geometries. (See Exercise 1 at the end of Section 6.10.)

2. Prove: In any ordered geometry, if two distinct lines meet, their intersection is a point.

3. Prove that $\mathbb{R}^n$ (Sections 5.6–5.9) is an ordered geometry. (You may use any of the exercises on the Join Geometry $\mathbb{R}^n$ at the end of Section 6.18.)

---

[2] For other examples similar to $\mathbb{R}^n$ but based on a partially ordered field, see Prenowitz [3], pp. 62–68.

4. Prove in an ordered geometry:
   (a) (pxq) and (pyq) imply exactly one of $x = y$, (pxy) or (pyx);
   (b) (pqx) and (pqy) imply exactly one of $x = y$, (pxy) or (pyx).

5. Prove: In an ordered geometry a line covers (Section 6.18) each of its points.

*6. Prove: In an ordered geometry no segment contains a proper ray.

12.3 A New Formula for Line *ab*

Line ab you recall was defined, if $a \neq b$, to be $\langle a, b \rangle$, and shown to be represented by the formula: line $ab = ab/ab$ (Section 6.10). In an ordered geometry a sharper formula can be obtained which is equivalent to a familiar Euclidean characterization of line.

Theorem 12.2. *In an ordered geometry*

$$\langle a, b \rangle = ab \cup a/b \cup b/a \cup a \cup b. \tag{1}$$

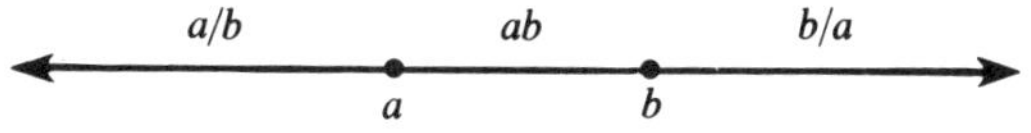

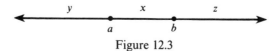

Figure 12.2

PROOF. If $a = b$ the theorem is trivial. Suppose $a \neq b$ (Figure 12.2). Certainly $\langle a, b \rangle \supset R$, where R is the right member of (1). To prove the reverse inclusion, let $x \subset \langle a, b \rangle$. If $x = a$ or $x = b$, then $x \subset R$. Suppose $x \neq a, b$. Then x, a, b are distinct points of the line $\langle a, b \rangle$. By Postulate O

$$x \subset ab, \qquad a \subset xb \quad \text{or} \quad b \subset xa.$$

In any case

$$x \subset ab \cup a/b \cup b/a \subset R.$$

Thus $R \supset \langle a, b \rangle$, and (1) follows. $\qquad\square$

Corollary 12.2. *In an ordered geometry*

$$\text{line } ab = ab \cup a/b \cup b/a \cup a \cup b, \qquad a \neq b. \tag{1}$$

Observe that (1) is a familiar characterization of line in Euclidean geometry expressed in join terminology [Section 6.1, (1)]. Interpreted in Euclidean geometry (Figure 12.3), it asserts that line ab, $a \neq b$, is the set

Figure 12.3

composed of a, b and all points x between a and b, all points y such that a is between y and b and all points z such that b is between a and z. It is, of course, intuitively obvious that none of the members $(ab, a/b, b/a, a, b)$ of the right side of (1) meets another. This is indeed the case in any join geometry (Exercise 1 at the end of Section 12.6 below). Thus by (1) we may say that $ab, a/b, b/a, a, b$ form a partition (Section 8.5) of line ab.

12.4 A Formula for a Ray

Theorem 12.3. *In an ordered geometry*

$$\overrightarrow{oa} = oa \cup a \cup a/o. \qquad (1)$$

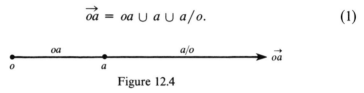

Figure 12.4

PROOF. By Corollary 8.4, $\overrightarrow{oa} \supset oa \cup a \cup a/o$ (Figure 12.4). Let $x \subset \overrightarrow{oa}$. By Theorem 12.2, $x \subset \langle o, a \rangle = oa \cup o/a \cup a/o \cup o \cup a$. If $x \subset oa \cup a \cup a/o$ for all x, then (1) holds. Otherwise $x \subset o/a \cup o$ for some x, so that $o \subset ax \cup x \subset \overrightarrow{oa}$. Hence $o = a$, and (1) is trivial. $\qquad \square$

Note by an observation above (the last sentence of Section 12.3), we see that if $a \neq o$, then $oa, a, a/o$ form a partition of the ray $\overrightarrow{oa}$.

Corollary 12.3. *In an ordered geometry let* $\overrightarrow{oa} = \overrightarrow{ob}$. *Then*

$$a \subset ob, \quad b \subset oa \quad or \quad a = b.$$

(See Figure 12.5.)

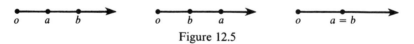

Figure 12.5

12.5 Another Formula for a Line

A line breaks down into two rays with a common endpoint.

Theorem 12.4. *In an ordered geometry let* $o \approx ab$. *Then*

$$\langle a, b \rangle = o/a \cup o/b \cup o. \qquad (1)$$

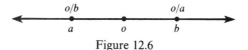
Figure 12.6

PROOF. If $a = b$, then $o = a$ and (1) is trivial. Suppose $a \neq b$ (Figure 12.6). Then $o \neq a$, and $\langle a, b \rangle$ is a line containing o, a. By Theorem 12.1,

$$\langle a, b \rangle = \langle o, a \rangle. \tag{2}$$

The condition $o \approx ab$ implies $a \subset o/b$, which is an o-ray. Thus $o/b = \overrightarrow{oa}$, and Theorem 12.3 implies

$$o/b = oa \cup a \cup a/o. \tag{3}$$

By Theorem 12.2,

$$\langle o, a \rangle = oa \cup o/a \cup a/o \cup o \cup a.$$

Then (2) and (3) yield (1), as desired.   □

Corollary 12.4.1. *In an ordered geometry let $o \approx ab$. Then*

$$\langle a, b \rangle = \overrightarrow{oa} \cup \overrightarrow{ob} \cup o.$$

PROOF. Apply Corollary 8.2.3.   □

Corollary 12.4.2. *In an ordered geometry let o be a point of line L. Then*

$$L = R \cup R' \cup o$$

where R and R' are proper opposite rays with endpoint o.

(See Figure 12.7.)

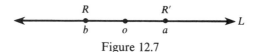
Figure 12.7

PROOF. Let $a \subset L$, $a \neq o$. Choose $b \subset o/a$. Then $o \approx ab$. Let $R = o/a$, $R' = o/b$. Then R, R' are opposite o-rays by Corollary 8.8. Note that o, a, b are distinct points of L. Thus R, R' are proper o-rays, and $L = \langle a, b \rangle$ by Theorem 12.1. Apply the theorem.   □

Note it follows from Theorem 8.6 that R, R', o form a partition of L.

Remark. The corollary says: In an ordered geometry any point of a line "divides" it into two rays. This does not hold for arbitrary join geometries —JG16 is a counterexample (see Exercise 2 below).

12.6 An Expansion Formula for a Segment

A familiar Euclidean representation of a segment in terms of two subsegments holds in an arbitrary ordered geometry.

Theorem 12.5 (Segment Expansion Theorem). *In an ordered geometry let* $p \subset ab$. *Then*

$$ab = pa \cup pb \cup p. \qquad (1)$$

PROOF. If $a = b$, (1) is trivial. Suppose $a \neq b$ (Figure 12.8). Let $S = pa \cup pb \cup p$. We show $S \subset ab$. Note that $p \subset ab$ implies $pa \subset aba = ab$. Similarly $pb \subset ab$. Thus $S \subset ab$.

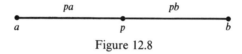

Figure 12.8

To show the reverse inclusion let $x \subset ab$. This implies by Theorem 12.4

$$\langle a, b \rangle = x/a \cup x/b \cup x.$$

Since $p \subset ab \subset \langle a, b \rangle$, we have

$$p \subset x/a \cup x/b \cup x.$$

Thus

$$x \subset pa \quad \text{or} \quad x \subset pb \quad \text{or} \quad x = p,$$

so that $x \subset S$ and the theorem follows.   □

Note that pa, pb, p form a partition of ab if $a \neq b$ (Exercise 3 below).

Remark. The theorem implies that in an ordered geometry any point of a proper segment "divides" it into two proper segments. This result does not hold for arbitrary join geometries. (See Remark above following Corollary 12.4.2.)

EXERCISES

1. Prove: In any join geometry, if $a \neq b$, then $ab, a/b, b/a, a, b$ are pairwise disjoint.

2. Show that Corollary 12.4.2 and Theorem 12.5 do not hold for an arbitrary join geometry and that JG16 is a counterexample in each case.

3. Prove: In any join geometry, if $a \neq b$ and $p \subset ab$, then pa, pb, p are pairwise disjoint.

4. Prove: If J is a join geometry which satisfies
$$\langle a, b \rangle = ab \cup a/b \cup b/a \cup a \cup b$$
for all $a, b \subset J$, then J is an ordered geometry.

5. Prove: If J is a join geometry which satisfies
$$\overrightarrow{ab} = ab \cup b \cup b/a$$
for all $a, b \subset J$, then J is an ordered geometry.

6. Must join geometry J be an ordered geometry if
   (a) $o \subset ab$ implies $\langle a, b \rangle = o/a \cup o/b \cup o$ for all $o, a, b \subset J$?
   (b) $p \subset ab$ implies $ab = pa \cup pb \cup p$ for all $p, a, b \subset J$?

7. Prove: In an ordered geometry, if A and B are convex subsets of a line and $A \approx B$, then $A \cup B$ is convex.

8. Prove: In an ordered geometry $C(ab) = [a, b]$.

9. Prove: In an ordered geometry $C(a/b) = a \cup a/b$.

10. Prove: In an ordered geometry $ab = cd$ implies $\{a, b\} = \{c, d\}$. This is essentially equivalent to: Any proper segment has a unique pair of endpoints. [Compare Exercise 4(b) at the end of Section 5.3.]

11. Prove: In an ordered geometry $a/b = c/d$ implies $a = c$, that is, any ray has a unique endpoint. (Compare Exercise 9 at the end of Section 8.7.)

*12. Prove: In an ordered geometry, if three proper rays are contained in a line, then one of the rays is a subset of a second.

*13. Prove: In an ordered geometry, if two segments meet, then their intersection is a segment (possibly degenerate).

*14. Prove: In an ordered geometry, if n points are given on a line, then the points can be labeled to form a sequence, $p_1, \ldots, p_n$, such that $i < j < k$ (i, j, k in the range $1, \ldots, n$) implies $p_j \subset p_i p_k$. Are there other such labelings of the points? If so, how many?

15. Prove: In an ordered geometry, if R and R' are opposite o-rays, then
$$\langle R \rangle = \langle R' \rangle = RR' = R \cup R' \cup o.$$

(Compare Theorem 8.12, and Exercises 6, 7 at the end of Section 8.7.)

12.7 An Expansion Formula for a Join of Points

A sequence of theorems is now developed which culminates in several important properties of polytopes.

First the segment expansion theorem is generalized from the join of two to the join of n points.

Theorem 12.6 (Join Expansion Theorem). *In an ordered geometry let* $p \subset a_1 \ldots a_n, \ n > 1$. *Then*

$$a_1 \ldots a_n = p(a_1 \cup \cdots \cup a_n \cup a_1 a_2 \cup \cdots \cup a_{n-1} a_n$$

$$\cup \cdots \cup a_1 \ldots a_{n-1} \cup \cdots \cup a_2 \ldots a_n) \cup p. \quad (1)$$

That is, $a_1 \ldots a_n$ *is obtained by joining p with the union of all joins of $n-1$ or fewer a's and adjoining p to the result.*

PROOF. To begin we express (1) in a more convenient form. Using Theorem 2.8 and J4, it can be shown that

$$\{a_1, \ldots, a_n\}^{n-1} = a_1 \cup \cdots \cup a_n \cup a_1 a_2 \cup \cdots \cup a_{n-1} a_n$$

$$\cup \cdots \cup a_1 \ldots a_{n-1} \cup \cdots \cup a_2 \ldots a_n.$$

Thus (1) becomes

$$a_1 \ldots a_n = p\{a_1, \ldots, a_n\}^{n-1} \cup p.$$

Restricting our attention to ordered geometries, the theorem may be stated:

$$\text{If} \quad p \subset a_1 \ldots a_n \quad \text{then} \quad a_1 \ldots a_n = p\{a_1, \ldots, a_n\}^{n-1} \cup p. \quad (A)$$

This will be proved by induction on n.

(A) holds for $n = 2$ by the Segment Expansion Theorem (Theorem 12.5), which yields

$$a_1 a_2 = pa_1 \cup pa_2 \cup p = p(a_1 \cup a_2) \cup p = p\{a_1, a_2\} \cup p.$$

Assume (A) for $n = k$: $r \subset a_1 \ldots a_k$ implies

$$a_1 \ldots a_k = r\{a_1, \ldots, a_k\}^{k-1} \cup r.$$

Now we suppose (Figure 12.9)

$$p \subset a_1 \ldots a_{k+1} \quad (2)$$

and prove

$$a_1 \ldots a_{k+1} = p\{a_1, \ldots, a_{k+1}\}^k \cup p. \quad (3)$$

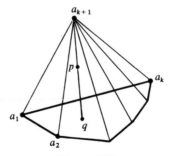

Figure 12.9

In order to apply the inductive assumption we "project" p from a_{k+1} onto $a_1 \ldots a_k$. Formally, the relation (2) implies the existence of q such that

$$p \subset qa_{k+1}, \qquad q \subset a_1 \ldots a_k. \tag{4}$$

By the Segment Expansion Theorem, (4) implies

$$qa_{k+1} = pq \cup pa_{k+1} \cup p. \tag{5}$$

By the inductive assumption for $r = q$, (4) implies

$$a_1 \ldots a_k = q\{a_1, \ldots, a_k\}^{k-1} \cup q. \tag{6}$$

Multiplying both members of (6) by a_{k+1}, we get

$$a_1 \ldots a_{k+1} = qa_{k+1}\{a_1, \ldots, a_k\}^{k-1} \cup qa_{k+1}. \tag{7}$$

Now using (7), (5) and (6) we have

$$a_1 \ldots a_{k+1} = (pq \cup pa_{k+1} \cup p)\{a_1, \ldots, a_k\}^{k-1}$$

$$\cup pq \cup pa_{k+1} \cup p$$

$$= pq\{a_1, \ldots, a_k\}^{k-1} \cup pa_{k+1}\{a_1, \ldots, a_k\}^{k-1}$$

$$\cup p\{a_1, \ldots, a_k\}^{k-1} \cup pq \cup pa_{k+1} \cup p$$

$$= p\big(\{a_1, \ldots, a_k\}^{k-1} \cup \{a_1, \ldots, a_k\}^{k-1}a_{k+1}$$

$$\cup a_{k+1} \cup q\{a_1, \ldots, a_k\}^{k-1} \cup q\big) \cup p$$

$$= p\big(\{a_1, \ldots, a_k\}^{k-1} \cup \{a_1, \ldots, a_k\}^{k-1}a_{k+1}$$

$$\cup a_{k+1} \cup a_1 \ldots a_k\big) \cup p$$

$$= p\{a_1, \ldots, a_{k+1}\}^{k} \cup p.$$

Thus (3) follows, and (A) holds by induction. $\qquad\square$

Remark. The theorem may be described as follows. Let P be the polytope $[a_1, \ldots, a_n]$, and let the union of all joins of $n-1$ or fewer a's be called the *quasifrontier* of P. Suppose p is a point of $\mathcal{J}(P) = a_1 \ldots a_n$. Then $\mathcal{J}(P)$ is obtained by joining p to the quasifrontier of P and adjoining p to the result. Note the application to a simplex (see Corollary 6.20.3).

Corollary 12.6. *In an ordered geometry let $p \subset [a_1, \ldots, a_n]$, $n > 1$. Then*

$$a_1 \ldots a_n \subset p\{a_1, \ldots, a_n\}^{n-1} \cup p. \tag{1}$$

PROOF. $p \subset a_{i_1} \ldots a_{i_r}$, some join of the a's. There is no loss of generality if we assume

$$p \subset a_1 \ldots a_r.$$

If $r = 1$ (that is, $p = a_1$), certainly (1) holds. Assume $r > 1$. By the theorem,

$$a_1 \ldots a_r = p\{a_1, \ldots, a_r\}^{r-1} \cup p.$$

Hence

$$a_1 \ldots a_n = \left(p\{a_1, \ldots, a_r\}^{r-1} \cup p\right)a_{r+1} \ldots a_n$$

$$= p\{a_1, \ldots, a_r\}^{r-1}a_{r+1} \ldots a_n \cup pa_{r+1} \ldots a_n$$

$$\subset p\{a_1, \ldots, a_n\}^{n-1} \cup p. \qquad \square$$

12.8 An Expansion for a Polytope

Now an elementary and very useful expansion for a polytope can be obtained.

Theorem 12.7 (Polytope Expansion Formula). *In an ordered geometry let* $p \subset [a_1, \ldots, a_n]$. *Then*

$$[a_1, \ldots, a_n] = [p, a_2, \ldots, a_n] \cup [a_1, p, a_3, \ldots, a_n]$$

$$\cup \cdots \cup [a_1, \ldots, a_{n-1}, p]. \qquad (1)$$

PROOF. If $n = 1$, then (1) is trivial. Assume $n > 1$. Let L and R be the left and right members of (1) respectively. Then $L \supset R$. We show $R \supset L$. Let T be the union of all joins of $n - 1$ or fewer a's; equivalently, $T = \{a_1, \ldots, a_n\}^{n-1}$. Then

$$R \supset p, T, pT.$$

This relation and the last corollary imply

$$R \supset pT \cup p \supset a_1 \ldots a_n.$$

Thus R contains all joins of the a's, so that $R \supset L$ and the theorem follows. $\qquad \square$

12.9 Modularity Principles

We begin with two theorems which are weak forms of distributive laws, so-called modular laws. (They are similar to Theorem 10.25.)

First a modular law for join and intersection.

Theorem 12.8 (Modular Law). *In any join geometry let* A *and* B *be sets of points,* L *linear and* $A \subset L$. *Then*

$$(AB) \cap L = A(B \cap L).$$

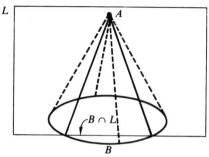

Figure 12.10

PROOF. $A(B \cap L) \subset AB$, L (Figure 12.10). Therefore

$$A(B \cap L) \subset (AB) \cap L.$$

Let $x \subset (AB) \cap L$. Then $x \subset ab$, where $a \subset A$, $b \subset B$. Hence

$$b \approx x/a \subset L.$$

Therefore

$$b \approx B \cap L$$

and

$$x \subset A(B \cap L).$$

Thus

$$(AB) \cap L \subset A(B \cap L),$$

and the theorem holds.   □

Remark. The conclusion of the theorem may be written, in view of $A \subset L$,

$$(AB) \cap L = (A \cap L)(B \cap L),$$

and so the theorem may be considered a weak form of distributive law for intersection with respect to join.

Now the following modular law involving convex hull and intersection is easily obtained.

Theorem 12.9 (Modular Law). *In any join geometry let A and B be convex, $A \subset L$, and L be linear. Then*

$$[A, B] \cap L = [A, B \cap L].$$

(See Figure 12.11.)

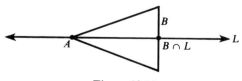

Figure 12.11

PROOF.

$$\begin{aligned}
[A, B] \cap L &= (A \cup B \cup AB) \cap L && \text{(Corollary 3.10)} \\
&= (A \cap L) \cup (B \cap L) \cup ((AB) \cap L) \\
&= A \cup (B \cap L) \cup A(B \cap L) && \text{(Theorem 12.8)} \\
&= [A, B \cap L] && \text{(Corollary 3.10).} \quad \square
\end{aligned}$$

12.10 The Section of a Polytope by a Linear Space

This section is devoted to a key result on polytopes: that a nonempty intersection of a polytope and a linear space must be a polytope. The proof makes use of the following lemma.

Lemma 12.10. *In any join geometry the union of a finite family of polytopes is a polytope, provided it is convex.*

PROOF. Let $P_1, \ldots, P_n$ be polytopes and $P_1 \cup \cdots \cup P_n$ be convex. Let Q be the polytope generated by the vertices of the P's. Then

$$Q \supset P_1 \cup \cdots \cup P_n.$$

Conversely $P_1 \cup \cdots \cup P_n \supset Q$, since it is convex and contains a set of generators of Q. Thus $P_1 \cup \cdots \cup P_n = Q$, and the lemma holds. $\quad \square$

Theorem 12.10 (The Polytope Section Theorem). *In an ordered geometry let P be a polytope and L a linear space. Then $P \cap L$ is $\varnothing$ or a polytope.*

PROOF. We apply induction on n, the number of elements in a set of generators of P. The theorem is restated to assert that the desired property holds for any polytope P that has a generating set composed of n elements.

The theorem certainly holds for $n = 1$.

Suppose it holds for $n = k$. Let polytope P have a generating set composed of $k + 1$ elements. Then

$$P = [a_1, \ldots, a_{k+1}], \quad \text{where the } a\text{'s are distinct.}$$

We may assume $P \cap L \neq \varnothing$. Let

$$p \subset P \cap L.$$

Since $p \subset P$ we have, by the Polytope Expansion Formula (Theorem 12.7),

$$P = [a_1, \ldots, a_{k+1}] = [p, a_2, \ldots, a_{k+1}] \cup \cdots \cup [a_1, \ldots, a_k, p].$$

Let

$$P_i = [a_1, \ldots, a_{i-1}, p, a_{i+1}, \ldots, a_{k+1}].$$

Then

$$P = P_1 \cup \cdots \cup P_{k+1},$$

and

$$P \cap L = (P_1 \cap L) \cup \cdots \cup (P_{k+1} \cap L). \qquad (1)$$

We show each of the intersections in the right member of (1) is a polytope. The case $P_1 \cap L$ is typical:

$$P_1 \cap L = [p, a_2, \ldots, a_{k+1}] \cap L$$

$$= [p, [a_2, \ldots, a_{k+1}]] \cap L$$

$$= [p, [a_2, \ldots, a_{k+1}] \cap L] \qquad \text{(Theorem 12.9)}$$

$$= [p, P'],$$

where $P' = [a_2, \ldots, a_{k+1}] \cap L$ and by the inductive assumption is $\varnothing$ or a polytope. If $P' = \varnothing$, then $P_1 \cap L = [p, \varnothing] = [p]$, a polytope. Otherwise $P' = [b_1, \ldots, b_r]$ for some points $b_1, \ldots, b_r$, hence

$$P_1 \cap L = [p, [b_1, \ldots, b_r]] = [p, b_1, \ldots, b_r],$$

and $P_1 \cap L$ is again a polytope. We infer from (1) that $P \cap L$ is a union of polytopes and so is a polytope by the preceding lemma. Hence the theorem holds by induction. $\qquad\qquad\qquad\qquad\qquad\qquad\qquad\qquad\qquad\square$

Counterexample. The theorem fails for an arbitrary join geometry. In JG16 let $a = (-1, -1)$, $b = (1, 1)$, and L be the line $y = 0$ (Figure 12.12). Then $[a, b] \cap L$ is not a polytope: it is the open segment with endpoints $(1, 0)$ and $(-1, 0)$. (See Exercise 1 at the end of Section 4.15.)

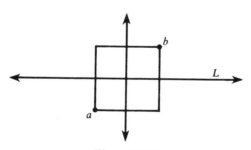

Figure 12.12

12.11 Consequences of the Polytope Section Theorem

We begin with a lemma whose proof does not depend on the Polytope Section Theorem.

Lemma 12.11. *In an ordered geometry any polytope P contained in a line must be a closed segment.*

PROOF. If *P* is a point, it is a (degenerate) closed segment. Suppose *P* not a point. Then it has two distinct extreme points, *a* and *b* (Corollary 4.21). Suppose *P* has a third extreme point *c*. By Postulate O,

$$a \subset bc, \qquad b \subset ac \quad \text{or} \quad c \subset ab,$$

contrary to the definition of extreme point. Thus *a* and *b* are the only extreme points of *P*, and $P = [a, b]$ by Theorem 4.21. □

Now several results on the intersection of a polytope or its frontier with a line, a ray and a closed segment are obtained.

Theorem 12.11. *In an ordered geometry let a line L meet a polytope P. Then their intersection is a closed segment. If the intersection is a proper closed segment, its endpoints are in $\mathfrak{F}(P)$.*

PROOF. The Polytope Section Theorem implies that $L \cap P$ is a polytope, and the lemma, that it is a closed segment (Figure 12.13). Suppose $L \cap P$ is the closed segment $[a, b]$, $a \neq b$. Does $a \subset \mathfrak{I}(P)$ hold? Suppose it does. Then $a \subset bc$, where $c \subset P$. Thus $c \subset a/b \subset L$ and $c \subset [a, b]$. But *a* is an extreme point of $[a, b]$ (Theorem 4.20). Hence $b = c$, so that $a = b$ which is impossible. Thus $a \not\subset \mathfrak{I}(P)$ and $a \subset \mathfrak{F}(P)$. By symmetry $b \subset \mathfrak{F}(P)$, and the proof is finished. □

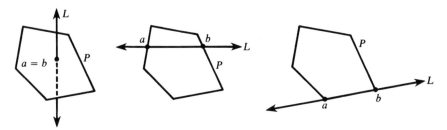

Figure 12.13

Corollary 12.11. *In an ordered geometry let a line intersect a polytope P in two distinct points. Then the line intersects $\mathfrak{F}(P)$ in at least two points.*

Next the section of a polytope by a closed segment. We prove an elementary characteristic property of a polytope.

Theorem 12.12. *In an ordered geometry let a closed segment meet a polytope. Then their intersection is a closed segment.*

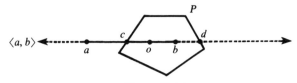

Figure 12.14

Proof. Suppose P is a polytope and $[a, b] \approx P$ (Figure 12.14). The conclusion is trivial if $a = b$. Suppose $a \neq b$. Let $S = [a, b] \cap P$. Then by Theorem 12.11

$$S = [a, b] \cap \langle a, b \rangle \cap P = [a, b] \cap [c, d],$$

where $[c, d] = \langle a, b \rangle \cap P$. Let $o \subset S$. By the Polytope Expansion Formula

$$S = [a, b] \cap [c, d] = ([o, a] \cup [o, b]) \cap ([o, c] \cup [o, d]),$$

so that

$$S = ([o, a] \cap [o, c]) \cup ([o, a] \cap [o, d]) \cup ([o, b] \cap [o, c])$$

$$\cup ([o, b] \cap [o, d]). \tag{1}$$

We assert that each term on the right is a closed segment. Consider $[o, a] \cap [o, c]$ as a typical case. The assertion is true if o, a, c are not distinct. Suppose then o, a, c are distinct. Note that $o, a, c \subset \langle a, b \rangle$. Hence by Postulate O

$$o \subset ac, \quad a \subset oc \quad \text{or} \quad c \subset oa. \tag{2}$$

If $a \subset oc$ then $[o, a] \subset [o, c]$ and $[o, a] \cap [o, c] = [o, a]$. Similarly, if $c \subset oa$, then $[o, a] \cap [o, c] = [o, c]$.

Finally suppose $a \not\subset oc$ and $c \not\subset oa$. Then (2) implies $o \subset ac$. We show

$$oa \not\approx oc. \tag{3}$$

Suppose $oa \approx oc$. We eliminate a between this relation and $o \subset ac$. We have

$$oc/o \approx a \approx o/c,$$

and, therefore, $oc \approx o$ and $o = c$, which is false. Thus (3) holds. Then

$$[o, a] \cap [o, c] = (o \cup a \cup oa) \cap (o \cup c \cup oc) = o.$$

In all cases then $[o, a] \cap [o, c]$ is a closed segment, and our assertion is justified. Now (1) implies that S is the union of four closed segments and so is a polytope (Lemma 12.10). Finally, $S \subset \langle a, b \rangle$ implies S is a closed segment (Lemma 12.11). ☐

Now another case where a line meets a polytope.

Theorem 12.13. *In an ordered geometry suppose line L contains an interior point o of polytope P and $L \subset \langle P \rangle$. Then L intersects $\mathcal{F}(P)$ in exactly two points, and these points are separated by o.*

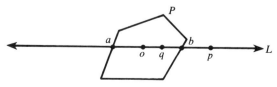

Figure 12.15

PROOF. Let $p \subset L$, $p \neq o$ (Figure 12.15). Then $o \subset \mathcal{I}(P)$, $p \subset \langle P \rangle$ imply $op \approx \mathcal{I}(P)$ (Theorem 6.27); say q is in both. Then $L \cap P \supset o, q$ and $o \neq q$. By Theorem 12.11, $L \cap P$ is a proper closed segment $[a, b]$ and $a, b \subset \mathcal{F}(P)$. Since $o \subset [a, b]$ and $o \neq a, b$, we have that o separates a, b; that is,

$$o \approx ab. \tag{1}$$

To complete the proof we argue as follows. Since ab is open, (1) implies $ab = oab$ (Corollary 4.25.1). But $o \subset \mathcal{I}(P)$ implies $oab \subset \mathcal{I}(P)$ (Corollary 2.15.1). Hence

$$ab \subset \mathcal{I}(P).$$

Then

$$L \cap \mathcal{F}(P) = L \cap P \cap \mathcal{F}(P) = [a, b] \cap \mathcal{F}(P) = \{a, b\}. \qquad \square$$

Finally the case where a ray emanates from an interior point of a polytope and pierces its frontier.

Theorem 12.14. *In an ordered geometry suppose P is a polytope, R a proper ray with endpoint o in $\mathcal{I}(P)$, and $R \subset \langle P \rangle$. Then R meets $\mathcal{F}(P)$ in a unique point.*

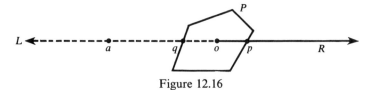

Figure 12.16

PROOF. Let $R = o/a$, $o \neq a$. Let $L = \langle R \rangle = \langle o/a \rangle = \langle o, a \rangle$ (Theorem 6.13). Then L is a line containing R, o. Moreover, $L \subset \langle P \rangle$, since $R \subset \langle P \rangle$. By Theorem 12.13,

$$L \cap \mathcal{F}(P) = \{p, q\}, \qquad p \neq q, \tag{1}$$

and

$$o \approx pq. \tag{2}$$

Theorem 12.1 and Corollary 12.4.1 applied to (1) and (2) respectively yield

$$L = \langle p, q \rangle = \overrightarrow{op} \cup \overrightarrow{oq} \cup o.$$

Since $R \subset L$ and R is a proper o-ray, we have $R \approx \overrightarrow{op}$ or $R \approx \overrightarrow{oq}$. Thus $R = \overrightarrow{op}$ or $R = \overrightarrow{oq}$ (Theorem 8.1), and $R \supset p$ or $R \supset q$. Note $R \supset p, q$ is impossible, since otherwise, $R \supset pq$ and (2) yields $R \supset o$, contrary to "R is proper." Hence R contains exactly one of p, q, say $R \supset p$. Then employing (1), we obtain

$$R \cap \mathcal{F}(P) = R \cap L \cap \mathcal{F}(P) = R \cap \{p, q\} = p,$$

and the theorem is proved. $\qquad \square$

12.12 Polytopes Are Closed

The last theorem has the consequence that polytopes are closed convex sets.

Theorem 12.15. *In an ordered geometry any polytope is closed.*

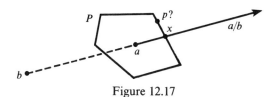

Figure 12.17

PROOF. Let P be a polytope (Figure 12.17). We must show $\mathcal{C}(P) \subset P$. It suffices to show

$$\mathcal{B}(P) \subset P. \tag{1}$$

Let $x \subset \mathcal{B}(P)$. Choose a in $\mathcal{I}(P)$. Then

$$xa \subset \mathcal{I}(P) \qquad \text{(Theorem 2.23)}.$$

Let $b \subset a/x$. Then $x \subset a/b$. Since $a, x \subset \mathcal{C}(P)$,

$$b \subset \langle \mathcal{C}(P) \rangle = \langle P \rangle \qquad \text{(Theorem 6.15)}.$$

Thus $a/b \subset \langle P \rangle$. Since $a \neq x$, we have $a \neq b$, and a/b is a proper ray. Then the preceding theorem implies that a/b meets $\mathcal{F}(P)$ in a point p. Thus $x, p \subset a/b$ and $\overrightarrow{ax} = \overrightarrow{ap}$. By Corollary 12.3

$$x \subset ap, \quad p \subset ax \quad \text{or} \quad x = p. \tag{2}$$

Consider the possibilities in (2). $x \subset ap$ is false, since it implies $x \subset \mathcal{I}(P)$ (Theorem 2.15). Similarly $p \subset ax$ is false. Thus (2) implies $x = p$ and $x \subset \mathcal{F}(P) \subset P$. This justifies (1), and the theorem follows. □

Corollary 12.15. *In an ordered geometry any closed segment is closed, that is,* $\mathcal{C}[a, b] = [a, b]$.

[Compare Exercise 4(a) at the end of Section 5.3.]

12.13 The Interior of a Polytope—Another View

Theorem 12.16. *In an ordered geometry let P be a polytope and $o \subset \mathcal{I}(P)$. Then*

$$\mathcal{I}(P) = o\mathcal{F}(P) \cup o.$$

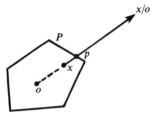

Figure 12.18

PROOF.

$$o\mathcal{F}(P) \cup o \subset \mathcal{I}(P) \qquad \text{(Corollary 2.15.1).}$$

To show the reverse inclusion let $x \subset \mathcal{I}(P)$ (Figure 12.18). If $x = o$, certainly

$$x \subset o\mathcal{F}(P) \cup o. \tag{1}$$

Suppose $x \ne o$. Then x/o is a proper ray with endpoint x in $\mathcal{I}(P)$. Moreover $x/o \subset \langle P \rangle$. By Theorem 12.14, $(x/o) \cap \mathcal{F}(P) = p$, a point. But $p \subset x/o$ implies

$$x \subset op \subset o\mathcal{F}(P),$$

which implies (1) and yields the result.   □

Remark. The theorem gives a simple geometric procedure for obtaining the interior of a polytope. It calls for comparison with Theorem 12.6, the Join Expansion Theorem, which can be stated so: Let $P = [a_1, \ldots, a_n]$, $n > 1$, $o \subset \mathcal{I}(P)$, and Q be the union of all joins of $n - 1$ or fewer a's. Then

$$\mathcal{I}(P) = oQ \cup o.$$

How is $\mathcal{F}(P)$ related to Q? The answer is: $\mathcal{F}(P) \subset Q$. For

$$\mathcal{F}(P) = (Q \cup a_1 \ldots a_n) - a_1 \ldots a_n \subset Q.$$

12.14 Separation Property of the Frontier of a Polytope

We make use of the definition of the separation of two sets by a set (Section 8.13) to define the separation of a set into two sets by a set.

Definition. Suppose S, T, A and B are sets of points (Figure 12.19) such that S (strictly) separates A and B and

$$T = S \cup A \cup B.$$

Then we say S (*strictly*) *separates* T *into* A *and* B or simply S (*strictly*) *separates* T.

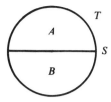

Figure 12.19

In Euclidean geometry, for example, a line of a plane separates the plane into two halfplanes, and a triangle separates its plane into its interior and its exterior.

Observation on Strict Separation. If S strictly separates A and B, then in addition to $S \not\approx A, B$ we have $A \not\approx B$ (Section 8.13). Thus if S strictly separates T into A and B, then S, A, B form a partition of T.

Theorem 12.17 (The Polytope Separation Theorem). *In an ordered geometry let P be a polytope which is not a point. Then the frontier of P strictly separates $\langle P \rangle$ into the interior of P and the exterior of P.*

PROOF. Clearly $\mathcal{I}(P) \neq \varnothing$. Note that $\mathcal{E}(P) \neq \varnothing$ by Corollary 6.28.2.

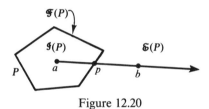

Figure 12.20

First we show $\mathcal{F}(P)$ separates $\mathcal{I}(P)$ and $\mathcal{E}(P)$ (Figure 12.20). Let $a \subset \mathcal{I}(P)$, $b \subset \mathcal{E}(P)$. Note that $a \neq b$, so that $\overrightarrow{ab}$ is a proper ray, and $\overrightarrow{ab} \subset \langle P \rangle$. By Theorem 12.14,

$$\overrightarrow{ab} \cap \mathcal{F}(P) = p.$$

Then $\overrightarrow{ab} = \overrightarrow{ap}$, and Corollary 12.3 yields

$$b \subset ap, \quad p \subset ab \quad \text{or} \quad p = b. \tag{1}$$

But $b \subset ap$ and $p = b$ each yield $b \subset P$, contrary to $b \subset \mathcal{E}(P)$. Hence (1) implies $p \subset ab$ and so $ab \approx \mathcal{F}(P)$. By definition $\mathcal{F}(P)$ separates $\mathcal{I}(P)$ and $\mathcal{E}(P)$. The separation is strict, since $\mathcal{F}(P) \not\approx \mathcal{I}(P)$ and $\mathcal{F}(P) \subset P \not\approx \mathcal{E}(P)$.

To complete the proof we have $\mathcal{E}(P) = \langle P \rangle - \mathcal{C}(P) = \langle P \rangle - P$ (Theorem 12.15), so that

$$\langle P \rangle = P \cup \mathcal{E}(P) = \mathcal{F}(P) \cup \mathcal{I}(P) \cup \mathcal{E}(P),$$

and the result holds by definition. □

This completes the development of the first constellation of results in the chapter, the elementary geometric properties of polytopes—elementary in the sense not of simplicity of proof but of ready accessibility to intuition.

EXERCISES

1. Verify Theorem 12.6, its corollary and Theorem 12.7 in several cases chosen in JG4 and JG5.

2. Prove: In an ordered geometry
$$\mathcal{C}(a_1 \ldots a_n) = [a_1, \ldots, a_n].$$

3. Prove: In any join geometry, if $a_1, \ldots, a_n$ are linearly independent and $p \subset a_1 \ldots a_n$, then every pair of
$$p, pa_1, \ldots, pa_n, pa_1a_2, \ldots, pa_{n-1}a_n, \ldots, \ pa_1 \ldots a_{n-1}, \ldots, pa_2 \ldots a_n$$
is a disjoint pair.

4. Prove: In an ordered geometry no polytope contains a proper ray.

5. Prove: In any join geometry, if K is convex and L is linear, then every extreme point of $K \cap L$ is in $\mathcal{F}(K)$ provided $K \cap L$ contains at least two points.

6. Prove: In an ordered geometry if K is convex and $p \subset \mathcal{I}(K)$, then segment pq and ray p/q cannot intersect $\mathcal{F}(K)$ in more than one point.

7. Prove: In an ordered geometry, if K is convex, $o \subset \mathcal{I}(K)$ and $x, y \subset \mathcal{F}(K)$, then $ox \approx oy$ implies $x = y$. How does this result relate to Theorem 12.16?

8. Prove: In an ordered geometry, if K is convex and has nonempty interior and exterior, then $\mathcal{I}(K) = o\mathcal{F}(K) \cup o$ for $o \subset \mathcal{I}(K)$ only if $\mathcal{F}(K)$ separates $\mathcal{I}(K)$ and $\mathcal{E}(K)$.

9. Prove: In an ordered geometry any point of
   (a) a line strictly separates the line into two rays.
   (b) a proper segment strictly separates the segment into two subsegments.
   (c) a proper ray strictly separates a ray into a segment and a subray.

10. Prove: In an ordered geometry any closed segment contained in
    (a) a line strictly separates the line into two rays.
    (b) a proper segment strictly separates the segment into two subsegments.
    (c) a proper ray strictly separates the ray into a segment and a subray.

11. Must the conclusion of Theorem 12.17 hold if P is
    (a) convex and has nonempty interior and exterior?
    (b) closed and has nonempty interior and exterior?

12.15 A New Formula for Linear Hulls

The object of this section is to derive a sharpened formula for the linear hull of a finite set. The result, Theorem 12.20, follows from two preliminary theorems.

Theorem 12.18. *In an ordered geometry*

$$ab/a = ab \cup b/a \cup b. \tag{1}$$

(Compare Theorem 12.3.)

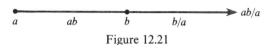

Figure 12.21

PROOF. By Theorem 8.5, $ab/a = \overrightarrow{ab}$. Then (1) is immediate by Theorem 12.3.   □

Remark. The formula (1) is easily recalled by a graphical mnemonic: To traverse the ray ab/a (see Figure 12.21), move from a to b, pass through b and continue moving indefinitely in the direction opposite to a. There is also a simple formal mnemonic: To obtain the three terms in the expansion of ab/a: (1) suppress a in the denominator; (2) suppress a in the numerator; (3) suppress a in both.

Corollary 12.18. *In an ordered geometry*

$$aB/a = aB \cup B/a \cup B.$$

The next theorem, which follows readily from Theorem 12.18, is the key to the structure of linear sets in an ordered geometry.

Theorem 12.19 (Common Factor Principle). *In an ordered geometry*

$$ab/ac = ab/c \cup b/ac \cup b/c.$$

(See Figure 12.22.)

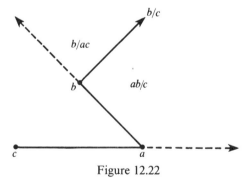

Figure 12.22

PROOF.

$$
\begin{aligned}
ab/ac &= (ab/a)/c \\
&= (ab \cup b/a \cup b)/c &&\text{(Theorem 12.18)} \\
&= ab/c \cup (b/a)/c \cup b/c &&\text{(Theorem 4.2)} \\
&= ab/c \cup b/ac \cup b/c.
\end{aligned}
$$
    □

Corollary 12.19 (Common Factor Principle). *In an ordered geometry*

$$aB/aC = aB/C \cup B/aC \cup B/C.$$

The formal mnemonic for Theorem 12.18 is valid for its corollary and for Theorem 12.19 and its corollary.

In an ordered geometry the formula for the linear hull of a finite nonempty set (Theorem 6.9) can be sharpened significantly.

Theorem 12.20. *In an ordered geometry* $\langle a_1, \dots, a_n \rangle$ *is the union of all sets expressible in the following forms*:

$$a_{i_1} \dots a_{i_r}, \qquad 1 \leqslant i_1 < \cdots < i_r \leqslant n, \tag{I}$$

$$a_{i_1} \dots a_{i_r}/a_{j_1} \dots a_{j_s}, \qquad 1 \leqslant i_1 < \cdots < i_r \leqslant n,$$

$$1 \leqslant j_1 < \cdots < j_s \leqslant n, \quad i_k \neq j_l. \tag{II}$$

(Compare Theorem 12.2.)

Thus in particular

$$\langle a_1, a_2, a_3 \rangle = a_1 \cup a_2 \cup a_3 \cup a_1 a_2 \cup a_1 a_3 \cup a_2 a_3 \cup a_1 a_2 a_3$$

$$\cup a_1/a_2 \cup a_1/a_3 \cup a_2/a_1 \cup a_2/a_3 \cup a_3/a_1 \cup a_3/a_2$$

$$\cup a_1/a_2 a_3 \cup a_2/a_1 a_3 \cup a_3/a_1 a_2$$

$$\cup a_1 a_2/a_3 \cup a_1 a_3/a_2 \cup a_2 a_3/a_1.$$

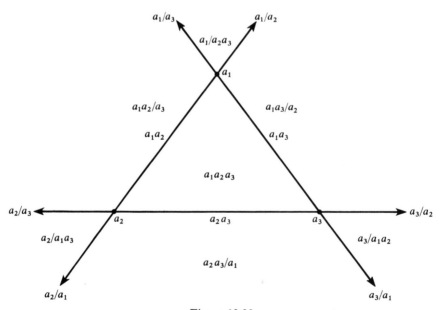

Figure 12.23

PROOF. Note that $\langle a_1, \ldots, a_n \rangle$ certainly contains all sets expressible in the forms (I) or (II). We have

$$\langle a_1, \ldots, a_n \rangle = a_1 \ldots a_n / a_1 \ldots a_n \qquad \text{(Theorem 6.9)}. \qquad (1)$$

We apply the Common Factor Principle (Corollary 12.19) to the right member of (1) in order to eliminate repetition of the factor a_1. We get

$$\langle a_1, \ldots, a_n \rangle = a_1 \ldots a_n / a_2 \ldots a_n \cup a_2 \ldots a_n / a_1 \ldots a_n$$

$$\cup a_2 \ldots a_n / a_2 \ldots a_n.$$

Similarly we eliminate repetitions of the factor a_2 in each term of this set union; for example,

$$a_1 \ldots a_n / a_2 \ldots a_n$$

is reduced to

$$a_1 \ldots a_n / a_3 \ldots a_n \cup a_1 a_3 \ldots a_n / a_2 \ldots a_n \cup a_1 a_3 \ldots a_n / a_3 \ldots a_n.$$

Continuing to eliminate repeated factors in this way, we eventually obtain an expansion for $\langle a_1, \ldots, a_n \rangle$ in which all terms are expressed in the forms (I) or (II). It should be noted that in dealing with an expression such as $a_1 \ldots a_n / a_1$, in which the repeated factor constitutes the denominator, the same process applies but is justified by Corollary 12.18. If the repeated factor is the numerator, the argument is even simpler; for example,

$$a_1 / a_1 \ldots a_n = (a_1 / a_1) / a_2 \ldots a_n = a_1 / a_2 \ldots a_n.$$

Thus $\langle a_1, \ldots, a_n \rangle$ is precisely the union of all expressions (I) and (II). $\square$

12.16 Convex Dependence and Independence

The notion of convex dependence is related to that of linear dependence and plays an important role in the theory of ordered geometries.

Definition. In any join geometry, points $a_1, \ldots, a_n$ are *convexly dependent* or *C-dependent* if there exist distinct integers $i_1, \ldots, i_r, j_1, \ldots, j_s$ in the range $1, \ldots, n$ such that

$$[a_{i_1}, \ldots, a_{i_r}] \approx [a_{j_1}, \ldots, a_{j_s}]. \qquad (1)$$

In the contrary case $a_1, \ldots, a_n$ are *convexly independent* or *C-independent*.

Note that if $a_1, \ldots, a_n$ are not distinct, they are C-dependent. Thus C-independent points are distinct, and it is natural to refer to a C-independent *set* of points. Therefore, we introduce the

Definition. Let S be a finite set of points. Suppose whenever S_1 and S_2 are disjoint subsets of S it follows that $[S_1] \not\approx [S_2]$. Then S is said to be *convexly independent* or *C-independent*.

Note that $\varnothing$ is C-independent and that any subset of a C-independent set is C-independent. (Compare Section 6.15.)

Note that a's which do not appear in (1) may be inserted into either member of (1) without affecting the validity of (1). Thus $a_1, \ldots, a_n$ are C-dependent if and only if there exists a permutation $1', \ldots, n'$ of $1, \ldots, n$ such that for some r, $1 \leqslant r < n$,

$$[a_{1'}, \ldots, a_{r'}] \approx [a_{(r+1)'}, \ldots, a_{n'}].$$

Figure 12.24 illustrates in Euclidean geometry the C-dependence of a, b, c and of a, b, c, d.

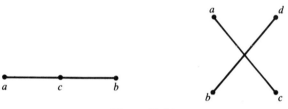

Figure 12.24

The term convexly independent is sometimes used in a different sense, but the present definition is the natural and useful one in the present context.

Hereafter the term linear independence will be abbreviated to L-*independence*, as convex independence was to C-*independence*, and similarly for linear dependence.

Independence concepts give rise naturally to notions of rank or dimension of linear spaces, of convex sets, or in general of sets of points.

Definition. In any join geometry let S be a set of points. Suppose there is a maximum m for the cardinal numbers of L-independent [C-independent] subsets of S. That is, S has an L-independent [C-independent] subset of cardinal number m, and no L-independent [C-independent] subset of S has cardinal number greater than m. Then S is said to be of *finite L-rank* [C-*rank*] and m is called the L-*rank* [C-*rank*] of S, denoted $r_L(S)$ [$r_C(S)$]. The L-*dimension* [C-*dimension*] of a set is its L-rank [C-rank] diminished by unity.[3]

Note that $\varnothing$ is of both finite L-rank and finite C-rank with $r_L(\varnothing) = r_C(\varnothing) = 0$.

C-dependence of points can be characterized without employing the convex hull concept.

---

[3] The L-rank of a set is defined exactly as the rank of a set in Section 11.11.

Theorem 12.21. *In any join geometry* $a_1, \ldots, a_n$ *are C-dependent if and only if there exist distinct integers* $p_1, \ldots, p_t, q_1, \ldots, q_u$ *in the range* $1, \ldots, n$ *such that*

$$a_{p_1} \cdots a_{p_t} \approx a_{q_1} \cdots a_{q_u}. \tag{1}$$

PROOF. Suppose $a_1, \ldots, a_n$ C-dependent. Then by definition there exist distinct integers $i_1, \ldots, i_r, j_1, \ldots, j_s$ in the range $1, \ldots, n$ such that

$$[a_{i_1}, \ldots, a_{i_r}] \approx [a_{j_1}, \ldots, a_{j_s}].$$

This relation and Theorem 3.2 imply that some join of the a_i's meets a join of the a_j's. Hence

$$a_{p_1} \cdots a_{p_t} \approx a_{q_1} \cdots a_{q_u},$$

where the subscripts are distinct and in the range $1, \ldots, n$.

Conversely, suppose (1), where the subscripts are distinct and in the range $1, \ldots, n$. Then

$$[a_{p_1}, \ldots, a_{p_t}] \approx [a_{q_1}, \ldots, a_{q_u}].$$

This relation asserts that $a_1, \ldots, a_n$ are C-dependent, and the proof is finished. ☐

12.17 How Are C-Dependence and L-Dependence Related?

C-dependence implies L-dependence.

Theorem 12.22. *In any join geometry, if* $a_1, \ldots, a_n$ *are C-dependent, they are L-dependent.*

PROOF. Suppose $a_1, \ldots, a_n$ are C-dependent. By Theorem 12.21 there exist distinct integers $p_1, \ldots, p_t, q_1, \ldots, q_u$ in the range $1, \ldots, n$ such that

$$a_{p_1} \cdots a_{p_t} \approx a_{q_1} \cdots a_{q_u}.$$

By Corollary 6.20.1, $a_1, \ldots, a_n$ are L-dependent. ☐

Remark. The result may be described by saying that the relation of C-dependence is at least as strong as L-dependence—whenever a sequence of points satisfies the first relation, it must satisfy the second. Actually the converse of Theorem 12.22 is not valid, and C-dependence is "stronger" than L-dependence (see Exercise 6 below).

Corollary 12.22.1. *In any join geometry, if* $a_1, \ldots, a_n$ *are L-independent, then they are C-independent.*

Corollary 12.22.2. *In any join geometry let S be of finite C-rank. Then S is of finite L-rank and $r_L(S) \leqslant r_C(S)$.*

PROOF. The result is trivial if $S = \emptyset$. Assume $S \neq \emptyset$, and let $r_C(S) = m$. Suppose points $a_1, \ldots, a_n$ of S are L-independent. By the previous corollary $a_1, \ldots, a_n$ are C-independent. Hence by definition of m, we have $n \leqslant m$. Thus there must be a maximum for the cardinal numbers of L-independent subsets of S, and this maximum cannot exceed m. By definition, S is of finite L-rank and $r_L(S) \leqslant r_C(S)$.   □

EXERCISES

1. Interpret Corollaries 12.18 and 12.19 in Euclidean geometry for various choices of sets B and C.

2. Verify Theorem 12.20 in Euclidean geometry for $n = 4$ if a_1, a_2, a_3, a_4 are vertices of
   (a) a tetrahedron;
   (b) a convex quadrilateral.

3. Prove: In any join geometry, if $a_1, \ldots, a_n$ are linearly independent, then any pair of expressions each of which occurs in (I) or (II) of Theorem 12.20 is a disjoint pair.

4. In a Euclidean plane let a, b, c, d be C-dependent distinct points that are noncollinear. Draw figures to illustrate how a, b, c, d are related. How many different types of figures can you find? Do the same for 5 noncoplanar points in Euclidean 3-space.

5. Find the L-rank and the C-rank of the basic set J in each of the join geometries JG1–JG16.

6. Show JG10 and JG16 are counterexamples for the converse of Theorem 12.22. What are the L-rank and the C-rank of the basic set J in JG10; in JG16?

12.18 Convex Dependence and Independence in an Ordered Geometry

In an ordered geometry convex dependence (independence) is equivalent to linear dependence (independence).

Theorem 12.23. *In an ordered geometry $a_1, \ldots, a_n$ are C-dependent if and only if they are L-dependent.*

PROOF. In view of Theorem 12.22 we need only show the L-dependence of $a_1, \ldots, a_n$ implies their C-dependence. Suppose $a_1, \ldots, a_n$ are L-dependent. Then one of the a's is in the linear hull of the remaining a's. Since the order of the a's is immaterial, let us say

$$a_n \subset \langle a_1, \ldots, a_{n-1} \rangle.$$

By Theorem 12.20 applied to $\langle a_1, \ldots, a_{n-1} \rangle$, we have

$$a_n \approx a_{i_1} \ldots a_{i_r}, \quad 1 \leqslant i_1 < \cdots < i_r \leqslant n-1; \tag{1}$$

or

$$a_n \approx a_{i_1} \ldots a_{i_r} / a_{j_1} \ldots a_{j_s}, \quad 1 \leqslant i_1 < \cdots < i_r \leqslant n-1,$$

$$1 \leqslant j_1 < \cdots < j_s \leqslant n-1, \quad i_k \neq j_l. \tag{2}$$

Relation (2) implies

$$a_n a_{j_1} \ldots a_{j_s} \approx a_{i_1} \ldots a_{i_r}. \tag{3}$$

Thus we have (1) or (3) and so an intersection relation of two joins of the a's with no repeated subscript. By Theorem 12.21, $a_1, \ldots, a_n$ are C-dependent, and the result follows. $\square$

Corollary 12.23. *In an ordered geometry $a_1, \ldots, a_n$ are C-independent if and only if they are L-independent.*

Theorem 12.24. *In an ordered geometry let S be of finite C-rank or finite L-rank. Then S is of both finite C-rank and finite L-rank, and $r_C(S) = r_L(S)$.*

PROOF. In view of Corollary 12.22.2 we need only show that if S has finite L-rank then S has finite C-rank and $r_C(S) \leqslant r_L(S)$. But since C-independent points are L-independent (Corollary 12.23), this is shown by the method used for Corollary 12.22.2. $\square$

The theorem suggests the following simplification in terminology.

Definition. In an ordered geometry the L-rank of a set S is called its *rank*.

The equivalence result in Theorem 12.23 for convex and linear dependence leads to the generalization to ordered geometries of several well-known theorems on convex sets in $\mathbb{R}^n$.

The first of these is due to Radon.

Theorem 12.25. *In an ordered geometry let linear space A have L-rank n. Then any $n+1$ points of A are C-dependent.*

PROOF. Since A has L-rank n, its C-rank is n by Theorem 12.24. Hence any $n+1$ points of A are C-dependent. $\square$

The theorem applies to Euclidean geometry and gives interesting information on Euclidean spaces of arbitrary dimension. Since a Euclidean

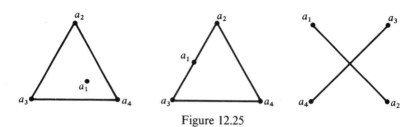

Figure 12.25

plane has L-rank 3, the theorem asserts: Any four points of a Euclidean plane are C-dependent. (See Figure 12.25; also Exercises 12 and 17(a) at the end of Section 3.4.)

Radon's Theorem as usually stated is now generalized.

Corollary 12.25 (Radon). *In an ordered join geometry if S is a set of $n + 1$ points in a linear space of rank n, then there exist disjoint subsets S_1 and S_2 of S such that $[S_1] \approx [S_2]$.*

12.19 Convex Dependence and Polytopes

Convex dependence plays a role in the theory of polytopes.

Theorem 12.26. *In any join geometry let $a_1, \ldots, a_n$ be C-dependent. Then the polytopes generated by sets of $n - 1$ of the a's,*

$$[a_1, \ldots, a_{n-1}], \ldots, [a_2, \ldots, a_n], \tag{1}$$

have a common point.

(See Figure 12.26.)

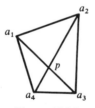

Figure 12.26

PROOF. Since $a_1, \ldots, a_n$ are C-dependent,

$$[a_{i_1}, \ldots, a_{i_r}] \approx [a_{j_1}, \ldots, a_{j_s}], \tag{2}$$

where $i_1, \ldots, i_r, j_1, \ldots, j_s$ are distinct integers in the range $1, \ldots, n$. Let p be a common point of the members of (2). Then each polytope in (1) contains one of the members of (2) and so contains p. □

The result may be restated thus: If a polytope is generated by n C-dependent points, then there is a point comon to the subpolytopes generated by sets of $n - 1$ of these points.

12.20 Helly's Theorem

Theorem 12.26 leads readily to a generalized form of a well-known theorem of Helly on convex sets in $\mathbb{R}^n$.

Theorem 12.27. *In any join geometry let $X_1, \ldots, X_{n+1}$ be convex sets in A, a linear space of C-rank $n \geqslant 1$. Let each n of $X_1, \ldots, X_{n+1}$ have a common point. Then there is a point common to $X_1, \ldots, X_{n+1}$.*

PROOF. We pick a point which is common to the members of each set of n X's. Specifically, let p_i, $1 \leqslant i \leqslant n + 1$, be in all of the X's with the possible exception of X_i:

$$p_i \subset X_j, \quad j \neq i, \quad 1 \leqslant j \leqslant n + 1.$$

Then

$$X_i \supset p_1, \ldots, p_{i-1}, p_{i+1}, \ldots, p_{n+1}.$$

Hence

$$X_i \supset [p_1, \ldots, p_{i-1}, p_{i+1}, \ldots, p_{n+1}]. \tag{1}$$

But the C-rank of A is n. Hence the points $p_1, \ldots, p_{n+1}$ must be C-dependent. Thus by Theorem 12.26 the sets

$$[p_1, \ldots, p_{i-1}, p_{i+1}, \ldots, p_{n+1}], \quad 1 \leqslant i \leqslant n + 1,$$

have a common point. Hence by (1) the X_i, $1 \leqslant i \leqslant n + 1$, have a common point. $\qquad\square$

Corollary 12.27 (Helly). *In an ordered geometry let linear space A have L-rank $n \geqslant 1$. Then if $X_1, \ldots, X_{n+1}$ are convex sets in A each n of which have a common point, all must have a common point.*

12.21 Decomposition of Polytopes into Subpolytopes: Caratheodory's Theorem

In an ordered geometry Theorem 12.26 yields a simple but nontrivial decomposition for a polytope.

Theorem 12.28. *In an ordered geometry let $a_1, \ldots, a_n$ be L-dependent. Then $[a_1, \ldots, a_n]$ is the union of the polytopes generated by sets of $n - 1$ of the* *a's:*

$$[a_1, \ldots, a_n] = [a_1, \ldots, a_{n-1}] \cup \cdots \cup [a_2, \ldots, a_n]. \qquad (1)$$

PROOF. For any $p \subset [a_1, \ldots, a_n]$, Theorem 12.7, the Polytope Expansion Formula, yields

$$[a_1, \ldots, a_n] = [p, a_1, \ldots, a_{n-1}] \cup \cdots \cup [p, a_2, \ldots, a_n]. \qquad (2)$$

We aim to make a judicious choice for p. By Theorem 12.23, $a_1, \ldots, a_n$ are C-dependent. Hence by Theorem 12.26, the terms in the right member of (1) have a common point. We choose such a point for p. Then p is redundant in each term of the right member of (2), and (2) reduces to (1). □

The following corollary is essentially a restatement of the theorem.

Corollary 12.28. *In an ordered geometry, if a polytope P has n vertices and is not a simplex, it is the union of the polytopes generated by sets of n − 1 of its vertices.*

The corollary suggests the following result.

Theorem 12.29. *In an ordered geometry any polytope P is expressible as a finite union of simplexes whose vertices are vertices of P.*

PROOF. First note that if a simplex is generated by vertices of P, its vertices are vertices of P (Theorem 4.19). Suppose P has n vertices. If P is a simplex the theorem holds. Suppose P is not a simplex. By the preceding corollary

$$P = P_1 \cup \cdots \cup P_n$$

where $P_1, \ldots, P_n$ are the polytopes generated by sets of $n - 1$ vertices of P. If each P_i is a simplex, we stop here. Suppose this is not so. Then P is the union of a family, possibly empty, of simplexes generated by $n - 1$ vertices of P and a family of polytopes generated by $n - 2$ vertices of P. Continuing in this way, we see that after at most $n - 1$ steps, P will be expressed as a finite union of simplexes generated by vertices of P. □

Corollary 12.29.1. *In an ordered geometry [S] is expressible as a union of simplexes whose vertices are points of S.*

(Compare Corollary 3.7.2.)

PROOF. Recall that $[S]$ is a union of polytopes generated by points of S (Corollary 3.7.2) and that the vertices of these polytopes must then be points of S (Theorem 4.19). Now apply the theorem. □

Corollary 12.29.2. *In an ordered geometry* $x \subset [S]$ *if and only if* x *is a point of a simplex whose vertices are in* S.

(Compare Corollary 3.7.3.)

In the theorem and its corollaries no information is given about the ranks of the simplexes involved. If however we impose a restriction on the rank of the convex set which is expressed as a union of simplexes, then information can be obtained on the ranks of these simplexes.

Theorem 12.30. *In an ordered geometry suppose* K *is a convex set of rank* r *and* $K = [S]$. *Then* K *is the union of a family of simplexes of ranks not exceeding* r *whose vertices are in* S.

PROOF. Apply Corollary 12.29.1, and observe since each simplex is a subset of K, it cannot have a linearly independent subset of more than r points. □

Corollary 12.30.1. *In an ordered geometry let* L *be a linear space of rank* r. *Suppose* $S \subset L$. *Then* $[S]$ *is the union of a family of simplexes of rank not exceeding* r *whose vertices are in* S.

A generalized form of Caratheodory's Theorem is immediate.

Corollary 12.30.2 (Caratheodory). *In an ordered geometry let* L *be a linear space of rank* r. *Suppose* $S \subset L$. *Then* $x \subset [S]$ *if and only if* x *is in a join of at most* r *points of* S.[4]

(Compare Corollary 3.7.1.)

Theorem 12.31. *In an ordered geometry suppose* $S \subset L$, *a nonempty linear space of rank* r. *Then* $[S] = S^r$.

(Compare Corollary 3.6.1 and Theorem 3.8.)

PROOF. Let $x \subset [S]$. By the preceding corollary

$$x \subset a_1 \ldots a_m, \tag{1}$$

where the a's are in S and $m \le r$. Note that $r - m$ factors a_m can be inserted in the right member of (1) so as to convert it into a join of r points of S. Then (1) implies $x \subset S^r$ and so $[S] \subset S^r$. The reverse inclusion follows by Theorem 3.8, and so the theorem follows. □

---

[4] It should be noted that Bryant and Webster [1] generalize the theorems of Radon, Helly and Caratheodory in $\mathbb{R}^n$ to a type of join geometry.

12.22 Equivalence of Linear and Convex Independence

Theorem 12.32. *A join geometry J is ordered if and only if L-independence and C-independence are equivalent in J.*

PROOF. In an ordered geometry the concepts of linear and convex independence are equivalent (Corollary 12.23). Suppose conversely that J is a join geometry in which these concepts are equivalent. We show every line in J is ordered. First we prove that the conclusion of Theorem 12.2 is valid in J:

$$\langle a, b \rangle = ab \cup a/b \cup b/a \cup a \cup b. \tag{1}$$

Let S denote the right member of (1). Then $S \subset \langle a, b \rangle$. Suppose

$$x \subset \langle a, b \rangle. \tag{2}$$

Then x, a, b are L-dependent, and so C-dependent. Thus

$$x \subset [a, b], \quad a \subset [x, b] \quad \text{or} \quad b \subset [x, a].$$

If $x \subset [a, b]$, then $x \subset S$. Suppose $a \subset [x, b]$. Then

$$a \subset x \cup b \cup xb. \tag{3}$$

If $a = b$, then (2) implies $x = a \subset S$. Suppose $a \neq b$. Then (3) implies $a = x$ or $a \subset xb$ and $x \subset S$ in either case. Similarly $b \subset [x, a]$ implies $x \subset S$. Thus $\langle a, b \rangle \subset S$, and (1) is verified.

Next we prove the following property:

$$p \subset \langle a, b \rangle, \quad p \neq a, \quad \text{implies} \quad \langle a, b \rangle = \langle a, p \rangle. \tag{E}$$

$p \subset \langle a, b \rangle$ implies $\langle a, p \rangle \subset \langle a, b \rangle$. By (1) and $p \neq a$ we have

$$p \subset ab \cup a/b \cup b/a \cup b.$$

This yields

$$b \subset p/a \cup a/p \cup ap \cup p \subset \langle a, p \rangle.$$

Hence $\langle a, b \rangle \subset \langle a, p \rangle$ and (E) is verified.

Now let L be any line of J, and suppose p, q, r are distinct points of L. We have to show

$$p \subset qr, \quad q \subset pr \quad \text{or} \quad r \subset pq. \tag{4}$$

This will follow readily from $L = \langle p, q \rangle$, which we proceed to prove. Let $L = \langle a, b \rangle$, $a \neq b$. Then $p, q \subset \langle a, b \rangle$. The relation $p \neq q$ implies $p \neq a$ or $q \neq a$; let us say $p \neq a$. Using (E) we have

$$L = \langle a, b \rangle = \langle a, p \rangle = \langle p, a \rangle. \tag{5}$$

Then $q \subset \langle p, a \rangle$ and $q \neq p$, so that (E) implies

$$\langle p, a \rangle = \langle p, q \rangle. \tag{6}$$

Equations (5) and (6) imply $L = \langle p, q \rangle$. Thus using (1),

$$r \subset \langle p, q \rangle = pq \cup p/q \cup q/p \cup p \cup q.$$

This relation implies, since $r \neq p, q$,

$$r \subset pq \cup p/q \cup q/p,$$

and (4) follows. Thus J is an ordered geometry, and the theorem holds. $\square$

Remark. Readers familiar with Chapter 11 will note that the property (E) in the proof is Postulate E of Section 11.1.

EXERCISES

1. Show that the results in Corollary 12.27, Theorem 12.28 and Theorem 12.31 do not hold in JG16.

2. Verify Theorem 12.31 for some of the sets presented in the set of exercises on Convex Hulls of Euclidean Sets at the end of Section 3.2.

3. Prove: S_1 and S_2 in Corollary 12.25 are uniquely determined if and only if every n points of S are linearly independent.

4. Prove the following extension of Theorem 12.27. In any join geomery let $X_1, \ldots, X_m$ be convex sets in A, a linear space of C-rank $n < m$. Let each n of $X_1, \ldots, X_m$ have a common point. Then there is a point common to $X_1, \ldots, X_m$.

5. In $\mathbb{R}^2$ let S_1, S_2, S_3, S_4 be vertical segments (i.e., each segment has endpoints with equal first coordinates). Suppose every three of S_1, S_2, S_3, S_4 meet a common line. Show that there is a line which meets each of S_1, S_2, S_3, S_4. [*Hint.* $\mathbb{R}^2$ has rank 3. Let C_i, $i = 1, 2, 3, 4$, be the set of all ordered pairs (m, b) of real numbers such that the line having equation $y = mx + b$ meets S_i. Prove each C_i is convex.]

6. Show that the conclusion of Theorem 12.31 cannot be sharpened. That is, show in any nonempty linear space L of rank r that there exists a set S such that $[S] \neq S^m$ for $m < r$.

12.23 The Separation of Linear Spaces by Linear Spaces

The final theme of the chapter is now treated: The separation of linear spaces by linear spaces in an ordered geometry. The results are typical of Euclidean geometry but not of arbitrary join geometries: a line is separated by any of its points, a plane by any of its lines, and in general a linear space by any linear space it covers.

First a formula for the linear hull of a nonempty linear space and a point.

Theorem 12.33. *In an ordered geometry let L be a nonempty linear space. Then*

$$\langle L, a \rangle = La/L \cup L/a \cup L. \tag{1}$$

(See Figure 12.27.)

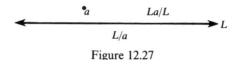

Figure 12.27

PROOF.

$$
\begin{array}{lll}
\langle L, a \rangle = \langle La \rangle & \text{(Theorem 6.13)} \\[4pt]
\quad = La/La & \text{(Corollary 6.11.1)} \\[4pt]
\quad = La/L \cup L/La \cup L/L & \text{(Corollary 12.19)} \\[4pt]
\quad = La/L \cup (L/L)/a \cup L \\[4pt]
\quad = La/L \cup L/a \cup L. & \qquad\qquad \square
\end{array}
$$

Remarks on the Theorem. If $a \not\subset L$, the case where the theorem has nontrivial content, then La/L and L/a are proper L-rays (halfspaces of L), indeed, they are opposite L-rays (Theorems 8.17, 8.19). Furthermore since no two of La/L, L/a and L meet, they form a partition of $\langle L, a \rangle$ (Section 8.5). Finally by employing Corollary 8.22.2 and the theorem, we obtain that L strictly separates $\langle L, a \rangle$ into La/L and L/a (Section 12.14).

Corollary 12.33. *In an ordered geometry let L be a nonempty linear space which is covered by linear space M. Then L strictly separates M into H and H', where H and H' are proper opposite halfspaces of L.*

(Compare Corollary 12.4.2.)

PROOF. Since M covers L, we have $M = \langle L, a \rangle$ for some $a \not\subset L$ (Theorem 6.21). Apply the above remarks. $\square$

Corollary 12.33 has a converse which holds in any join geometry.

Theorem 12.34. *In any join geometry suppose L is a nonempty linear space which strictly separates the linear space M into sets A and B. Then M covers L, and A and B are proper opposite halfspaces of L.*

PROOF. Each of L, A, B is nonempty, and these sets form a partition of M (Figure 12.28). Let a be a point of A, and b a point of B. Then for any

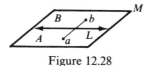

Figure 12.28

$x \subset A$ we have $xb \approx L$, so that $x \approx L/b$. Hence

$$A \subset L/b \subset M, \tag{1}$$

and similarly

$$B \subset L/a \subset M. \tag{2}$$

Next $a, b \not\subset L$ imply L/a, L/b are proper halfspaces of L. Since $ab \approx L$, Corollary 8.20 yields that L/a and L/b are opposite halfspaces of L. Then no two of L, L/a, L/b meet, so that (1) and (2) together with

$$M = L \cup A \cup B \tag{3}$$

imply $A = L/b$ and $B = L/a$. Thus A and B are proper opposite half-spaces of L.

It remains to be shown that M covers L. Clearly M properly contains L. Suppose X is linear, $X \neq L$ and $L \subset X \subset M$. Then by (3), $X \approx A$ or $X \approx B$. Suppose $X \approx A = L/b$. Then $b \approx L/X \subset X$, so that

$$A = L/b \subset L/X \subset X. \tag{4}$$

Moreover, since $ab \approx L$, we have by (4) $a \approx L/b \subset X$. Hence

$$B = L/a \subset L/X \subset X. \tag{5}$$

Similarly $X \approx B = L/a$ implies (4) and (5). By (3), (4) and (5), $X = M$. Thus M covers L and the theorem is proven.   □

We may combine results.

Corollary 12.34. *In an ordered geometry let L and M be nonempty linear spaces. Then L strictly separates M if and only if M covers L.*

Our final result is a point spanning principle in ordered geometries for the linear covering relation.

Theorem 12.35. *In an ordered geometry let A and B be linear. Then A covers B if and only if $A = \langle B, a \rangle$ for some $a \not\subset B$.*

(Compare Theorem 6.21.)

PROOF. Suppose A covers B. Then Theorem 6.21 certainly yields $A = \langle B, a \rangle$ for some $a \not\subset B$. Conversely suppose $A = \langle B, a \rangle$ and $a \not\subset B$. If $B = \varnothing$ then $A = a$, so that A covers B. If $B \neq \varnothing$, then the Remarks on Theorem 12.33 imply that B strictly separates A. Thus Corollary 12.34 implies A covers B.   □

EXERCISES

1. Prove: In an ordered geometry, if a, b, c are linearly independent and L is a line contained in $\langle a, b, c \rangle$, then $\langle a, b, c \rangle$ covers L.

2. Prove Pasch's Postulate: In an ordered geometry, if a, b, c are linearly independent and L is a line such that $L \subset \langle a, b, c \rangle$ but $L \not\supset a, b, c$, then $L \approx ab$ implies $L \approx ac$ or $L \approx bc$ but not both. Interpret the result in Euclidean geometry.

3. Prove: In an ordered geometry, if L is linear, then
   (a) $\mathcal{C}(L/a) = L \cup L/a$.
   (b) $\mathcal{C}(L \cup L/a) = L \cup L/a$.
   (Compare the Remark in Section 8.15.)

4. Prove: In an ordered geometry, if L is a nonempty linear space, and H and H' are opposite L-rays, then
$$\langle H \rangle = \langle H' \rangle = HH' = H \cup H' \cup L.$$
   If H is a proper L-ray, does $\langle H \rangle$ cover L? Explain. (Compare Exercise 15 at the end of Section 12.6.)

*5. Prove: In an ordered geometry if five points, no three of which lie on a line, are contained in a linear space of rank 3, then there exist at least four of these points which are vertices of some polytope. (For a generalization see Grünbaum [1], p. 22, Exercise 12.)

The Structure of Polytopes in an Ordered Geometry **13**

This chapter falls into three parts. First the theory of extreme sets and extremal linear spaces for convex sets is employed to elicit basic facial structural properties of polytopes in ordered geometries. The chief result is that a nontrivial polytope can be represented as the intersection of a certain finite family of closed halfspaces. These halfspaces have edges which are linear hulls of the maximal proper faces, or *facets*, of the polytope. Secondly, conditions are obtained, involving intersection properties with lines and segments, for a convex set to be the convex hull of its extreme points. Thirdly two characterizations of polytopes as types of convex set are obtained. One is based on the intersection properties mentioned above, the other on the representation of a polytope as an intersection of closed halfspaces. The second characterization yields the result that two polytopes which meet intersect in a polytope.

13.1   Enclosing Convex Sets in Closed Halfspaces

Given a linear space L and a convex set K not contained in L (Figure 13.1), we seek conditions under which K will be contained in a proper closed halfspace of L. Our first result observes that L must be extremal to K (Section 7.27).

Theorem 13.1. *In any join geometry let K be convex, L a nonempty linear space and H a proper halfspace of L. Then $K \subset L \cup H$ implies L is extremal to K.*

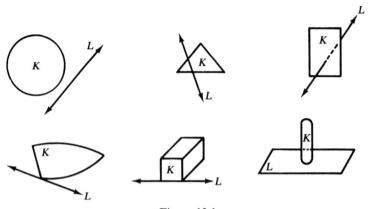

Figure 13.1

PROOF. Suppose

$$L \approx xy \quad \text{for } xy \subset K. \tag{1}$$

Then $K \subset L \cup H$ implies

$$xy \subset L \cup H. \tag{2}$$

We show

$$xy \subset L. \tag{3}$$

Suppose (3) fails. Then (2) yields $xy \approx H$. Let $p \subset xy$, H. Since xy is open, Corollary 4.25.1 yields

$$xy = pxy \subset H(L \cup H) = HL \cup H = H \quad \text{(Theorem 8.16).} \tag{4}$$

But (1) and (4) imply $L \approx H$, which is impossible given that H is a proper halfspace of L. Thus (3) holds, and L is extremal to K by definition. $\square$

Suppose L is extremal to K. Is there a closed halfspace of L which contains K? Here is an answer.

Theorem 13.2. *In an ordered geometry let K be convex and L a nonempty extremal to K which does not contain K. Then L has a proper halfspace H such that $K \subset L \cup H$ if and only if $\langle K, L \rangle$ covers L (in the family of linear spaces).*

PROOF. Suppose

$$K \subset L \cup H, \tag{1}$$

where H is a proper halfspace of L. Since $L \not\supset K$, there exists

$$a \subset K - L. \tag{2}$$

Then (1) and (2) yield $a \subset H$, so that

$$H = La/L \quad \text{(Theorem 8.17).} \tag{3}$$

Then (1), (3) and (2) imply

$$\langle K, L \rangle \subset \langle L \cup H, L \rangle = \langle La/L, L \rangle = \langle a, L \rangle \subset \langle K, L \rangle,$$

so that

$$\langle K, L \rangle = \langle a, L \rangle. \tag{4}$$

By Theorem 12.35, (2) and (4) imply

$$\langle K, L \rangle \text{ covers } L. \tag{5}$$

Conversely suppose (5) holds. Then since $L \neq \varnothing$, Corollary 12.33 implies

$$\langle K, L \rangle = L \cup H \cup H', \tag{6}$$

where H and H' are proper halfspaces of L which are separated by L. Then certainly (6) gives

$$K \subset L \cup H \cup H'. \tag{7}$$

We show that it is impossible for both $K \approx H$ and $K \approx H'$ to hold. Suppose otherwise. Then there exist points p and q with p common to K and H and q common to K and H'. Since L separates H and H', we obtain $L \approx pq$. But this implies $L \supset p, q$, since L is extremal to K (Corollary 7.40.1). Then $L \approx H$ and $L \approx H'$, which is in contradiction to the fact that H and H' are proper halfspaces of L. Thus either $K \not\approx H$ or $K \not\approx H'$; say $K \not\approx H'$. Then (7) implies (1) and the theorem is proven.   □

13.2 Interpreting the Results

Let us interpret Theorems 13.1 and 13.2 in Euclidean geometry. First we take K to be a closed convex planar region and L to be a line in the plane of K (Figure 13.2). In Theorem 13.1, the situation in the right hand drawing is ruled out. In both the left hand drawing and the center drawing, L is extremal to K and $K \subset L \cup H$. But the situation depicted in the center seems desirable. Here, in a certain sense, $L \cup H$ "just covers" K—or contains it "most economically". For if L and H were slid upward, $L \cup H$ would fail to contain K, and if downward, $L \cup H$ would be unnecessarily

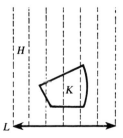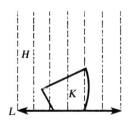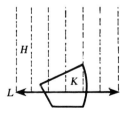

Figure 13.2

"large". In this Euclidean case the relation of $L \cup H$ to K can be described so: There is no closed halfspace *between* the sets $L \cup H$ and K. Intuitively speaking, we require L to be in "contact" with K.

Secondly, we take K to be a closed convex planar region and L to be an extremal linear space in contact with K. In Theorem 13.2, the situation in the right hand part of Figure 13.3 is ruled out—$\langle K, L \rangle$ does not cover L. In both the left hand part and the center part $K \subset L \cup H$, but the situation depicted in the center is again the desirable one. In the left hand part there is too much "space" in $L \cup H$; $L \cup H$ can contain solid regions. Thus it seems desirable for L to be contained in the plane of K, and in general for $L \subset \langle K \rangle$ to hold. Notice that in this Euclidean example the desirable situation, the center drawing, can be obtained from the left hand drawing by finding the intersections of $\langle K \rangle$ with L and with H. The resulting linear space $L \cap \langle K \rangle$ and its halfspace $H \cap \langle K \rangle$ are precisely those of the center drawing. Hence to require $L \subset \langle K \rangle$ is not unduly restrictive.

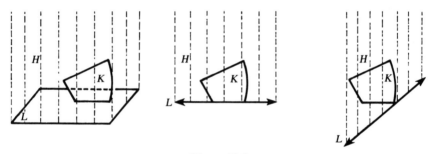

Figure 13.3

Under the assumption $L \subset \langle K \rangle$ the condition "$\langle K, L \rangle$ covers L" of Theorem 13.2 reduces to "$\langle K \rangle$ covers L", or equivalently, "L is a hyperplane of $\langle K \rangle$". Having observed first that L is extremal to K and $L \approx \mathcal{C}(K)$ should also hold, we see that L should be a supporting hyperplane to K (Section 7.31).

Hence we are led to study halfspaces of supporting hyperplanes to convex sets.

Theorem 13.3. *In an ordered geometry let L be a supporting hyperplane to the convex set K. Then there exists a unique proper halfspace H of L such that*

$$K \subset L \cup H. \tag{1}$$

Moreover $\langle K \rangle \supset L \cup H$.

PROOF. The existence of a proper halfspace H of L that satisfies (1) is a consequence of Theorem 13.2.

To show uniqueness suppose H' is a proper halfspace of L that satisfies (1). The relation (1) implies

$$K - L \subset H, H'. \tag{2}$$

Since $\langle K \rangle$ covers L, $K \not\subset L$. Thus $K - L \neq \varnothing$. Then (2) implies $H \approx H'$ and $H = H'$ (Theorem 8.13).

Finally we show $\langle K \rangle \supset L \cup H$. Certainly $\langle K \rangle \supset L$. The relation (2) and $K - L \neq \varnothing$ imply $K \approx H$. Let $p \subset K, H$. Then using Theorem 8.17, $\langle K \rangle \supset Lp/L = \overrightarrow{Lp} = H$ and the theorem holds.   □

13.3   Supporting Halfspaces to Convex Sets

The theorem suggests the following

Definition. In any join geometry let K be a convex set and L a supporting hyperplane to K. Suppose H is a proper halfspace of L such that $K \subset L \cup H$. Then we call H a *supporting halfspace* to K and $L \cup H$ a *supporting closed halfspace* to K. We also say H or $L \cup H$ *supports* K.

The following corollary is essentially a restatement of Theorem 13.3.

Corollary 13.3. *In an ordered geometry let L be a supporting hyperplane to the convex set K. Then there exist a unique halfspace H of L and a unique closed halfspace H' of L that support K. Moreover*

$$H' = L \cup H \subset \langle K \rangle.$$

Remarks on the Definition.

(1) The notion could have been introduced in Chapter 8 when halfspaces were defined. It can be studied in any join geometry, but has a richer aggregate of properties in an ordered geometry.

(2) We call H a supporting halfspace to K, even though it may not contain all points of K, for convenience of expression and for the substantive reason that $H \supset \mathcal{G}(K)$, the "core" of K. This we now prove.

Theorem 13.4. *In any join geometry let H be a supporting halfspace to convex set K. Then $H \supset \mathcal{G}(K)$.*

PROOF. H is a proper halfspace of a supporting hyperplane L to K, and

$$K \subset L \cup H. \tag{1}$$

Suppose $\mathcal{G}(K) \approx L$. Then certainly $\mathcal{G}(K) \approx L \cap K$. But L is extremal to K, so that $L \cap K$ is a face of K. Then by Theorem 7.28, $K \subset L \cap K$; that is, $K \subset L$. This contradicts the fact that L is a hyperplane of $\langle K \rangle$. Thus $\mathcal{G}(K) \not\approx L$. Then (1) yields $\mathcal{G}(K) \subset H$, as desired.   □

Remark I. The proof shows that a supporting hyperplane L to K does not meet the interior of K. Since $L \approx \mathcal{C}(K)$ (by definition), we must have $L \approx \mathcal{B}(K)$, and so L meets K in points of $\mathcal{F}(K)$, if at all.

Remark II. The theorem implies that the interior of a convex set K is contained in the intersection of the halfspaces that support K. This immediately raises the question: Is $\mathcal{F}(K)$ the intersection of the supporting halfspaces to K? Similarly: Is K the intersection of its supporting *closed* halfspaces? In general, and even in ordered geometries, the answer to both questions is no (see Exercises 5 and 7 below). (The question of the existence of supporting hyperplanes and halfspaces to K is in many join geometries related to the topological notion of completeness of the space— a notion we do not develop here.) Still there are interesting and important situations where the answers are yes. One of these, the case of polytopes (which are not simply points) in an ordered geometry, will be taken up next.

EXERCISES

1. Show that JG16 provides a counterexample for the contention that the forward implication in Theorem 13.2 holds in an arbitrary join geometry.

2. Prove: In an ordered geometry, if K is convex and L is a supporting hyperplane to K, then there exists a unique proper halfspace H of L satisfying $H \subset \mathcal{S}(K)$ (Section 6.21).

3. Prove: In an ordered geometry, if K is convex, L is linear and H is a proper halfspace of L, then H is a supporting halfspace to K if and only if $L \approx \mathcal{C}(K)$ and

$$K \subset L \cup H \subset \langle K \rangle.$$

4. In an ordered geometry suppose K is convex and L is a supporting hyperplane to K. Prove:
   (a) LK/L is a supporting closed halfspace to K if $K \approx L$.
   (b) LK/L is a supporting halfspace to K if $K \napprox L$.

5. In JG15, let K be the set of all rational numbers between $\sqrt{2}$ and $\sqrt{3}$. Show that K is convex and both open and closed. Use this to conclude that there do not exist any supporting hyperplanes to K.

*6. In JG16 let J' consist of the origin $(0, 0)$ together with all points not on the x-axis or y-axis (Figure 13.4), that is, J' is the set of points $x = (x_1, x_2)$ satisfying $x_1 = 0$ if and only if $x_2 = 0$. A join operation $\cdot$ is defined in J' by relativizing the join operation in JG16 to J'. That is, for $a, b \subset J'$,

$$a \cdot b = (ab) \cap J',$$

where the join on the right is that in JG16. Show that $(J', \cdot)$ is a join geometry, and that it provides a counterexample to the contention that Theorem 13.3 holds for an arbitrary join geometry. Does the reverse implication in Theorem 13.2 hold in an arbitrary join geometry? (Compare Exercise 1.)

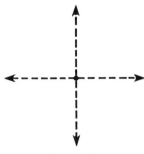

Figure 13.4

*7. In an ordered geometry let p be an extreme point of convex set K. Show that there does not necessarily exist a supporting hyperplane to K that contains p. [*Hint.* Let $J = \mathbb{Q}^2$ (Exercise 14 at the end of Section 5.9), $p = (0, 0)$ and K be the set of points $x = (x_1, x_2)$ satisfying $(x_1 - \sqrt{2})^2 + (x_2 - 1)^2 \leqslant 3$.]

13.4 Supporting Hyperplanes to Polytopes Exist[1]

Given a polytope, we wish to produce supporting hyperplanes to it. First let us examine the case of a polyhedron P in Euclidean 3-space (Figure 13.5). We see that $\mathcal{F}(P)$, the surface of P, is composed of planar faces. These planar faces have linear hulls which are supporting hyperplanes to P. Thus in the general setting of a polytope in an ordered geometry, the example suggests a search for faces of the polytope which have supporting hyperplanes as linear hulls. Our search is somewhat indirect, but it is rewarded. We begin with a lemma which seems unrelated to our problem.

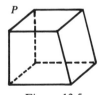

Figure 13.5

Lemma 13.5. *In any join geometry suppose K is convex and $L_1, \ldots, L_n$ are linear. Then $K \subset L_1 \cup \cdots \cup L_n$ implies $K \subset L_i$ for some i in the range $1, \ldots, n$.*

PROOF. We first consider the case where K is a segment. Suppose the lemma is false for segments. Then there exists a segment ab and linear

---

[1]This section is an adaptation to join geometries of Grünbaum [1], Theorem 3.1.1 and its proof.

spaces $L_1, \ldots, L_n$ such that

$$ab \subset L_1 \cup \cdots \cup L_n,$$

but $ab \not\subset L_i$ for each i in the range $1, \ldots, n$. Furthermore such a counterexample can be chosen to be minimal in the sense that, if $a'b'$ is a segment and $L_1', \ldots, L_m'$ are linear spaces with

$$a'b' \subset L_1' \cup \cdots \cup L_m',$$

then $m < n$ implies $a'b' \subset L_j'$ for some j in the range $1, \ldots, m$. Let $ab, L_1, \ldots, L_n$, as given above, furnish such a minimal counterexample.

Clearly $n \geqslant 2$, and there exists $p \subset ab$ satisfying

$$p \subset L_1 - (L_2 \cup \cdots \cup L_n). \tag{1}$$

Consider segment ap. We claim

$$ap \subset L_2 \cup \cdots \cup L_n. \tag{2}$$

Let $x \subset ap$, and suppose

$$x \subset L_1. \tag{3}$$

Then $a \subset x/p$, so that (1) and (3) yield $a \subset L_1$. Then $b \subset p/a \subset L_1$, and the contradiction $ab \subset L_1$ results. Hence (3) fails. But

$$x \subset ap \subset aab = ab,$$

so that

$$x \subset L_1 \cup \cdots \cup L_n.$$

This and the failure of (3) yield

$$x \subset L_2 \cup \cdots \cup L_n,$$

so that (2) holds as claimed. But (2), in view of the minimality condition, implies $ap \subset L_i$ for some i in the range $2, \ldots, n$. Then

$$p \subset \mathcal{C}(L_i) = L_i \qquad \text{(Corollary 6.15.3)},$$

a contradiction to (1). Thus no counterexample exists, and the lemma holds for segments.

We now consider the case of an arbitrary convex set K. We have the relation

$$K \subset L_1 \cup \cdots \cup L_n.$$

We may assume in this relation that n is minimal, but $n \neq 1$. Then there exists

$$a \subset K - L_1. \tag{4}$$

Let $x \subset K$. We show

$$x \subset L_2 \cup \cdots \cup L_n. \tag{5}$$

Certainly $ax \subset K$. Hence

$$ax \subset L_1 \cup \cdots \cup L_n.$$

By the case just considered, $ax \subset L_i$ for some i in the range $1, \dots , n$. Hence

$$a, x \subset \mathcal{C}(L_i) = L_i. \qquad (6)$$

The relations (4) and (6) yield $i \neq 1$. Then (6) yields (5). Thus

$$K \subset L_2 \cup \cdots \cup L_n.$$

This is impossible, and the lemma holds.   □

The next theorem enables us to produce supporting hyperplanes to polytopes in ordered geometries. It is the key to the representation of polytopes as intersections of supporting closed halfspaces.

Theorem 13.5. *In an ordered geometry let P be a polytope and let $a \subset \mathcal{E}(P)$. Then there exists a face F of P such that $\langle F \rangle$ is a supporting hyperplane to P and $a\mathcal{I}(P) \approx F$.*

(See Figure 13.6.)

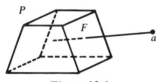

Figure 13.6

PROOF. Note that P is not a point, since $\mathcal{E}(P) \neq \varnothing$. P has only a finite number of faces (Corollary 7.36.2). Let $A_1, \dots , A_n$ be the *proper* faces of P. Let

$$x \subset \mathcal{I}(P). \qquad (1)$$

Theorem 12.17 applies and yields $ax \approx \mathcal{F}(P)$. By Theorem 7.34,

$$ax \approx A_1 \cup \cdots \cup A_n.$$

Thus

$$x \approx (A_1 \cup \cdots \cup A_n)/a = A_1/a \cup \cdots \cup A_n/a,$$

so that (1) yields

$$\mathcal{I}(P) \subset A_1/a \cup \cdots \cup A_n/a. \qquad (2)$$

In the right side of (2) we eliminate unnecessary terms and obtain

$$\mathcal{I}(P) \subset F_1/a \cup \cdots \cup F_r/a, \qquad (3)$$

where each F is an A and

$$\mathcal{I}(P) \approx F_i/a \quad \text{for } i = 1, \dots , r. \qquad (4)$$

From (3) we obtain

$$\mathcal{I}(P) \subset \langle F_1, a \rangle \cup \cdots \cup \langle F_r, a \rangle.$$

By the lemma,

$$\mathcal{G}(P) \subset \langle F, a \rangle, \tag{5}$$

where F is one of $F_1, \ldots, F_r$. From (5) we obtain, by Theorem 6.16,

$$\langle P \rangle = \langle \mathcal{G}(P) \rangle \subset \langle F, a \rangle.$$

But $F, a \subset \langle P \rangle$, so that

$$\langle P \rangle = \langle F, a \rangle = \langle \langle F \rangle, a \rangle. \tag{6}$$

Suppose $a \subset \langle F \rangle$. Then (6) becomes $\langle P \rangle = \langle F \rangle$, a contradiction to Theorem 7.42. Thus $a \not\subset \langle F \rangle$, and (6) and Theorem 12.35 yield that $\langle P \rangle$ covers $\langle F \rangle$. Finally, since F is an F_i, (4) yields $a\mathcal{G}(P) \approx F$. Note $F \neq \emptyset$. Since F is a face of P, $\langle F \rangle$ is extremal to P (Corollary 7.39). But $\langle F \rangle \approx P$, and so is a supporting hyperplane to P.   □

Corollary 13.5. *In an ordered geometry let P be a polytope which is not a point. Then there exists a face F of P such that $\langle F \rangle$ is a supporting hyperplane to P.*

PROOF. Since P is not a point, $\mathcal{E}(P)$ is nonempty (Corollary 6.28.2). Apply the theorem.   □

13.5 Polytopes Are Intersections of Closed Halfspaces

Consider Corollaries 13.3 and 13.5. The latter ensures that a polytope P has a face whose linear hull L is a supporting hyperplane to P. Then the former ensures that L has a unique halfspace that supports P. Together they enable us to represent a polytope as an intersection of supporting closed halfspaces.

Theorem 13.6. *In an ordered geometry let P be a polytope which is not a point. Let $F_1, \ldots, F_n$ be the faces of P whose linear hulls $L_1, \ldots, L_n$ are supporting hyperplanes to P. Let $H_1, \ldots, H_n$ be the halfspaces of $L_1, \ldots, L_n$ which support P. Then*

(a) $P = (L_1 \cup H_1) \cap \cdots \cap (L_n \cup H_n);$

(b) $\mathcal{G}(P) = H_1 \cap \cdots \cap H_n;$

(c) $\mathcal{F}(P) \subset L_1 \cup \cdots \cup L_n.$

PROOF. By definition of supporting halfspace

$$P \subset L_i \cup H_i, \qquad i = 1, \ldots, n. \tag{1}$$

By Theorem 13.4

$$\mathcal{G}(P) \subset H_i, \qquad i = 1, \ldots, n. \tag{2}$$

Thus by (1), $P \subset U$, where U denotes the right member of (a), and by (2), $\mathcal{G}(P) \subset V$, where V denotes the right member of (b).

To show the reverse inclusions, first suppose $x \subset U$. Then $x \subset \langle P \rangle$, since for each i, $L_i \cup H_i \subset \langle P \rangle$ by Corollary 13.3. Suppose $x \not\subset P$. Then $x \subset \mathcal{E}(P)$, since P is closed (Theorem 12.15). By Theorem 13.5 there exists a face A of P such that $\langle A \rangle$ is a supporting hyperplane to P and

$$x \mathcal{G}(P) \approx A. \tag{3}$$

Then $A = F_j$ for some j, and (3) becomes

$$x \mathcal{G}(P) \approx F_j. \tag{4}$$

But (4) in view of $x \subset U$ and (2) yields

$$(L_j \cup H_j)H_j \approx L_j,$$

which implies, by Theorem 8.16, $H_j \approx L_j$. But this is contrary to the fact that H_j is a proper halfspace of L_j. Thus our supposition is false, and $x \subset P$ holds, so that $U = P$ and (a) is proven.

Next observe that $V \subset U = P$. But V is the intersection of the open sets $H_1, \ldots, H_n$ and so is open (Corollary 4.26). Thus the relation $\mathcal{G}(P) \subset V$ and the fact that $\mathcal{G}(P)$ is a component of P imply $\mathcal{G}(P) = V$ (Theorem 7.15). Hence (b) is proven.

Finally, for (c), suppose $x \subset \mathcal{F}(P)$ and $x \not\subset L_1 \cup \cdots \cup L_n$. Then (a) yields $x \subset H_i$ for each i, so that (b) gives $x \subset \mathcal{G}(P)$, a contradiction. Thus (c) is proven. $\square$

Corollary 13.6. *In an ordered geometry let P be a polytope which is not a point. Then there exist faces of P such that P is the intersection of closed halfspaces of the linear hulls of these faces, $\mathcal{G}(P)$ is the intersection of these halfspaces, and $\mathcal{F}(P)$ is contained in the union of these linear hulls.*

13.6 Covering of Faces of a Polytope and Their Linear Hulls

A study of the relationship between the covering concept for faces of a convex set (Section 7.26) and the covering concept for linear spaces, as it applies to the linear hulls of these faces, was begun in Section 7.29. We obtain an improved result for a polytope in an ordered geometry.

Theorem 13.7 (The Polytope Facial Covering Theorem). *In an ordered geometry let A and B be faces of a polytope P. Then A covers B (in the* *family of faces of P*) *if and only if* $\langle A \rangle$ *covers* $\langle B \rangle$ (*in the family of linear spaces*).

(See Figure 13.7.)

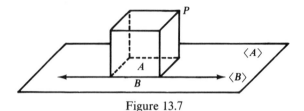

Figure 13.7

PROOF. By Theorem 7.43, $\langle A \rangle$ covers $\langle B \rangle$ implies A covers B. Suppose conversely that A covers B. We consider two cases. First assume $B = \varnothing$. Then A is a polytope whose vertices are vertices of P (Corollary 7.36.1) and so faces of P. Thus A has only one vertex, say a, and $A = a$. Then $\langle A \rangle = a$, which covers $\langle B \rangle = \varnothing$. Secondly assume $B \ne \varnothing$. Again A is a polytope, but this time A cannot be a point. By Theorem 13.6,

$$\mathscr{F}(A) \subset L_1 \cup \cdots \cup L_n, \tag{1}$$

where each L_i is a hyperplane of $\langle A \rangle$ and the linear hull of a face F_i of A. Observe that B is a face of A, indeed a proper face, so that $B \subset \mathscr{F}(A)$ by Theorem 7.34. Then (1) yields

$$B \subset L_1 \cup \cdots \cup L_n,$$

and consequently Lemma 13.5 implies $B \subset L_i$ for some i. Furthermore by Corollary 7.38,

$$B \subset L_i \cap A = \langle F_i \rangle \cap A = F_i.$$

Note that F_i is properly contained in A and that F_i is a face of P. By applying the covering hypothesis we obtain $F_i = B$. Thus $L_i = \langle B \rangle$ is a hyperplane of $\langle A \rangle$, that is, $\langle A \rangle$ covers $\langle B \rangle$.   □

13.7 Facets of a Polytope

Theorem 13.6 calls our attention to an important family of faces of a polytope P. Can these faces be characterized intrinsically in P? The Polytope Facial Covering Theorem suggests faces of P which are covered by P. This notion can be defined for any convex set.

Definition. In any join geometry, if A is a face of a convex set K and K covers A, then we call A a *facet* of K.

The definition is used to formulate an important corollary of Theorem 13.7.

Corollary 13.7. *In an ordered geometry let F be a face of polytope P. Then F is a facet of P if and only if* $\langle F \rangle$ *is a hyperplane of* $\langle P \rangle$.

Theorem 13.8. *In an ordered geometry let F be a face of a polytope P. Then* $\langle F \rangle$ *is a supporting hyperplane to P if and only if F is a nonempty facet of P.*

PROOF. Since $\langle F \rangle$ is an extremal to P, which meets $\mathcal{C}(P) = P$ if and only if $F \neq \varnothing$, the result is a consequence of Corollary 13.7. □

Corollary 13.8. *In an ordered geometry let F be a nonempty facet of a polytope P. Then P is supported by a unique halfspace of* $\langle F \rangle$.

PROOF. Apply Corollary 13.3. □

The treatment of supporting halfspaces to polytopes culminates in the next theorem, which is essentially a recasting of Theorem 13.6 based on the results of this section.

Theorem 13.9. *In an ordered geometry let P be a polytope which is not a point. Then*:

(a) *P is the intersection of the supporting closed halfspaces to P of the linear hulls of its facets.*

(b) $\mathcal{I}(P)$ *is the intersection of the supporting halfspaces to P of the linear hulls of its facets.*

(c) *Each point of* $\mathcal{F}(P)$ *is in a supporting hyperplane to P which is the linear hull of a facet of P.*

PROOF. Since P is not a point, Corollary 13.5 yields faces of P whose linear hulls are supporting hyperplanes to P. By Theorem 13.8 each such face is a facet of P. On the other hand, since P is not a point, no facet of P can be empty. By Theorem 13.8 each facet of P has a linear hull which is a supporting hyperplane to P. Thus the facets of P are precisely those faces of P whose linear hulls are supporting hyperplanes to P. Then (a), (b) and (c) follow from Theorem 13.6. □

Corollary 13.9. *In an ordered geometry let P be a polytope which is not a point. Then* P ($\mathcal{I}(P)$) *is the intersection of a finite family of closed (open) halfspaces of hyperplanes in* $\langle P \rangle$.

EXERCISES

1. Prove in an ordered geometry that a polytope which is not a point is the intersection of the family of all its supporting closed halfspaces. Show by examples in Euclidean geometry that this family may contain closed halfspaces of hyperplanes which are not linear hulls of facets of the polytope.

2. Prove: In an ordered geometry, if P is a polytope which is not a point, then $\mathfrak{S}(P)$ is the union of a finite family of halfspaces of hyperplanes in $\langle P \rangle$.

3. Prove: In an ordered geometry, any polytope not a point contains at least two facets.

4. Using Theorem 12.16 instead of Theorem 12.17, convert the proof of Theorem 13.5 into a proof of Corollary 13.5.

5. Prove: In any join geometry, if A is a nonpathological convex set contained in a union of faces of a convex set K, then A is contained in one of these faces. Is the result still true if A is not assumed nonpathological? Justify your assertion.

6. Show that Corollary 13.5 and Theorem 13.7 are false if the join geometry is not assumed to be ordered.

7. Prove: In an ordered geometry, if F is a facet of polytope P and $L_1, \ldots, L_n$ are linear spaces properly contained in $\langle P \rangle$ such that $\mathfrak{F}(P) \subset L_1 \cup \cdots \cup L_n$, then there is an i in the range $1, \ldots, n$ such that $L_i = \langle F \rangle$.

8. Prove: In an ordered geometry, if F is a nonempty facet of polytope P, then
   (a) P/F is a supporting closed halfspace to P;
   (b) $\mathfrak{S}(P)/F$ is a supporting halfspace to P.

*9. Prove: In an ordered geometry the intersection of all the facets of a polytope is empty. Show that the result is false if the join geometry is not assumed to be ordered.

10. Prove: In an ordered geometry, if the polytope P is the intersection of a finite family F of supporting closed halfspaces to P, and A is a facet of P, then some member of F is a closed halfspace of $\langle A \rangle$.

13.8   Facial Structure of a Polytope

We begin a brief investigation into the facial structure of a polytope in an ordered geometry. First we obtain formulas which express a facet and its interior in terms of halfspaces and hyperplanes.

Theorem 13.10. *In an ordered geometry let P be a polytope which is not a point. Let $F_1, \ldots, F_n$ be the facets of P; $L_1, \ldots, L_n$ the linear hulls of $F_1, \ldots, F_n$; $H_1, \ldots, H_n$ the halfspaces of $L_1, \ldots, L_n$ that support P. Then*

(a) $$F_i = (L_1 \cup H_1) \cap \cdots \cap (L_{i-1} \cup H_{i-1}) \cap L_i$$
$$\cap (L_{i+1} \cup H_{i+1}) \cap \cdots \cap (L_n \cup H_n),$$

(b) $$\mathfrak{S}(F_i) = H_1 \cap \cdots \cap H_{i-1} \cap L_i \cap H_{i+1} \cap \cdots \cap H_n.$$

(See Figure 13.8.)

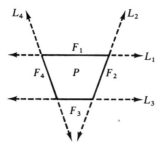

Figure 13.8

PROOF. (a)

$$F_i = P \cap \langle F_i \rangle = P \cap L_i$$

$$= (L_1 \cup H_1) \cap \cdots \cap (L_n \cup H_n) \cap L_i \qquad \text{(Theorem 13.9)}$$

$$= (L_1 \cup H_1) \cap \cdots \cap (L_{i-1} \cup H_{i-1}) \cap (L_i \cup H_i) \cap L_i$$

$$\cap (L_{i+1} \cup H_{i+1}) \cap \cdots \cap (L_n \cup H_n)$$

$$= (L_1 \cup H_1) \cap \cdots \cap (L_{i-1} \cup H_{i-1}) \cap L_i$$

$$\cap (L_{i+1} \cup H_{i+1}) \cap \cdots \cap (L_n \cup H_n).$$

(b) Let

$$R = H_1 \cap \cdots \cap H_{i-1} \cap L_i \cap H_{i+1} \cap \cdots \cap H_n.$$

Suppose $x \subset \mathcal{G}(F_i)$. Then since $L_i = \langle F_i \rangle$,

$$x \subset L_i. \qquad (1)$$

Now suppose

$$x \subset L_j, \quad \text{where } j \neq i. \qquad (2)$$

Then

$$x \subset L_j \cap P = \langle F_j \rangle \cap P = F_j.$$

Thus $F_j \approx \mathcal{G}(F_i)$, and Theorem 7.28 gives

$$F_j \supset F_i. \qquad (3)$$

But since F_j and F_i are distinct facets of P, (3) is impossible. Thus (2) is false, so that

$$x \not\subset L_j \quad \text{for each } j \neq i. \qquad (4)$$

But it follows from (a) that

$$x \subset L_j \cup H_j \quad \text{for each } j \neq i.$$

Thus by (4),

$$x \subset H_j \quad \text{for each } j \neq i. \qquad (5)$$

The relations (1) and (5) imply $x \subset R$, and we have

$$\mathcal{G}(F_i) \subset R. \tag{6}$$

Since P is not a point, $F_i \neq \emptyset$, so that $\mathcal{G}(F_i) \neq \emptyset$. Thus (6) yields $\mathcal{G}(F_i) \approx R$. But by (a), $R \subset F_i$, moreover, since R is an intersection of open sets, it is open. Hence Theorem 7.11(b) yields $\mathcal{G}(F_i) \supset R$, and the proof is finished.   □

Theorem 13.11. *In an ordered geometry let P be a polytope. Let A be a proper face of P which is not a facet of P. Then there are at least two facets of P which contain A.*

PROOF. Note the hypothesis implies that P is not a point. Let then $F_1, \ldots, F_n$ be the facets of P; $L_1, \ldots, L_n$ their linear hulls; $H_1, \ldots, H_n$ the halfspaces of these linear hulls which support P.

Since A is not P and not a facet of P, A is properly contained in some facet of P, say F_i. We first consider the possibility that $A = \emptyset$. Since F_i is not P, there exists a vertex of P not in F_i. This vertex must also be contained in a facet of P, and clearly this facet must be distinct from F_i. Thus A is contained in at least two facets of P.

Now assume $A \neq \emptyset$. Then $\mathcal{G}(A) \neq \emptyset$; say $p \subset \mathcal{G}(A)$. Suppose $p \subset \mathcal{G}(F_i)$. Then $\mathcal{G}(A) \approx \mathcal{G}(F_i)$, and Corollary 7.28 implies $A = F_i$, which is impossible. Thus $p \not\subset \mathcal{G}(F_i)$. By Theorem 13.10,

$$p \not\subset H_1 \cap \cdots \cap H_{i-1} \cap L_i \cap H_{i+1} \cap \cdots \cap H_n. \tag{1}$$

Note that $p \subset L_i$ holds, so that (1) yields

$$p \not\subset H_j \quad \text{for some } j \neq i. \tag{2}$$

But $p \subset F_i$, so that Theorem 13.10 implies

$$p \subset (L_1 \cup H_1) \cap \cdots \cap (L_{i-1} \cup H_{i-1}) \cap L_i$$
$$\cap (L_{i+1} \cup H_{i+1}) \cap \cdots \cap (L_n \cup H_n). \tag{3}$$

Thus $p \subset L_j \cup H_j$, and (2) gives $p \subset L_j$. Then

$$p \subset L_j \cap P = F_j.$$

Thus $\mathcal{G}(A) \approx F_j$, so that $A \subset F_j$. Since $F_j \neq F_i$ the theorem is proved.   □

13.9 Facets of a Facet of a Polyope

We improve on Theorem 13.11 for the case where a face of P is covered by a facet of P.

Theorem 13.12. *In an ordered geometry let F be a face of a polytope P which is covered by a facet A of P. Then there is exactly one proper face B of P*

distinct from A which properly contains F. Moreover, B is also a facet of P, B covers F, and F = A ∩ B.

(See Figure 13.9.)

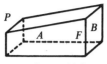

Figure 13.9

PROOF. We first show that at most one proper face of P other than A can properly contain F. Suppose B and C are proper faces of P distinct from A which properly contain F. Let

$$b \subset B - F, \quad c \subset C - F. \tag{1}$$

In order to show $B = C$, it suffices to show

$$b \subset C, \quad c \subset B. \tag{2}$$

We first observe that $A \cap B$ and $A \cap C$ are properly contained in A and contain F. Since A covers F, we have

$$A \cap B = F = A \cap C. \tag{3}$$

Since A is certainly a nonempty facet of P, we have by Corollary 13.8 that $P \subset L \cup H$, where $L = \langle A \rangle$ and H is a halfspace of L. Thus

$$b, c \subset L \cup H. \tag{4}$$

Suppose $b \subset L$. Then by (3),

$$b \subset L \cap B = L \cap P \cap B = \langle A \rangle \cap P \cap B = A \cap B = F.$$

This contradicts (1), and we conclude $b \not\subset L$. Similarly $c \not\subset L$. Then (4) yields

$$b, c \subset H. \tag{5}$$

By Theorem 8.17, $H = Lc/L$, and so (5) gives

$$b \subset Lc/L. \tag{6}$$

However, since A covers F, Theorem 13.7 implies $L = \langle A \rangle$ covers $\langle F \rangle$. Choose

$$a \subset A - F. \tag{7}$$

Then $a \not\subset \langle F \rangle$, since $\langle F \rangle \cap P = F$. By Theorem 6.21,

$$L = \langle \langle F \rangle, a \rangle = \langle F, a \rangle. \tag{8}$$

We consider the possibility that $F = \varnothing$. In this case (8) gives $L = a$, so that (6) becomes $b \subset ac/a$. By Theorem 12.18,

$$b \subset ac \cup c/a \cup c,$$

so that $b \approx ac$ or $c \approx ab$ or $b = c$. But $b \approx ac$ implies $B \approx ac$. Since B is a face of P, we get $B \supset a$, which is impossible given (3) and (7). Hence $b \approx ac$ fails, and by symmetry $c \approx ab$ fails. Thus $b = c$ and (2) is verified.

Next suppose $F \neq \emptyset$. Then by Theorem 6.13 and Corollary 6.11.1,

$$\langle F, a \rangle = \langle Fa \rangle = Fa / Fa,$$

so that (8) and (6) yield

$$b \subset (Fa/Fa)c/(Fa/Fa) \subset (Fac/Fa)/(Fa/Fa) \subset Fac/Fa. \quad (9)$$

But (9) in view of Corollary 12.19 becomes

$$b \subset Fac/F \cup Fc/Fa \cup Fc/F. \quad (10)$$

Suppose $b \approx Fac/F$. Then

$$ac \approx Fb/F \subset \langle B \rangle,$$

so that

$$ac \approx \langle B \rangle \cap P = B.$$

Again $B \supset a$ follows, contrary to (3) and (7). Similarly the supposition $b \approx Fc/Fa$ yields

$$ab \approx Fc/F \subset \langle C \rangle,$$

which by symmetry is also impossible. Thus (10) yields

$$b \approx Fc/F \subset \langle C \rangle, \quad (11)$$

from which $b \subset \langle C \rangle \cap P = C$ follows. But in addition, (11) gives

$$c \approx Fb/F \subset \langle B \rangle$$

from which $c \subset B$ follows. Thus (2) holds and $B = C$.

Hence we have shown that there is at most one proper face of P other than A which properly contains F. But Theorem 13.11 applies and yields the existence of a proper face, indeed a facet, B of P which is not A but which contains F. Certainly B properly contains F. It remains to be shown that B covers F and that $A \cap B = F$. But both of these follow immediately from the status of B as the only proper face of P other than A that properly contains F. □

EXERCISES

1. In an ordered geometry let P be a polytope which is not a point. Suppose a is in each supporting closed halfspace to P of the linear hull of a facet of P except for the supporting closed halfspace to P of the linear hull of the facet F of P. Suppose $b \subset P - F$. Prove that $ab \approx F$ in a unique point.

2. Prove: In an ordered geometry each proper face of a polytope P is the intersection of facets of P.

3. In an ordered geometry let P be a polytope and A a proper face of P. Suppose there exists a sequence

$$A = A_0, A_1, \ldots, A_n = P$$

of faces of P such that

$$A_i \text{ covers } A_{i-1}, \qquad i = 1, \ldots, n.$$

Prove that A is covered by at least n faces of P.

13.10 Generation of Convex Sets by Extreme Points

A polytope in any join geometry is generated by (that is, is the convex hull of) its vertices or extreme points (Theorem 4.21). The focus of interest is shifted now from the structure of polytopes to a broader question of when a convex set is generated by its set of extreme points. Elementary geometrical conditions are obtained (Theorem 13.22) which ensure that a convex set is generated by its extreme points. The conditions are satisfied by polytopes in ordered geometries, and the development concludes with applications of these conditions to convex sets to characterize polytopes independently of the usual defining property of finiteness of generation.

We adapt to the theory of join geometries concepts and results on the generation of convex sets by extreme points given by Klee [1]. Klee's aim in this paper is to generalize a classical result of Minkowski: In $\mathbb{R}^n$ a closed and bounded convex set is generated by its set of extreme points. [Recall that a set of points is *bounded* if it is contained in a polytope (Section 3.11)]. The point of departure in the paper is the introduction of a criterion that indicates how dispensable a point is in forming a set of generators of a convex set.

Let K be convex and S a subset of K which is to generate K. Suppose we start by taking $S = K$ and inquire: When can a point of K be deleted from S without destroying the property $[S] = K$? A simple answer: If

$$p \subset ab, \qquad a, b \subset K, \qquad a \neq b, \tag{1}$$

then $[S - p] = K$. In other words, if p, a, b satisfy (1), then p is dispensable as a member of S if a and b are retained in S. But a (or b or both) may be just as dispensable as p. For example, if $a \subset pq$, then a is dispensable if p and q are retained as members of S. So we must find a condition for p to be "more" dispensable than a, and perhaps even better, a condition for p to be "more" dispensable than a and b.

First a condition involving a and p is stated (Figure 13.10):

$$p \subset aK \quad \text{but} \quad a \not\subset pK. \tag{2}$$

This condition implies that p is dispensable if a and some point b also of K are retained—but we cannot say a is dispensable if p and another point x of K are retained, that is, $a \subset px$ is false for any $x \subset K$. Observe that (2) is concerned with the notion of precedence in K (Section 4.26). The condition (2) can be restated: p precedes a with respect to K, but a does not precede p with respect to K.

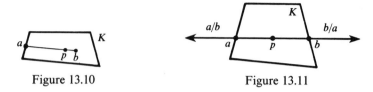

Figure 13.10   Figure 13.11

Next a condition involving a, b and p is given (Figure 13.11):

$$p \subset ab, \qquad a/b \not\approx K, \qquad b/a \not\approx K. \qquad (3)$$

The condition (3) implies that p is dispensable if a and b are retained, but (3) also implies (2). [For suppose $a \subset pK$. Then $K \approx a/p$, so that

$$K \approx a/ab = (a/a)/b = a/b,$$

contrary to (3).] Since (3) is symmetrical in a and b, (3) also implies (2) for b in place of a. Thus (3) does indeed ensure that p is "more" dispensable than a and b.

13.11 Terminal Segments and Dispensability of Generating Points

Our discussion suggests two definitions.

Definition. In any join geometry let K be convex and $a, b \subset K$ (Figure 13.12). Then ab is called a *terminal segment* of K if a/b, $b/a \not\approx K$.

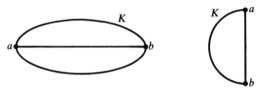

Figure 13.12

Note that if ab is a terminal segment, then $a \neq b$. Observe also that, since $a/b \not\approx K$ is equivalent to $a \not\approx bK$, we have (i) if a and b are distinct extreme points of K then ab is a terminal segment of K, and (ii) if ab is a terminal segment of K then a, b are in $\mathcal{F}(K)$.

Definition. In any join geometry let K be convex and $x, y \subset K$. If $x \subset yK$ but $y \not\subset xK$, then we say x is *more dispensable than* y (in K) or x is *deeper than* y (in K) and write $x \prec y$.

The second reading given for $x \prec y$ is intended to signalize that x is embedded in K "more deeply" than y: x is "more interior" to K than y, y

is "more peripheral" than x. This property is characterized in terms of the precedence relation on points in K as indicated above.

Relative dispensability and terminal segments are strongly related notions.

Theorem 13.13. *In a join geometry let K be convex, $a, b \subset K$ and $p \subset ab$. Then ab is a terminal segment of K if and only if $p \prec a$ and $p \prec b$.*

PROOF. Suppose ab is a terminal segment of K. Then $p \prec a$ and $p \prec b$ follow from the observations in the last section concerning conditions (2) and (3). Conversely if $p \prec a$ and $p \prec b$, then $a \not\gtrsim pK$ and $b \not\gtrsim pK$, or equivalently, $a/p \not\gtrsim K$ and $b/p \not\gtrsim K$. But $p \subset ab$ yields

$$a/p \subset a/ab = (a/a)/b = a/b.$$

Since a/p and a/b are a-rays, $a/p = a/b$. Thus $a/b \not\gtrsim K$. Similarly, $b/a \not\gtrsim K$, and ab is a terminal segment of K by definition.   □

EXERCISES

1. Can you find in Euclidean geometry examples of convex sets K such that
   (a) no point of K is in a terminal segment of K?
   (b) each point of K is in a terminal segment of K?
   (c) neither (a) nor (b) holds?
   (d) (a) holds but some points of K are more dispensable than others?

2. Prove: In a join geometry, if K is a convex set, then for points in K,
   (a) $x \not\prec x$.
   (b) $x \prec y$ implies $y \not\prec x$.
   (c) $x \prec y$ and $y \prec z$ imply $y \prec z$.

3. Choose a few convex sets in a Euclidean plane and find a few sequences

$$x_1 \prec x_2, \quad x_2 \prec x_3, \ldots, x_{n-1} \prec x_n$$

in each. Make a conjecture as to how many terms such a sequence can have. Do the same for a Euclidean 3-space.

4. In a join geometry suppose K is convex and ab is a terminal segment of K. Must

$$[a, b] = K \cap \langle a, b \rangle?$$

Justify your answer. If your answer was no, reconsider the exercise for an ordered geometry.

5. In a join geometry suppose K is convex, L is linear and

$$K \cap L = [a, b]$$

for distinct points a and b. Must ab be a terminal segment of K? Justify your answer. If your answer was no, reconsider the exercise for an ordered geometry.

6. In a join geometry suppose K is convex and ab is a terminal segment of K. Must ab be a maximal member of the family of all segments contained in K? Justify your answer. If your answer was no, reconsider the exercise for an ordered geometry.

7. In a join geometry suppose K is convex and $[a, b]$, $a \neq b$, is a maximal member of the family of all closed segments contained in K. Must ab be a terminal segment of K? Justify your answer. If your answer was no, reconsider the exercise for an ordered geometry.

8. In a join geometry suppose K is a convex set in which $p \prec q$. Let A be the least face of K that contains p (Section 7.22). Prove $q \subset A$. Which of $q \subset \mathcal{I}(A)$, $q \subset \mathcal{F}(A)$ holds? Justify your answer.

9. Prove: In a join geometry, if K is convex and $a, b \subset K$, then ab is a terminal segment of K if and only if there exist faces A, B of K such that $a \subset A - B$ and $b \subset B - A$.

*10. Prove: In a join geometry, if K is a convex set in which

$$x_1 \prec x_2, \qquad x_2 \prec x_3, \ldots, \quad x_{n-1} \prec x_n,$$

then $x_1, \ldots, x_n$ are convexly independent (Section 12.16).

13.12 Dispensability Sequences

Now we can describe a program for finding a set of generators for a given convex set K. Let $x \subset K$. We seek a terminal segment $x_1 x_2$ which contains x. Suppose there is one. Then by the last theorem

$$x \prec x_1, \qquad x \prec x_2.$$

Similarly we seek containing terminal segments for x_1 and x_2, say $x_{11} x_{12}$ and $x_{21} x_{22}$. If they exist, we have

$$x_1 \prec x_{11}, \qquad x_1 \prec x_{12};$$

$$x_2 \prec x_{21}, \qquad x_2 \prec x_{22}.$$

Continuing in this way, we obtain a sequence of points $x, x_1, x_2, \ldots$ and a series of pairs of relations $u \prec v$, $u \prec v'$ connecting the points, such that vv' is a terminal segment of K containing u.

This process can be pictured by the use of a tree diagram (Figure 13.13). In the tree each branch at any given stage has stopped growing or splits into exactly two new branches. Consider the case where the whole tree stops growing—where the sequence of points terminates after a finite

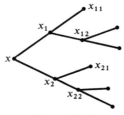

Figure 13.13

number of steps. Then x will be in the join of the "final points" in the sequence, that is, the points of the sequence which are not contained in a terminal segment of K. In this case the final points will be included in the desired set S of generators of K. If the sequence does not terminate, there is no satisfactory way to construct S.

Hence we must develop conditions which ensure that the sequence for each point $x \subset K$ will terminate. We find that it pays to consider the facial structure of K.

Theorem 13.14. *In a join geometry let K be convex and $a, b \subset K$. Suppose the least face of K that contains a is A, and the least face that contains b is B. Then in K, $a \prec b$ if and only if A properly contains B.*

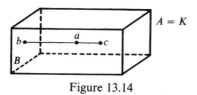

Figure 13.14

PROOF. By Theorem 7.30

$$a \subset \mathcal{G}(A) \tag{1}$$

and

$$b \subset \mathcal{G}(B). \tag{2}$$

Preparatory to proving the theorem we show

$$a \subset bK \quad \text{is equivalent to} \quad A \supset B. \tag{3}$$

To establish (3), first assume

$$a \subset bK. \tag{4}$$

Then since $a \subset A$, (4) implies $A \approx bc$ for some $c \subset K$. Since A is a face of K, Theorem 7.21 yields $A \supset b$. Then (2) implies

$$A \approx \mathcal{G}(B),$$

which yields (Theorem 7.28)

$$A \supset B. \tag{5}$$

Next assume (5) holds. Then $b \subset B$ gives $b \subset A$. Hence (1) implies $a \subset bc$ for some $c \subset A$. Clearly $c \subset K$ and (4) holds. Therefore we have established (3).

By the symmetry in the hypothesis of the theorem we have the validity of the statement which is symmetrical to (3):

$$b \subset aK \quad \text{is equivalent to} \quad B \supset A,$$

or

$$b \not\subset aK \quad \text{is equivalent to} \quad B \not\supset A. \tag{6}$$

From (3) and (6) it is immediate that in K, $a \prec b$ if and only if A properly contains B. Thus the theorem is proven. $\qquad \square$

We employ the last theorem to give two conditions under which dispensability sequences will terminate.

Theorem 13.15. *In a join geometry let K be convex and have m faces. Then in K,*

$$x_1 \prec x_2, \qquad x_2 \prec x_3, \ldots, x_{n-1} \prec x_n$$

implies $n < m$.

PROOF. For $i = 1, \ldots, n$, let A_i be the least face of K that contains x_i. Then Theorem 13.14 implies

$$A_1, \ldots, A_n, \varnothing$$

is a sequence of faces of K each member of which properly contains its successor. In particular the members of the sequence are distinct, and $n < m$ follows.   □

Theorem 13.16. *In a join geometry let K be convex and $\langle K \rangle$ be of height h. Then in K,*

$$x_1 \prec x_2, \qquad x_2 \prec x_3, \ldots, x_{n-1} \prec x_n$$

implies $n \leqslant h$.

PROOF. Again for $i = 1, \ldots, n$, let A_i be the least face of K that contains x_i. Then again

$$A_1, \ldots, A_n, \varnothing$$

is a sequence of faces of K each member of which properly contains its successor. By Theorem 7.42,

$$\langle A_1 \rangle, \ldots, \langle A_n \rangle, \varnothing$$

is a decreasing sequence of linear subsets of $\langle K \rangle$ (Section 6.19). Hence its length n can not exceed the height of $\langle K \rangle$, that is, $n \leqslant h$.   □

Compare the theorem with Exercise 3 at the end of Section 13.11.

13.13 Generation Conditions

Now we obtain a set of generators for a convex set K that depends on the family of terminal segments contained in K.

Theorem 13.17. *In a join geometry let K be convex. Suppose there is a positive integer m such that in K,*

$$x_1 \prec x_2, \qquad x_2 \prec x_3, \ldots, x_{n-1} \prec x_n$$

implies $n \leqslant m$. *Then the set* S *of points of* K *that do not belong to any terminal segment of* K *generates* K, *that is,* $K = [S]$.

PROOF. Obviously $[S] \subset K$. We show

$$K \subset [S]. \tag{1}$$

Suppose (1) is false. Then there exists x_1 satisfying

$$x_1 \subset K, \qquad x_1 \not\subset [S]. \tag{2}$$

Thus $x_1 \not\subset S$, so that

$$x_1 \subset x_2 y_2, \tag{3}$$

where $x_2 y_2$ is a terminal segment of K. Now if both $x_2, y_2 \subset [S]$, then (3) implies

$$x_1 \subset [S][S] \subset [S],$$

contrary to (2). Thus one of x_2, y_2 is not in $[S]$; say x_2. Then x_2 satisfies

$$x_2 \subset K, \qquad x_2 \not\subset [S]. \tag{4}$$

By Theorem 13.13,

$$x_1 \prec x_2. \tag{5}$$

Therefore (2) implies the existence of x_2 satisfying (4) and (5). By a comparison of (2) and (4), it is evident that (4) implies the existence of x_3 satisfying

$$x_3 \subset K, \qquad x_3 \not\subset [S]$$

and

$$x_2 \prec x_3.$$

The process may be repeated any number of times. In particular, for the positive integer $m + 1$ we have a sequence of points $x_1, \ldots, x_{m+1}$ of K satisfying

$$x_1 \prec x_2, \qquad x_2 \prec x_3, \ldots, x_m \prec x_{m+1}.$$

This clearly contradicts the hypothesis of the theorem, so (1) is not false, and the theorem follows. $\square$

Corollary 13.17.1. *In a join geometry let* K *be convex and have finitely many faces. Let* S *be the set of points of* K *that do not belong to any terminal segment of* K. *Then* $K = [S]$.

PROOF. Apply Theorem 13.15. $\square$

Corollary 13.17.2. *In a join geometry let* K *be convex and* $\langle K \rangle$ *be of finite height. Let* S *be the set of points of* K *that do not belong to any terminal segment of* K. *Then* $K = [S]$.

PROOF. Apply Theorem 13.16. $\square$

Remark. The theorem and its corollaries do not automatically yield nontrivial sets of generators for convex sets K satisfying their hypotheses. They are completely hypothetical as regards the existence of terminal segments in K. For example, if K is open, it certainly has finitely many faces, but it also has no terminal segments, and so $S = K$. To obtain significant applications of the theorem and its corollaries, conditions must be imposed on K to ensure the existence of terminal segments.

13.14 Segmental Closure and Linear Boundedness

Following Klee [1], we introduce two elementary geometrical properties of a convex set which imply an existence theorem (Theorem 13.21) for terminal segments.

Definition. In a join geometry a convex set K is said to be *segmentally closed* if its intersection with each closed segment that meets it is a closed segment: that is,

$$[p, q] \approx K$$

implies

$$[p, q] \cap K = [u, v]$$

for some points u and v (Figure 13.15).

Definition. In a join geometry a convex set K is said to be *linearly bounded* provided its intersection with each line L that meets it is contained in a closed segment of L: that is,

$$L \approx K,$$

where L is a line, implies

$$L \cap K \subset [a, b],$$

where $a, b \subset L$ (Figure 13.16).

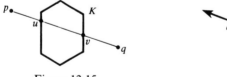

Figure 13.15

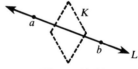

Figure 13.16

Remark. Trivially, in any join geometry, $\varnothing$ and any point are both segmentally closed and linearly bounded. In an ordered geometry every polytope is segmentally closed (Theorem 12.12) and linearly bounded (Theorem 12.11).

To continue we prove some elementary results on segmental closure and linear boundedness.

Theorem 13.18. *In a join geometry, if convex sets A and B are segmentally closed, then $A \cap B$ is segmentally closed.*

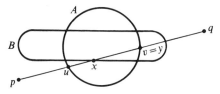

Figure 13.17

PROOF. Suppose (Figure 13.17)

$$[p, q] \approx A \cap B. \tag{1}$$

Then $[p, q] \approx A$, and the segmental closure of A implies

$$[p, q] \cap A = [u, v] \tag{2}$$

for some points u and v. Then by (1) and (2),

$$\emptyset \neq [p, q] \cap A \cap B = [u, v] \cap B. \tag{3}$$

Hence $[u, v] \approx B$, and the segmental closure of B gives

$$[u, v] \cap B = [x, y] \tag{4}$$

for some points x and y.
By (3) and (4),

$$[p, q] \cap A \cap B = [x, y],$$

and $A \cap B$ is by definition segmentally closed.   □

Corollary 13.18. *In a join geometry the intersection of finitely many segmentally closed convex sets is segmentally closed.*

Theorem 13.19. *In a join geometry, if A and B are convex, $A \supset B$ and A is linearly bounded, then B is linearly bounded.*

PROOF. Suppose L is a line and $L \approx B$ (Figure 13.18). Then since $A \supset B$, we have $L \approx A$. Since A is linearly bounded, we have

$$L \cap A \subset [p, q], \quad \text{where } p, q \subset L. \tag{1}$$

But

$$L \cap B \subset L \cap A. \tag{2}$$

The relations (1) and (2) yield

$$L \cap B \subset [p, q], \quad \text{where } p, q \subset L,$$

and so B is by definition linearly bounded.   □

Figure 13.18   Figure 13.19

Theorem 13.20. *In a join geometry suppose convex set K is segmentally closed and linearly bounded. Then for any line L that meets K,*

$$L \cap K = [a, b] \tag{1}$$

for some a and b.

PROOF. Suppose $L \approx K$ (Figure 13.19). Then since K is linearly bounded, there exist points p, q such that

$$\varnothing \neq L \cap K \subset [p, q] \subset L. \tag{2}$$

Then (2) implies $K \approx [p, q]$. Since K is segmentally closed, there exist points a, b such that

$$[p, q] \cap K = [a, b]. \tag{3}$$

But by (2),

$$L \cap K \subset [p, q] \cap K \subset L \cap K. \tag{4}$$

The relations (3) and (4) yield (1), as desired.   □

13.15 Convex Sets Generated by Their Extreme Points

A convex set K which satisfies the intersection properties of the last section has—with trivial exceptions—an abundance of terminal segments.

Theorem 13.21. *In a join geometry let the convex set K be segmentally closed and linearly bounded. Then any nonextreme point x of K is in a terminal segment of K.*

PROOF. Since x is a nonextreme point of K, we have (Figure 13.20)

$$x \subset ab, \tag{1}$$

where a and b are distinct points of K. Let $L = \langle a, b \rangle$. Then L is a line which meets K. By Theorem 13.20,

$$L \cap K = [p, q] \tag{2}$$

for some points p and q. Note that $p, q \subset K$.

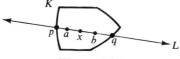

Figure 13.20

We show pq is a terminal segment of K. Suppose

$$p/q \approx K. \tag{3}$$

In view of (2), $p/q \subset L$, and so (3) yields

$$p/q \approx L \cap K.$$

By (2),

$$p/q \approx [p, q],$$

which implies

$$p \approx [p, q]q = (p \cup q \cup pq)q = pq \cup q.$$

Thus $p = q$ results. But note that $a, b \subset L \cap K$, so that by (2),

$$a, b \subset [p, q]. \tag{4}$$

Since $a \neq b$, it is impossible for $p = q$ to hold. Thus (3) fails; $p/q \not\approx K$. By symmetry $q/p \not\approx K$, and we infer pq is a terminal segment of K.

Finally we show

$$x \subset pq. \tag{5}$$

The relation (4) gives

$$a, b \subset p \cup q \cup pq,$$

so that $a \neq b$ is seen, through simple case by case set theoretical considerations, to imply

$$ab \subset pq.$$

Then (1) yields (5), and the theorem is proved.   □

Now we come to the key theorem in the development.

Theorem 13.22. *In a join geometry let the convex set K be segmentally closed and linearly bounded. Let K have finitely many faces, or let $\langle K \rangle$ have finite height. Then K is the convex hull of its set of extreme points.*

PROOF. Applying either Corollary 13.17.1 or Corollary 13.17.2 as the case may be, we have

$$K = [S], \tag{1}$$

where S is the set of points of K that are in no terminal segment of K. By Theorem 4.19, S contains each extreme point of K. On the other hand, by Theorem 13.21, no point of S is other than an extreme point of K. Hence S is precisely the set of extreme points of K, and the theorem follows from (1).   □

EXERCISES

1. Find in Euclidean geometry examples of convex sets K such that the set S of points of K that do not belong to any terminal segment of K satisfies
   (a) $S = K$;   (b) $S = \mathcal{F}(K)$;
   (c) $S = \mathcal{I}(K)$;   (d) $S = K$, $\mathcal{F}(K) \neq \emptyset$.

2. Prove: In a join geometry, if K is convex, $\langle K \rangle$ is of finite height and every point of K that does not belong to a terminal segment of K is in $\mathcal{I}(K)$, then K is open.

3. In $\mathbb{R}^*$ (Section 5.12) let K be the convex set which consists of all elements of $\mathbb{R}^*$ in which there are no negative entries. For each positive integer i, let x_i be that element of K whose first i entries are 0 and whose other entries are each 1, i.e., $x_1 = (0, 1, 1, 1, \ldots)$, $x_2 = (0, 0, 1, 1, 1, \ldots)$, $\ldots$. Show that for each i, $x_i < x_{i+1}$.

4. Prove: In a join geometry, the requirement in the definition of segmental closure that K be a convex set is actually implied by the intersection condition.

5. Prove that in a join geometry a segmentally closed convex set is closed. Show JG15 provides a counterexample to the converse of this result.

6. Prove: In an ordered geometry a bounded (Section 3.11) convex set is linearly bounded.

7. Prove: In an ordered geometry a convex set is linearly bounded if and only if it does not contain a proper ray.

8. Prove: In an ordered join geometry a convex set is segmentally closed if its intersection with each line that meets it is a closed segment. Find an example in JG10 which shows that the result is false if the assumption that the join geometry is ordered is deleted.

9. In a join geometry, is the intersection of any family of segmentally closed convex sets segmentally closed? Justify your answer. If your answer was no, reconsider the exercise for an ordered geometry.

13.16 A Characterization of a Polytope

The last theorem is employed to characterize polytopes in an ordered geometry. First a condition is given under which a convex set in any join geometry is a polytope.

Theorem 13.23. *In a join geometry let K be a nonempty convex set. Then K is a polytope if it is segmentally closed, is linearly bounded and has a finite number of faces.*

PROOF. By Theorem 13.22, K is the convex hull of its set of extreme points. Since K has a finite number of faces, it certainly has a finite number of extreme points, and so K, being nonempty, is a polytope by definition. $\square$

We have observed (Section 13.14, Remark) that in an ordered geometry every polytope is segmentally closed and linearly bounded. Hence we have in ordered geometries a new characterization of a polytope.

Theorem 13.24. *In an ordered geometry let K be a nonempty convex set. Then K is a polytope if and only if it is segmentally closed, is linearly bounded and has a finite number of faces.*

13.17 Polytopes and Intersections of Closed Halfspaces

The characterization of polytopes in Theorem 13.24 may be described as in the spirit of elementary geometry since it leans heavily on intersection properties with lines and closed segments. A different type of characterization of polytopes—as intersections of closed halfspaces—is suggested by Theorem 13.9 and its corollary. The characterization (Theorem 13.28 below) is a consequence of the following results on linear spaces and closed halfspaces.

Theorem 13.25. *In an ordered geometry a linear space is segmentally closed.*

PROOF. Let L be a linear space, and suppose $[a, b] \approx L$. Then by Theorem 12.10, $[a, b] \cap L$ is a polytope. If the polytope is a point, it is a closed segment. Thus we may assume $[a, b] \cap L$ is not a point. Then $\langle a, b \rangle$ is a line and

$$[a, b] \cap L \subset \langle a, b \rangle.$$

By Lemma 12.11, $[a, b] \cap L$ is a closed segment.   □

Remark. It may be tempting to conjecture that the theorem holds in any join geometry. However, this is false, for the counterexample following Theorem 12.10 is also a counterexample here.

Theorem 13.26. *In an ordered geometry a closed halfspace is segmentally closed.*

PROOF. Let L be a nonempty linear space, and let H be a halfspace of L, so that $L \cup H$ is a closed halfspace (Figure 13.21). By the previous theorem we may assume H is a proper halfspace of L. Suppose $[p, q] \approx L \cup H$. We have to show

$$[p, q] \cap (L \cup H) \text{ is a closed segment.} \tag{1}$$

It is therefore not restrictive to suppose

$$[p, q] \cap (L \cup H) \supset u, v, \quad u \neq v. \tag{2}$$

Figure 13.21

Note then that $p \neq q$. Let H' be the halfspace of L which is opposite to H. Then (2) implies

$$u, v \subset L \cup H \cup H'. \tag{3}$$

But since $H = La/L$ and $H' = L/a$ for some a, Theorem 12.33 implies that $L \cup H \cup H'$ is linear. Then (3) implies

$$\langle u, v \rangle \subset L \cup H \cup H'. \tag{4}$$

But (2) implies $u, v \subset \langle p, q \rangle$, so that Theorem 12.1 gives $\langle p, q \rangle = \langle u, v \rangle$. Then (4) implies

$$p, q \subset L \cup H \cup H'. \tag{5}$$

We now consider alternatives. If $p, q \subset L \cup H$, then (1) holds. Thus we may assume one of p, q is not in $L \cup H$; say q. Then by (5),

$$q \subset H'. \tag{6}$$

We show

$$p \subset H. \tag{7}$$

Suppose $p \subset L \cup H'$. Then

$$pq \subset (L \cup H')H' = LH' \cup H' = H' \quad \text{(Theorem 8.16).} \tag{8}$$

Using (6), (8) and the disjointness of H' and $L \cup H$ we have

$$[p, q] \cap (L \cup H) = (p \cup q \cup pq) \cap (L \cup H) \subset (p \cup H') \cap (L \cup H)$$
$$= p \cap (L \cup H) \subset p,$$

a contradiction to (2). Thus $p \not\subset L \cup H'$, and so by (5) we infer (7). Since H and H' are opposite halfspaces of L, they are separated by L. Hence (6) and (7) give $pq \approx L$; say r is a common point. Then by Theorem 12.5,

$$pq = r \cup pr \cup qr. \tag{9}$$

But $r \subset L$, and so $pr \subset HL = H$; similarly $qr \subset H'$. Then using (6), (7), (9) and the disjointness of $L \cup H$ and H',

$$[p, q] \cap (L \cup H) = (p \cup q \cup r \cup pr \cup qr) \cap (L \cup H)$$
$$\subset (p \cup r \cup pr \cup H') \cap (L \cup H)$$
$$= (p \cup r \cup pr) \cap (L \cup H)$$
$$= p \cup r \cup pr$$
$$= [p, r].$$

Hence again (1) holds, and the theorem follows.   □

Corollary 13.26. *In an ordered geometry, if $A_1, \ldots, A_n$ are closed halfspaces, then $A_1 \cap \cdots \cap A_n$ is segmentally closed.*

PROOF. Apply Corollary 13.18. □

Theorem 13.27. *In any join geometry let H be a halfspace of the nonempty linear space L. Then the components of the closed halfspace $L \cup H$ are $\varnothing$, L and H.*

PROOF. Note that L is open. If $H = L$, the result is trivial. If H is proper, then Theorem 13.1 may be applied with K chosen as $L \cup H$ to conclude that L is extremal to $L \cup H$, and so a component of $L \cup H$. Since H is open, it must be contained in a component of $L \cup H$, and this component must be disjoint to L. Thus H is itself a component of $L \cup H$. The result now follows. □

Corollary 13.27. *In any join geometry, if $A_1, \ldots, A_n$ are closed halfspaces, then $A_1 \cap \cdots \cap A_n$ has finitely many faces.*

PROOF. Apply repeatedly Corollary 7.17.2—that if each of K_1, K_2 is a convex set with finitely many components, then so is $K_1 \cap K_2$—to conclude that $A_1 \cap \cdots \cap A_n$ has finitely many components. Since any face of $A_1 \cap \cdots \cap A_n$ is a union of a family of these components (Theorem 7.14), the result follows. □

We have now developed enough information to characterize polytopes in ordered geometries in a new way.

Theorem 13.28. *In an ordered geometry let K be a nonempty convex set. Then K is a polytope if and only if K is a linearly bounded intersection of finitely many closed halfspaces.*

PROOF. Suppose K is a polytope. Then K is linearly bounded, by Theorem 13.24. By Corollary 13.9, K is the intersection of finitely many closed halfspaces, provided K is not a point. But if K is a point p, then K may be considered to be the closed (improper) halfspace p of linear space p. Thus in any case the desired conclusion holds.

Conversely, suppose K is linearly bounded and $K = A_1 \cap \cdots \cap A_n$, where each A is a closed halfspace. By Corollary 13.26, K is segmentally closed. By Corollary 13.27, K has finitely many faces. Therefore K is a polytope by Theorem 13.24. □

The theorem can be applied to the intersection of two polytopes.

Theorem 13.29. *In an ordered geometry let P_1 and P_2 be polytopes. Then $P_1 \cap P_2$ is $\varnothing$ or a polytope.*

PROOF. We may assume $P_1 \cap P_2 \neq \varnothing$. Since $P_1 \cap P_2 \subset P_1$ and P_1 is linearly bounded, Theorem 13.19 implies $P_1 \cap P_2$ is linearly bounded. Since P_1 and

P_2 are both intersections of finitely many closed halfspaces (Theorem 13.28), so is $P_1 \cap P_2$. Therefore, $P_1 \cap P_2$ is a polytope by Theorem 13.28. $\square$

Compare the theorem with Exercise 6 at the end of Section 7.17.

EXERCISES

1. Prove: In an ordered geometry, if L is linear and $L \approx [a, b]$ in more than a single point, then $L \supset [a, b]$. Use this result to construct a different proof of Theorem 13.25.

2. In any join geometry determine the faces of a closed halfspace. Prove your assertion.

3. Prove: In an ordered geometry, if $a_1, \ldots, a_n$ are linearly dependent, then

$$\lfloor a_1, \ldots, a_{n-1}] \cap \cdots \cap [a_2, \ldots, a_n]$$

is a polytope.

References

W. Barnes [1]. *Introduction to Abstract Algebra*. Boston, Heath, 1963.

V. W. Bryant and R. J. Webster [1]. Generalizations of the theorems of Radon, Helly, and Caratheodory. *Monatsh. Math.* **73**, 309–315, 1969.

C. Chevalley [1]. *Fundamental Concepts of Algebra*. New York, Academic Press, 1956.

Euclid [1]. *Euclid's Elements*. New York, Dover, 1956.

B. Grünbaum [1]. *Convex Polytopes*. New York, Interscience, 1967.

D. Hilbert [1]. *Foundations of Geometry, 2nd ed.* La Salle, Open Court, 1971.

N. Jacobson [1]. *Lectures in Abstract Algebra*. New York, Van Nostrand, 1956, Vol. 1.

V. Klee [1]. Extreme points of convex sets without completeness of the scalar field. *Mathematika* **11**, 59–63, 1964.

S. Mac Lane and G. Birkhoff [1]. *Algebra*. New York, Macmillan, 1967.

H. Paley and P. M. Weichsel [1]. *A First Course in Abstract Algebra*. New York, Holt, 1966.

W. Prenowitz [1]. Descriptive geometries as multigroups. *Trans. Am. Math. Soc.* **59**, 333–380, 1946.

[2]. Spherical geometries and multigroups. *Can. J. Math.* **2**, 100–119, 1950.

[3]. A contemporary approach to classical geometry. *Am. Math. Month.* **68** (No. 1, Part II), 1–67, 1961.

and J. Jantosciak [1]. Geometries and join spaces. *J. Reine Angew. Math.* **257**, 100–128, 1972.

and M. Jordan [1]. *Basic Concepts of Geometry*. New York, Blaisdell, 1965.

R. T. Rockafellar [1]. *Convex Analysis*. Princeton, NJ. Princeton University Press, 1970.

B. L. van der Waerden [1]. *Modern Algebra, 2nd ed.* Heidelberg, Springer, 1941.

Index

Undergraduate Texts in Mathematics

Apostol: Introduction to Analytic
Number Theory.
1976. xii, 370 pages. 24 illus.

Childs: A Concrete Introduction to
Higher Algebra.
1979. xiv, 338 pages. 8 illus.

Chung: Elementary Probability Theory
with Stochastic Processes. Third Edition
1979. 336 pages.

Croom: Basic Concepts of Algebraic
Topology.
1978. x, 177 pages. 46 illus.

Fleming: Functions of Several Variables.
Second edition.
1977. xi, 411 pages. 96 illus.

Halmos: Finite-Dimensional Vector
Spaces. Second edition.
1974. viii, 200 pages.

Halmos: Naive Set Theory.
1974. vii, 104 pages.

Kemeny/Snell: Finite Markov Chains.
1976. ix, 210 pages.

Lax/Burstein/Lax: Calculus with
Applications and Computing,
Volume 1.
1976. xi, 513 pages. 170 illus.

LeCuyer: College Mathematics with
A Programming Language.
1978. xii, 420 pages. 126 illus. 64 diagrams.

Malitz: Introduction to Mathematical
Logic.
Set Theory - Computable Functions -
Model Theory.
1979. 255 pages. 2 illus.

Prenowitz/Jantosciak: The Theory of
Join Spaces.
A Contemporary Approach to Convex
Sets and Linear Geometry.
1979. Approx. 350 pages. 404
illus.

Priestley: Calculus: An Historical
Approach.
1979. 400 pages. 300 illus.

Protter/Morrey: A First Course in Real
Analysis.
1977. xii, 507 pages. 135 illus.

Sigler: Algebra.
1976. xi, 419 pages. 32 illus.

Singer/Thorpe: Lecture Notes on
Elementary Topology and Geometry.
1976. viii, 232 pages. 109 illus.

Smith: Linear Algebra
1978. vii, 280 pages. 21 illus.

Thorpe: Elementary Topics in
Differential Geometry.
1979. 256 pages. Approx. 115 illus.

Whyburn/Duda: Dynamic Topology.
1979. Approx. 175 pages. Approx. 20
illus.

Wilson: Much Ado About Calculus.
A Modern Treatment with Applications
Prepared for Use with the Computer.
1979. Approx. 500 pages. Approx. 145
illus.